# A-Level Mathematics

## A Comprehensive and Supportive Companion to the Unified Curriculum

STUDENT BOOK

$$2ab \le a^2 + b^2$$

$$+ \frac{f(x_2)}{2} + \ldots + \frac{f(x_N)}{2}$$

$$\frac{d\sin(x)}{dx} = \cos(x)$$

$$\sin^2\left(\frac{\pi}{24}\right)$$

Tarquin

YEAR ONE

Edited by Tom Bennison and Edward Hall

## Editors/Authors:
Dr. Tom Bennison;
Dr. Edward Hall.

## Contributing Authors:
Rob Beckett;
Katie Binks;
Kieran Fitness;
Dr. Jenny Gladstone;
Dr. David Marles;
Stuart Price;
Matilde Warden;
Dr. Tom Wicks.

## Figure Acknowledgements:
Figure 3.1, Page 23, Shutterstock Image;
Figure 4.1(b), Page 39, Shutterstock Image;
Figure 5.1, Page 79, Shutterstock Image;
Figure 6.1, Page 116, Public Domain:
https://civilityandtruth.com/2017/01/07/creating-the-chrysalis-design/;
Figure 8.1, Page 147, Shutterstock Image;
Figure 11.1(a), Page 193, Shutterstock Image;
Figure 11.1(b), Page 193, Shutterstock Image;
Figure 11.2, Page 194, Courtesy of Luke Jolley;
Figure 14.2, Page 260, Shutterstock Image;
Figure 17.1, Page 336, Modified Shutterstock Image;
Figure 18.1(b), Page 352, Modified Shutterstock Image;
Figure 18.2(a), Page 353, Shutterstock Image;
Figure 18.2(b), Page 353, Modified Shutterstock Image;
Figure 19.1, Page 374, Shutterstock Image;
Figure 20.1, Page 388, Modified Shutterstock Image;
Figure 25.1 Figure 26.1(a), Page 503, Shutterstock Image;
Figure 26.1(b), Page 503, Shutterstock Image;
Figure 27.1, Page 527, Modified Shutterstock Image;
Figure 28.1(a), Page 556, Shutterstock Image;
Figure 28.1(b), Page 556, Courtesy of Edward Hall.

A Tarquin EReader Licence may be free for
purchasers of this book and subscribers to
Tarquin's ALevel Service: contact
info@tarquingroup.com for more details.

Tarquin Publications
Suite 74, 17 Holywell Hill
St Albans
AL1 1DT
UK
www.tarquingroup.com

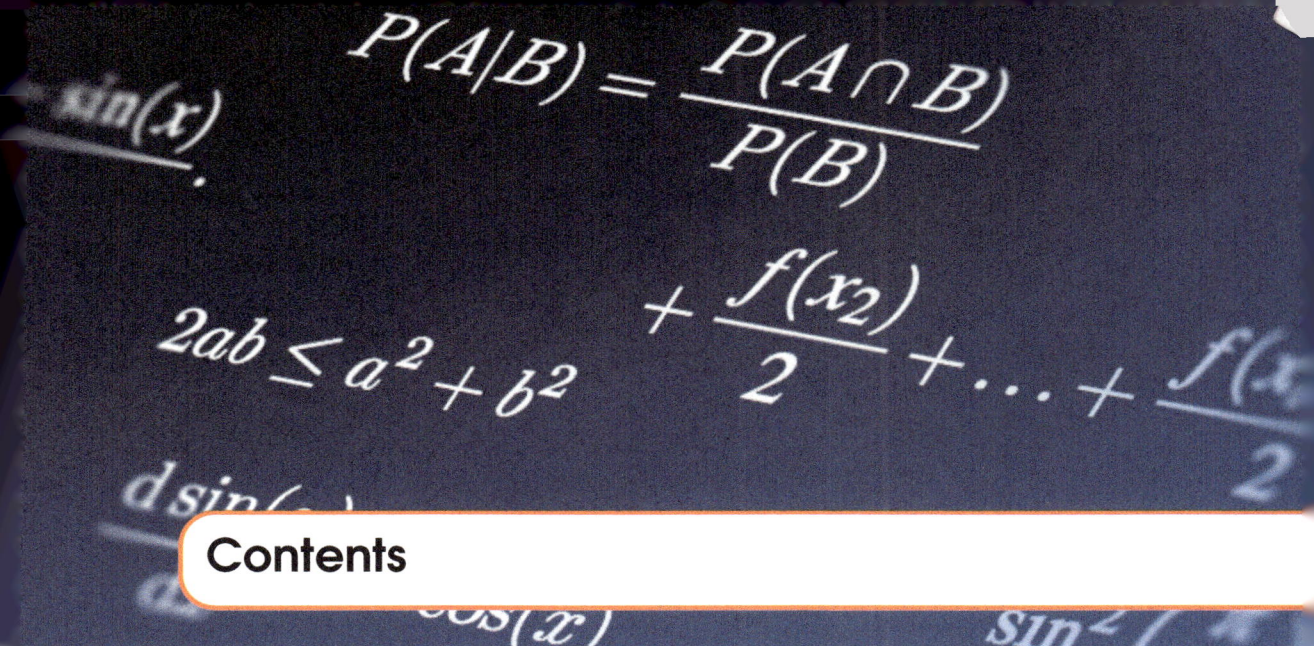

$$P(A|B) = \frac{P(A \cap B)}{P(B)}$$

$$2ab \le a^2 + b^2$$

$$+ \frac{f(x_2)}{2} + \ldots + \frac{f(x}{2}$$

# Contents

# 1. Introduction

The introduction of the new A-Level specification (Sept. 2017) has motivated the writing of this mathematics textbook. It has been written not only by experienced teachers, but also by professional mathematicians who are experts in the field. As the content of the new A-Level syllabus should be consistent across all examination boards, it is intended that this book could be the only textbook that is needed: questions in the style of the each examination board are included in the downloadable assessment exercises at the end of each chapter. However, the book could be used as a companion to the examination boards' own offerings, especially where more explanation of techniques and theory is required. The book can be used for self-study, or as part of a school based syllabus.

The new A-level specification aims to raise awareness of the importance of proof in mathematics. The concept of *proof* is fundamental in mathematics. In other sciences, theories are subject to experimental evidence: as the evidence changes, so does the theory. One example of this are the theories within mechanics. For a number of centuries, Newton's laws of motion were assumed to be correct, but under the weight of evidence, notable scientists such as Albert Einstein and Niels Bohr developed theories of *relativity* and *quantum mechanics* showing that Newton's laws break down when an object's speed is close to that of light, or when the object is very small. These theories are unlikely to be the final word in mechanics and may change as our understanding grows in the future. In mathematics, once something has been proved once, it has been proved for all time. Hence, Pythagoras' theorem remains as relevant today as it did 2500 years ago, the square root of 2 can never be rational, the quadratic equation formula will hold for any quadratic equation forever and there will always be an infinite number of prime numbers. Of course, many things still remain to be proved and a large part of a professional mathematician's work is spent developing new proofs. In this book, we have provided proofs of all of the most important results, or given guidance on how they could be proved. In some cases we have proved theorems in alternative ways, in an attempt to link together separate parts of the course. Of course, these proofs have been developed before, but exposure to these proofs fosters a deeper understanding of mathematics and will greatly assist those who are planning on continuing their studies at degree level.

The A-level specification also states that the use of technology should be central to the course. An e-book version of the book comes with the subscription. This provides the perfect platform to host Geogebra and Tarquin interactive activities and it is available through web browsers and as native iOS and Android apps. The Geogebra activities (see the figure below) have been specifically designed to enable students to quickly see how new ideas, techniques and methods work.

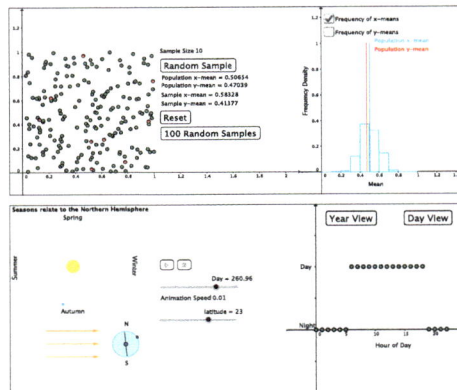

The Tarquin activities include electronic 'card sorts', 'fill in the gaps', 'ordering' activities, with automatic marking to give instantaneous feedback on understanding of a particular topic. In addition to this, there is support for the use of both scientific and graphical Casio and Texas Instruments calculators.

Emphasis has been placed on an explanation of why the mathematical techniques presented in the syllabus are useful. When studying mathematics, it is often important to understand why a particular topic is being studies and how useful it can be. As this book has been written not just by teachers, but by professional mathematicians and scientists, an emphasis has been placed on applications of the techniques.

In summary, the main features of this book are:
- Specifically written and designed for the new A-level syllabus (Sept. 2017), not a reworking of previous material.
- Problems in the style of each examination board.
- Teacher version offering more explanation of topics and suggestions for activities to promote understanding by students.
- Discussion of misconceptions
- Detailed worked solutions for all exercises.
- Special emphasis placed on the importance of proof in mathematics.
- Interactive activities in Geogebra and the Tarquin learning environment.
- Support for scientific and graphical calculators.
- Discussion of application areas of the mathematics.
- A dedicated chapter exploring ways to use the large data set.

$$b^m \div b^n = b^{m-n},$$
$$(b^m)^n = b^{mn},$$
$$b^0 = 1, \qquad b \neq 0,$$
$$b^{-m} = \frac{1}{b^m}, \qquad b \neq 0,$$

# 2. Indices and Surds

In this chapter we shall explore the concepts of indices and surds and how they can be used for simplifying algebraic manipulation. As algebra is prevalent throughout the whole of mathematics, understanding the concepts introduced in this chapter is essential.

Figure 2.1: A Pythagoras Tree.

Figure 2.1 shows a fractal based on Pythagoras' Theorem and is called a *Pythagoras Tree*. In this construction, each square (except the largest) is connected to three others, the ratio of the sides of these squares, which will contain surds, is determined by the properties of the triangle enclosed between squares. For example, in Figure 2.1 sides are in the ratio $\sqrt{3} : \sqrt{2} : 1$. The total number of squares in the fractal can be expressed using index notation as $2^{n+1} - 1$, where $n$ is the number of colours.

## 2.1 Indices

Index notation was first used to reduce printing costs by representing repeated multiplication such as $b \times b \times b$ in the more compact form $b^3$. Here, we call $b$ the base and 3 the exponent or power. It turns out that this shorthand notation has many useful properties which extend beyond positive integer powers. We shall start by exploring properties for integer powers.

---

**Formulae 2.1 — Index Laws**
Below are some important properties of indices.

$$b^m \times b^n = b^{m+n}, \tag{2.1}$$
$$b^m \div b^n = b^{m-n}, \tag{2.2}$$
$$(b^m)^n = b^{mn}, \tag{2.3}$$
$$b^0 = 1, \qquad b \neq 0, \tag{2.4}$$
$$b^{-m} = \frac{1}{b^m}, \qquad b \neq 0. \tag{2.5}$$

---

We can prove property (2.1) simply by using the definition of the notation as follows:

$$b^m \times b^n = \overbrace{b \times b \times \cdots \times b}^{m \text{ times}} \times \overbrace{b \times b \times \cdots \times b}^{n \text{ times}} = b^{m+n}.$$
$$\underbrace{\phantom{b \times b \times \cdots \times b \times b \times b \times \cdots \times b}}_{m+n \text{ times}}$$

---

**Remark**
To understand the meaning of negative powers, consider the following sequence. Each row can be obtained by dividing the preceding row by 3.

$$
\begin{aligned}
3^4 &= 3 \times 3 \times 3 \times 3 = 81 \\
3^3 &= 3 \times 3 \times 3 = 27 \\
3^2 &= 3 \times 3 = 9 \\
3^1 &= 3 = 3 \\
3^0 &= 3 \div 3 = 1 \\
3^{-1} &= 1 \div 3 = \tfrac{1}{3} \\
3^{-2} &= \tfrac{1}{3} \div 3 = \tfrac{1}{9} \\
3^{-3} &= \tfrac{1}{9} \div 3 = \tfrac{1}{27}
\end{aligned}
$$

---

**Activity 2.1 — Proving index laws**
Write out a simple proof for each of the index laws showing clear reasoning.

---

**Example 2.1**
Simplify the following expressions.
(a) $(3xy^3)^2 \times (2x^4y^5)^3$,
(b) $\frac{8x^2 \times 3x^2 y^7}{xy^3}$,
(c) $\frac{x^{-3} \times x^{-5}}{x^4 y^3}$.

**Solution:**

(a)

$$(3xy^3)^2 \times (2x^4y^5)^3 = 9x^2y^6 \times 8x^{12}y^{15} \qquad \text{(by (2.3))}$$
$$= 72x^{14}y^{21}. \qquad \text{(by (2.1))}$$

(b)

$$\frac{8x^2 \times 3x^2y^7}{xy^3} = \frac{24x^4y^7}{xy^3} \qquad \text{(by (2.1))}$$
$$= 24x^3y^4. \qquad \text{(by (2.2))}$$

(c)

$$\frac{x^{-3} \times x^{-5}}{x^4y^3} = \frac{x^{-8}}{x^4y^3} \qquad \text{(by (2.1))}$$
$$= x^{-12}y^{-3}. \qquad \text{(by 2.2 and 2.5)}$$

So far, we have used integer powers only, but these rules also hold for rational powers. By using only the rules already established, we can understand the meaning of rational powers. If we try to think of $x^{\frac{1}{2}}$ as a repeated multiplication it makes little sense. However, if we write

$$x^{\frac{1}{2}} \times x^{\frac{1}{2}} = x^{\frac{1}{2}+\frac{1}{2}} = x,$$

it becomes clear that $x^{\frac{1}{2}} = \sqrt{x}$.

---

**Formulae 2.2 — Further Index Laws**
Index laws for rational exponents:

$$x^{\frac{1}{2}} = \sqrt{x}, \tag{2.6}$$
$$x^{\frac{1}{q}} = \sqrt[q]{x}, \tag{2.7}$$
$$x^{\frac{p}{q}} = \sqrt[q]{x^p} = (\sqrt[q]{x})^p. \tag{2.8}$$

---

**Example 2.2**
Evaluate or simplify the following expressions:
  (a) Evaluate $16^{\frac{3}{2}}$;
  (b) Evaluate $4^{-\frac{1}{2}}$;
  (c) Evaluate $\left(\frac{3}{4}\right)^{-3}$;
  (d) Simplify $9x^{\frac{4}{5}} \times 4x^{\frac{6}{5}}$.

**Solution:**

(a)

$$16^{\frac{3}{2}} = \left(\sqrt{16}\right)^3 \qquad\qquad \text{(by (2.8))}$$
$$= (4)^3$$
$$= 64.$$

(b)

$$4^{-\frac{1}{2}} = \frac{1}{4^{\frac{1}{2}}} \qquad\qquad \text{(by (2.5))}$$
$$= \frac{1}{\sqrt{4}} \qquad\qquad \text{(by (2.7))}$$
$$= \frac{1}{2}.$$

(c)

$$\left(\frac{3}{4}\right)^{-3} = \frac{1}{\left(\frac{3}{4}\right)^3} \qquad\qquad \text{(by (2.5))}$$
$$= \frac{1}{\left(\frac{27}{64}\right)}$$
$$= \frac{64}{27}.$$

(d)

$$9x^{\frac{4}{5}} \times 4x^{\frac{6}{5}} = 36x^{\frac{10}{5}} \qquad\qquad \text{(by (2.1))}$$
$$= 36x^2.$$

---

**Example 2.3**

(a) Express $\sqrt[5]{x^2} \times \sqrt[3]{x^4}$ in the form $x^n$,

(b) Evaluate $\left(\frac{125}{27}\right)^{-\frac{2}{3}}$,

(c) Express $\dfrac{12}{(\sqrt{64x})^{\frac{2}{3}}}$ in the form $ax^n$.

**Solution:**

(a)

$$\sqrt[5]{x^2} \times \sqrt[3]{x^4} = x^{\frac{2}{5}} \times x^{\frac{4}{3}} \qquad\qquad \text{(by(2.8))}$$
$$= x^{\left(\frac{2}{5}+\frac{4}{3}\right)} \qquad\qquad \text{(by (2.1))}$$
$$= x^{\frac{26}{15}}.$$

(b)

$$\left(\frac{125}{27}\right)^{-\frac{2}{3}} = \left(\frac{27}{125}\right)^{\frac{2}{3}} \qquad \text{(by (2.5))}$$

$$= \left(\sqrt[3]{\frac{27}{125}}\right)^{2} \qquad \text{(by (2.8))}$$

$$= \left(\frac{3}{5}\right)^{2} \qquad \text{(cube root evaluated)}$$

$$= \frac{9}{25}.$$

(c)

$$\frac{12}{(\sqrt{64x})^{\frac{2}{3}}} = \frac{12}{\left((64x)^{\frac{1}{2}}\right)^{\frac{2}{3}}} \qquad \text{(by (2.7))}$$

$$= \frac{12}{(64x)^{\frac{1}{3}}} \qquad \text{(by (2.3))}$$

$$= \frac{12}{4x^{\frac{1}{3}}} \qquad \text{(cube root of 64 evaluated)}$$

$$= 3x^{-\frac{1}{3}}. \qquad \text{(by (2.5))}$$

---

**Example 2.4**

Solve the equation $3^{2x-6} = 81$.

**Solution:**

First we manipulate each term until they all have the same base.

$$3^{2x-6} = 81$$
$$\Rightarrow \quad 3^{2x-6} = 3^{4}.$$

Since the base is the same on both sides their powers must be equal.

$$2x - 6 = 4$$
$$\Rightarrow \qquad 2x = 10$$
$$\Rightarrow \qquad x = 5.$$

**Example 2.5**
Solve the equation $2^{x+1} = 8^x$.

**Solution:**
First we ensure each term has the same base.

$$
\begin{aligned}
& 2^{x+1} = 8^x \\
\Rightarrow\ & 2^{x+1} = (2^3)^x \\
\Rightarrow\ & 2^{x+1} = 2^{3x}.
\end{aligned}
$$

Since the base is the same on both sides their powers must be equal.

$$
\begin{aligned}
& x + 1 = 3x \\
\Rightarrow\ & \quad 1 = 2x \\
\Rightarrow\ & \quad x = \frac{1}{2}.
\end{aligned}
$$

---

**Exercise 2.1**

Q1. Evaluate the following.
   (a) $2^3 \times 2^5$;
   (b) $275^0$;
   (c) $49^{\frac{1}{2}}$;
   (d) $8^{-2}$;
   (e) $25^{-\frac{1}{2}}$;
   (f) $32^{\frac{4}{5}}$;
   (g) $\left(\frac{625}{81}\right)^{\frac{3}{4}}$;
   (h) $36^{1.5}$;
   (i) $\left(\frac{4}{9}\right)^{-2}$;
   (j) $\left(\frac{32}{243}\right)^{-\frac{3}{5}}$.

Q2. Simplify the following.
   (a) $x^3 \times x^5$;
   (b) $(2y)^4 \times 3y^2$;
   (c) $20x^8 \div 30x^2$;
   (d) $\frac{12xy^3 \times 3x^5}{2x^2 y}$;
   (e) $\left(\frac{2}{x^2}\right)^{-2}$;
   (f) $(7x^2 y^{-1})^{-2}$;
   (g) $\frac{1}{x^{-3}}$;
   (h) $(8x^6)^{\frac{1}{3}} \times 2x^{\frac{1}{2}}$;
   (i) $\sqrt{x} \times \sqrt[4]{x}$;
   (j) $\frac{\sqrt[3]{x} \times x^2}{\sqrt{x}}$.

Q3. Solve the following equations.
   (a) $7^x = \frac{1}{49}$;
   (b) $3^x \times 27^{x-1} = 243$;
   (c) $5^{2x} = 125^{1-3x}$;

(d) $9^{2x+1} = 27^{5-x}$;

(e) $16 - 8^{x-4} = 0$;

(f) $\left(\frac{1}{4}\right)^{x+2} = \sqrt[3]{8}$;

(g) $\frac{2^{3x+1}}{4^{2-x}} = \frac{8^{2x}}{16}$.

## 2.2 Surds

Here we take a deeper look at roots, in particular, *surds*. A surd provides a way of denoting a root when there is no rational expression available. For example, numbers such as $\sqrt{4} = 2$ or $\sqrt[3]{\frac{8}{27}} = \frac{2}{3}$ are *rational* but $\sqrt{2}$ and $\sqrt{3}$ are *surds*. Some useful properties of surds arise naturally by using the index laws discussed earlier.

**Formulae 2.3 — Surd Laws**

Properties of surds for $x, y \geq 0$.

$$\sqrt{x} \times \sqrt{y} = \sqrt{x \times y}; \tag{2.9}$$

$$\sqrt{x} \div \sqrt{y} = \sqrt{\frac{x}{y}}. \tag{2.10}$$

**Remark**

A common misconception when working with surds is to assume that $\sqrt{x+y} = \sqrt{x} + \sqrt{y}$. In general, this is not true unless one or both of $x$ and $y$ are zero. However, it can be shown that

$$\sqrt{x+y} \leq \sqrt{x} + \sqrt{y}.$$

This is an example of a *Triangle Inequality*.

**Activity 2.2 — Proving Surd Laws**

Property 2.9 can be proven in the following way. Let $a = \sqrt{x} \times \sqrt{y}$ and $b = \sqrt{x \times y}$. Squaring these we obtain,

$$a^2 = (\sqrt{x}\sqrt{y})^2 = (\sqrt{x}\sqrt{y})(\sqrt{x}\sqrt{y}) = \sqrt{x}\sqrt{x}\sqrt{y}\sqrt{y} = xy$$
$$b^2 = (\sqrt{xy})^2 = xy.$$

Equating these we have $a^2 = b^2$. Since $x, y \geq 0$ we also have $a, b \geq 0$. Hence $a = b$ and $\sqrt{x}\sqrt{y} = \sqrt{xy}$.

Q1. Write a similar proof for (2.10).

Q2. Prove that $\sqrt{x+y} \leq \sqrt{x} + \sqrt{y}$ holds for all $x, y \geq 0$.

Using the properties from Formulae 2.3 we can manipulate surds to make them simpler or more convenient to work with. We can simplify surds such as $\sqrt{48}$ by applying the multiplication property 2.9. This can be done in a number of ways, some are shown below.

$$\sqrt{48} = \sqrt{24 \times 2} = \sqrt{24} \times \sqrt{2},$$
$$\sqrt{48} = \sqrt{12 \times 2 \times 2} = \sqrt{12} \times \sqrt{2} \times \sqrt{2} = 2\sqrt{12},$$
$$\sqrt{48} = \sqrt{16 \times 3} = \sqrt{16} \times \sqrt{3} = 4\sqrt{3}.$$

The form $\sqrt{48} = 4\sqrt{3}$ is the simplest, since 16 is the largest square number that is also a factor of 48. Writing surds in this form allows us to simplify numbers such as $\sqrt{48} + 2\sqrt{3}$ to $6\sqrt{3}$.

**Example 2.6**
Simplify $\sqrt{180} + \sqrt{20}$.

**Solution:**

$$\sqrt{180} + \sqrt{20} = \sqrt{36 \times 5} + \sqrt{4 \times 5}$$
$$= 6\sqrt{5} + 2\sqrt{5}$$
$$= 8\sqrt{5}.$$

**Example 2.7**
Expand and simplify $(7\sqrt{3} + 6)(2\sqrt{3} - 2)$.

**Solution:**

$$(7\sqrt{3} + 2)(2\sqrt{3} - 6) = 14 \times 3 - 14\sqrt{3} + 12\sqrt{3} - 12$$
$$= 30 - 2\sqrt{3}.$$

**Tip**
It can sometimes be difficult to find the largest square factor of large numbers without a calculator. We can write any number as a product of its prime factors (using a factor tree if necessary) as shown below.

$$\sqrt{180} = \sqrt{\underbrace{2 \times 2 \times 3 \times 3}_{36 \text{ is a factor}} \times 5} = 6\sqrt{5}.$$

All pairs of prime factors will give square factors.

When thinking about fractions involving surds, such as $\frac{1}{\sqrt{2}}$, it can be difficult to understand what it means to divide by an irrational number ($\sqrt{2} = 1.41421...$). We can simplify fractions of this form by *rationalising the denominator*. To do this, we multiply the fraction by 1 in such a way that the denominator becomes a rational number as shown below.

$$\frac{1}{\sqrt{2}} = \frac{1}{\sqrt{2}} \times \frac{\sqrt{2}}{\sqrt{2}} = \frac{\sqrt{2}}{2}.$$

Now, rather than trying to think what $1 \div 1.41421...$ is, we can see this is actually the same as $1.41421... \div 2 = 0.7071....$.

---

**Example 2.8**

Simplify $\frac{8}{3\sqrt{2}}$.

**Solution:**

$$\frac{8}{3\sqrt{2}} = \frac{8}{3\sqrt{2}} \times \frac{\sqrt{2}}{\sqrt{2}}$$
$$= \frac{8\sqrt{2}}{3 \times 2}$$
$$= \frac{4\sqrt{2}}{3}.$$

---

**Example 2.9**

Simplify $\frac{14}{3+\sqrt{2}}$.

**Solution:**

$$\frac{14}{3 + \sqrt{2}} = \frac{14}{3 + \sqrt{2}} \times \frac{3 - \sqrt{2}}{3 - \sqrt{2}}$$
$$= \frac{14(3 - \sqrt{2})}{9 + 3\sqrt{2} - 3\sqrt{2} - 2}$$
$$= \frac{14(3 - \sqrt{2})}{7}$$
$$= 6 - 2\sqrt{2}.$$

---

**Example 2.10**

Simplify $\frac{4\sqrt{2}}{\sqrt{5}-\sqrt{2}}$.

**Solution:**

$$\frac{4\sqrt{2}}{\sqrt{5} - \sqrt{2}} = \frac{4\sqrt{2}}{\sqrt{5} - \sqrt{2}} \times \frac{\sqrt{5} + \sqrt{2}}{\sqrt{5} + \sqrt{2}}$$
$$= \frac{4\sqrt{2}(\sqrt{5} + \sqrt{2})}{5 + \sqrt{10} - \sqrt{10} - 2}$$
$$= \frac{4\sqrt{10} + 8}{3}.$$

---

**Tip**

If the denominator of a fraction is of the form $a \pm b\sqrt{c}$ or $a\sqrt{b} \pm c\sqrt{d}$, we can use the difference of two squares identity to identify a suitable multiplier. Some examples are shown below.

$$(1 + 2\sqrt{3})(1 - 2\sqrt{3}) = 1 - 2\sqrt{3} + 2\sqrt{3} - 4\sqrt{9} = -11,$$
$$(\sqrt{2} - 3\sqrt{5})(\sqrt{2} + 3\sqrt{5}) = \sqrt{4} + 3\sqrt{10} - \sqrt{10} - 9\sqrt{25} = -43.$$

**Interactive Activity 2.1 — Order the Surd Expressions**
In the Tarquin interactive activity below (in the digital book), place the surd expressions in order.

**Exercise 2.2**
Q1. Simplify the following.
  (a) $\sqrt{5} \times \sqrt{5}$;
  (b) $\sqrt{8} \times \sqrt{2}$;
  (c) $2\sqrt{7} \times 3\sqrt{7}$;
  (d) $\sqrt{12} \times \sqrt{3}$;
  (e) $\frac{\sqrt{30}}{\sqrt{5}}$;
  (f) $\frac{\sqrt{24}}{\sqrt{3}}$;
  (g) $\frac{\sqrt{180}}{\sqrt{5}}$;
  (h) $\frac{\sqrt{5}}{\sqrt{20}}$.
Q2. Simplify the following.
  (a) $\sqrt{18}$;
  (b) $\sqrt{120}$;
  (c) $\sqrt{72}$;
  (d) $\sqrt{24}$;
  (e) $\sqrt{20} + \sqrt{5}$;
  (f) $\sqrt{12} + \sqrt{12}$;
  (g) $\sqrt{8} + \sqrt{50}$;
  (h) $3\sqrt{27} + \sqrt{48}$.
Q3. Expand and simplify the following.
  (a) $(2 + \sqrt{3})(2 - \sqrt{3})$;
  (b) $(1 - 2\sqrt{5})^2$;
  (c) $(2\sqrt{2} + \sqrt{3})(3\sqrt{2} + 2\sqrt{3})$;
  (d) $(\sqrt{3} + \sqrt{5})(\sqrt{3} + \sqrt{7})$.
Q4. Rationalise the denominator of the following fractions.
  (a) $\frac{1}{\sqrt{7}}$;
  (b) $\frac{20}{\sqrt{5}}$;
  (c) $\frac{\sqrt{7}}{\sqrt{2}}$;
  (d) $\frac{3\sqrt{10}}{2\sqrt{15}}$;
  (e) $\frac{1}{4 + \sqrt{2}}$;

(f) $\frac{12}{\sqrt{5}-1}$;

(g) $\frac{2+\sqrt{2}}{\sqrt{2}-2}$;

(h) $\frac{5+2\sqrt{3}}{4-\sqrt{3}}$.

## Key Facts — Indices and Surds

- The following are properties of indices:

$$b^m \times b^n = b^{m+n},$$
$$b^m \div b^n = b^{m-n},$$
$$(b^m)^n = b^{mn},$$
$$b^0 = 1, \qquad b \neq 0,$$
$$b^{-m} = \frac{1}{b^m}, \qquad b \neq 0,$$
$$x^{\frac{1}{2}} = \sqrt{x},$$
$$x^{\frac{1}{q}} = \sqrt[q]{x},$$
$$x^{\frac{p}{q}} = \sqrt[q]{x^p} = (\sqrt[q]{x})^p.$$

- The following are properties of surds:

$$\sqrt{x} \times \sqrt{y} = \sqrt{x \times y},$$
$$\sqrt{x} \div \sqrt{y} = \sqrt{\frac{x}{y}}.$$

- When simplifying $\sqrt{a}$, first seek any square factors of $a$. For example,

$$\sqrt{432} = \sqrt{16 \times 9 \times 3}$$
$$= (4 \times 3)\sqrt{3}$$
$$= 12\sqrt{3}.$$

- Simplifying surds of the form $\frac{\sqrt{a}+\sqrt{b}}{\sqrt{c}\pm\sqrt{d}}$ can be performed as follows

$$\frac{\sqrt{a}+\sqrt{b}}{\sqrt{c}\pm\sqrt{d}} = \frac{\sqrt{a}+\sqrt{b}}{\sqrt{c}\pm\sqrt{d}} \times \frac{\sqrt{c}\mp\sqrt{d}}{\sqrt{c}\mp\sqrt{d}} = \frac{(\sqrt{a}+\sqrt{b})(\sqrt{c}\mp\sqrt{d})}{c^2-d^2}.$$

## Chapter Assessment — Indices and Surds

Download and sit the 30 minute assessment for this chapter from the digital book.

$$-\sin(x)$$

$$R = \sqrt{1^2 + 2^2}$$
$$= \sqrt{1+4}$$
$$= \sqrt{5}$$
$$= 2.24N \text{ to 2 decimal places}$$

$$d\sin(\phantom{x})$$

$$+ \frac{f(x}{2}$$

# 3. Vectors

The word "vector" is from the Latin word for "carrier". A vector is what is needed to "carry" the point $A$ to the point $B$. Vectors are used in many applications of mathematics, particularly for solving problems in physics and engineering. Examples include:

- Problems in geometry, such as finding the point of intersection between straight lines and finding the distance between points.
- Problems in mechanics, such as particle motion and resolving forces in statics.

Vectors are incredibly useful for modelling mechanical systems in three dimensions, which can be geometrically complicated otherwise. The basic rules of vector algebra apply regardless of the complexity of the problem being studied.

Most measurable quantities fall into one of two different categories:

- *Scalars* have a *size* or *magnitude*, *i.e.* they have a numerical value, but no specified direction. Examples of scalar quantities are mass, density, frequency, speed and time. A scalar can be represented by a (single) number.
- *Vectors* have both a *magnitude* **and** a *direction*. Examples of vector quantities are force, velocity, acceleration, and position.

The principal difference between vector and a scalar is that a vector quantity involves a movement in a particular direction, whereas a scalar is simply a quantity. Figure 3.1 shows the magnetic force field around a bar magnet. At every point around the magnet, the magnetic force has both a direction and a magnitude. Lines of equal force magnitude are shown and the direction of the force is then given by the direction of these lines.

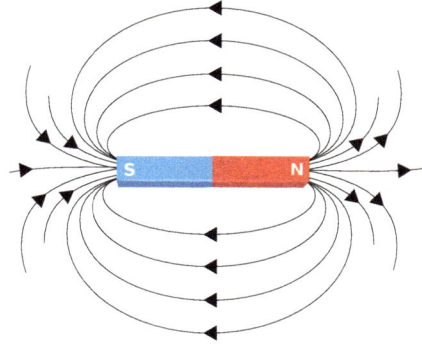

Figure 3.1: Force field around a bar magnet.

It is important to distinguish clearly between scalars and vectors in written work.
- In books, vectors are written in bold type, *e.g.* **a**.
- When writing mathematics by hand, the notation $\underline{a}$ or $\underset{\sim}{a}$ is used.

## 3.1 Vectors and their Geometrical Interpretation

A vector **a** is represented diagrammatically by a directed line with an arrow at the end as shown in Figure 3.2.

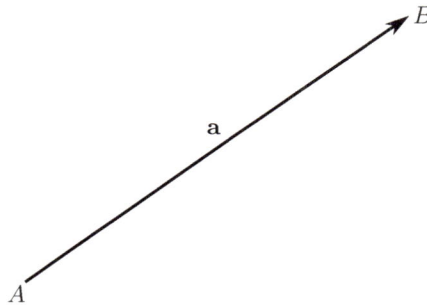

Figure 3.2: The vector $\mathbf{a} = \overrightarrow{AB}$.

The *magnitude* of the vector **a** is represented by the length of the line; its *direction* is that of the arrow. If the points $A$ and $B$ are at the ends of the arrow, then an alternative notation for the vector is $\overrightarrow{AB}$ (the directed line segment from $A$ to $B$).

### 3.1.1 Reversing a Vector

If $\mathbf{a} = \overrightarrow{AB}$, then $\overrightarrow{BA} = -\mathbf{a}$ is the *reverse vector*, of **a**. This is geometrically intuitive, since **a** is the vector from the point $A$ to the point $B$, while $-\mathbf{a}$ is the vector from $B$ to $A$.

### 3.1.2 Equality

Two vectors **a** and **b**, are said to be *equal*, written **a** = **b**, if **a** and **b** have both the same length and direction. An examples of this is shown in Figure 3.3.

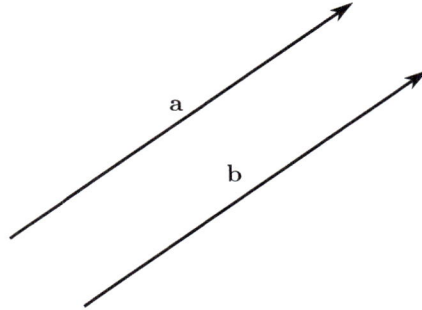

Figure 3.3: The vectors **a** and **b** are parallel and have equal length, so **a** = **b**.

### 3.1.3 Addition of Vectors

The sum (or *resultant*) of two vectors **a** and **b**, written **a**+**b**, is constructed by first drawing the vector **a**, then drawing the vector **b** with its tail end starting at the arrow end of **a**. The vector **c** = **a** + **b** is given by the vector starting at the tail end of **a** with its arrow end at the arrow end of **b**. This is known as the triangle law of vector addition, as **c** is the third side of a triangle, as can be seen in Figure 3.4.

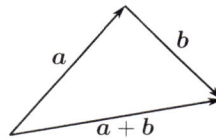

Figure 3.4: The triangle law of vector addition.

Similarly, we can determine **b** + **a**.

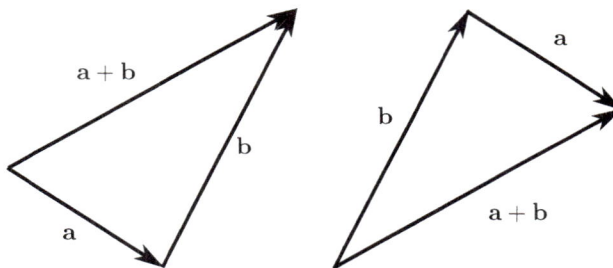

Figure 3.5

We see that **a** + **b** = **b** + **a** so that the order of adding vectors does not matter, as shown in Figure 3.5.

Alternatively, we use the *parallelogram law of vector addition* (Figure 3.6). This law is constructed by combining the triangles in Figure 3.5.

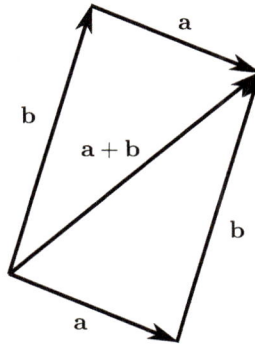

Figure 3.6: The parallelogram law of vector addition.

In addition, we also have that

$$\mathbf{a} + (\mathbf{b} + \mathbf{c}) = (\mathbf{a} + \mathbf{b}) + \mathbf{c}.$$

It follows that $\mathbf{a} + \mathbf{b} + \mathbf{c}$ can be evaluated without ambiguity by first adding any two of the three vectors $\mathbf{a}$, $\mathbf{b}$, $\mathbf{c}$, and then adding the third.

**Activity 3.1 — Associativity of Vector Addition**
Prove that $\mathbf{a} + (\mathbf{b} + \mathbf{c}) = (\mathbf{a} + \mathbf{b}) + \mathbf{c}$.

**Interactive Activity 3.1 — Vector Addition**
Using the vectors provided in the Geogebra applet below (in the digital book) show the vectors $2\mathbf{u}$, $\mathbf{u} + \mathbf{v}$, $\mathbf{v} + \mathbf{u}$, $\mathbf{u} - 2\mathbf{v}$, $-\mathbf{v}$ and $\mathbf{w} + \mathbf{u}$ with reference to the blue dot. The vectors at the top of the activity can be moved and placed in the appropriate location.

### 3.1.4   The zero vector

The *zero vector* is denoted by $\mathbf{0}$ and has zero magnitude and arbitrary direction. Vector addition includes the case where one of the vectors begins and ends at the same point so that

$$\mathbf{a} + \mathbf{0} = \mathbf{a}.$$

On the other hand, if $\mathbf{a} + \mathbf{b} = \mathbf{0}$, then it is natural, using basic arithmetic, to write

$$\mathbf{b} = -\mathbf{a}.$$

The vector $-\mathbf{a}$ has the same magnitude as $\mathbf{a}$ but points in the opposite direction to $\mathbf{a}$, as defined in Section 3.1.1.

**Example 3.1**

Consider the diagram below.

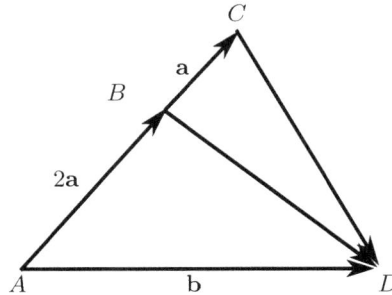

Find the vectors $\overrightarrow{BD}$ and $\overrightarrow{CD}$ in terms of **a** and **b**.

**Solution:**

$$\overrightarrow{BD} = \overrightarrow{BA} + \overrightarrow{AD}$$
$$= -2\mathbf{a} + \mathbf{b},$$
$$\overrightarrow{CD} = \overrightarrow{CB} + \overrightarrow{BA} + \overrightarrow{AD}$$
$$= -\mathbf{a} - 2\mathbf{a} + \mathbf{b} = -3\mathbf{a} + \mathbf{b}.$$

**Example 3.2**

In the diagram below, point $E$ is the midpoint of $CB$, and the point $D$ is such that $BD : DA$ is in the ratio $4 : 3$. Given that $\overrightarrow{CA} = \mathbf{a}$ and $\overrightarrow{CB} = \mathbf{b}$, show that $ED$ is not parallel to $CA$.

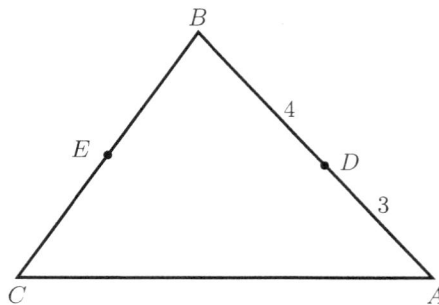

**Solution:**

We find the vector $\overrightarrow{ED}$.

$$\overrightarrow{ED} = \overrightarrow{EB} + \overrightarrow{BD}.$$

Since $E$ is the midpoint of $CB$,

$$\overrightarrow{EB} = \frac{1}{2}\,\overrightarrow{CB},$$
$$= \frac{1}{2}\mathbf{b}.$$

As $BD : DA$ is in the ratio $4 : 3$,

$$\overrightarrow{BD} = \frac{4}{7}\,\overrightarrow{BA},$$
$$= \frac{4}{7}(-\mathbf{b} + \mathbf{a}).$$

Hence,

$$\overrightarrow{ED} = \overrightarrow{EB} + \overrightarrow{BD},$$
$$= \frac{1}{2}\mathbf{b} + \frac{4}{7}(-\mathbf{b} + \mathbf{a}),$$
$$= \frac{4}{7}\mathbf{a} - \frac{1}{14}\mathbf{b}.$$

Since $\overrightarrow{ED}$ is not a scalar multiple of $\overrightarrow{CA}$, $ED$ is not parallel to $CA$.

---

**Example 3.3**

A ferry takes people from a terminal on one bank of a straight stretch of river to a spot directly opposite on the other bank. On one particular day, the river is in flood and flowing at $4\,\mathrm{m\,s^{-1}}$; the boat has a top speed of $5\,\mathrm{m\,s^{-1}}$ in still water.

Ignoring any acceleration and deceleration, calculate the direction (as an acute angle to the river bank) the boat must head to make the crossing in the shortest time and the resultant crossing speed.

**Solution:**

The boat must head upstream at an angle $\theta$ to the bank to overcome being swept downstream by the motion of the water. The resultant direction must be perpendicular to the bank. Using the parallelogram law of vector addition, we obtain the following vector diagram.

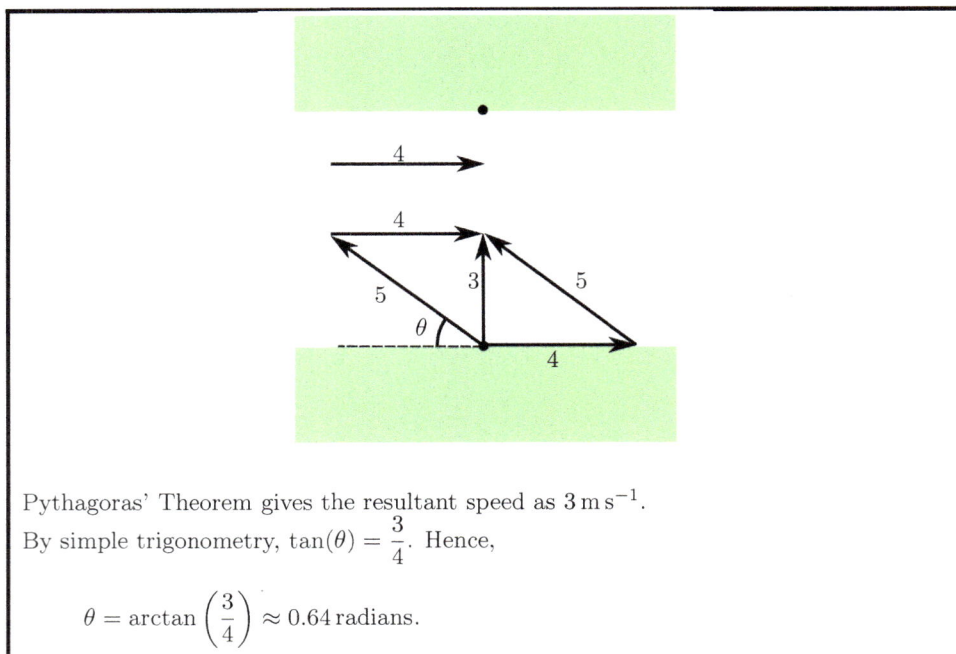

Pythagoras' Theorem gives the resultant speed as $3\,\mathrm{m\,s^{-1}}$.

By simple trigonometry, $\tan(\theta) = \dfrac{3}{4}$. Hence,

$$\theta = \arctan\left(\frac{3}{4}\right) \approx 0.64\,\text{radians}.$$

### 3.1.5    Position Vectors

We define points using vectors by defining their position relative to a fixed origin $O$. The point $A$ is located at the end point of the *position vector* **a**, which starts at $O$. In general, we use $\overrightarrow{AB}$ to denote the vector from the point $A$ to the point $B$. Hence, the position vector **a** can be denoted $\overrightarrow{OA}$. Using the triangle law of addition, we define $\overrightarrow{AB}$ in terms of the position vectors of **a** and **b**.

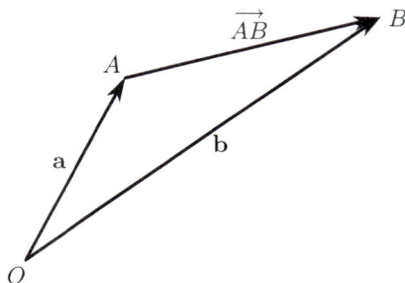

Figure 3.7: The vector $\overrightarrow{AB}$ is the vector between the points $A$ and $B$, which have position vectors **a** and **b** respectively.

Using Figure 3.7 as a guide, we see that

$$\vec{OB} = \vec{OA} + \vec{AB},$$

$$\Rightarrow \quad \mathbf{b} = \mathbf{a} + \vec{AB},$$

$$\Rightarrow \quad \vec{AB} = \mathbf{b} - \mathbf{a}.$$

> **Remark**
> Note that the vector $\vec{AB}$ is fixed regardless of where the origin is, since the points $A$ and $B$ have fixed positions relative to each other.

### 3.1.6 Multiplication of a Vector by a Scalar

It is natural to consider multiples of vectors. For example, we might wish to consider the effect of doubling some applied force. The following rules apply, where $k$ is an arbitrary nonzero scalar and $k_1$ and $k_2$ are nonzero scalars with $k_1 \neq k_2$.

1. $k\mathbf{a}$ denotes a vector having magnitude $|k|$ times the magnitude of $\mathbf{a}$ which is parallel to $\mathbf{a}$. If $k < 0$, the direction is reversed (by Section 3.1.1).
2. $0\mathbf{a} = \mathbf{0}$.
3. $k\mathbf{0} = \mathbf{0}$.
4. $1\mathbf{a} = \mathbf{a}$.
5. $(-k)\mathbf{a} = -(k\mathbf{a})$.
6. $(-1)\mathbf{a} = -(\mathbf{a})$.
7. $k(\mathbf{a} + \mathbf{b}) = k\mathbf{a} + k\mathbf{b}$.
8. $(k_1 + k_2)\mathbf{a} = k_1\mathbf{a} + k_2\mathbf{a}$.
9. $(k_1 k_2)\mathbf{a} = k_1(k_2\mathbf{a})$.

> **Definition 3.1 — Magnitude Notation**
> The magnitude (often referred to as the *length* or *modulus*) of a vector is denoted by $|\mathbf{a}|$. In handwritten work, the magnitude is denoted by $a$, $|\underline{a}|$ or $|\underset{\sim}{a}|$.

> **Activity 3.2 — Vector Diagrams**
> Use vector diagrams or the rules for vector addition to prove the above rules for scalar multiplication.

**Exercise 3.1**

Q1. Consider the diagram below.

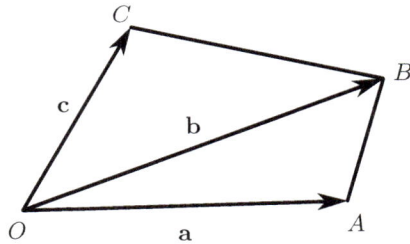

Write the following in terms of **a**, **b** and **c**.

(a) $\overrightarrow{AB}$;

(b) $\overrightarrow{BA}$;

(c) $\overrightarrow{BC}$;

(d) $\overrightarrow{AC}$.

Check that the same answer is obtained in terms of **a**, **b** and **c** regardless of which path between the two points is chosen.

Q2. The diagram below shows a triangle $OAB$. The point $M$ is the midpoint of the line $AB$.

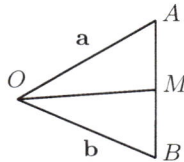

Determine $\overrightarrow{OM}$ in terms of **a** and **b**.

Q3. The diagram below shows two forces $\mathbf{F}_1$ and $\mathbf{F}_2$ acting on an object $O$.

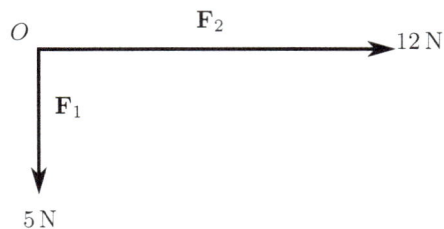

(a) Add the vector representing the resultant force $\mathbf{R} = \mathbf{F}_1 + \mathbf{F}_2$ to the diagram.

(b) Find the magnitude of **R**.

(c) Find the angle between **R** and $\mathbf{F}_2$.

Q4. If **a** and **b** are the position vectors of two points $A$ and $B$, find the position vector of the point $C$ on $\overrightarrow{AB}$ such that $\overrightarrow{AC} = \frac{4}{5}\overrightarrow{AB}$.

Q5. Suppose that the vectors **a** and **b** describe the horizontal and sloped sides of a parallelogram respectively. There exists $c$ such that the vector $\mathbf{v} = \mathbf{b} - c\mathbf{a}$ is perpendicular to **a**. What is the geometric interpretation of $|\mathbf{v}|$?

Q6. If **a**, **b** and **c** are the position vectors of the vertices $A$, $B$ and $C$ respectively of the triangle $ABC$, prove, using a vector based argument, that the straight line joining the midpoints of the lines $AB$ and $AC$ is parallel to $BC$ and half the length of $BC$.

Q7. Suppose that $\alpha\mathbf{a} + \beta\mathbf{b} = \mathbf{0}$, and **a** and **b** are non-zero vectors which are not parallel. Show that $\alpha = \beta = 0$.

Q8. In the trapezium below, $DA = 2CB$. $FB$ is parallel to $DE$ and $|\overrightarrow{FB}| = \frac{5}{7}|\overrightarrow{DE}|$. The lengths $BE$ and $EA$ are in the ratio $2:3$ Find the ratio $DF:FC$.

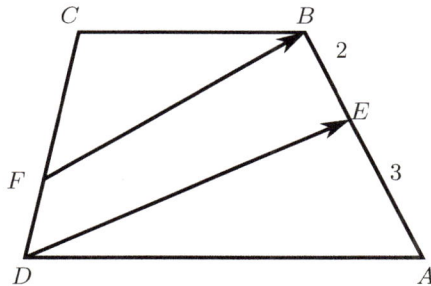

## 3.2 Components of a Vector

So far, we have dealt with vectors as geometric objects and explored the algebraic results that apply to them. To perform complex calculations using vectors, we also need to describe them numerically. In the *Cartesian* coordinate system, we break a vector down into its components in the $x$ and $y$ direction.

**Definition 3.2 — 2D Cartesian Basis Vectors**

We write 2D vectors in terms of the basis vectors **i** and **j**, which are vector of unit length in the direction of the $x$ and $y$ axis respectively (see the diagram below).

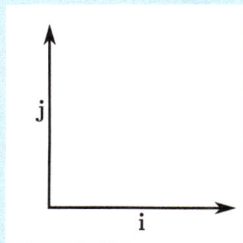

If the vector $\mathbf{a} = a_1\mathbf{i} + a_2\mathbf{j}$, then the scalars $a_1$ and $a_2$ are the *components* of the vector $\mathbf{a}$ relative to $\mathbf{i}$ and $\mathbf{j}$, respectively. We may also refer to the components $a_1$ and $a_2$ as the $x$ and $y$ components of $\mathbf{a}$, respectively.

### 3.2.1 Vector Notation

When writing the vector $\mathbf{a}$ with components $a_1$ and $a_2$, we can use any of the following equivalent notations.
- Cartesian form: $\mathbf{a} = a_1\mathbf{i} + a_2\mathbf{j}$.
- Row form: $\mathbf{a} = (a_1, a_2)$.
- Column form: $\mathbf{a} = \begin{pmatrix} a_1 \\ a_2 \end{pmatrix}$.

### 3.2.2 Position Vectors

Using the component form of vectors, we can interpret coordinates of points in the $xy$-plane as vectors from the origin $O$ to the point, as shown in Figure 3.8.

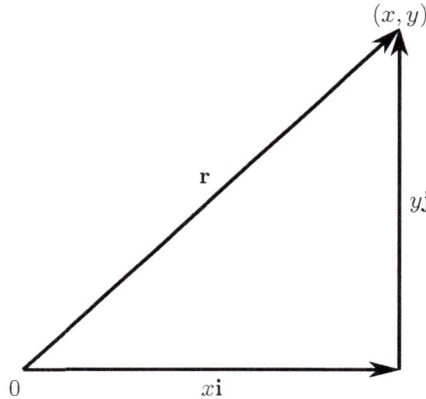

Figure 3.8: Position vector in two dimensions.

The point with coordinates $(x, y)$ relative to a fixed origin $O$ has position vector $\mathbf{r} = x\mathbf{i} + y\mathbf{j}$.

**Formula 3.1 — Addition of 2D Vectors**
In component form, we add vectors by simply adding their components, so if $\mathbf{a} = (a_1, a_2)$ and $\mathbf{b} = (b_1, b_2)$, then

$$\mathbf{a} + \mathbf{b} = (a_1 + b_1, a_2 + b_2).$$

**Formula 3.2 — Magnitude of 2D vectors**
The magnitude of the vector $\mathbf{a}$ is equal to its geometric length. Using Pythagoras' theorem in Figure 3.8 and recalling that $|\mathbf{i}| = |\mathbf{j}| = 1$, the vector $\mathbf{a} = (a_1, a_2)$ has

magnitude

$$|\mathbf{a}| = \sqrt{a_1^2 + a_2^2}.$$

We can use right angled trigonometry to find the direction of a vector relative to the unit vectors $\mathbf{i}$ or $\mathbf{j}$.

**Remark**

If we are working in a Cartesian framework then this is equivalent to finding the direction relative to the $x$-axis or $y$-axis respectively.

---

**Example 3.4**

Find the angle between the vector $5\mathbf{i} + 3\mathbf{j}$ and the positive $x$-axis.

**Solution:**

We first sketch the vector to ensure that we find the correct angle.

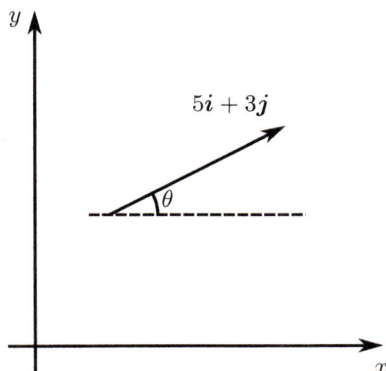

Using right angled trigonometry, which should be familiar from GCSE, we can find the angle $\theta$ using the tangent ratio.

$$\tan(\theta) = \frac{3}{5},$$

$$\Rightarrow \qquad \theta = \left(\frac{3}{5}\right),$$

$$= 30.96375653.$$

Hence, the vector $5\mathbf{i} + 3\mathbf{j}$ makes an angle of approximately $31°$ with the $x$-axis.

**Interactive Activity 3.2 — Magnitude and Direction of Vectors**

Use the Geogebra applet below (in the digital book) to find the magnitude and direction of vectors. The blue circle representing the end of the vector can be moved around.

**Definition 3.3 — The unit vector**

A *unit vector* in the direction of the vector $\mathbf{a}$, denoted $\hat{\mathbf{a}}$, has the same direction as $\mathbf{a}$, but has unit magnitude, *i.e.* $|\hat{\mathbf{a}}| = 1$. The unit vector in the direction of $\mathbf{a}$ is found using the formula

$$\hat{\mathbf{a}} = \frac{\mathbf{a}}{|\mathbf{a}|}.$$

**Example 3.5**

Let $\mathbf{a} = (2, -1)$ and $\mathbf{b} = (3, 2)$. Find
  (a) $2\mathbf{a} + 3\mathbf{b}$.
  (b) $|\mathbf{a}|$ and $|\mathbf{b}|$.
  (c) The unit vectors $\hat{\mathbf{a}}$ and $\hat{\mathbf{b}}$.

**Solution:**
  (a)

$$2\mathbf{a} + 3\mathbf{b} = 2(2, -1) + 3(3, 2),$$
$$= (4, -2) + (9, 6),$$
$$= (13, 4).$$

  (b) $|\mathbf{a}| = \sqrt{2^2 + (-1)^2} = \sqrt{5}$, $|\mathbf{b}| = \sqrt{3^2 + 2^2} = \sqrt{13}$.
  (c) Using the magnitudes computed in part (b),

$$\hat{\mathbf{a}} = \frac{\mathbf{a}}{|\mathbf{a}|} = \frac{1}{\sqrt{5}}(2, -1) = \left( \frac{2}{\sqrt{5}}, -\frac{1}{\sqrt{5}} \right),$$
$$\hat{\mathbf{b}} = \frac{\mathbf{b}}{|\mathbf{b}|} = \frac{1}{\sqrt{13}}(3, 2) = \left( \frac{3}{\sqrt{13}}, \frac{2}{\sqrt{13}} \right).$$

**Example 3.6**

The points $A$ and $B$ have coordinates $(1, 2)$ and $(-2, 5)$ respectively. Find the vector $\overrightarrow{AB}$ and hence compute the distance between $A$ and $B$.

**Solution:**
$A$ has position vector $\mathbf{a} = (1, 2)$ and $B$ has position vector $\mathbf{b} = (-2, 5)$. Then

$$\overrightarrow{AB} = \mathbf{b} - \mathbf{a} = (-2, 5) - (1, 2) = (-3, 3).$$

The distance between $A$ and $B$ is given by the length of $\overrightarrow{AB}$, where

$$| \overrightarrow{AB} | = \sqrt{(-3)^2 + 3^2} = \sqrt{18} = 3\sqrt{2}.$$

**Example 3.7**

Find the direction between the vectors $\mathbf{u} = 3\mathbf{i} + 4\mathbf{j}$ and $\mathbf{v} = 2\mathbf{i} + 2\mathbf{j}$.

**Solution:**

We first find the angle, $\theta_1$ between $\mathbf{u}$ and the unit vector $\mathbf{i}$.

$$\theta_1 = \arctan\left(\frac{4}{3}\right),$$
$$\approx 53.1°.$$

Similarly, the angle $\theta_2$ between $\mathbf{v}$ and the unit vector $\mathbf{i}$ is given by

$$\theta_2 = \arctan\left(\frac{2}{2}\right),$$
$$= 45°.$$

Hence, the angle between $\mathbf{u}$ and $\mathbf{v}$ is approximately $53.1 - 45 = 8.13°$.

**Use of Technology 3.1 — Vector Operations**

Some scientific calculators can calculate the magnitude and direction of vectors.

**Interactive Activity 3.3 — Magnitude and Direction of Vectors - Matching Pairs**

The matching pairs interactive in the digital book can be used to practice finding the magnitude and direction of vectors.

**Exercise 3.2**

Q1. Write the following vectors in (i) row form; and (ii) column form.

(a) $\mathbf{i} + \mathbf{j}$;

(b) $2\mathbf{i}$;

(c) $-3\mathbf{j}$;

(d) $3\boldsymbol{i} + 2\boldsymbol{j}$;

(e) $\mathbf{j} - 4\mathbf{i}$;

(f) $\frac{1}{2}\mathbf{i} - \frac{3}{2}\mathbf{j}$.

Q2. For each of the following vectors, $\mathbf{a}$, find (i) $|\mathbf{a}|$, the magnitude of $\mathbf{a}$; and (ii) $\hat{\mathbf{a}}$, the unit vector in the direction of $\mathbf{a}$, writing your answer in the same form as that given.

(a) $3\mathbf{i} + 4\mathbf{j}$;

(b) $(1, -2)$;

(c) $\begin{pmatrix} -2 \\ 4 \end{pmatrix}$;

(d) $(-1, 1)$.

+

Q3. For the following pairs of points $A$ and $B$ with position vectors $\mathbf{a}$ and $\mathbf{b}$ respectively, find (i) the vector $\overrightarrow{AB}$, writing your answer in row form; and (ii) the distance between $A$ and $B$.

(a) $\mathbf{a} = (1, 1)$, $\mathbf{b} = (0, 1)$;

(b) $\mathbf{a} = (3, 2)$, $\mathbf{b} = (1, -2)$;

(c) $\mathbf{a} = (0, 4)$, $\mathbf{b} = (4, 0)$;

(d) $\mathbf{a} = (5, 4)$, $\mathbf{b} = (-2, 3)$.

Q4. Find vectors with magnitude $k$ in the direction $\mathbf{a}$, where

(a) $k = 9$, $\mathbf{a} = (3, 4)$;

(b) $k = 3$, $\mathbf{a} = (1, -2)$;

(c) $k = \dfrac{1}{3}$, $\mathbf{a} = (-4, 2)$;

(d) $k = \sqrt{2}$, $\mathbf{a} = (2, 2)$.

Q5. Find the direction between the vectors given below and the $y$-axis.

(a) $\mathbf{u} = 7\mathbf{i} + 2\mathbf{j}$;

(b) $\mathbf{u} = 2\mathbf{i} + 4\mathbf{j}$;

(c) $\mathbf{u} = -4\mathbf{i} + 3\mathbf{j}$;

**Learning Resource 3.1 — Categorise the Vectors**

From the digital book download the resource below. Sort the vectors into categories and justify why they are suitable categories.

**Key Facts — Vectors**

- A vector has both *magnitude* and *direction*.
- For two vectors $\mathbf{a}$ and $\mathbf{b}$, $\mathbf{a} + \mathbf{b} = \mathbf{b} + \mathbf{a}$.
- The magnitude of a vector $\mathbf{a}$ is written as $|\mathbf{a}|$.
- The following rules apply, where $k$ is an arbitrary nonzero scalar and $k_1$ and $k_2$ are nonzero scalars with $k_1 \neq k_2$.
    1. $k\mathbf{a}$ denotes a vector having magnitude $|k|$ times the magnitude of $\mathbf{a}$ which is parallel to $\mathbf{a}$. If $k < 0$, the direction is reversed.
    2. $0\mathbf{a} = \mathbf{0}$.
    3. $k\mathbf{0} = \mathbf{0}$.
    4. $1\mathbf{a} = \mathbf{a}$.

5. $(-k)\mathbf{a} = -(k\mathbf{a})$.
6. $(-1)\mathbf{a} = -(\mathbf{a})$.
7. $k(\mathbf{a} + \mathbf{b}) = k\mathbf{a} + k\mathbf{b}$.
8. $(k_1 + k_2)\mathbf{a} = k_1\mathbf{a} + k_2\mathbf{a}$.
9. $(k_1 k_2)\mathbf{a} = k_1(k_2\mathbf{a})$.

- $\mathbf{i}$ and $\mathbf{j}$ are vectors of unit magnitude in the $x$- and $y$-directions, respectively.
- A vector $\mathbf{a}$ can be written either in terms of $\mathbf{i}$ and $\mathbf{j}$:

$$\mathbf{a} = a_1\mathbf{i} + a_2\mathbf{j},$$

or as row or column vector:

$$\mathbf{a} = (a_1, a_2), \quad \mathbf{a} = \begin{pmatrix} a_1 \\ a_2 \end{pmatrix}.$$

- Addition of vectors is performed componentwise. For example, if $\mathbf{a} = (a_1, a_2)$ and $\mathbf{b} = (b_1, b_2)$, then

$$\mathbf{a} + \mathbf{b} = (a_1 + b_1, a_2 + b_2).$$

- The magnitude of a vector can be found using Pythagoras' theorem, for example, for $\mathbf{a} = (a_1, a_2)$

$$|\mathbf{a}| = \sqrt{a_1^2 + a_2^2}.$$

- A unit vector $\hat{\mathbf{a}}$ in the direction of $\mathbf{a}$ is

$$\hat{\mathbf{a}} = \frac{\mathbf{a}}{|\mathbf{a}|}.$$

**Chapter Assessment — Vectors**

Download and sit the 30 minute assessment for this chapter from the digital book.

# 4. Linear Functions

Many quantities can be related to other quantities, for example, the height and weight of a person are linked. In some instances, the relationship between the two can be represented by a linear function. The study of such relationships forms the basis for this chapter. In economics, the supply and demand for a quantity can be modelled by a linear relationship. For both the supply and demand functions, price is the independent variable. The supply function describes the relationship between the quantity of a product produced by a manufacturer and its price. As the price goes up, one would expect the customer demand to go down, since customers are less likely to be prepared to pay a high price for a product. Conversely, as price goes up, manufacturers are more likely to be willing to produce products, increasing the supply, since their profit margins will be higher. From a business perspective, it is important that supply is not significantly higher than the demand, since there would be a cost implication to the business. As an example, consider an English vineyard modelling the supply and demand for a new sparkling wine they are producing. We use $S(p)$ to denote the supply at price $p$ and $D(p)$ to denote the demand at price $p$. We model the supply function and the demand function as follows

$$S(p): \quad 5875p - 2q - 55750 = 0,$$
$$D(p): \quad 5000p + 3q - 115000 = 0.$$

The equilibrium point for the supply and demand of sparkling wine is the intersection point of the two lines shown in Figure 4.1(a). The equilibrium point is approximately $(14.38, 14\,400)$, meaning that a production of $14\,400$ bottles at a sale price of £14.40 would satisfy both the manufacturer and the consumer. In this chapter, we study the equations of such straight lines and model real-life situations with straight line functions. Finding the intersection points of two lines will be explored in more detail in Chapter 11.

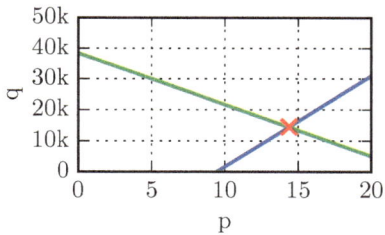 

(a) (b)

Figure 4.1: Supply and Demand lines for a sparkling wine (a) and an English vineyard (b).

## 4.1 The Cartesian Coordinate System

In 1637, the French philosopher and Mathematician René Descartes published the idea of referring to points in the plane with reference to a fixed line or axis. The Cartesian system of coordinates we use today is named after Descartes, however, the two axes for representing points in two dimensional space that are familiar today were developed later. In a Cartesian coordinate system, points are represented by an ordered pair of numbers $(a, b)$, which denote the perpendicular distance from the point to the $y$-axis and $x$-axis respectively. For example, the coordinate $(1, 5)$ represents the point $A$ shown in Figure 4.2. This point is a distance 5 away from the $y$-axis and so the $x$-coordinate is 5. Similarly it is a distance of 1 away from the $x$-axis and so the $y$ coordinate is 1.

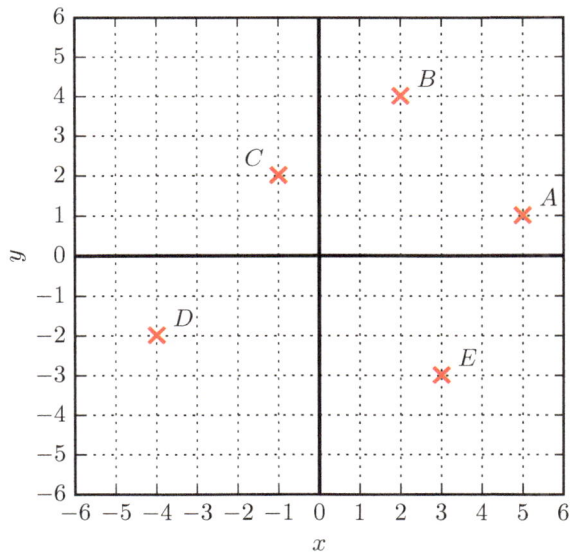

Figure 4.2: A collection of points represented on a Cartesian coordinate system.

**Remark**

The $x$-coordinate is sometimes referred to as the *abscissa* and the $y$-coordinate the *ordinate*.

The four quadrants shown in Figure 4.2 are often numbered in the following way.

| Quadrant | $x$-coordinate | $y$-coordinate |
|----------|----------------|----------------|
| 1        | $+$ve          | $+$ve          |
| 2        | $-$ve          | $+$ve          |
| 3        | $-$ve          | $-$ve          |
| 4        | $+$ve          | $-$ve          |

**Remark**

In Figure 4.2, the points $A$, $B$ lie in Quadrant 1, $C$ in Quadrant 2, $D$ in Quadrant 3 and $E$ in Quadrant 4.

In this book we use the notation $A(x_A, y_A)$ to denote the point $A$ with $x$-coordinate $x_A$ and $y$-coordinate $y_A$.

**Tip**

It is also common to see notation such as $A(x_1, y_1)$ when referring to a general point in a coordinate system.

**Interactive Activity 4.1 — Identifying Quadrants**

Use the activity in the digital book to practise identifying the quadrant points lie in.

### 4.1.1  Distance Between Two Points

Suppose we wish to move from point $B$ to point $A$ in Figure 4.2. We can move vertically downwards 3 units and then to the right 3 units, moving a total distance of 6 units. However, it is visually clear that is shorter to move diagonally from $B$ to $A$. This situation is shown in Figure 4.3.

**Remark**

In fact, for any triangle, the sum of the length of two sides is always of a smaller magnitude than the length of the third side. This is a known as the *triangle inequality* and can be generalised to more abstract settings. For example, in an inner product space,

$$||x + y||^2 \leq ||x||^2 + ||y||^2,$$

where $|| \cdot ||^2$ denotes a norm induced by the inner product.

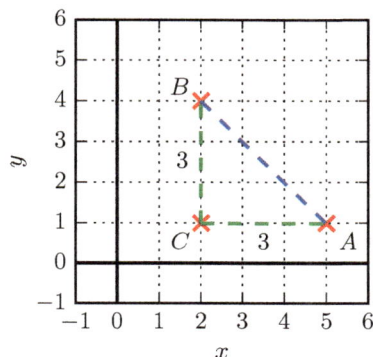

Figure 4.3: Finding the distance between two points.

As the axes in a Cartesian coordinate system are perpendicular, triangle $ABC$ shown in Figure 4.3 is right angled and so we can use Pythagoras' Theorem to calculate the distance between $A$ and $B$. Note, we use $|AB|$ to denote the distance of the line segment between $A$ and $B$.

$$
\begin{aligned}
|AB| &= \sqrt{|BC|^2 + |CA|^2} \\
&= \sqrt{3^2 + 3^2} \\
&= \sqrt{9 + 9} \\
&= \sqrt{18} \\
&= \sqrt{9 \times 2} \\
&= 3\sqrt{2}.
\end{aligned}
$$

**Formula 4.1 — Distance Between Two Points**

The distance between two points $A(x_A, y_A)$ and $B(x_B, y_B)$ is given by

$$|AB| = \sqrt{(x_B - x_A)^2 + (y_B - y_A)^2}. \tag{4.1}$$

**Tip**

In Formula 4.1, the order of the coordinates is unimportant, as long as we are consistent. So, we could also find the distance $|AB|$ using

$$|AB| = \sqrt{(x_A - x_B)^2 + (y_A - y_B)^2}.$$

**Remark**

Formula 4.1 can be generalised to find the distance between two points in a higher dimensional space. For example, if $A(x_A, y_A, z_A)$ and $B(x_B, y_B, z_B)$ are in three dimensional space, then

$$|AB| = \sqrt{(x_B - x_A)^2 + (y_B - y_A)^2 + (z_B - z_A)^2}.$$

**Example 4.1**

Find the distance between the points $A(3, 4)$ and $B(-2, -2)$.

**Solution:**

We first sketch the points in a Cartesian plane.

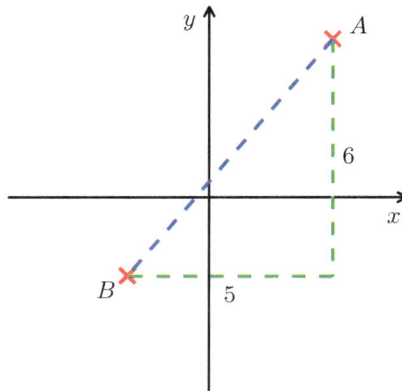

Using Formula 4.1,

$$\begin{aligned} |AB| &= \sqrt{(-2 - 3)^2 + (-2 - 4)^2} \\ &= \sqrt{(-5)^2 + (-6)^2} \\ &= \sqrt{25 + 36} \\ &= \sqrt{61}. \end{aligned}$$

**Tip**

The calculations in Example 4.1 are straightforward, however they are prone to error in exam situations. Ensure every line of working is checked.

We can also use Formula 4.1 to find the coordinates of one end of a line segment given the other coordinate and length of the segment.

**Example 4.2**

Given that the distance between $A(6,6)$ and $B(11,y)$ is 13, find the value of $y$.

**Solution:**

Using Formula 4.1,

$$13 = \sqrt{(11-6)^2 + (y-6)^2},$$
$$\Rightarrow \quad 13^2 = 5^2 + (y-6)^2,$$
$$\Rightarrow \quad 169 = 25 + y^2 - 12y + 36,$$
$$\Rightarrow \quad 0 = y^2 - 12y - 133.$$

This quadratic can be factorised using techniques familiar from GCSE study:

$$y^2 - 12y - 133 = 0,$$
$$\Rightarrow \quad (y-19)(y+7) = 0.$$

Thus, we have two possible values of $y$, namely, $y = -7$ and $y = 19$. The points from these values of $y$ are shown in the sketch below.

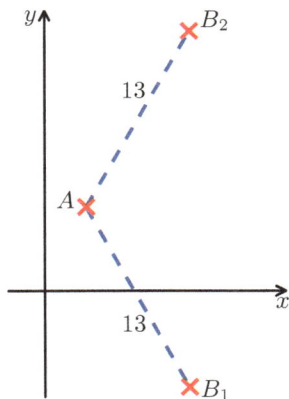

**Tip**

Factorising quadratics is discussed in full in Chapter 5.

Recall that when factorising a quadratic of the form $y^2 + by + c = (y+d)(y+e)$ the values of $d$ and $e$ are chosen so that $de = c$ and $.d + e = b$.

We can look ahead if these factorisation skills are unfamiliar.

**Example 4.3**
Find all points, such that the $y$-coordinate is one greater than the $x$-coordinate, which are a distance 5 away from the point $A(2,2)$.

**Solution:**
We seek points of the form $(x, x+1)$ which are 5 units away from the point $(2,2)$. Using Formula 4.1 we have

$$5 = \sqrt{(2-x)^2 + (2-(x+1))^2},$$
$$\Rightarrow \quad 5 = \sqrt{(2-x)^2 + (1-x)^2},$$
$$\Rightarrow \quad 25 = (2-x)^2 + (1-x)^2,$$
$$\Rightarrow \quad 25 = 4 - 4x + x^2 + 1 - 2x + x^2,$$
$$\Rightarrow \quad 25 = 2x^2 - 6x + 5,$$
$$\Rightarrow \quad 0 = 2x^2 - 6x - 20.$$

As in the previous example, we factorise this quadratic,

$$\Rightarrow \quad 0 = 2(x^2 - 3x - 10),$$
$$\Rightarrow \quad 0 = 2(x-5)(x+2).$$

Hence, the possible $x$-coordinates are 5 or $-2$. Using the relation between $x$- and $y$-coordinates, the points that satisfy the given conditions are $(5,6)$ and $(-2,-1)$, as shown below.

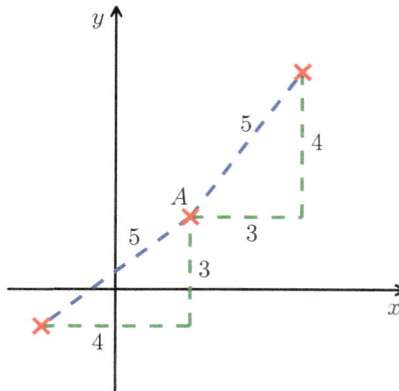

The points we have found do, in fact, lie on the circumference of a circle, radius 5 centre $(2,2)$. Circle geometry is studied in Chapter 13.

**Interactive Activity 4.2 — Distance Between Points**

Using the interactive contained in the digital book, practise finding the distance between two points.

## 4.1.2 The Gradient of a Line Segment

The slope of a line segment is known as the gradient and describes the change in the $y$-direction for a unit change in the $x$-direction.

**Definition 4.1 — Gradient of a Line Segment**

For the points $A(x_A, y_A)$ and $B(x_B, y_B)$ shown in Figure 4.4, the gradient of the line segment joining them is given by

$$\frac{y_B - y_A}{x_B - x_A}. \tag{4.2}$$

The gradient of a line segment (or line as we shall see later) is commonly denoted $m$.

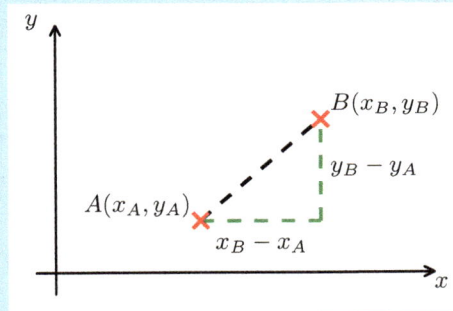

Figure 4.4: How to compute the gradient of a straight line segment.

**Remark**

As with the distance between two points, we can label the points in the opposite order and calculate the gradient as

$$m = \frac{y_A - y_B}{x_A - x_B}.$$

The first recorded use of the letter "m" to represent the gradient of a straight line was in O'Brians' 1844 book on geometry.

**Tip**

A useful way to recall the definition of the gradient is to remember that

$$\text{Gradient} = \frac{\text{change in } y}{\text{change in } x}.$$

This may sometimes be expressed as

$$\text{Gradient} = \frac{\Delta y}{\Delta x},$$

where $\Delta x$ denotes the change in $x$, for example.

---

**Example 4.4**

Find the gradient of the line segment joining the points $A(1, 8)$ to $B(7, 20)$.

**Solution:**

We calculate the gradient as follows.

$$
\begin{aligned}
m &= \frac{y_B - y_A}{x_B - x_A} \\
&= \frac{20 - 8}{7 - 1} \\
&= \frac{12}{6} \\
&= 2.
\end{aligned}
$$

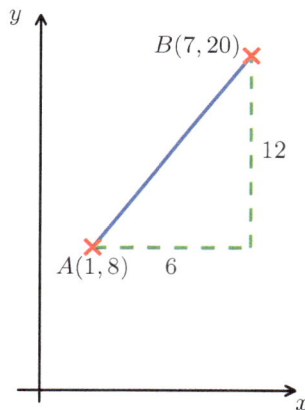

**Example 4.5**

Find the gradient of the line segment joining $A(2,1)$ to $B(6,5)$ and the gradient of the line segment joining $C(1,3)$ to $D(3,1)$. What can be observed?

**Solution:**

For the segment AB, let $m_1$ denote the gradient, then

$$
\begin{aligned}
m_1 &= \frac{y_B - y_A}{x_B - x_A} \\
&= \frac{5 - 1}{6 - 2} \\
&= \frac{4}{4} \\
&= 1.
\end{aligned}
$$

Letting $m_2$ denote the gradient of the line segment $CD$,

$$
\begin{aligned}
m_2 &= \frac{y_C - y_D}{x_C - x_D} \\
&= \frac{3 - 1}{1 - 3} \\
&= \frac{2}{-2} \\
&= -1.
\end{aligned}
$$

With reference to the figure below, we note that the line segments appear to be perpendicular to each other, and that $m_1 m_2 = -1$.

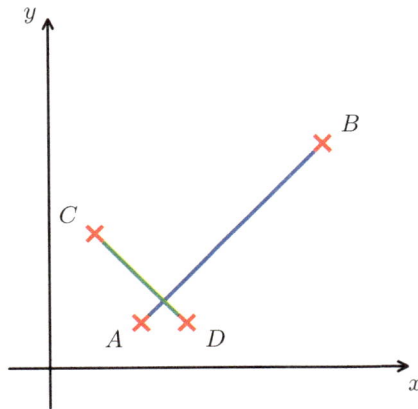

**Example 4.6**

The gradient of the line segment joining $A(-4, -2)$ to $B(x_B, 3)$ is $\frac{4}{7}$. What is the value of $x_B$?

**Solution:**

We first sketch the situation.

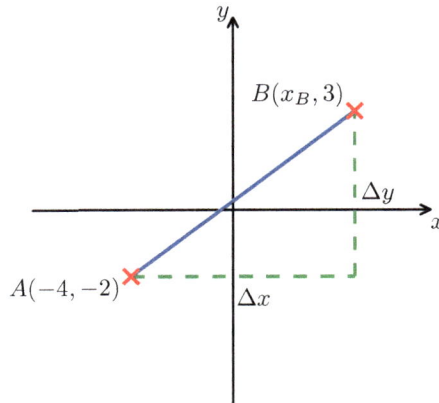

From the information given $\frac{\Delta y}{\Delta x}$ must simplify to $\frac{4}{7}$. Using Formula

$$\frac{4}{7} = \frac{3 - (-2)}{x_B - (-4)},$$

$$\Rightarrow \quad 4(x_B - (-4)) = 7(3 - (-2)),$$

$$\Rightarrow \quad 4x_B + 16 = 7(5),$$

$$\Rightarrow \quad 4x_B = 35 - 16,$$

$$\Rightarrow \quad 4x_B = 19,$$

$$\Rightarrow \quad x_B = \frac{19}{4}.$$

**Remark**

For the examples above, the sketch may not seem that useful, however, for more complicated questions, a sketch provides a useful way to spot any mistakes and check that a solution is "sensible".

### 4.1.3  The Midpoint of a Line Segment

Working in a Cartesian framework allows us to easily work out the location of points that split the distance $AB$ into a given ratio. In particular, we can easily compute the midpoint of a line segment. In Figure 4.5, the point $C$ is halfway between the points $A$ and $B$ (*i.e.* the distance $|AC|$ equals the distance $|CB|$) and so is known as the midpoint of the line segment $AB$.

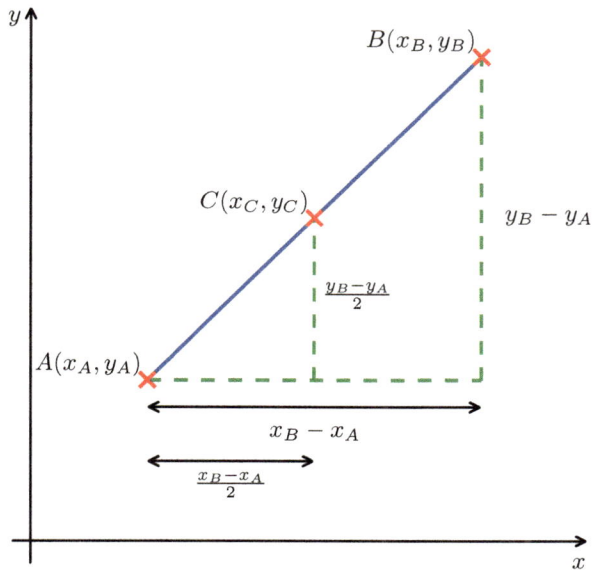

Figure 4.5: Finding the midpoint of two points.

The distance between the $x$-coordinates of $A$ and $B$ is $x_B - x_A$ and so the $x$-coordinate of the point $C$ can be obtained by adding half of this length to the $x$-coordinate of $A$. Thus,

$$x_C = x_A + \frac{x_B - x_A}{2}$$
$$= \frac{x_A + x_B}{2}.$$

Similarly, the $y$-coordinate of the point $C$ is given by

$$y_C = y_A + \frac{y_B - y_A}{2}$$
$$= \frac{y_A + y_B}{2}.$$

---

**Formula 4.2 — The Midpoint of a Line Segment**

The coordinates of the midpoint of the line segment joining the points $A(x_A, y_A)$ and $B(x_B, y_B)$ are

$$\left( \frac{x_A + x_B}{2}, \frac{y_A + y_B}{2} \right).$$

It is not immediately obvious that defining the point $C$ with these coordinates results in $|AC| = |CB|$; this can be shown by an application of Formula 4.1.

$$\begin{aligned}
|AC| &= \sqrt{\left(\frac{x_A + x_B}{2} - x_A\right)^2 + \left(\frac{y_A + y_B}{2} - y_A\right)^2} \\
&= \sqrt{\left(\frac{x_B - x_A}{2}\right)^2 + \left(\frac{y_B - y_A}{2}\right)^2} \\
&= \sqrt{\left(x_B - \frac{x_A + x_B}{2}\right)^2 + \left(y_B - \frac{y_A + y_B}{2}\right)^2} \\
&= |CB|.
\end{aligned}$$

---

**Example 4.7**

Find the midpoint of the line segment with endpoints $A(1,1)$ and $B(5,9)$.

**Solution:**

Let $C$ denote the midpoint of $AB$, then using Formula 4.2

$$\begin{aligned}
(x_C, y_C) &= \left(\frac{x_A + x_B}{2}, \frac{y_A + y_B}{2}\right) \\
&= \left(\frac{1+5}{2}, \frac{1+9}{2}\right) \\
&= \left(\frac{6}{2}, \frac{10}{2}\right) \\
&= (3, 5).
\end{aligned}$$

---

**Example 4.8**

The line segment $AB$ has midpoint $C(5.5, 4)$. Given that the point $A$ has coordinates $(9,3)$, find the coordinates of $B$.

**Solution:**

We again use Formula 4.2 and compare $x$- and $y$-coordinates.

$$(5.5, 4) = \left(\frac{9 + x_B}{2}, \frac{3 + y_B}{2}\right).$$

Comparing the $x$-coordinates

$$\begin{aligned}
\frac{11}{2} &= \frac{9 + x_B}{2}, \\
\Rightarrow \quad 11 &= 9 + x_B, \\
\Rightarrow \quad x_B &= 2.
\end{aligned}$$

Similarly, comparing the $y$-coordinates

$$4 = \frac{3 + y_B}{2},$$

$$\Rightarrow \quad 8 = 3 + y_B,$$

$$\Rightarrow \quad y_B = 5.$$

Hence, the coordinates of $B$ are $(2, 5)$.

---

**Example 4.9**

Consider four points in the Cartesian plane $A(3, 1)$, $B(10, 3)$, $C(12, 7)$ and $D(5, 5)$.
  (a) Show that the midpoint of the line segment $AC$ has the same coordinates as the midpoint of the line segment $BD$.
  (b) Explain how this lets us identify the quadrilateral $ABCD$.
  (c) Find the perimeter of this quadrilateral.

**Solution:**

  (a) Let $E$ denote the midpoint of $AC$ and $F$ denote the midpoint of $BD$. We use Formula 4.2 to find the midpoint of $AC$ and $BD$.

$$(x_E, y_E) = \left( \frac{x_A + x_C}{2}, \frac{y_A + y_C}{2} \right)$$

$$= \left( \frac{3 + 12}{2}, \frac{1 + 7}{2} \right)$$

$$= (7.5, 4),$$

$$(x_F, y_F) = \left( \frac{x_B + x_D}{2}, \frac{y_B + y_D}{2} \right)$$

$$= \left( \frac{10 + 5}{2}, \frac{3 + 5}{2} \right)$$

$$= (7.5, 4).$$

Hence, the points $E$ and $F$ are coincident.
  (b) Since $E$ is the midpoint of $AC$, $|AE| = |EC|$ and similarly $|BE| = |ED|$. Since the diagonals bisect each other, we can deduce that the quadrilateral is a parallelogram.
  (c) To find the area of this parallelogram, we first calculate the lengths $|AB|$ and $|BC|$ using Formula 4.1:

$$|AB| = \sqrt{(x_B - x_A)^2 + (y_B - y_A)^2}$$

$$= \sqrt{(10 - 3)^2 + (3 - 1)^2}$$

$$= \sqrt{7^2 + 2^2}$$

$$= \sqrt{53}.$$

$$|BC| = \sqrt{(x_C - x_B)^2 + (y_C - y_B)^2}$$
$$= \sqrt{(12 - 10)^2 + (7 - 3)^2}$$
$$= \sqrt{2^2 + 4^2}$$
$$= \sqrt{20}.$$

Hence, the perimeter is $2\sqrt{53} + 2\sqrt{20}$.
The figure below shows this graphically.

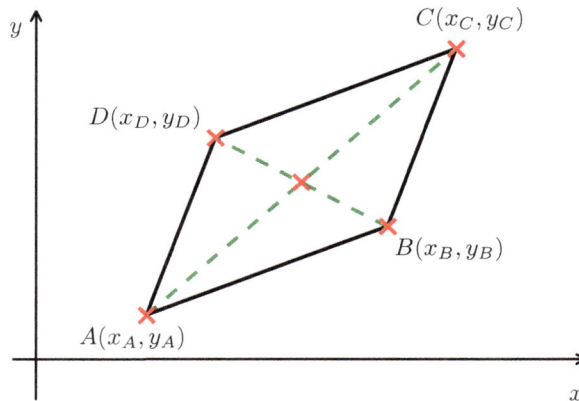

### 4.1.4  Perpendicular Line Segments

**Definition 4.2 — Perpendicular Line Segments**
Line segment $AB$ is said to be perpendicular to the line segment $CD$ if the angle between them, at the point they intersect, is $90°$.

**Formula 4.3 — Gradient Condition for Perpendicular Line Segments**
For the points $A(x_A, y_A)$, $B(x_B, y_B)$, $C(x_C, y_C)$ and $D(x_D, y_D)$, the line segments $AB$ and $CD$ are perpendicular if and only if

$$m_1 m_2 = -1, \tag{4.3}$$

where $m_1$ is the gradient of $AB$ and $m_2$ is the gradient of $CD$.

**Proof** Without loss of generality, we consider the situation where the points $B$ and $D$ coincide, and so the line segments join each other at one end as shown in Figure 4.6.

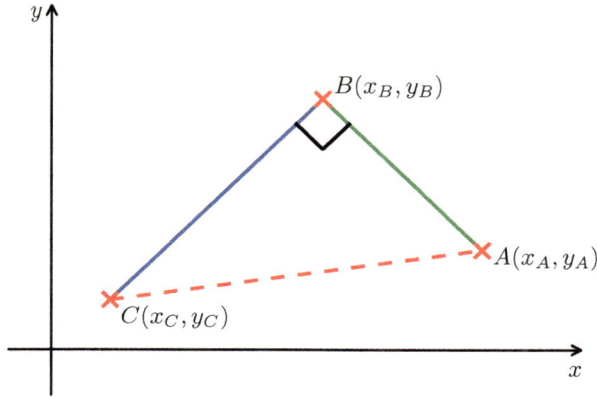

Figure 4.6: Line segment $AB$ is perpendicular to line segment $BC$.

The gradients $m_1$, $m_2$ of the line segments $AB$, and $BC$ respectively, are,

$$m_1 = \frac{y_B - y_A}{x_B - x_A},$$
$$m_2 = \frac{y_C - y_B}{x_C - x_B}.$$

Now, since $AB$ is perpendicular to $BC$, triangle $ABC$ is right-angled and so Pythagoras' theorem can be applied:

$$|AB|^2 + |BC|^2 = |AC|^2. \tag{4.4}$$

Considering the left hand side of Equation (4.4),

$$|AB|^2 + |BC|^2 = (x_B - x_A)^2 + (y_B - y_A)^2 + (x_C - x_B)^2 + (y_B - y_C)^2. \tag{4.5}$$

Considering the right hand side of Equation (4.4),

$$
\begin{aligned}
|AC|^2 &= (x_C - x_A)^2 + (y_C - y_A)^2, \\
&= (x_C + x_B - x_B - x_A)^2 + (y_C + y_B - y_B - y_A)^2, \\
&= [(x_C - x_B) + (x_B - x_A)]^2 + [(y_C - y_B) + (y_B - y_A)]^2, \\
&= (x_C - x_B)^2 + (x_B - x_A)^2 + 2(x_C - x_B)(x_B - x_A) + (y_C - y_B)^2 \\
&\quad + (y_B - y_A)^2 + 2(y_C - y_B)(y_B - y_A).
\end{aligned} \tag{4.6}
$$

Since Equation (4.4) holds,

$$0 = 2(x_C - x_B)(x_B - x_A) + 2(y_C - y_B)(y_B - y_A),$$
$$\Rightarrow \quad -2(y_C - y_B)(y_B - y_A) = 2(x_C - x_B)(x_B - x_A),$$
$$\Rightarrow \quad \frac{(y_C - y_B)(y_B - y_A)}{(x_C - x_B)(x_B - x_A)} = \frac{2}{-2},$$
$$\Rightarrow \quad \frac{(y_C - y_B)}{(x_C - x_B)} \frac{(y_B - y_A)}{(x_B - x_A)} = -1,$$
$$\Rightarrow \quad m_1 m_2 = -1.$$

Thus, if the two line segments $AB$ and $BC$ are perpendicular then their gradients multiply to give negative one.

All the steps in the above proof can be reversed and so we can also conclude that, if the gradients of two line segments multiply together to give $-1$, then they are perpendicular.

---

**Question**

Why can we assume, without loss of generality, that the points $B$ and $D$ coincide?

---

**Remark**

In deriving Equation (4.6), we have twice used a technique that can often be applied when trying to prove an assertion - the addition of zero. For example, $(x_C - x_A)$ became $(x_C + x_B - x_B - x_A)$ as $+x_B - x_B = 0$ and so adding this leaves the original expression unchanged.

A similar technique makes use of the multiplicative identity to multiply a term by 1 in a clever way to enable progress to be made in a proof. Look out for this in other proofs.

---

**Tip**

The algebraic manipulation in deriving Equation (4.6) could have been presented in a better way by defining, for example, $\delta x_{AC} = x_C - x_A$, *etc.* In this way, the argument would have been more concise and less prone to error making. When performing difficult algebraic manipulations, always consider if similar definitions are appropriate.

---

**Example 4.10**

Show that the line segment joining $A(-2, 6)$ and $B(8, -4)$ is perpendicular to the line segment joining $C(2, -2)$ and $D(10, 6)$.

**Solution:**

To show $AB$ is perpendicular to $CD$, we first find the gradients of these line segments. Let $m_1$ denote the gradient of $AB$ and $m_2$ denote the gradient of $CD$. We have

$$
\begin{aligned}
m_1 &= \frac{-4 - 6}{8 - -2} \\
&= \frac{-10}{10} \\
&= -1, \\
m_2 &= \frac{6 - -2}{10 - 2} \\
&= \frac{8}{8} \\
&= 1.
\end{aligned}
$$

Hence, $m_1 m_2 = -1$ and so the line segments are perpendicular.

**Example 4.11**

Prove that $A(8,2)$, $B(11,6)$ and $C(7,9)$ are the vertices of a right-angled triangle and show that the area of this triangle can be expressed as $\frac{p}{q}$ squared units, where $p, q \in \mathbb{Z}$.

**Solution:**

We proceed by calculating the gradient of each line segment. Let $m_1$ be the gradient of $AB$, $m_2$ be the gradient of $BC$ and $m_3$ be the gradient of $CA$. We have

$$
\begin{aligned}
m_1 &= \frac{6-2}{11-8} \\
&= \frac{4}{3}, \\
m_2 &= \frac{9-6}{7-11} \\
&= -\frac{3}{4}, \\
m_3 &= \frac{2-9}{8-1} \\
&= -7.
\end{aligned}
$$

Since these three gradients are different, the points $A, B, C$ are not collinear and we conclude that $ABC$ is a triangle. To show that it is a right-angled triangle, two of the line segments must be perpendicular. We notice that,

$$
\begin{aligned}
m_1 m_2 &= \frac{4}{3} \times -\frac{3}{4} \\
&= -1.
\end{aligned}
$$

Hence, $AB$ is perpendicular to $BC$ and $ABC$ is, therefore, a right angled triangle. To find the area of this triangle we first find the lengths of the line segments $AB$ and $BC$.

$$
\begin{aligned}
|AB| &= \sqrt{(11-8)^2 + (6-2)^2} \\
&= \sqrt{3^2 + 4^2} \\
&= 5, \\
|BC| &= \sqrt{(7-11)^2 + (9-6)^2} \\
&= \sqrt{(-4)^2 + 3^2} \\
&= 5.
\end{aligned}
$$

The area is then $\frac{1}{2} \times 5 \times 5 = \frac{25}{2}$ squared units.

### 4.1.5 Parallel Line Segments

Two line segments are said to be parallel if their gradients are equal. This notion will be formalised in Section 4.2.2 when we discuss general parallel straight lines.

---

**Example 4.12**

Consider the points $A(4,9)$, $B(6,5)$, $C(8,7)$ and $D(10,3)$. Show that the line segments $AB$ and $CD$ are parallel.

**Solution:**

We use Equation (4.2) to calculate the gradients of both $AB$ and $CD$.

$$\text{Gradient of } AB = \frac{5-9}{6-4}$$
$$= \frac{-4}{2}$$
$$= -2,$$
$$\text{Gradient of } CD = \frac{3-7}{10-8}$$
$$= \frac{-4}{2}$$
$$= -2.$$

Since the gradients are the same, $AB$ and $CD$ are parallel.

---

**Example 4.13**

For the points $A(3,6)$, $B(8,8)$, $C(6,3)$ and $D(x_D, y_D)$, find a relationship between $x_D$ and $y_D$ so that the line segment $CD$ is parallel to the line segment $AB$.

**Solution:**

We first calculate the gradient of the line segments $AB$ and $CD$, using Equation (4.2).

$$\text{Gradient of } AB = \frac{8-6}{8-3}$$
$$= \frac{2}{5},$$
$$\text{Gradient of } CD = \frac{y_D - 3}{x_D - 6}.$$

Since $CD$ is parallel to $AB$, we can equate their gradients:

$$\frac{2}{5} = \frac{y_D - 3}{x_D - 6},$$
$$\Rightarrow \quad 2x_D - 12 = 5y_D - 15,$$
$$\Rightarrow \quad 2x_D = 5y_D - 3.$$

Hence,

$$y_D = \frac{2}{5}x_D + \frac{3}{5},$$

is a relationship between $x_D$ and $y_D$ for $CD$ to be parallel to $AB$.

**Exercise 4.1**

Q1. For each of the pairs of points listed below, find the distance $|AB|$, the gradient of $AB$ and the midpoint of $AB$.
  (a) $A(2,2)$ and $B(4,4)$;
  (b) $A(8,5)$ and $B(11,2)$;
  (c) $A(6,-3)$ and $B(6,3)$;
  (d) $A(12,7)$ and $B(2,5)$;
  (e) $A(-3,6)$ and $B(3,-3)$;
  (f) $A(11,-2)$ and $B(15,6)$;
  (g) $A(3,8)$ and $B(-1,5)$;
  (h) $A(-2,2)$ and $B(1,-1)$;

Q2. Find the gradient between the following pairs of points.
  (a) $(a,a)$ and $(2a,2a)$;
  (b) $(10a,0)$ and $(0,5a)$;
  (c) $(4t,t^2)$ and $(8t,2t^2)$.

Q3. The line segment joining $A(1,1)$ to $B(6,p)$ has gradient $\frac{4}{5}$. Find $p$.

Q4. The line segment joining $A(5,3)$ to $B(z,-3)$ has gradient $-1$. Find $z$.

Q5. Show that the points $A(-2,8)$, $B(4,4)$ and $C(13,-2)$ can be joined by a straight line.

Q6. Show that the midpoint of $AB$ is equal to the midpoint of $CD$ where $A(8,2)$, $B(14,6)$, $C(7,5)$ and $D(15,3)$.

Q7. The midpoint of $A(1,3)$ and $B(x_B,y_B)$ has coordinates $C\left(\frac{13}{2},5\right)$, find the coordinates of $B$.

Q8. The midpoint of $A(1,y_A)$ and $B(x_B,-4)$ is $C\left(\frac{13}{2},-\frac{5}{2}\right)$ find $y_A$ and $x_B$.

Q9. Consider all possible pairings for the coordinates $A(3,4)$, $B(9,7)$, $C(3,1)$, $D(3,6)$, $E(7,3)$ and $F(10,-3)$. Which line segments are parallel and which are perpendicular?

## 4.2 Linear Relationships and Straight Lines

Since our primary schooling we have been familiar with completing tables of input-outputs for a given mapping.

| Set $X$ | Set $Y$ |
|---|---|
| 1 | 4 |
| 2 | 7 |
| 3 | 10 |
| 10 | 31 |

Table 4.1: A linear relationship.

For example, the table of values shown in Table 4.1 can be described using the mapping $x \mapsto 3x + 1$.

**Tip**
This notation is read "$x$ is mapped to $3x + 1$".

This mapping is known as a linear mapping, since the highest power of $x$ appearing in the right hand side of the mapping definition is one.

If we let $y$ represent any element of Set $Y$, then we can also write this mapping in the form $y = \alpha x + \beta$. This form is known as the equation of a straight line.

**Activity 4.1 — Linear Mappings**
This activity concerns the use of mappings to represent linear relationships between two variables.

(a) Complete the missing information in each table, based on the relationships shown.

| Relationship A | |
|---|---|
| Set X | Set Y |
| 2 | 3 |
| 6 | 11 |
| 5 | 9 |
| 3 | |
| $x$ | |

| Relationship B | |
|---|---|
| Set X | Set Y |
| 2 | 4 |
| 6 | 6 |
| 5 | 5.5 |
| 3 | |
| $x$ | |

| Relationship C | |
|---|---|
| Set X | Set Y |
| 2 | 5 |
| 6 | 1 |
| 5 | 2 |
| 3 | |
| $x$ | |

| Relationship D | |
|---|---|
| Set X | Set Y |
| 2 | 6 |
| 6 | 14 |
| 5 | 11 |
| 3 | |
| $x$ | |

(b) Which of these mappings are linear? For those which are linear, write the mapping in the form $x \mapsto \alpha x + \beta$.

(c) Define three more mappings, one of which is non-linear.

## 4.2.1 Defining Straight Lines

If we consider the mapping $x \mapsto 2x - 1$, the equivalent representation $y = 2x - 1$ allows us to represent this mapping on a pair of Cartesian axes as shown in Figure 4.7. It is clear that any one of the three points shown on this graph in Figure 4.7 could be removed and there still be a unique line through the remaining two points.

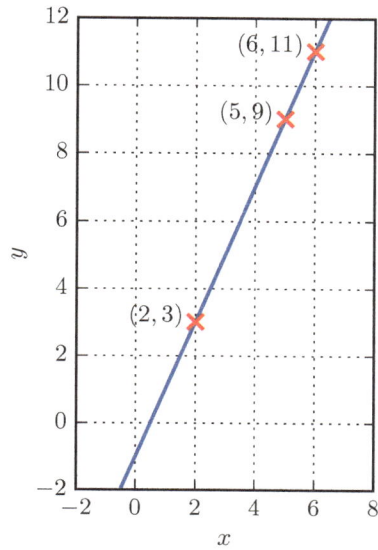

Figure 4.7: The line representing the equation $y = 2x - 1$.

Figure 4.8 below shows that a single point alone is not enough to uniquely define a line.

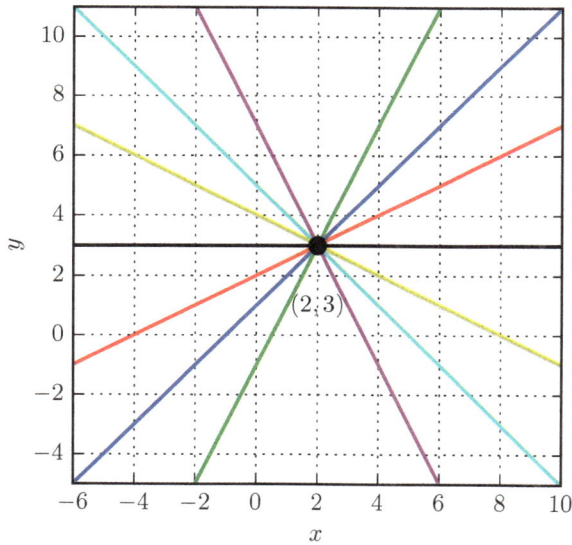

Figure 4.8: Many lines passing through the point $(2, 3)$.

A natural question is to ask "what is the least amount of information required to uniquely define a line?".

Before we address this question, we relate our mapping representation to the previous section on the gradient of a line segment.

With reference to Figure 4.7, we can calculate the gradient of the line segment joining $(2, 3)$ to $(5, 9)$,

$$
\begin{aligned}
m &= \frac{9 - 3}{5 - 2} \\
&= \frac{6}{3} \\
&= 2.
\end{aligned}
$$

This is the same as the value of $\alpha$ in the mapping notation. The value of $\beta$ is known as the $y$-intercept and is the value taken by $y$ when $x = 0$. Since we commonly use $m$ to represent a gradient, instead of using the letters $\alpha$, $\beta$ to represent the general equation of a straight line we use $m$ and $c$, respectively.

---

**Formula 4.4 — Equation of a Straight Line 1**

The equation of the line with gradient $m$ and $y$-intercept $c$ can be written in the form

$$y = mx + c. \tag{4.7}$$

---

**Remark**

This form of the equation of a straight line should be familiar from study at GCSE.

---

**Example 4.14**

Write down the equation of the straight line with gradient 4 and $y$-intercept 2.

**Solution:**

Since $m = 4$ and $c = 2$, using Equation 4.7, we have

$$y = 4x + 2.$$

---

The equation of a straight line can also be written in the following alternative form.

---

**Formula 4.5 — Equation of a Straight Line 2**

The equation of a straight line can also be written in the form

$$ax + by + c = 0. \tag{4.8}$$

**Example 4.15**

Write the equation $y = 5x - 6$ in the form $ax + by + c = 0$.

**Solution:**

We rearrange to give

$$-5x + y + 6 = 0,$$

which is in the required form with $a = -5$, $b = 1$ and $c = 6$.

**Example 4.16**

Write the equation $y = \frac{5}{3} - \frac{3}{2}x$ in the form $ax + by + c = 0$. Hence, find where the line crosses both coordinate axes.

**Solution:**

We first employ equivalent fractions so that the fractions appearing in $y = \frac{5}{3} - \frac{3}{2}x$ all have the same denominator. The lowest common multiple of 2 and 3 is 6, so we have

$$y = \frac{5}{3} - \frac{3}{2}x,$$
$$\Rightarrow \qquad y = \frac{10}{6} - \frac{9}{6}x,$$
$$\Rightarrow \qquad 6y = 10 - 9x,$$
$$\Rightarrow \quad 9x + 6y - 10 = 0.$$

To find where the line crosses the $y$-axis we let $x = 0$ and then

$$6y - 10 = 0,$$
$$\Rightarrow \qquad 6y = 10,$$
$$\Rightarrow \qquad y = \frac{10}{6}.$$

Similarly, to find where the line crosses the $x$-axis we let $y = 0$ and then solve the resulting equation:

$$9x - 10 = 0,$$
$$\Rightarrow \qquad 9x = 10,$$
$$\Rightarrow \qquad x = \frac{10}{9}.$$

Hence, the $y$-intercept is $\frac{10}{6}$ and the $x$-intercept is $\frac{10}{9}$.

**Remark**

If we plot the line $9x + 6y - 10 = 0$ (from Example 4.16), we notice that the coefficients 9, 6 and 10 appear in the coordinates of the intercepts of both axes.

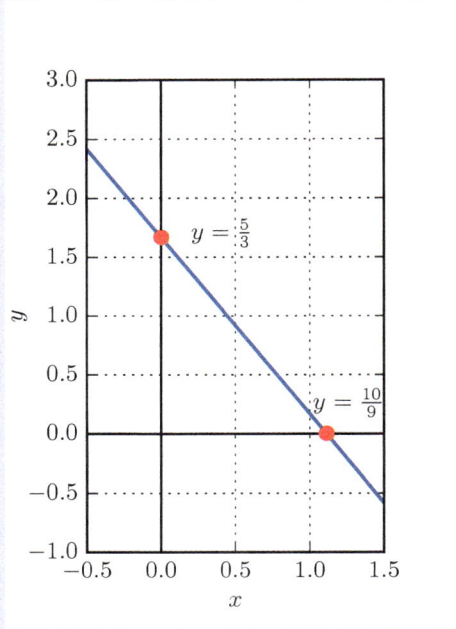

Figure 4.9: The intercepts of the line $9x + 6y - 10 = 0$.

For a straight line in the form $ax + by + c = 0$, then:

- The $y$-intercept has coordinates $\left(0, \frac{-c}{b}\right)$ since it occurs when $x$ is zero.

$$ax + by + c = 0,$$
$$\Rightarrow \qquad by + c = 0,$$
$$\Rightarrow \qquad x = -\frac{c}{b}.$$

- The $x$-intercept has coordinates $\left(0, \frac{-c}{a}\right)$ since it occurs when $y$ is zero.

$$ax + by + c = 0,$$
$$\Rightarrow \qquad ax + c = 0,$$
$$\Rightarrow \qquad x = -\frac{c}{a}.$$

- We also note that the gradient of the line can be calculated as $m = \frac{-a}{b}$ by rearranging the equation of the line into the form $y = mx + c$.

Each of the lines in Figure 4.8 has a different gradient as shown in Figure 4.10.

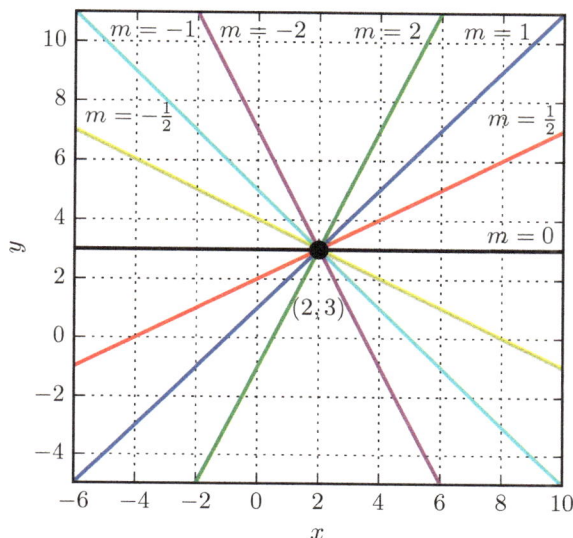

Figure 4.10: Figure 4.8 with gradients marked.

We can therefore uniquely define a line given its gradient and a point on the line.

> **Formula 4.6 — The Equation of a Line Given the Gradient and One Point**
> Suppose the gradient of the line we wish to find the equation for is $m$ and that it passes through the point $(x_1, y_1)$, then the equation of the line can be found by rearranging
>
> $$y - y_1 = m(x - x_1). \tag{4.9}$$

---

**Example 4.17**
Find the equation of the line passing through $(2, 3)$ which has a gradient of $m = 2$.

**Solution:**
**Method 1:** Using Formula 4.4 we can substitute $m = 2$, $x = 2$ and $y = 3$ to find the value of the $y$-intercept $c$.

$$
\begin{aligned}
& y = mx + c, \\
\Rightarrow \quad & 3 = 2 \times 2 + c, \\
\Rightarrow \quad & c = 3 - 4 \\
& \phantom{c} = -1.
\end{aligned}
$$

Hence, the equation of the straight line is $y = 2x - 1$ or $2x - y - 1 = 0$.

**Method 2:** We use Formula 4.6 with $m = 2$, $x_1 = 2$, and $y_1 = 3$.

$$\begin{aligned} y - y_1 &= m(x - x_1), \\ \Rightarrow \quad y - 3 &= 2(x - 2), \\ \Rightarrow \quad y - 3 &= 2x - 4, \\ \Rightarrow \quad y &= 2x - 1. \end{aligned}$$

Hence, the equation of the straight line is $y = 2x - 1$ or $2x - y - 1 = 0$.

---

### Example 4.18

The line with equation $2x + 3y - 24 = 0$ passes through the point $(9, b)$. Find the point $b$ and the gradient, $m$, of the line.

**Solution:**

We first rearrange into the form $y = mx + c$.

$$\begin{aligned} 2x + 3y - 24 &= 0, \\ \Rightarrow \quad 3y &= -2x + 24, \\ \Rightarrow \quad y &= -\frac{2}{3}x + \frac{24}{3}. \end{aligned} \tag{4.10}$$

From this we find that the gradient is $-\frac{2}{3}$. To find $b$, substitute $x = 9$ into Equation (4.10):

$$\begin{aligned} y &= -\frac{2}{3} \times 9 + \frac{24}{3} \\ &= 2. \end{aligned}$$

Hence, $b = 2$ and $m = -\frac{2}{3}$.

---

### Example 4.19

Show that the line with gradient 3 passing through $(-2, -10)$ also passes through the point $(4, 8)$.

**Solution:**

**Method 1:** Using Formula 4.4 we can substitute $m = 3$, $x = -2$ and $y = -10$ to find the value of the $y$-intercept $c$.

$$\begin{aligned} y &= mx + c, \\ \Rightarrow \quad -10 &= 3 \times (-2) + c, \\ \Rightarrow \quad c &= -10 + 6 \\ &= -4. \end{aligned}$$

Hence, the equation of the straight line is $y = 3x - 4$. Substituting $x = 4$ into this gives $y = 8$, confirming that the point $(4, 8)$ also lies on the line.

**Method 2:** We note that in moving from an $x$-coordinate of $-2$ to an $x$-coordinate of 4 we travel 6 units in the positive $x$-direction. Since the gradient is the change in $y$-direction per unit change in the $x$-direction we travel $6 \times 3 = 18$ units in the $y$-direction and so arrive at the point $(4, 8)$. This confirms that the point $(4, 8)$ lies on the line.

Since we can find the gradient of a line by considering the gradient between two points, we can also uniquely define a straight line given two points on the line.

**Formula 4.7 — Equation of a Line Given Two Points**
Given two points $A(x_A, y_A)$ and $B(x_B, y_B)$, the equation of the line that passes through $A$ and $B$ is given by

$$\frac{y - y_A}{y_B - y_A} = \frac{x - x_A}{x_B - x_A} \tag{4.11}$$

**Proof:**
To prove this formula, we use Formula 4.6 with the Definition 4.1 for the gradient between $A$ and $B$. Therefore,

$$m = \frac{y_B - y_A}{x_B - x_A},$$

and

$$y - y_A = \frac{y_B - y_A}{x_B - x_A}(x - x_A),$$

$$\Rightarrow \quad \frac{y - y_A}{y_B - y_A} = \frac{x - x_A}{x_B - x_A}.$$

**Tip**
It is common to see Formula 4.7 written for the equation of a straight line between the points $(x_1, y_1)$ and $(x_2, y_2)$. In which case,

$$\frac{y - y_1}{y_2 - y_1} = \frac{x - x_1}{x_2 - x_1}$$

**Example 4.20**
Find the equation of the line passing through the points $A(1, 2)$ and $B(13, 8)$.

**Solution:**
**Method 1:** First we calculate the gradient between $A$ and $B$. Using Formula 4.1, we have

$$m = \frac{8 - 2}{13 - 1}$$
$$= \frac{6}{12}$$
$$= \frac{1}{2}.$$

We then use Formula 4.6 and one of the points to find the equation of the line. If we choose to use the point $A$ then

$$y - y_A = m(x - x_A),$$
$$\Rightarrow \quad y - 2 = \frac{1}{2}(x - 1),$$
$$\Rightarrow \quad y = \frac{1}{2}x + \frac{3}{2}.$$

**Method 2:** We use Formula 4.7 directly:

$$\frac{y - y_A}{y_B - y_A} = \frac{x - x_A}{x_B - x_A},$$
$$\Rightarrow \quad \frac{y - 2}{8 - 2} = \frac{x - 1}{13 - 1},$$
$$\Rightarrow \quad 12(y - 2) = 6(x - 1),$$
$$\Rightarrow \quad 12y - 24 = 6x - 6,$$
$$\Rightarrow \quad 12y = 6x + 18,$$
$$\Rightarrow \quad y = \frac{1}{2}x + \frac{3}{2}.$$

**Tip**

Both methods are acceptable in the exam. However, when using Method 2, it is important to be consistent with which point is labelled as $A$ and which is $B$.

## 4.2.2 Parallel Straight Lines

Now that we can find the equation of a straight line, we return to the properties of parallel straight lines, extending the discussion in Section 4.1.5.

**Definition 4.3 — Parallel Straight Lines**
Two straight lines in two-dimensional Euclidean space are said to be parallel if they do not intersect. They are a constant distance apart for their entire length.

**Remark**
In three-dimensional space, for two lines to be considered parallel they must not only never intersect but also lie in a common plane. Two lines that do not intersect but do not lie in a common plane are known as *skew lines*.

**Definition 4.4 — Parallel Straight Lines**
Two straight lines, $l_1 : y = m_1 x + c_1$ and $l_2 : y = m_2 x + c_2$ are parallel if their gradients are equal, *i.e.* $m_1 = m_2$.

**Proof:** Figure 4.11 shows two parallel lines $l_1 : y = m_1 x + c_1$ and $l_2 : y = m_2 x + c_2$. By definition, the gradients of the two lines are

$$m_1 = \frac{|BA|}{|CA|},$$
$$m_2 = \frac{|ED|}{|FD|}.$$

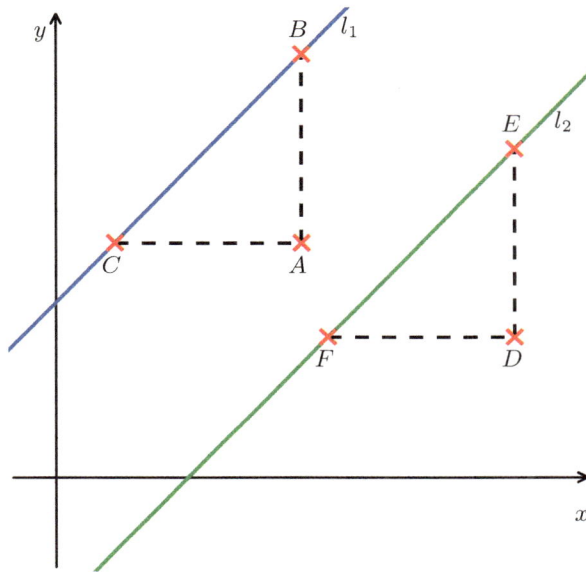

Figure 4.11: Two parallel lines.

Since $l_1$ and $l_2$ are parallel, the line segments $CB$ and $FE$ are parallel. The line segments $CA$ and $FD$ are parallel to the $x$-axis and the line segments $AB$ and $DE$ are parallel to the $y$-axis. Because of these three facts, triangle $ABC$ is similar to triangle $DEF$. Due to similarity,

$$\frac{|BA|}{|ED|} = \frac{|CA|}{|FD|},$$
$$\Rightarrow \quad \frac{|BA|}{|CA|} = \frac{|ED|}{|FD|},$$
$$\Rightarrow \quad m_1 = m_2.$$

For the converse, assume we have two lines $l_1$ and $l_2$ with the same gradient. If they have the same gradient then they make the same angle with the $x$-axis, and due to corresponding angles must, therefore, be parallel.

**Example 4.21**
Find the equation of the straight line parallel to $l_1 : 4x - 3y + 12 = 0$ which passes through the point $(5, 8)$.

**Solution:**
We rearrange $4x - 3y + 12 = 0$ into the form $y = mx + c$ so that we know the gradient.

$$4x - 3y + 12 = 0,$$
$$\Rightarrow \qquad y = \frac{4}{3}x + 4.$$

Hence the line we wish to find has gradient $\frac{4}{3}$, since it is parallel to $l_1$.
We use Formula 4.6 to find the equation of the line we seek,

$$y - 8 = \frac{4}{3}(x - 5),$$
$$\Rightarrow \qquad y = \frac{4}{3}x - \frac{20}{3} + 8,$$
$$\Rightarrow \qquad y = \frac{4}{3}x + \frac{4}{3}.$$

**Remark**
We could alternatively give our answer to Example 4.21 as $4x - 3y - 4 = 0$.

### 4.2.3 Perpendicular Straight Lines

Similarly to Section 4.1.4 two lines $l_1 : y = m_1 x + c_1$ and $l_2 : y = m_2 x - c_2$ are perpendicular if one gradient is a negative reciprocal of the other, that is,

$$m_1 m_2 = -1. \tag{4.12}$$

**Question**
Why does the proof in Section 4.1.4 hold in this case?

**Example 4.22**
Find the equation of the straight line which is perpendicular to $y = -\frac{1}{2}x + \frac{13}{2}$ and passes through the point $(7, 6)$.

**Solution:**
Let $m_1$ be the gradient of the line we are finding, then

$$-\frac{1}{2} \times m_1 = -1,$$
$$\Rightarrow \qquad m_1 = 2.$$

Using Formula 4.6,

$$y - 6 = 2(x - 7),$$
$$\Rightarrow \quad y = 2x - 14 + 6,$$
$$\Rightarrow \quad y = 2x - 8.$$

**Exercise 4.2**

Q1. Find the gradients of the following lines:
   (a) $y = 2x - 2$;
   (b) $y = 6$;
   (c) $y = -4x + 1$;
   (d) $-y = 2x - 4$;
   (e) $y = \frac{1}{3}x + 4$;
   (f) $2x + 3y - 12 = 0$;
   (g) $9x - 3y + 2 = 0$;
   (h) $6y + 4 = 13x$.

Q2. Write the lines below in the form $ax + by + c = 0$, where $a$, $b$, $c$ are integers.
   (a) $y = 3x - 2$;
   (b) $y = \frac{1}{2}x - 6$
   (c) $y = -3x - 7$
   (d) $y = \frac{1}{3}x + \frac{5}{4}$

Q3. Find the equation of the line parallel to the line $l_1 : x + 2y + 3 = 0$, that passes through the point $(-2, 5)$.

Q4. Find the equation of the line that cuts the $y$-axis at $(0, 7)$ and the $x$-axis at $(12, 0)$.

Q5. Find the area of the triangle enclosed by the $x$-axis, $y$-axis and the line with equation $l_1 : 3x + 2y - 18 = 0$.

Q6. The straight line with equation $y = mx + 5$ passes through the point $(1, 9)$. Find the value of $m$.

Q7. A straight line has $y$-intercept $(0, 2)$ and gradient $\frac{1}{2}$. This line has equation $ax + by - 4 = 0$; find the values of $a$ and $b$.

Q8. Find the equation of the line with gradient $m$ passing through the point $A(x_A, y_A)$ for the given gradients and points.
   (a) $m = 2$, $A(0, 0)$;
   (b) $m = 3$, $A(3, 2)$;
   (c) $m = -1$, $A(1, 4)$;
   (d) $m = -2$, $A(-3, 7)$;
   (e) $m = \frac{3}{2}$, $A(4, 12)$;
   (f) $m = -\frac{2}{3}$, $A\left(5, -\frac{25}{3}\right)$;
   (g) $m = 6$, $A(2, 25)$;
   (h) $m = -\frac{2}{13}$, $A\left(3, \frac{1}{2}\right)$.

Q9. Find the equation of the line, in terms of $a$, with gradient $\frac{5}{2}$ which passes through the point $(2a, 4a)$.

Q10. Find the equation passing through each pair of points:
   (a) $A(-1, 1)$, $B(2, 4)$;
   (b) $A(-1, 6)$, $B(0, 5)$;
   (c) $A(3, 1)$, $B(9, 9)$;

(d) $A(2,8)$, $B(7,7)$;
(e) $A(2,1)$, $B(5,8)$;
(f) $A(-5,9)$, $B(1,4)$;
(g) $A(3,-8)$, $B(8,-3)$;
(h) $A(2,-5)$, $B(15,1)$.

Q11. Line $l_1 : 5x+4y-30$ meets the $x$-axis at the point $P$. The line $l_2 : -2x-5y+20 = 0$ meets the $y$-axis at the point $Q$. Find the equation of the line joining $P$ to $Q$.

Q12. The lines $y = x+1$ and $x+2y-14 = 0$ intersect at the point $P$. Find the equation of the line joining $P$ to $Q(9,7)$.

Q13. Find the equations of the lines perpendicular to the given line and passing through the given point. The equations of the lines should be given in the form $ax+by+c = 0$ for $a,b$, $c \in \mathbb{Z}$.
(a) $y = 3x - 2$ and $(5,3)$;
(b) $y = \frac{5}{7}x + 4$ and $(1,8)$;
(c) $3x + 8y - 7 = 0$, and $(5,3)$;
(d) $6x - 11y + 28 = 0$ and $(8,-2)$.

## 4.3  Exploring Straight Lines

The techniques discussed in the previous two sections can be used to explore situations involving one or more straight lines.

**Example 4.23**

The line $l_1$ passes through the point $(9,-4)$, has gradient $\frac{1}{3}$ and crosses the $y$-axis at the point $C$. The line $l_2$ passes through the origin, $O$, has gradient $-2$ and intersects $l_1$ at the point $P$.
(a) Find an equation for $l_1$ in the form $ax + by + c = 0$, where $a,b$ and $c$ are integers.
(b) Calculate the coordinates of $P$.
(c) Calculate the exact area of the triangle $OCP$

**Solution:**

(a) We can use Formula 4.6 to find the equation of $l_1$ since we know the gradient and a point on the line.

$$y - (-4) = \frac{1}{3}(x - 9),$$
$$\Rightarrow \qquad y + 4 = \frac{x}{3} - 3,$$
$$\Rightarrow \qquad y = \frac{x}{3} - 7,$$
$$\Rightarrow \qquad 3y = x - 21,$$
$$\Rightarrow \quad x - 3y - 21 = 0.$$

This is in the required form with $a = 1$, $b = -3$ and $c = -21$.
(b) To find the coordinates of $P$ we first need an equation of $l_2$.
Since $l_2$ passes through the origin, the equation of $l_2$ is $y = -2x$. To find the point

of intersection we can substitute this into the equation for $l_1$. Hence,

$$x - 3y - 21 = 0,$$
$$\Rightarrow \quad x - 3(-2x) - 21 = 0,$$
$$\Rightarrow \quad x + 6x - 21 = 0,$$
$$\Rightarrow \quad 7x = 21,$$
$$\Rightarrow \quad x = 3.$$

When $x = 3$, by substituting into the equation of line $l_2$, $y = -6$. Hence, the point of intersection is $P(3, -6)$.

(c) To find the area of the triangle $OCP$, we first need to find the coordinates of $C$. As this is the $y$-intercept of $l_1$, substituting $x = 0$ into the equation of line $l_1$ gives $C(0, -7)$. To find the area of the triangle we use the area formula for a right angled triangle with height, $h = 3$ and base, $b = 7$ as shown below.

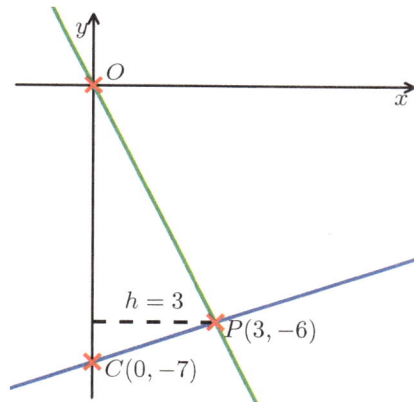

Thus, the area is $A = \frac{1}{2} \times 3 \times 7 = \frac{21}{2}$ square units.

**Example 4.24**

The line $l_1$ has equation $2y - 3x - k = 0$, where $k$ is a constant. Given that the point $A(1, 4)$ lies on $l_1$, find

(a) the value of $k$;

(b) the gradient of $l_1$.

Given that the line $l_2$ passes through $A$ and is perpendicular to $l_1$.

(c) Find an equation of $l_2$, giving your answer in the form $ax + by + c = 0$, where $a$, $b$ and $c$ are integers.

The line $l_2$ crosses the $x$-axis at the point $B$.

(d) Find the coordinates of $B$.

(e) Find the exact length of $AB$.

**Solution:**

(a) Substituting the coordinates of $A$ into the equation for $l_1$ lets us find the value of $k$.

$$2y - 3x - k = 0,$$
$$\Rightarrow \quad 2 \times 4 - 3 \times 1 - k = 0,$$
$$\Rightarrow \quad 5 - k = 0,$$
$$\Rightarrow \quad k = 5.$$

(b) Using this value of $k$, we can rearrange the equation of line $l_1$ into the form $y = m_1 x + c$ to find the gradient:

$$2y - 3x - 5 = 0,$$
$$\Rightarrow \quad 2y = 3x + 5,$$
$$\Rightarrow \quad y = \frac{3}{2}x + \frac{5}{2}.$$

Hence, the gradient, $m_1$ of $l_1$ is $m_1 = \frac{3}{2}$.

(c) As $l_2$ is perpendicular to $l_1$, letting the gradient of $l_2$ be $m_2$, we have

$$m_1 m_2 = -1,$$
$$\Rightarrow \quad \frac{3}{2} \times m_2 = -1,$$
$$\Rightarrow \quad m_2 = -\frac{2}{3}.$$

As this line also passes through $A(1, 4)$, we can use Formula 4.6 to find the equation of the line $l_2$.

$$y - 4 = -\frac{2}{3}(x - 1),$$
$$\Rightarrow \quad y - 4 = -\frac{2}{3}x + \frac{2}{3},$$
$$\Rightarrow \quad y = -\frac{2}{3}x + 4 + \frac{2}{3},$$
$$\Rightarrow \quad y = -\frac{2}{3}x + \frac{14}{3},$$
$$\Rightarrow \quad 3y = -2x + 14,$$
$$\Rightarrow \quad 2x + 3y - 14 = 0.$$

Hence, line $l_2$ has the equation $2x + 3y - 14 = 0$.

(d) Substituting $y = 0$ into this, leads to $x = 7$. Hence, $B$ is the point $B(7, 0)$.

(e) Since we have the coordinates of $A(1,4)$ and $B(7,0)$ we can use Pythagoras' theorem to find the length of $AB$.

$$\begin{aligned} |AB| &= \sqrt{(7-1)^2 + (0-4)^2} \\ &= \sqrt{6^2 + (-4)^2} \\ &= \sqrt{36 + 16} \\ &= \sqrt{52}. \end{aligned}$$

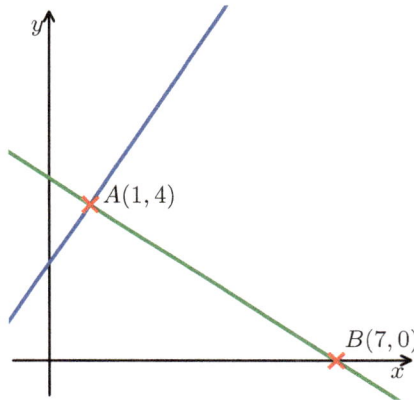

Straight lines can be used to investigate the relationship between two quantities. Many physical quantities can be related with a linear function, for example, the relationship between a temperature measured in Fahrenheit and its measurement in Celsius.

**Example 4.25**

The maximum temperature in Barcelona in July 2016 was $31°$C, or $88°$F. In December the maximum temperature was $16°$C, or $61°$F.

(a) Let $C$ represent the temperature in Celsius and $F$ represent the temperature in Fahrenheit. Derive an equation relating $C$ and $F$ in the form $C = aF + b$.

(b) What do the values $a$ and $b$ mean physically?

(c) If the temperature in May is $73°$F, what is the temperature in Celsius?

(d) If the temperature in April is $19°$C, what is the temperature in Fahrenheit?

(e) Comment on any sources of error for this model.

(f) Can the equation derived in part (a) be used to convert temperatures measured in Moscow?

**Solution:**

(a) With $F$ as our independent variable and $C$ as the dependent variable, we have two points, $A(f_1, c_1) = (88, 31)$ and $B(f_2, c_2) = (61, 16)$ that lie on a straight line, hence, we can use Formula 4.7 to find the equation of that line.

$$\frac{C - C_1}{C_2 - C_1} = \frac{F - F_1}{F_2 - F_1},$$

$$\Rightarrow \quad \frac{C - 31}{16 - 31} = \frac{F - 88}{61 - 88},$$

$$\Rightarrow \quad \frac{C - 31}{-15} = \frac{F - 88}{-27},$$

$$\Rightarrow \quad C - 31 = \frac{15}{27}(F - 88),$$

$$\Rightarrow \quad C - 31 = \frac{5}{9}F - \frac{440}{9},$$

$$\Rightarrow \quad C = \frac{5}{9}F - \frac{161}{9}.$$

Hence, our equation relating $C$ to $F$ is

$$C = \frac{5}{9}F - \frac{161}{9}, \tag{4.13}$$

with $a = \frac{5}{9}$ and $b = -\frac{161}{9}$.

(b) The quantity $a$ is the increase in temperature, measured in degrees Celsius per unit increase in temperature as measured in Fahrenheit. The quantity $b$ is the temperature in Celsius when the temperature is $0°F$.

(c) When the temperature is $73°F$, we can use Equation (4.13) to find the temperature in Celsius.

$$C = \frac{5}{9}F - \frac{161}{9},$$

$$= \frac{5}{9} \times 73 - \frac{161}{9},$$

$$= \frac{68}{3},$$

$$= 23°C \quad \text{to the nearest degree.}$$

(d) Rearranging Equation (4.13), we can find the temperature in Fahrenheit, given a temperature in Celsius.

$$C = \frac{5}{9}F - \frac{161}{9},$$

$$\Rightarrow \quad 19 = \frac{5}{9}F - \frac{161}{9},$$

$$\Rightarrow \quad 19 + \frac{161}{9} = \frac{5}{9}F,$$

$$\Rightarrow \quad \frac{332}{9} = \frac{5}{9}F,$$

$$\Rightarrow \quad F = 66.4,$$

$$\Rightarrow \quad F = 66°F \quad \text{to the nearest degree.}$$

(e) The temperatures we were given were only measured to the nearest degree, this will result in a slight error in our coefficients appearing in Equation 4.13. In fact the relationship between Fahrenheit and Celsius, derived by Anders Celsius in 1742, is $C = \frac{5}{9}(F - 32)$.

(f) The conversion between units of temperature is independent of location, and so this relationship can be used to convert temperatures in Moscow.

**Exercise 4.3**

Q1. (a) Find the equation of the line parallel to the line with equation $3x + 2y + 1 = 0$ that passes through the point $(0, 5)$.

(b) Find the equation of the line perpendicular to the line with equation $3x + 2y + 1 = 0$ that passes through the point $(0, 5)$.

(c) What can be said about the solutions to (a) and (b) if they are written in the form $ax + by + c = 0$, where $a$, $b$ and $c$ are integers.

Q2. Ew are given one pair in a linear relationship: $2 \mapsto 3$. Which of these mappings could be correct?

(a) $x \mapsto 2x + 1$;
(b) $x \mapsto \frac{1}{2}x + 2$;
(c) $x \mapsto 3x - 2$;
(d) $x \mapsto -\frac{1}{2}x + 4$.

Q3. A straight line has gradient $-\frac{2}{3}$ and passes through the point $M(3, 2)$. It crosses the $y$-axis at point $A$ and the $x$-axis at point $B$. $O$ is the origin. Show that the line through $OM$ cuts the area of triangle $AOB$ in half.

Q4. Consider the two straight lines drawn below.

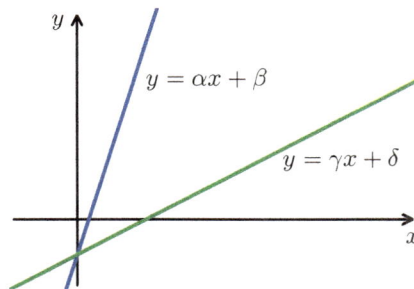

(a) Which is true?
i. $\alpha > \gamma$;
ii. $\alpha = \gamma$;
iii. $\alpha < \gamma$.

(b) Which is true?
i. $\beta > \delta$;
ii. $\beta = \delta$;
iii. $\beta < \delta$.

Q5. $A(13, -2)$, $B(6, 6)$, $C(4, -2)$ are the vertices of a triangle.

(a) Show that this is a scalene triangle.
(b) Find the area of the triangle.

    (c) Find equations of the lines
        i. $AB$;
        ii. $BC$;
        iii. $CA$.

Q6. The straight line $l_1$ has equation $y = 2x - 1$. The straight line $l_2$ has equation $5x + 6y - 45 = 0$. The lines $l_1$ and $l_2$ intersect at the point $A$.

    (a) Find the coordinates of $A$.

    (b) The straight line $l_2$ crosses the $x$-axis at $B$. Find the coordinates of $B$.

    (c) The straight line $l_1$ crosses the $x$-axis at $C$. Find the coordinates of $C$.

    (d) Show that the area of the triangle $OBA$ (where $O$ is the origin) is 1.25 square units bigger than the area of the triangle $BAC$.

Q7. A commercial printer charges £15 per 100 business cards, plus an initial setup charge of £30.

    (a) Write a linear model relating the cost, $c$, and the number, $n$ of business cards, where $n$ is measured in hundreds.

    (b) What would the cost of 2300 business cards be?

    (c) A customer is charged £1365, how many cards did they receive?

Q8. A new school has opened in Yorkshire, starting with an intake of 240 Year 7 students. Each year a new intake will join the school until it is an 11-16 school. The expected number of students per year is 225.

    (a) Derive a linear model for the population of the school, where $n$ is the number of students, $y$ is the number of years after opening.

    (b) Explain why the $y$-intercept is non-physical?

    (c) How many students would we expect four years after opening?

    (d) Why would this model not be valid in 10 years time?

Q9. Consider the square shown below.

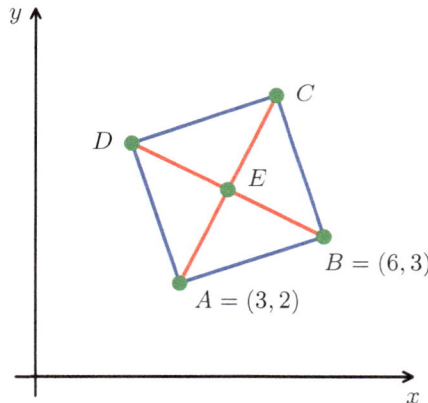

Find the coordinates of points $C$, $D$ and $E$, and the equations of all the edges and diagonals. Also find the area of the square.

**Key Facts — Linear Functions**

- The location of points in a two dimensional Cartesian plane can be described in terms of the quadrant they lie in, based on the values of their coordinates.

| Quadrant | $x$-coordinate | $y$-coordinate |
|----------|-----------|-----------|
| 1 | +ve | +ve |
| 2 | −ve | +ve |
| 3 | −ve | −ve |
| 4 | +ve | −ve |

- The $x$-coordinate of a point is known as the *abscissa* and the $y$-coordinate the *ordinate*

- The distance between two points $A(x_A, y_A)$ and $B(x_B, y_B)$ is given by

$$|AB| = \sqrt{(x_B - x_A)^2 + (y_B - y_A)^2}.$$

- For the points $A(x_A, y_A)$ and $B(x_B, y_B)$ shown in Figure 4.4, the gradient of the line segment joining them is given by

$$\frac{y_B - y_A}{x_B - x_A}.$$

- In the context of straight line graphs and linear functions the gradient is often denoted $m$.

- The coordinates of the midpoint of the line segment joining the points $A(x_A, y_A)$ and $B(x_B, y_B)$ are

$$\left( \frac{x_A + x_B}{2}, \frac{y_A + y_B}{2} \right).$$

- For the points $A(x_A, y_A)$, $B(x_B, y_B)$, $C(x_C, y_C)$ and $D(x_D, y_D)$, the line segments $AB$ and $CD$ are perpendicular if and only if

$$m_1 m_2 = -1,$$

where $m_1$ is the gradient of $AB$ and $m_2$ is the gradient of $CD$.

- Two line segments $AB$ and $CD$ are parallel if they have the same gradient.

- The equation of the line with gradient $m$ and $y$-intercept $c$ can be written in the form

$$y = mx + c.$$

- The equation of a straight line can also be written in the form

$$ax + by + c = 0.$$

- If we know the gradient, $m$, of a line and that it passes through the point $(x_1, y_1)$, then the equation of the line can be found by rearranging

$$y - y_1 = m(x - x_1).$$

- Given two points $A(x_A, y_A)$ and $B(x_B, y_B)$, the equation of the line that passes through $A$ and $B$ is given by

$$\frac{y - y_A}{y_B - y_A} = \frac{x - x_A}{x_B - x_A}$$

- Two straight lines, $l_1 : y = m_1 x + c_1$ and $l_2 : y = m_2 x + c_2$ are parallel if their gradients are equal, *i.e.* $m_1 = m_2$.
- Two lines $l_1 : y = m_1 x + c_1$ and $l_2 : y = m_2 x - c_2$ are perpendicular if one gradient is a negative reciprocal of the other, that is $m_1 m_2 = -1$.

### Chapter Assessment — Linear Functions

Download and sit the 30 minute assessment for this chapter from the digital book.

$$\ldots + 1 = (2x + d)(3x + e),$$
$$= 6x^2 + 2xe + 3xd + de,$$
$$= 6x^2 + (2e + 3d)x + de.$$

# 5. Quadratic Functions

The use of quadratic functions is widespread throughout both high level pure and applied mathematics. Quadratic functions often arise when we model physical situations mathematically, for example, if we consider a projectile moving through the air (assuming no air-resistance) then a quadratic function can be used to model its path. Figure 5.1 shows a cannon ball projected from a launcher positioned on a table top. The ball leaves the cannon with velocity $u\,\mathrm{m\,s^{-1}}$ at a height $h$ above the floor and follows a parabolic path until it hits the floor a horizontal distance $s$ away from the end of the cannon barrel. The modelling of projectiles moving through the air is studied in Year Two of A-Level Mathematics.

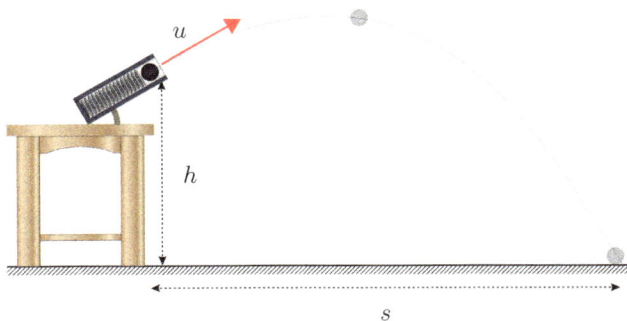

Figure 5.1: Parabolic path of a projectile launched from a spring cannon.

> **Remark**
> The shape of the graph of a quadratic function is known as a parabola and is one of
> four *conic sections* which can be obtained by slicing through a cone in different ways.
> Conic sections are studied in further detail in the Further Mathematics A-Level.

Quadratic functions are also the lowest order polynomials which exhibit interesting behaviour; for example, they possess turning points. Because of this, quadratic functions,
and working with then, form a large part of the first year A-Level syllabus.

A quadratic function is a polynomial which maps a set of numbers (the domain) to another
set of numbers (the range). For example, the set $\{x : -2 \leq x \leq 3\}$ is mapped to the set
$\{y : -\frac{1}{4} \leq y \leq 20\}$ by the function $f(x) = x^2 + 3x + 2$. More formally, we can define a
quadratic function in the following way.

> **Definition 5.1 — Quadratic Function**
> A function of $x$, $f(x)$, is said to be a quadratic function if it is of the form $f(x) =
> ax^2 + bx + c$, where $a, b, c \in \mathbb{R}$ and $a \neq 0$.

> **Interactive Activity 5.1 — Exploring Quadratic Functions**
> Using the Geogebra applet below as a visual aid, describe what happens when the values
> of $a, b$ and $c$ are changed in the general formula for a quadratic function.

> **Exercise 5.1**
> Q1. Which of the following are quadratic functions? Please explain your answer.
>     a) $f(x) = x^2 + 3x + 2$;
>     b) $y = 3x + 2$;
>     c) $x = y^2 + 2y - 15$;
>     d) $f(x) = (x + 3)^2$;
>     e) $f(x) = 5x^2 + 2x$;
>     f) $f(x) = (2x - 3)^2 - 4x^2$.

## 5.1 Plotting Quadratic Functions

We begin the exploration of quadratic functions by considering how to plot them. Plotting
a function is distinct from sketching a function and we shall discuss sketching quadratics
towards the end of this chapter.

> **Example 5.1**
> To plot the graph of $y = x^2 - 4x + 3$, we can construct the table below.

| $x$ | $x^2$ | $-4x$ | $+3$ | $y$ |
|----|------|------|-----|-----|
| $-1$ | 1 | $+4$ | $+3$ | 8 |
| 0 | 0 | 0 | $+3$ | $+3$ |
| 1 | 1 | $-4$ | $+3$ | 0 |
| 2 | 4 | $-8$ | $+3$ | $-1$ |
| 3 | 9 | $-12$ | $+3$ | 0 |
| 4 | 16 | $-16$ | $+3$ | $+3$ |
| 5 | 25 | $-20$ | $+3$ | 8 |

Plotting these points on a pair of coordinate axes and joining them up with a smooth curve leads to the plot shown in Figure 5.2.

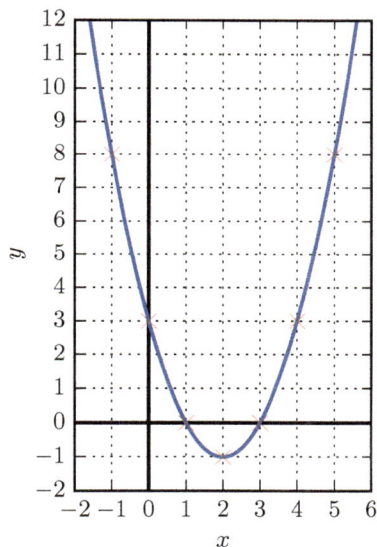

Figure 5.2: The graph of $y = x^2 - 4x + 3$.

The shape of this quadratic (and indeed the graph of any quadratic function) is known as a parabola. Notice that this curve is symmetrical about the line $x = 2$ and that this line of symmetry passes through the vertex $(2, -1)$ of the parabola, which in this case is a minimum.

Note, in the above example we have stated the coordinate of the vertex by reading the values from the graph. Later in this chapter we shall study algebraic methods that can be used to find this point and the equation of the line of symmetry.

**Remark**

The working shown in Example 5.1 is incredibly verbose. Most of the time we do not need to be this detailed and a table such as the one below will suffice.

| $x$ | $-1$ | 0 | 1 | 2 | 3 | 4 | 5 |
|---|---|---|---|---|---|---|---|
| $y$ | 8 | 3 | 0 | $-1$ | 0 | 3 | 8 |

**Exercise 5.2**

Q1. Which of the functions in the figure below are quadratic functions?

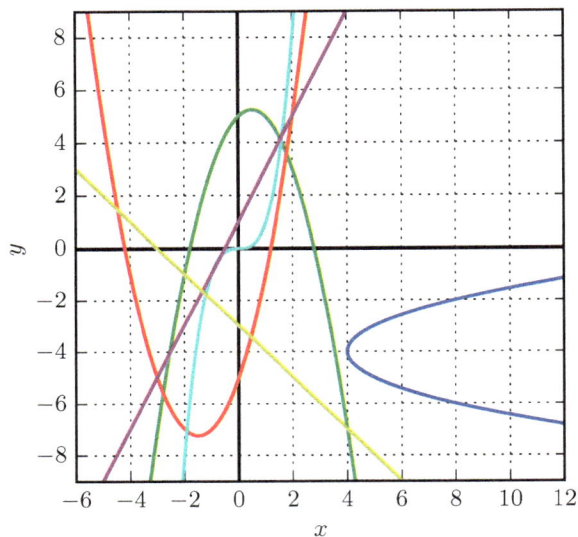

**Interactive Activity 5.2 — Investigating $y = x^2$**

Consider the Geogebra activity located below. Here the function $y = x^2$ is plotted for $x \in [-12, 12]$. Using the sliders for $a$ and $b$, answer the following questions.

Q1. Write down the coordinates of the points $A$ and $B$.

Q2. What can be said about the value of the $y$-coordinate of the point where the chord $AB$ crosses the $y$-axis?

Q3. Prove the conjecture made in Question 2.

**Exercise 5.3**

Q1. Plot the graphs of the following quadratics between $x = -3$ and $x = 3$. State the line of symmetry of the curve and the vertex of the parabola.

(a) $y = x^2 + 1$;
(b) $y = x^2 + 2x$;
(c) $y = x^2 + 2x + 1$;
(d) $y = x^2 + 2x + 5$;
(e) $y = (x + 2)^2$;
(f) $y = -x^2 + 4x + 7$;
(g) $y = -x^2 + 4x - 4$;
(h) $x = y^2 + y + 2$;

Q2. Plot on the same pair of axes the functions $f(x) = x^2 - 4x + 3$ and $f(x) = -x^2 + 4x - 3$. What can be said about these curves?

## 5.2  Factorising Quadratic Expressions

As we know from previous studies, factorising involves expressing a function (often a polynomial) as a product of simpler expressions contained in brackets in some way. Where possible, it is often advantageous to express a quadratic in terms of a pair of linear expressions that are multiplied together. The general quadratic can be expressed in factorised form as

$$ax^2 + bx + c = (fx + d)(gx + e),$$

where $d, e, f$ and $g$ are to be determined.

The ease of solving equations that are in factorised form is just one of the reasons why it is often advantageous to factorise expressions. Many of the quadratics encountered in the A-Level course will factorise with $d, e, f, g \in \mathbb{Z}$, if they factorise at all. If there is to be further significant algebraic manipulation, it is always worth investigating whether a given quadratic will factorise.

**Remark**

Of course, in general, most quadratics are not factorable with $d, e, f$ and $g$ being integers.

**Use of Technology 5.1**

Factorising quadratics is an ideal opportunity to introduce the use of Computer Algebra System (CAS) packages into the classroom. We can use any CAS (such as Mathematica, Maple or SymPy) or a website such as WolframAlpha to check the factorisation of a quadratic. Some graphical calculators will also have this symbolic facility. When using a CAS, it is important to learn the correct syntax to ensure that the software works as expected.

### 5.2.1  Factorising Monic Quadratics

A monic quadratic expression is of the the form $x^2 + bx + c$, *i.e.*, the coefficient $a$ of the general quadratic is 1. To factorise quadratics of this form, first consider the result of multiplying two binomials:

$$(x + d)(x + e) = x^2 + dx + ex + de = x^2 + (d + e)x + de. \tag{5.1}$$

Suppose we wish to express the quadratic $x^2 + bx + c$ in the form $(x + d)(x + e)$, then, using (5.1) we have

$$x^2 + bx + c = x^2 + (d + e)x + de. \tag{5.2}$$

Comparing coefficients in (5.2), we find that $b = d + e$ and $c = de$. To put it another way, the numbers we seek to find as factors are such that their product is equal to the constant term and their sum is equal to the coefficient of $x$ in the original quadratic. Forming equations such as these allows us to efficiently factorise any monic quadratic.

The easiest quadratics to factorise are those where $c$ is a prime number, as shown in the example below.

---

**Question**

Suppose the quadratic $y = x^2 + bx + c$ factorises into $(x + d)(x + e)$ where both $d$ and $e$ are integers. What can be said about the coefficients $b$ and $c$ if $c$ is prime?

---

**Example 5.2**

Factorise the quadratic $y = x^2 + 8x + 7$.

**Solution:**
Let $x^2 + 8x + 7 = (x + d)(x + e)$, then, since 7 is prime, we know that $(d, e) = (1, 7)$, and so,

$$x^2 + 8x + 7 = (x + 1)(x + 7).$$

---

**Remark**
Note, we could equivalently write $x^2 + 8x + 7 = (x + 7)(x + 1)$.

---

If $c$ is non-prime, an efficient way to obtain the factorisation of a monic quadratic is to consider the factor pairs of $c$ in turn.

---

**Example 5.3**

Factorise the quadratic $y = x^2 + 5x + 6$.

**Solution:**
Let $x^2 + 5x + 6 = (x + d)(x + e)$, then we seek $d, e$ such that,

$$de = 6,$$
$$d + e = 5.$$

Now, considering the possible factor pairs we could have $(d, e) = (1, 6)$ or $(d, e) = (2, 3)$. It is clear that the only pair such that $d + e = 5$ is $(d, e) = (2, 3)$. Hence,

$$x^2 + 5x + 6 = (x + 3)(x + 2).$$

**Remark**

Note that, in the above example, we can discount any pairs of negative numbers that multiply to give 6 as the coefficient $b$ is positive. This kind of observation will often reduce the amount of time taken to factorise a quadratic, and we shall come back to this later.

**Example 5.4**

Factorise the quadratic $y = x^2 + x - 6$.

**Solution:**

Let $x^2 + x - 6 = (x + d)(x + e)$, then we seek $d, e$ such that,

$$de = -6,$$
$$d + e = 1.$$

Since $c = -6$, we now have to consider the possible signed factor pairs. There are four such pairs:

$$
\begin{array}{rcr}
1 & \text{and} & -6, \\
-1 & \text{and} & 6, \\
2 & \text{and} & -3, \\
-2 & \text{and} & 3.
\end{array}
$$

Considering these in turn, we find that the only pair which sum to 1 is $(d, e) = (-2, 3)$. Hence,

$$x^2 + x - 6 = (x - 2)(x + 3).$$

**Example 5.5**

Factorise the quadratic $y = x^2 - 7x + 12$.

**Solution:**

Let $x^2 - 7x + 12 = (x + d)(x + e)$, then we seek $d, e$ such that

$$de = 12,$$
$$d + e = -7.$$

As $c$ is positive and $b$ is negative, we consider the negative signed factor pairs of 12.

Namely, $(d, e) = (-1, -12)$, $(d, e) = (-2, -6)$ or $(d, e) = (-3, -4)$. Of these, $(d, e) = (-3, -4)$ sum to 1 and so

$$x^2 - 7x + 12 = (x - 3)(x - 4).$$

**Remark**

From the examples above it should be clear that the signs of the coefficients $b$ and $c$ in the quadratic $y = x^2 + bx + c$ can be used to determine the signs of the factors.

| Sign of $b$ | Sign of $c$ | Signs of the factors |
| :---: | :---: | :---: |
| + | + | + and +. |
| + | − | + and − |
| − | + | − and − |
| − | − | + and − |

When the signs of $b$ and $c$ are different, we can also deduce information about which factor will contain the negative. For example, consider the factor pair $(d, e)$: when the sign of $b$ is negative and the sign of $c$ is positive, we can deduce that the sign of the factor which has the largest magnitude must be negative.

**Learning Resource 5.1 — Sum-Products**

As we have seen, a key skill when factorising quadratics is finding a pair of numbers which multiply to one value and add together to another value. The puzzles in the sheet below (in the digital book) can be used to practise this skill.

These factorising skills are useful for factorising other expressions; these are often referred to as "hidden quadratics", for example $y = x^4 + 5x^2 + 6$.

**Example 5.6**

Factorise the expression $y = x^4 + 5x^2 + 6$.

**Solution:**

Let $w = x^2$, then we have

$$y = w^2 + 5w + 6.$$

This is a quadratic in $w$ and, in fact, is the same as that factorised in Example 5.4. Hence, $y = (w + 3)(w + 2)$, now substituting $w = x^2$ back into our factorised expression, we have the factorised form of the original expression,

$$x^4 + 5x^2 + 6 = (x^2 + 2)(x^2 + 3).$$

**Tip**

These hidden quadratics make an appearance throughout mathematics and can be easily identified by comparing a function to the general form of a quadratic, $ax^2 + bx + c$.
If the power of one of the variables is twice that of the power of the other then it is possible the function is a hidden quadratic. The following are all examples of hidden quadratics:

$$y = x^{12} + 7x^6 + 12,$$
$$y = \sin^2(2x) - 3\sin(2x) - 10.$$

**Example 5.7**

Factorise the expression $f(t) = t^{\frac{2}{3}} - t^{\frac{1}{3}} - 6$.

**Solution:**

If we make the substitution $x = t^{\frac{1}{3}}$, then we have

$$t^{\frac{2}{3}} - t^{\frac{1}{3}} - 6 = x^2 - x - 6,$$
$$= (x + 2)(x - 3).$$

Substituting $x = t^{\frac{1}{3}}$ back into our factorised expression,

$$t^{\frac{2}{3}} - t^{\frac{1}{3}} - 6 = (t^{\frac{1}{3}} + 2)(t^{\frac{1}{3}} - 3)$$

**Remark**

The factorisation of quadratics can be explained diagrammatically. For example, consider the quadratic $y = x^2 + 4x + 3$. Using algebra tiles we can represent this as one square tile of area $x^2$, four rectangular tiles of area $x$ and 3 unit tiles of area 1, as shown below.

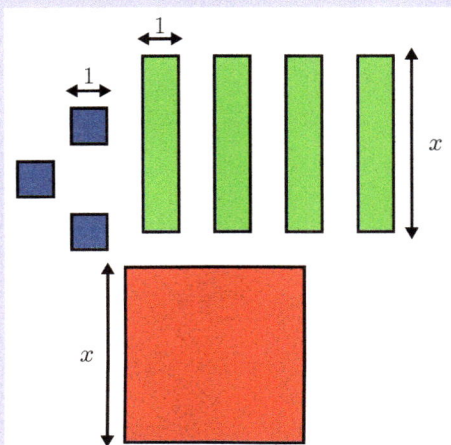

To factorise this quadratic, we seek to arrange these tiles in a rectangular array; this can be done in the following way.

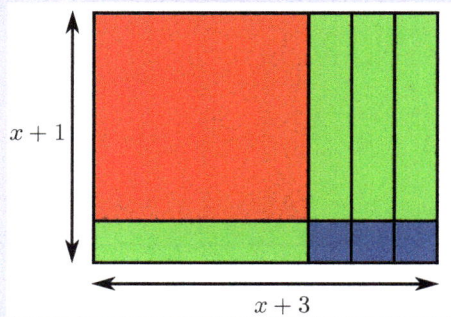

In the above picture we have a rectangle of area $x^2 + 4x + 3$ with side lengths $x + 1$ and $x + 3$. Hence, $y = x^2 + 4x + 3 = (x + 1)(x + 3)$.

As a further example, consider $y = x^2 + 3x + 2$ which factorises to $y = (x + 2)(x + 1)$. Algebra tiles can be used to represent this factorisation, as shown below.

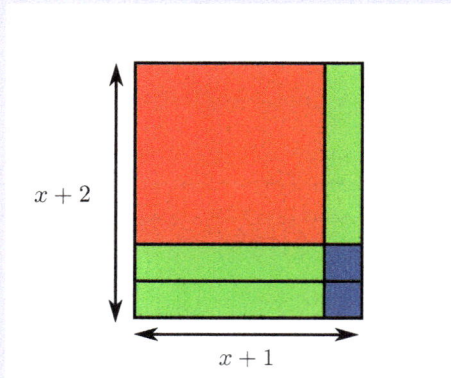

This pictorial representation of factorising can be extremely powerful, however there are limitations. For example, it is tricky to represent the factorisation $x^2 + 2x - 8 = (x - 2)(x + 4)$ due to the difficulties of displaying a quantity with negative area.

**Exercise 5.4**

Q1. Show pictorially the factorisations of the following quadratics.
 (a) $x^2 + 4x + 4$;
 (b) $x^2 + 5x + 6$.

Q2. Factorise the following quadratics:

 (a) $y = x^2 + 12x + 11$;          (d) $y = t^2 + 24t + 23$;
 (b) $y = x^2 + 18x + 17$;          (e) $y = x^2 - 12x - 13$;
 (c) $y = x^2 + 42x + 41$;          (f) $y = x^2 - 24x + 23$.

Q3. Factorise the following:

(a) $y = x^2 + 25x + 24$;
(b) $y = x^2 + 11x + 28$;
(c) $y = x^2 + 6x + 9$;
(d) $y = x^2 + 9x + 20$;
(e) $y = x^2 + 10x + 21$;
(f) $y = x^2 + 6x + 8$;
(g) $y = x^2 + x - 6$;
(h) $y = x^2 + 2x - 35$;
(i) $y = x^2 + 3x - 4$;
(j) $y = x^2 - 2x - 15$;
(k) $y = x^2 + 2x - 8$;

(l) $y = x^2 - 2x - 8$;
(m) $y = x^2 - 4x - 5$;
(n) $y = x^2 + x - 12$;
(o) $y = x^2 - 5x + 6$;
(p) $y = x^2 - 7x + 6$;
(q) $y = x^2 - 11x + 24$;
(r) $y = x^2 - 12x + 35$;
(s) $y = x^2 - 14x + 24$;
(t) $y = x^2 - 13x + 42$;
(u) $y = x^2 - x - 42$;
(v) $y = x^2 + 4x - 21$.

### 5.2.2  Factorising the General Quadratic

For the general quadratic $ax^2 + bx + c$ with $a \neq 1$ the situation is a little more complicated. For $d, e, f, g \in \mathbb{R}$, consider the following expansion,

$$(fx + d)(gx + e) = fgx^2 + de + dgx + fex$$
$$= fgx^2 + (dg + fe)x + de.$$

So,

$$ax^2 + bx + c = fgx^2 + (dg + fe)x + de. \tag{5.3}$$

Comparing coefficients in (5.3), we have

$$a = fg,$$
$$b = dg + fe,$$
$$c = de.$$

Here we have three equations and four unknowns, so we cannot simply solve these to obtain the values of $d, e, f$ and $g$.

In practice, a good way to proceed is to decide on values for $f$ and $g$ and see if a factorisation can be made to work. If $a$ is prime this is straightforward.

Proceeding in this way leads to a semi-deterministic method for factorising a general non-monic quadratic. In practise the selection of values for $f$ and $g$ is often chosen based on experience - for this reason practising factorising quadratics of this form is essential.

**Remark**

The fact that above we have to solve for 4 unknowns with only 3 equations indicates that, if there is one factorisation, then there are an infinite number of possible factorisations. However, we are specifically looking for factorisations with integer coefficients. Note that when $a = 1$, we could also seek a factorisation of the form $(fx + d)(gx + e)$ and we would be led to the same set of equations as above. Again this hints that there could be an infinite number of solutions, but, by fixing $f = g = 1$, the three equations are reduced to two, with now only two unknowns $d$ and $e$, hence a single factorisation is possible.

For example, consider the quadratic function $f(x) = x^2 + 5x + 6$.

$$x^2 + 5x + 6 = (x+2)(x+3)$$
$$= \left(\frac{1}{2}x + 1\right)(2x + 6)$$
$$= \left(\frac{1}{\pi}x + \frac{2}{\pi}\right)(2\pi x + 6\pi).$$

There are many such factorisations, however $x^2 + 5x + 6 = (x+2)(x+3)$ is generally recognised as the factorisation.

**Question**
Suggest some reasons why we prefer factorisations with integer coefficients.

**Example 5.8**
Factorise the quadratic $y = 3x^2 + 16x + 5$.

**Solution:**
The coefficient of $x^2$, 3, is prime and so we can take $f = 3$ and $g = 1$ to obtain

$$(3x + d)(x + e) = 3x^2 + 16x + 5.$$

Comparing coefficients, we see that

$$de = 5,$$
$$3e + d = 16.$$

Solving these, again using techniques discussed in Chapter 11 if necessary, we find that $e = 5$ and $d = 1$. Hence,

$$3x^2 + 16x + 5 = (3x + 1)(x + 5).$$

**Example 5.9**
Factorise the quadratic $y = 3x^2 + 6x + 3$.

**Solution:**
The coefficient of $x^2$, 3, is prime and so we can take $f = 3$ and $g = 1$ to obtain

$$(3x + d)(x + e) = 3x^2 + 6x + 3.$$

Comparing coefficients, we see that

$$de = 3,$$
$$3e + d = 3.$$

To solve the above equations, we can take $d = 3$ and $e = 1$. Hence,

$$3x^2 + 6x + 5 = (3x + 3)(x + 1).$$

**Remark**

If, in an exam, we were asked to 'Factorise completely the quadratic $y = 3x^2 + 6x + 3$' the solution given in Example 5.9 would not be sufficient. This is due to the terms in the first bracket having a common factor of 3; so $3x^2 + 6x + 3 = 3(x + 1)(x + 1) = 3(x + 1)^2$. Of course, this potential for a mistake can be avoided by looking for any common factors before attempting the factorisation. In the above example we could factorise out a common factor of three before having to factorise a quadratic. Factorising $x^2 + 2x + 1$ is certainly easier than factorising $3x^2 + 6x + 3$.

To factorise the quadratic $ax^2 + bx + c$ where $a \neq 1$ and not prime is more challenging. It is often best to combine a heuristic approach with coefficient matching to perform the factorisation efficiently, as the next two examples show.

**Example 5.10**

Factorise the quadratic $y = 6x^2 + 5x + 1$.

**Solution:**

We seek to write $6x^2 + 5x + 1$ in the form $(fx + d)(gx + e)$. To do this, we first choose the numbers $f$ and $g$ by noticing that $fg = 6$. Considering factor pairs of 6 we have two possibilities for $f$ and $g$, namely $(f, g) = (1, 6)$ or $(f, g) = (2, 3)$. Often, picking the factor pair where the absolute difference between the factors is minimised works; and so we choose $(f, g) = (2, 3)$. Consider

$$6x^2 + 5x + 1 = (2x + d)(3x + e),$$
$$= 6x^2 + 2xe + 3xd + de,$$
$$= 6x^2 + (2e + 3d)x + de.$$

Hence,

$$de = 1,$$
$$2e + 3d = 5.$$

Solving these we obtain $(d, e) = (1, 1)$. Therefore, the factorised form of $y = 6x^2 + 5x + 1$ is $y = (2x + 1)(3x + 1)$.

Sometimes the first choice of $f$ and $g$ will not result in a successful factorisation.

**Example 5.11**

Factorise $y = 8x^2 - 37x - 15$.

**Solution:**

As in Example 5.10, we first choose values for $f$ and $g$ using (5.3). Observe that the possible factor pairs for 8 are $(f,g) = (1,8)$ or $(f,g) = (2,4)$. As before, we choose the "middle" pair and consider

$$8x^2 - 37x - 15 = (2x + d)(4x + e).$$

Comparing coefficients, we have that

$$de = -15,$$
$$2e + 4d = -37.$$

However, this pair of simultaneous equations has no solutions; we therefore need to consider the other factor pair for $f$ and $g$, namely, $(f,g) = (1,8)$. Now,

$$8x^2 - 37x - 15 = (8x + d)(x + e)$$
$$= 8x^2 + (d + 8e)x + de.$$

Once again, comparing coefficients, we have that,

$$de = -15,$$
$$8e + d = -37.$$

These do have a solution pair: $(d,e) = (3, -5)$, and so, in factorised form,

$$8x^2 - 37x - 15 = (8x + 3)(x - 5).$$

---

**Remark**

As we did for monic quadratics, we can use algebra tiles to represent the factorisation of a non-monic quadratic. Consider the quadratic $y = 2x^2 + 5x + 2$; this can be represented as a rectangle with side widths $2x + 1$ and $x + 2$ as shown below.

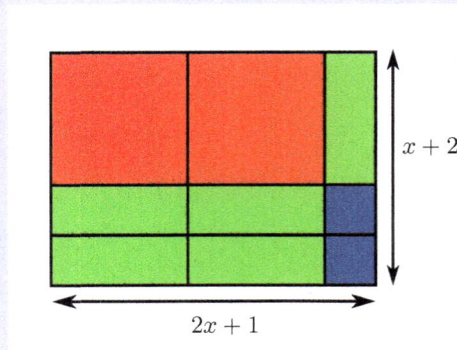

Hence, we have

$$2x^2 + 5x + 2 = (2x + 1)(x + 2).$$

---

With practise, the methods for factorising quadratics described in the above examples will become easier. It is also possible to systematically reduce the problem of factorising a general quadratic to that of factorising a monic quadratic. The next two examples show how this can be done.

---

**Example 5.12**

Factorise $y = 3x^2 - 5x - 12$.

**Solution:**

We first aim to make the leading term a perfect square. To this end, we multiply the quadratic by 3:

$$3y = 9x^2 - 15x - 36.$$

Let $a = 3x$, then

$$3y = a^2 - 5a - 36.$$

We can then factorise this monic quadratic in $a$ using the techniques described in Section 5.2.1:

$$3y = (a + 4)(a - 9). \tag{5.4}$$

Replacing $a$ with $3x$ and factorising out any common factors in the brackets of Equation (5.4), we obtain,

$$3y = (3x + 4)(3x - 9)$$
$$= 3(3x + 4)(x - 3).$$

Hence,

$$y = (3x + 4)(x - 3).$$

---

**Example 5.13**

Factorise $y = 12x^2 + 23x + 10$.

**Solution:**

As in the above example, we first aim to make the leading term a perfect square. To this end, we multiply the quadratic by 12,

$$12y = 144x^2 + 276x + 120.$$

Let $a = 12x$, then

$$12y = a^2 + 23a + 120.$$

We can then factorise this monic quadratic in $a$ using the techniques described in Section 5.2.1:

$$12y = (a + 15)(a + 8) \tag{5.5}$$

Replacing $a$ with $12x$ and factorising out any common factors in the brackets of (5.5) we obtain,

$$12y = (12x + 15)(12x + 8)$$
$$= 3(4x + 5)4(3x + 2)$$
$$= 12(4x + 5)(3x + 2).$$

Hence,

$$y = (4x + 5)(3x + 2).$$

### Exercise 5.5

Q1. Show pictorially the factorisations of the following quadratics.
    (a) $2x^2 + 7x + 3$;
    (b) $3x^2 + 4x + 1$.

Q2. Factorise the following quadratics:

    (a) $y = 2x^2 + 3x + 1$;        (e) $y = 5x^2 + 14x + 8$;
    (b) $y = 2x^2 + 5x + 3$;        (f) $y = 7x^2 + 20x - 3$;
    (c) $y = 3x^2 + 6x + 3$;        (g) $y = 7x^2 - 33x - 10$;
    (d) $y = 5x^2 + 8x + 3$;        (h) $y = 13x^2 - 29x + 6$.

Q3. Factorise the following quadratics:

    (a) $y = 4x^2 + 8x + 4$;        (f) $y = 12x^2 - 10x + 2$;
    (b) $y = 4x^2 + 10x + 6$;      (g) $y = 10x^2 - x - 2$;
    (c) $y = 6x^2 + 8x + 2$;        (h) $y = 10x^2 + 32x + 6$;
    (d) $y = 4x^2 + 7x + 3$;        (i) $y = 12x^2 + 32x + 5$;
    (e) $y = 8x^2 - 2x - 15$;      (j) $y = 30x^2 + 47x + 14$.

## 5.3  Completing the Square

For some quadratics, $y = x^2 + bx + c$, it is possible to factorise them using the identities below.

$$(x + d)^2 \equiv x^2 + 2dx + d^2$$
$$(x - d)^2 \equiv x^2 - 2dx + d^2$$

When this is possible, the quadratic is said to be a "perfect square".

### Remark

This generalises the idea of a "perfect square" integer, such as 9. 9 is sometimes referred to as "perfect square" since $9 = 3 \times 3$.

**Example 5.14**

Express $y = x^2 + 8x + 4$ in the form $(x \pm d)^2$.

**Solution:**

Since $16 = 4^2$ and $8 = 4 \times 2$, we can write

$$x^2 + 8x + 16 = (x + 4)^2.$$

This process is called *completing the square.*

**Remark**

If we visualise this process using algebra tiles it becomes clear why it is known as *completing the square*. Consider the example above for $y = x^2 + 8x + 4$. It is clear from the diagram above that the components of $x^2 + 8x + 16$ can be arranged in a square.

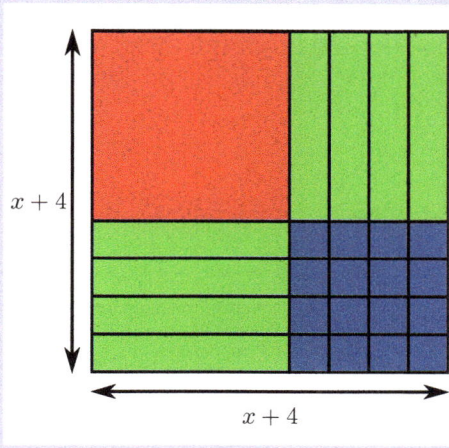

This process also works when the coefficients of the quadratic are rational numbers.

**Example 5.15**

Express $y = x^2 + \frac{4}{5}x + \frac{4}{25}$ in the form $(x \pm d)^2$.

**Solution:**

We can find $d$ by considering the coefficient of $x$ in the quadratic and halving it. Hence,

$$x^2 + \frac{4}{5}x + \frac{4}{25} = \left(x + \frac{2}{4}\right)^2.$$

**Exercise 5.6**

Q1. Using algebra tiles show the completed square form of the following quadratic functions.

  (a) $y = x^2 + 6x + 9$;

  (b) $y = x^2 + 2x + 1$.

Q2. Express in completed square form the following:

(a) $y = x^2 + 4x + 4$;  (d) $y = x^2 + 6x + 9$;
(b) $y = x^2 + 14x + 49$;  (e) $y = x^2 - 2x + 1$;
(c) $y = x^2 + 16x + 64$;  (f) $y = x^2 - 10x + 25$.

Q3. Express in completed square form the following:

(a) $y = x^2 + \frac{2}{3}x + \frac{1}{9}$;  (d) $y = x^2 + \frac{6}{7}x + \frac{9}{49}$;
(b) $y = x^2 - \frac{2}{3}x + \frac{1}{9}$;  (e) $y = x^2 - \frac{2}{5}x + \frac{1}{25}$;
(c) $y = x^2 + \frac{8}{5}x + \frac{16}{25}$;  (f) $y = x^2 + \frac{8}{9}x + \frac{16}{81}$.

In general, a given quadratic $y = x^2 + bx + c$ is not going to be of the form $y = x^2 + 2dx + d^2$ for some $d$. Consequently, there will not be a relation between the coefficients $b$ and $c$. However, it is still possible to *complete the square* on any quadratic function.

**Formula 5.1 — The Completed Square Form of a Quadratic Function**
We can express the general monic quadratic $y = x^2 + bx + c$ in completed square form as

$$x^2 + bx + c = \left( x + \frac{b}{2} \right)^2 + q,$$

where $q$ is chosen so that $\left(\frac{b}{2}\right)^2 + q = c$.

**Remark**
The $q$ appearing in Formula 5.1 can be found by subtracting $\left(\frac{b}{2}\right)^2$ and then adding $c$. That is, write the quadratic as $y = \left(x + \frac{b}{2}\right)^2 - \left(\frac{b}{2}\right)^2 + c$, and then simplify.

**Example 5.16**
Express, in completed square form, the quadratic $y = x^2 + 6x + 15$.

**Solution:**
Using Formula 5.1, we can write this quadratic in completed square form in the following way.

$$x^2 + 6x + 15 = (x + 3)^3 - 3^2 + 15$$
$$= (x + 3)^2 + 6.$$

**Example 5.17**
Express, in completed square form, the quadratic $y = -x^2 - 4x + 8$.

**Solution:**
The presence of a $-x^2$ makes this slightly more complicated than the above example.

Using Formula 5.1,

$$\begin{aligned}
y &= -x^2 - 4x + 8 \\
&= -(x^2 + 4x) + 8 \\
&= -[(x+2)^2 - 2^2] + 8 \\
&= -[(x+2)^2 - 4] + 8 \\
&= -(x+2)^2 + 4 + 8 \\
&= -(x+2)^2 + 12.
\end{aligned}$$

Expressing a quadratic with a leading coefficient that is not equal to one remains straightforward (and is often more straightforward than factorising, even if that is possible), as shown in the example below. In these cases we start by factorising the leading coefficient out of the first two terms of the quadratic.

**Example 5.18**

Express, in completed square form, the quadratic $y = 2x^2 + 6x + 10$.

**Solution:**

We begin by taking out a factor of 2 from the fist two terms and then proceed as before.

$$\begin{aligned}
y &= 2x^2 + 6x + 10 \\
&= 2\left[x^2 + 3x\right] + 10 \\
&= 2\left[\left(x + \frac{3}{2}\right)^2 - \frac{9}{4}\right] + 10 \\
&= 2\left(x + \frac{3}{2}\right)^2 - \frac{18}{4} + 10 \\
&= 2\left(x + \frac{3}{2}\right)^2 + \frac{11}{2}.
\end{aligned}$$

**Remark**

Algebra tiles can be used to show pictorially the method of completing the square for a general quadratic function. Consider again the quadratic $x^2 + 4x + 3$. To complete the square, we seek to arrange the algebra tiles in a square as shown below. From the diagram below, due to the missing square in the the top right corner, it is clear that,

$$x^2 + 4x + 3 = (x+2)^2 - 1.$$

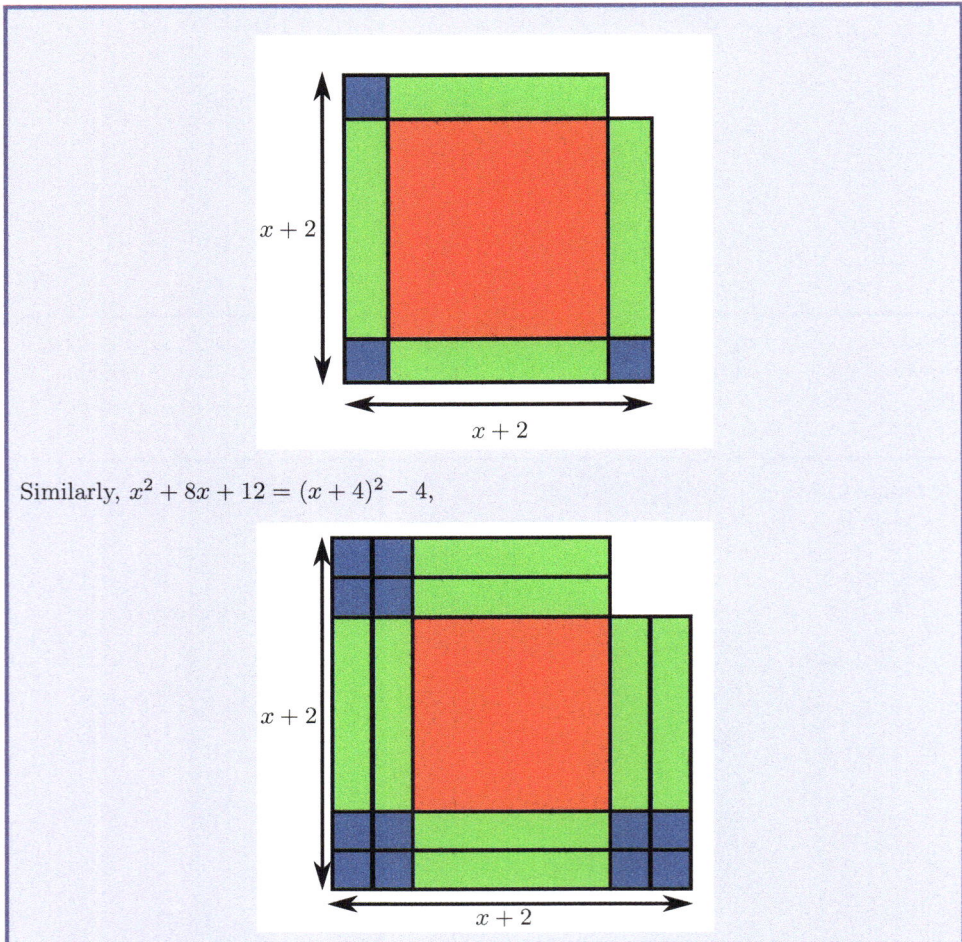

Similarly, $x^2 + 8x + 12 = (x+4)^2 - 4$,

Recall that in Section 5.1 we identified the minimum point and line of symmetry of a quadratic. Completing the square enables us to do this algebraically without plotting the quadratic.

**Example 5.19**

Find the location of the minimum point of the quadratic $y = x^2 + 4x - 6$. Find also the line of symmetry of this quadratic.

**Solution:**

Completing the square, we have that

$$y = x^2 + 4x - 3$$
$$= (x+2)^2 - 7.$$

Hence, the minimum point has coordinate $(-2, -7)$ and the line of symmetry is $x = -2$. The minimum point is shown with a red cross and the line of symmetry with a dashed green line in the plot below.

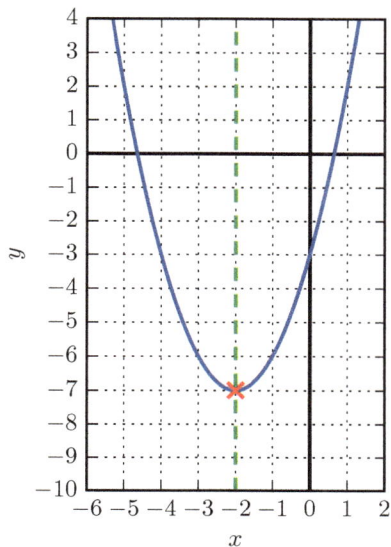

**Exercise 5.7**

Q1. Show pictorially the method of completing the square for the quadratics below.
   (a) $y = x^2 + 2x + 2$;
   (b) $y = x^2 + 6x + 6$.
   (c) $y = x^2 + 4x + 3$.

Q2. Express the following in completed square form:

   (a) $y = x^2 + 2x + 2$;        (f) $y = x^2 + 4x - 4$;
   (b) $y = x^2 + 2x + 5$;        (g) $y = x^2 + 2x - 5$;
   (c) $y = x^2 + 4x + 9$;        (h) $y = x^2 - 4x + 9$;
   (d) $y = x^2 + 6x + 11$;       (i) $y = x^2 - 8x + 22$;
   (e) $y = x^2 + 6x + 4$;        (j) $y = x^2 - 2x + 16$.

Q3. Express the following in completed square form

   (a) $y = x^2 + 3x + 4$;        (f) $y = x^2 - 3x - 3$;
   (b) $y = x^2 + 3x + 7$;        (g) $y = x^2 - 11x + 4$;
   (c) $y = x^2 + x + 4$;         (h) $y = x^2 - 5x - 3$;
   (d) $y = x^2 + 5x - 7$;        (i) $y = x^2 - x - 1$;
   (e) $y = x^2 + 5x + 9$;        (j) $y = x^2 - 5x + 6$.

Q4. Express the following in completed square form:

(a) $y = -x^2 - 4x + 8$;
(b) $y = -x^2 + 2x + 1$;
(c) $y = -x^2 + 4x - 3$;
(d) $y = -x^2 - 8x + 4$;

(e) $y = -x^2 + 5x - 2$;
(f) $y = -x^2 - 3x$;
(g) $y = -x^2 - 14x - 49$;
(h) $y = -x^2 + 7x - 4$.

Q5. Express the following in completed square form:

(a) $y = 2x^2 + 4x + 6$;
(b) $y = 2x^2 - 8x + 2$;
(c) $y = 3x^2 + 12x + 1$;
(d) $y = 5x^2 - 20x - 2$;

(e) $y = 2x^2 - 3x + 1$;
(f) $y = 3x^2 + 4x + 5$;
(g) $7x^2 - 3x + 14$;
(h) $5x^2 - 12x + 3$.

Q6. Find the coordinate of the minimum of the quadratics below and show this on a plot of the function.

(a) $y = x^2 + 2x - 5$;
(b) $y = x^2 + 6x + 3$;
(c) $y = x^2 - 6x + 11$;
(d) $y = x^2 - x + 11$;
(e) $y = x^2 - 3x - 4$;

(f) $y = 3x^2 - 3x - 4$;
(g) $y = 4x^2 + 5x - 26$;
(h) $y = 2x^2 - x + 5$;
(i) $y = 5x^2 - 7x + 1$;
(j) $y = 2x^2 + 16x - 7$.

## 5.4 Solving Quadratic Equations

Solving quadratic equations will occur frequently throughout both the applied and pure topics in A-Level Mathematics; in this section we describe four methods to do this.

### Use of Technology 5.2

In the exam it is possible to use a scientific or graphical calculator to check your solutions to a quadratic equation. In fact, unless you are asked to show working, or equivalent, this will be the most efficient way to find the solutions to a quadratic equation.

### 5.4.1 Finding Approximate Solutions Graphically

Sometimes a solution that is only accurate to one decimal place, say, is acceptable. For many quadratics, if this is the case then the solutions (or *roots* of the quadratic equation) can be found by plotting the quadratic function and observing where the graph crosses the $x$-axis.

### Example 5.20

Find approximate solutions (accurate to one decimal place) for the quadratic equation $x^2 - 6x + 4 = 0$.

**Solution:**
Plotting the quadratic function $y = x^2 - 6x + 4$ we obtain the graph shown in Figure 5.3.

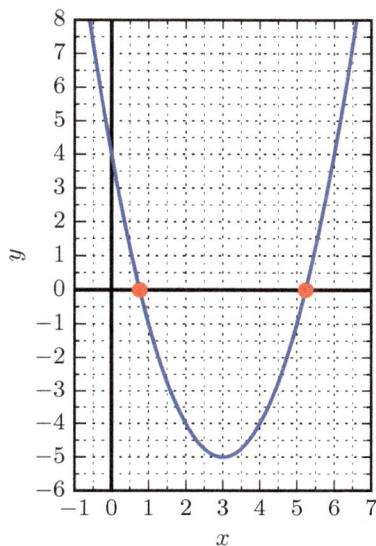

Figure 5.3: The graph of $y = x^2 - 6x + 4$.

From the graph it can be seen that the quadratic $x^2 - 6x + 4$ has two real roots, $x_1 \approx 0.8$ and $x_2 \approx 5.2$.

### 5.4.2 Solution by Factorising

Once a quadratic has been factorised, it is easy to find the roots of the associated quadratic equation.

**Example 5.21**
Solve $x^2 - 9x + 20 = 0$ by factorising the quadratic.

**Solution:**
Factorising the quadratic gives

$$x^2 - 9x + 20 = 0,$$
$$\Rightarrow \quad (x - 5)(x - 4) = 0.$$

From this we can see that either $x - 5 = 0$ or $x - 4 = 0$, hence the roots of the equation are $x = 5$ and $x = 4$.

**Example 5.22**

Solve $x^2 - x - 6 = 0$ by factorising the quadratic.

**Solution:**

Factorising the quadratic gives

$$x^2 - x - 6 = 0,$$
$$\Rightarrow \quad (x - 3)(x + 2) = 0.$$

From this we can see that either $x - 3 = 0$ or $x + 2 = 0$, hence the roots of the equation are $x = 3$ and $x = -2$.

---

**Example 5.23**

Solve by factorising $8x^2 + 26x + 21 = 0$.

**Solution:**

Factorising the quadratic,

$$8x^2 + 26x + 21 = 0,$$
$$\Rightarrow \quad (2x + 3)(4x + 7) = 0.$$

From this, we have that $2x + 3 = 0$ or $4x + 7 = 0$ and so the roots of the equation are $x = -\frac{3}{2}$ and $x = -\frac{7}{4}$.

---

**Interactive Activity 5.3 — Finding Solutions to Quadratic Equations By Factorising**

In the Tarquin activity below match quadratics with their solutions.

---

**Learning Resource 5.2 — Transforming Roots**

The resource contained in the digital book below explores how a quadratic with given roots can be transformed to a quadratic with different roots.

### 5.4.3 Solution by Completing the Square

In addition to being a quick method for locating the minima or maxima of a quadratic equation, it is possible to solve a quadratic by completing the square.

---

**Example 5.24**

Solve $x^2 + 6x + 3 - 0$ by completing the square.

**Solution:**

Completing the square we have that

$$x^2 + 6x + 3 = 0,$$
$$\Rightarrow \quad (x+3)^2 - 6 = 0,$$
$$\Rightarrow \quad (x+3)^2 = 6,$$
$$\Rightarrow \quad x+3 = \pm\sqrt{6},$$
$$\Rightarrow \quad x = -3 \pm \sqrt{6}.$$

Hence, the two roots of the quadratic equation $x^2 + 6x + 3 = 0$ are

$$x = -3 + \sqrt{6} \quad \text{and} \quad x = -3 - \sqrt{6}.$$

---

**Example 5.25**

Solve $x^2 - 7x + 3 = 0$ by completing the square.

**Solution:**

Completing the square, we have

$$x^2 - 7x + 3 = 0,$$
$$\Rightarrow \quad \left(x - \frac{7}{2}\right)^2 - \frac{37}{4} = 0,$$
$$\Rightarrow \quad \left(x - \frac{7}{2}\right)^2 = \frac{37}{4},$$
$$\Rightarrow \quad x - \frac{7}{2} = \pm\sqrt{\frac{37}{4}},$$
$$\Rightarrow \quad x = \frac{7}{2} \pm \sqrt{\frac{37}{4}}.$$

Hence, the two roots of the quadratic equation $x^2 - 7x + 3 = 0$ are

$$x = \frac{7}{2} + \sqrt{\frac{37}{4}} \quad \text{and} \quad x = \frac{7}{2} - \sqrt{\frac{37}{4}}.$$

### 5.4.4  Solution by the Quadratic Formula

By completing the square on the general quadratic function $ax^2 + bx + c$, we can derive a formula which allows us to solve any equation of the form $ax^2 + bx + c = 0$, provided that we know the values of $a$, $b$ and $c$.

$$ax^2 + bx + c = 0,$$
$$\Rightarrow \quad x^2 + \frac{b}{a}x + \frac{c}{a} = 0,$$
$$\Rightarrow \quad x^2 + \frac{b}{a}x = -\frac{c}{a},$$
$$\Rightarrow \quad x^2 + \frac{b}{a}x + \left(\frac{b}{2a}\right)^2 = -\frac{c}{a} + \left(\frac{b}{2a}\right)^2,$$
$$\Rightarrow \quad \left(x + \frac{b}{2a}\right)^2 = -\frac{c}{a} + \left(\frac{b}{2a}\right)^2,$$
$$\Rightarrow \quad x + \frac{b}{2a} = \pm\sqrt{-\frac{c}{a} + \left(\frac{b}{2a}\right)^2},$$
$$\Rightarrow \quad x = -\frac{b}{2a} \pm \sqrt{-\frac{c}{a} + \left(\frac{b}{2a}\right)^2},$$
$$\Rightarrow \quad x = -\frac{b}{2a} \pm \sqrt{-\frac{c}{a} + \frac{b^2}{4a^2}},$$
$$\Rightarrow \quad x = -\frac{b}{2a} \pm \sqrt{\frac{-4a^2c + b^2a}{4a^2}},$$
$$\Rightarrow \quad x = -\frac{b}{2a} \pm \frac{1}{2a}\sqrt{-4ac + b^2},$$
$$\Rightarrow \quad x = \frac{-b \pm \sqrt{b^2 - 4ac}}{2a}.$$

This derivation allows us to define formulae for the two roots of the general quadratic formula.

---

**Formula 5.2 — The Quadratic Formula**

The two roots $x_1$ and $x_2$ of the quadratic equation $ax^2 + bx + c = 0$ can be computed using

$$x_1 = \frac{-b + \sqrt{b^2 - 4ac}}{2a},$$
$$x_2 = \frac{-b - \sqrt{b^2 - 4ac}}{2a}.$$

---

**Remark**

For brevity, we often see the quadratic formula stated as,

$$x = \frac{-b \pm \sqrt{b^2 - 4ac}}{2a}.$$

**Example 5.26**

Solve, using the quadratic formula, $3x^2 + 6x + 2 = 0$.

**Solution:**

Comparing coefficients with the general quadratic $ax^2 + bx + c$, we have that $a = 3$, $b = 6$ and $c = 2$. Using Formula 5.2 we have

$$
\begin{aligned}
x &= \frac{-b \pm \sqrt{b^2 - 4ac}}{2a} \\
&= \frac{-6 \pm \sqrt{6^2 - 4 \times 3 \times 2}}{2 \times 3} \\
&= \frac{-6 \pm \sqrt{36 - 24}}{6} \\
&= \frac{-6 \pm \sqrt{12}}{6} \\
&= \frac{-6 \pm 2\sqrt{3}}{6} \\
&= -1 \pm \frac{\sqrt{3}}{3}.
\end{aligned}
$$

Hence,

$$x_1 = -1 + \frac{\sqrt{3}}{3}$$

and

$$x_2 = -1 - \frac{\sqrt{3}}{3}.$$

**Activity 5.1 — An Alternative Quadratic Formula**

In certain situations (when programmed on a computer) the quadratic formula will incorrectly solve a quadratic equation due to how computer processors handle numbers. In this case an alternative quadratic formula can be used:

$$
\begin{aligned}
x_1 &= \frac{2c}{-b - \sqrt{b^2 - 4ac}}, \\
x_2 &= \frac{2c}{-b + \sqrt{b^2 - 4ac}}.
\end{aligned}
$$

Q1. Show how the above formulae can be derived from Formula 5.2.

Q2. Do the above formulae always give the same answers (assuming exact arithmetic) as Formula 5.2?

**Example 5.27**
Solve, by using the alternative formula, the quadratic equation $3x^2 + 6x + 2 = 0$.

**Solution:**

$$x = \frac{2c}{-b \mp \sqrt{b^2 - 4ac}}$$
$$= \frac{2 \times 2}{-6 \mp \sqrt{6^2 - 4 \times 3 \times 2}}$$
$$= \frac{4}{-6 \mp \sqrt{36 - 24}}$$
$$= \frac{4}{-6 \mp \sqrt{12}}$$
$$= \frac{4}{-6 \mp 2\sqrt{3}}.$$

Hence, $x_1 = \frac{4}{-6-2\sqrt{3}}$ and $x_2 = \frac{4}{-6+2\sqrt{3}}$.

**Exercise 5.8**

Q1. Find approximate solutions, accurate to one decimal place, for the following quadratic equations using a graphical method.

(a) $x^2 - 4x - 1 = 0$;
(b) $x^2 - 2x - 4 = 0$;
(c) $x^2 + 2x - 4 = 0$;
(d) $x^2 + 6x + 1 = 0$;

(e) $x^2 + 6x + 5 = 0$;
(f) $x^2 - 2x - 5 = 0$;
(g) $x^2 + 3x - 8 = 0$;
(h) $x^2 + 3x - 2 = 0$.

Q2. Solve the following quadratics by factorising.

(a) $x^2 - 2x - 3 = 0$;
(b) $x^2 + 11x - 28 = 0$;
(c) $x^2 - x - 12 = 0$;
(d) $2x^2 + 7x - 3 = 0$;

(e) $2x^2 - x - 3 = 0$;
(f) $6x^2 - 11x + 3 = 0$;
(g) $6x^2 + 7x - 5 = 0$;
(h) $12x^2 - 25x + 7 = 0$.

Q3. Find the roots of the following quadratics by completing the square.

(a) $x^2 + 4x - 1$;
(b) $x^2 - 8x + 9$;
(c) $x^2 + 8x - 4$;
(d) $x^2 + 2x - 1$;

(e) $x^2 - 3x - 3$;
(f) $x^2 + 4x + 3$;
(g) $x^2 - 2x - 4$;
(h) $x^2 + 6x + 3$.

Q4. Use the quadratic formula to solve the following.

(a) $x^2 + 3x + 1 = 0$;
(b) $x^2 + 5x - 2 = 0$;
(c) $x^2 - 3x + 1 = 0$;

(d) $x^2 + 2x + 1 = 0$;
(e) $2x^2 + 3x - 4 = 0$;
(f) $5x^2 + 5x - 2 = 0$;

> Q5. Solve the following quadratics using any method that results in an exact answer.
>    (a) $3x^2 + 8x - 4$;
>    (b) $3x^2 - 5x - 2$.

## 5.5 The Role of the Discriminant

From the previous sections it should be apparent that the graphs of all quadratic functions have a similar 'look'. Hence, we should be able to categorise quadratic functions in some way. To do this, we refer back to Formula 5.2 (the Quadratic Formula) and make the following definition.

**Definition 5.2 — The Discriminant of a Quadratic**
The discriminant of the quadratic function $f(x) = ax^2 + bx + c$ is defined to be $\Delta = b^2 - 4ac$.

This definition allows us to compute a numerical quantity associated with every possible quadratic.

**Example 5.28**
We compute the discriminant of the quadratic $y = 3x^2 + 4x - 2$ in the following way, noting that, in this instance, $a = 3$, $b = 4$ and $c = -2$. Hence,

$$\Delta = b^2 - 4ac$$
$$= 4^2 - 4 \times 3 \times (-2)$$
$$= 40.$$

For the general quadratic $f(x) = ax^2 + bx + c$, we can use the discriminant together with the sign of the coefficient $a$ to determine the shape of the graph of $f(x)$. The table above illustrates the six possibilities. From Table 5.1 we can also classify the algebraic properties of the roots of the quadratic equation $ax^2 + bx + c = 0$ using the discriminant.

- $b^2 - 4ac > 0$ implies that the quadratic equation $ax^2 + bx + c = 0$ has two real and distinct roots.
- $b^2 - 4ac = 0$ implies that the quadratic equation $ax^2 + bx + c = 0$ has one repeated root.
- $b^2 - 4ac < 0$ implies that the quadratic equation $ax^2 + bx + c = 0$ has no real roots.

If we are only interested in classifying the roots of a quadratic equation and not finding them, then it suffices to just investigate the discriminant.

**Activity 5.2 — Why the Discriminant?**
Explain why the discriminant allows us to categorise the roots of a quadratic equation.

|  | $b^2 - 4ac > 0$ | $b^2 - 4ac = 0$ | $b^2 - 4ac < 0$ |
|---|---|---|---|
| $a > 0$ | | | |
| $a < 0$ | | | |

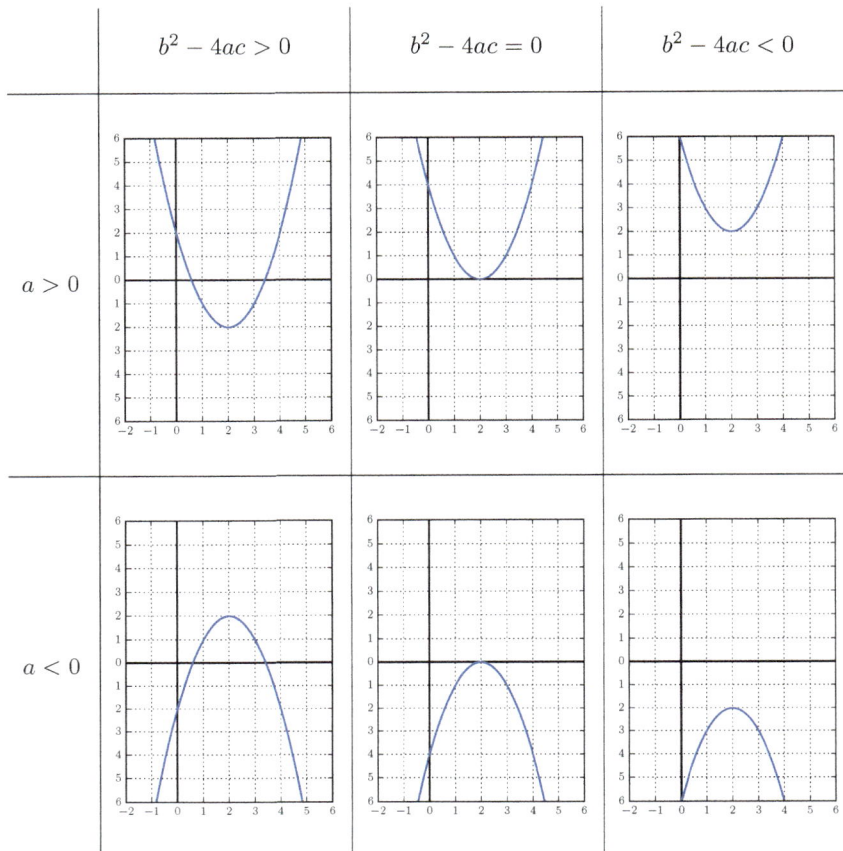

Table 5.1: Graphical meaning of the discriminant.

**Example 5.29**

Consider the quadratic equation $x^2 + 2x - 2 = 0$, then the discriminant is given by:

$$\begin{aligned} \Delta &= b^2 - 4ac \\ &= 2^2 - 4 \times 1 \times (-2) \\ &= 4 - (-8) \\ &= 12. \end{aligned}$$

Thus, this quadratic equation has two distinct real roots.

**Activity 5.3 — Properties of the Discriminant**

What can be said if the discriminant of the quadratic equation $ax^2 + bx + c = 0$ is greater than zero and a whole number? What can be said if the discriminant of the quadratic equation $ax^2 + bx + c = 0$ is greater than zero and a perfect square?

**Remark**

The use of the discriminant for a quadratic is well known, however the formula for the discriminant of a cubic equation is less well known.

For the cubic polynomial $ax^3 + bx^2 + cx + d$ the discriminant can be calculated using

$$\Delta = b^2 c^2 - 4ac^3 - 4b^3 d - 27a^2 d^2 + 18abcd.$$

This can be used in a similar way to the discriminant of a quadratic equation to classify the roots of a cubic equation:

- $\Delta > 0$: the equation has 3 distinct real roots.
- $\Delta = 0$: at least two roots coincide and they are all real.
- $\Delta < 0$: the equation has 1 real root and two complex roots.

This is not required knowledge for A-Level mathematics, but can be useful for STEP and Challenge style problems.

**Interactive Activity 5.4 — Calculating the Discriminant**

Using the interactive matching pairs game, match the quadratic with the value of the corresponding discriminant.

**Exercise 5.9**

Q1. Find the discriminant of the quadratic functions below.

(a) $y = x^2 + 3x - 2$;
(b) $y = x^2 + x + 5$;
(c) $y = 2x^2 + 2x - 1$;
(d) $y = x^2 - 2x + 6$;
(e) $y = -x^2 - 2x + 5$;

(f) $y = -x^2 - 6x + 1$;
(g) $y = x^2 + 8x + 4$;
(h) $y = x^2 + 3x + 4$;
(i) $y = 2x^2 + 4x + 2$;
(j) $y = x^2 + 2x - 5$.

Q2. Classify the roots of the following quadratic equations.

(a) $x^2 + 5x + 6$;
(b) $x^2 + 5x + 1$;
(c) $x^2 + 1$;
(d) $3x^2 + 6x + 3$;

(e) $x^2 - 3x + 2$;
(f) $x^2 - 2x - 1$;
(g) $x^2 - 2x + 8$;
(h) $5x^2 + 10x + 5$.

Q3. Show that, for all $k \in \mathbb{R}$, the quadratic $2x^2 + 4kx - 2$ will have two real and distinct roots.

Q4. Find the $p$ such that $x^2 + 3x + p = 0$ has a repeated root.

Q5. Find the range of $k$ for which $x^2 + 2kx - k = 0$ has real roots.

## 5.6  Sketching Quadratics

In Chapter 9 we will discuss curve sketching in general; here we only look at two ways to sketch quadratic functions.

In Section 5.1 we plotted quadratic functions. In practice, if we wanted to do this, some graph plotting software would be employed. It is much more common, however, to sketch a quadratic function as this provides a visual description of its behaviour.

Typically a sketch of a quadratic function will include the following key points:

- The general shape of the graph.
- Where the graph crosses the $y$-axis.
- The location of any points where the graph crosses the $x$-axis.
- The location of the vertex of the quadratic.

**Remark**

The above list is comprehensive; the information shown should be dictated by the use of the sketch. If we are not interested in the locations of the roots, it is not necessary to show them on the sketch (or calculate them).

**Example 5.30**

Sketch the graph of $y = x^2 + 6x + 7$.

**Solution:**

To sketch the graph, we complete the square on the quadratic function to easily calculate all the information required. To this end,

$$x^2 + 6x + 7 = (x + 3)^2 - 9 + 7$$
$$= (x + 3)^2 - 2.$$

The coefficient of $x^2$ tells us that this quadratic has a minimum point, which, from our completed square expression above, has coordinates $(-3, -2)$. The curve crosses the $x$-axis at the roots which can be found in the following way:

$$(x + 3)^2 - 2 = 0,$$
$$\Rightarrow \qquad (x + 3)^2 = 2,$$
$$\Rightarrow \qquad x + 3 = \pm\sqrt{2},$$
$$\Rightarrow \qquad x = -3 \pm \sqrt{2}.$$

Hence we can plot the points $(-3 - \sqrt{2}, 0)$ and $(-3 + \sqrt{2}, 0)$. Finally, when $x = 0$,

$y = 7$, hence the point $(0, 7)$ is also known. With this information, we can sketch the quadratic as below.

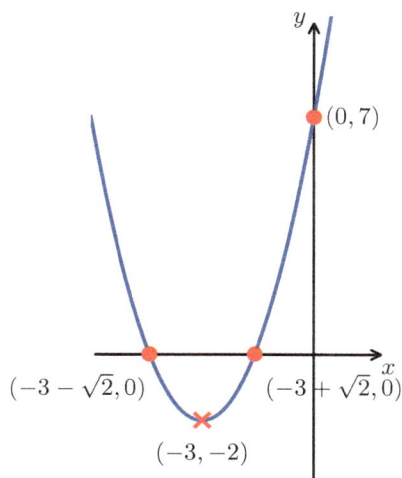

---

**Example 5.31**

Sketch the graph of $y = x^2 + 3x - 5$.

**Solution:**

In this approach to sketching, we exploit the symmetry of the quadratic. First, factorise the first two terms so that we have

$$x^2 + 3x - 5 = x(x + 3) - 5.$$

From the above we can make the following deductions.

- When $x = 0$ and $x = -3$ the function has the same value, namely, $-5$.
- By symmetry the curve must have 'turned' (more precisely, the gradient has changed sign) between $x = 0$ and $x = -3$.
- The positive coefficient of $x^2$ indicates that this quadratic has a minimum point, and that the $x$-coordinate of this is half way between $x = 0$ and $x = -3$.
- Hence, the minimum value occurs when $x = -1.5$ and at this point $y = -\frac{29}{4}$.

Using this information we can sketch the quadratic.

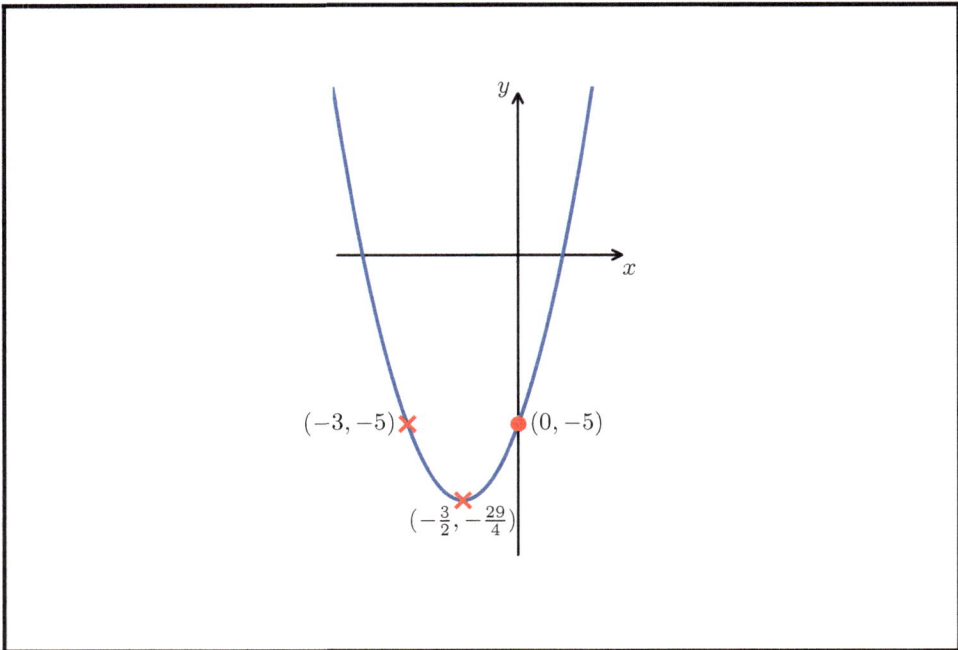

The curve passes through $(-3,-5)$ and $(0,-5)$ with vertex $\left(-\frac{3}{2}, -\frac{29}{4}\right)$.

**Exercise 5.10**

Q1. Sketch the following quadratics, showing all points of intersection with the axes and the minimum value.
   (a) $y = x^2 + 4x - 5$;
   (b) $y = x^2 - 6x + 5$;
   (c) $y = x^2 - x - 8.75$;
   (d) $y = -x^2 + 6x - 5$.

Q2. Show, on a sketch, the location of the vertex of the quadratics below:
   (a) $y = x^2 - 2x - 3$;
   (b) $y = -x^2 - 2x + 4$;
   (c) $y = x^2 + x - 6$;
   (d) $y = x^2 - 5x + 4$.

The exercises below are intended to consolidate understanding of this topic.

**Interactive Activity 5.5 — Identify the Quadratic**

In the interactive activity below match the graph with the correct equation.

**Learning Resource 5.3 — Quadratic Equations Revision**

The file linked below can be used at the end of this topic to check understanding and fluency. The cards should be sorted into groups based on the underlying quadratic. For example, one group could contain an quadratic in expanded form, the quadratic in completed square form and a plot of this quadratic. Some of the cards have information missing from them.

**Exercise 5.11**

Q1. The two quadratics below have the same minimum value; find the value of $b$.

$$y_1 = x^2 + 8x - 2,$$
$$y_2 = x^2 + 6x + b.$$

Q2. The two quadratics below have the same minimum value; find the value of $a$ and hence the second quadratic.

$$y_1 = x^2 - 4x + 8,$$
$$y_2 = x^2 - 2ax + 20.$$

Q3. Both the $x$ and $y$ coordinates of the minima of the three quadratics below form a different increasing arithmetic sequence. Determine the common difference. Given that both arithmetic sequences have the same common difference, determine the value of $b$.

$$y_1 = x^2 + 2x - 1,$$
$$y_2 = x^2 - 2x + 1,$$
$$y_3 = x^2 + bx + 11.$$

Q4. The minima of the quadratics $A$ and $B$ form two vertices of a square.

$$A \quad : \quad y = x^2 - 4x + 2,$$
$$B \quad : \quad y = x^2 - 12x + 38.$$

Find two quadratics that could form the other two vertices and determine the area of the square.

Q5. Consider the quadratic below for $k \in \mathbb{R}$. Find the value of $k$ which will maximise the minimal value of the quadratic. For this value of $k$, find the co-ordinates of the minimal point.

$$y = x^2 + 4kx + 2k,$$

Q6. Find all solutions to the following equation

$$\left(x^2 - 3x + 3\right)^{x^2 + 7x + 12} = 1,$$

Q7. What possible integers can be placed in the $\square$ in the quadratics below so that they are factorable with integer roots?

(a)
$$x^2 + \square x + 6.$$

(b)
$$x^2 + \square x + 24.$$

(c)
$$x^2 + 10x + \square.$$

Q8. The sides of a right angled triangle are, given in order of increasing size, $x$, $x+7$ and $x+8$. Determine the lengths of the sides.

**Key Facts — Quadratic Functions**

- A quadratic function is one of the form $f(x) = ax^2 + bx + c$, where $a, b, c \in \mathbb{R}$ and $a \neq 0$.
- A monic quadratic function has the form $f(x) = x^2 + bx + c$, i.e., the coefficient $a$ of the general quadratic is 1.
- The signs of the coefficients $b$ and $c$ in the quadratic $y = x^2 + bx + c$ can be used to determine the signs of the factors. This is summarised in the table below.

| Sign of $b$ | Sign of $c$ | Signs of the factors |
|:---:|:---:|:---:|
| + | + | + and +. |
| + | − | + and − |
| − | + | − and − |
| − | − | + and − |

- The completed square form of a general monic quadartic equation is

$$x^2 + bx + c = \left(x + \frac{b}{2}\right)^2 + q,$$

where $q$ is chosen so that $\left(\frac{b}{2}\right)^2 + q = c$.
- To complete the square of a general quadratic, first take out the factor $a$.
- The roots of a quadratic equation are the values of $x$ such that $ax^2 + bx + c = 0$.
- The quadratic equation formula for finding the roots of a quadratic is

$$x = \frac{-b \pm \sqrt{b^2 - 4ac}}{2a}$$

- The discriminant of a general quadratic function is $\Delta = b^2 - 4ac$.
- The sign of the discriminant reveals important information about the roots:
  - $\Delta > 0$ implies that the quadratic equation has two real and distinct roots.
  - $\Delta = 0$ implies that the quadratic equation has one repeated root.

- $\Delta < 0$ implies that the quadratic equation has no real roots.
- When sketching quadratic functions, always show any roots and points of intersection with the $y$-axis.

**Chapter Assessment — Quadratic Functions**

Download and sit the 30 minute assessment for this chapter from the digital book.

$$A'(x) = \lim_{h \to 0} \frac{A(x+h) - A(x)}{h} + \frac{f(x)}{2}$$

# 6. Polynomials

In Chapters 4 and 5, we were introduced to linear and quadratic functions, these are both examples of polynomials, of first and second order, respectively. In this chapter we focus on polynomials of higher order. Polynomials have many applications, partly due to their ability to approximate complicated shapes very accurately. They are commonly used in computer aided design (CAD) software packages and in the computer gaming industry to draw realistic graphics. Figure 6.1 shows the design and construction of The Chrysalis pavilion at Merriweather Park at Symphony Woods in Maryland. A CAD package has been used which implements a NURBS (Non-Uniform Rational B-Spline) drawing tool, which is based on polynomials.

Figure 6.1: Design and Construction of The Chrysalis pavilion.

When we combine two numbers using addition, subtraction or multiplication, the result is another number. For example, $12 + 37 = 49$ and $-3 \times 6 = -18$.

If we perform one of these operations using a number and a variable quantity $x$, then we obtain a linear expression such as $3x$ or $(x - 7)$. Although multiplying two numbers gives us another number, multiplying two linear expressions does not give another linear expression. Instead we get a quadratic expression, such as $3x^2 - 4x + 12$.

If we repeatedly multiply these expressions, we might ask how complicated they can become. The answer is that, however many operations we perform, all of our expressions will be *polynomials*.

---

**Definition 6.1 — Polynomials**

A polynomial in a single variable $x$ is any expression of the form

$$a_0 + a_1 x + a_2 x^2 + \cdots + a_n x^n,$$

where $n$ is some positive integer, called the *degree* of the polynomial, and $a_0, a_1, \ldots, a_n$ are constants, with $a_n \neq 0$.

---

**Remark**

When we write out a polynomial, such as $x^2 - 3x + 2$, the distinction between the coefficients and the variable quantity is obvious. The coefficients are numbers, such as $-3$ and $2$, and the variable is shown as a letter, $x$.

The distinction becomes more subtle when we do not know the value of every coefficient. If we have a polynomial $f(x) = x^2 - ax + 2$, then the distinction between $a$ and $x$ is not as clear. We may wonder, since both quantities are denoted by letters whose values are not known, why they are treated as belonging to different categories.

This difficulty is compounded by questions which appear to treat the two types of quantity in equivalent ways. In the example above, we are told that $f(4) = 6$ and so can substitute $x = 4$ into the polynomial and solve the equation to find $a = 3$. If we are now asked to find the roots of the polynomial, we must solve the equation $f(x) = 0$ to find $x = 1$ and $x = 2$. These two processes are similar, and appear to treat the terms $a$ and $x$ in the same way.

The difference is that, whereas $a$ is considered to have a fixed value, possibly unknown, $x$ is considered to be a variable. We may find the value of $x$ under certain conditions (such as when $f(x) = 0$) but we can never find *the* value of $x$, as it can take any value.

---

## 6.1 Graphs of Polynomials

---

**Activity 6.1 — Polynomial Behaviour at Large Values**

Consider the polynomial $f(x) = 2x^3 - 20x^2 + 3x + 12$.

What value, approximately, does it take at $x = 100$ or $x = -100$?

What about at $x = 1000$ or $x = -1000$?

In general, describe the behaviour of the polynomial as $x$ becomes very large.

Now describe the behaviour of $g(x) = -3x^3 + 10x^2 + 5x + 300$ as $x \to \infty$ and as $x \to -\infty$.

---

In some situations, polynomial expressions behave in the same way as numbers, and in many cases it is helpful to compare the methods we use for polynomials with the familiar methods from standard arithmetic. However, a polynomial is much more interesting than a single number, because it has a variable quantity, $x$. This means that we can also consider the behaviour of the polynomial as $x$ varies over all possible values. Some examples of graphs of polynomials are shown in Figure 6.2.

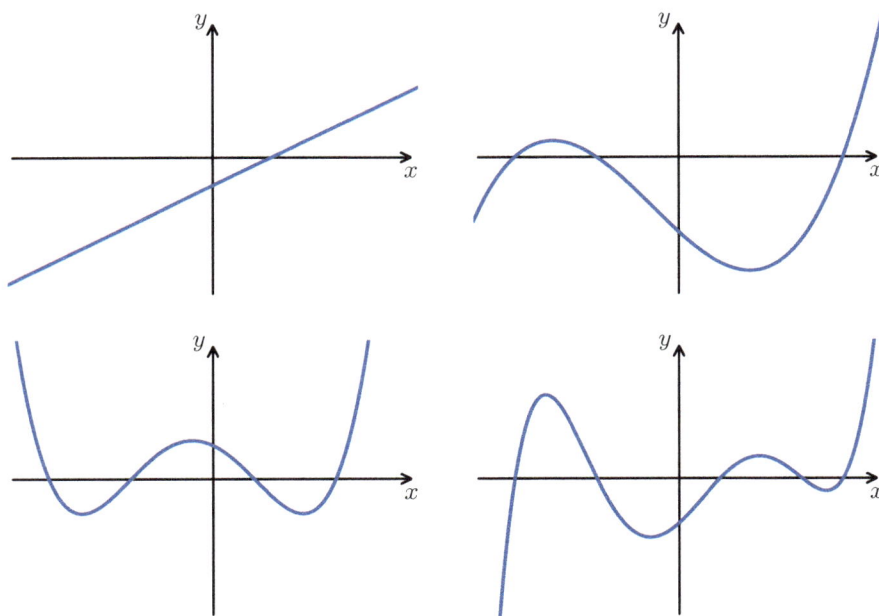

Figure 6.2: Examples of graphs of polynomial expressions.

We see that, for a given level of accuracy, if we choose a large enough value of $x$, only the highest order term is relevant. Eventually, all of the other terms are too small to affect the result at the chosen level of accuracy. This means that we can determine the behaviour of the polynomial at extreme values of $x$ simply by examining the highest order term, $a_n x^n$. This is helpful when sketching out graphs of polynomials.

> **Tip**
> If $n$ is even, then $x^n \to \infty$ both as $x \to \infty$ and $x \to -\infty$.
> If $n$ is odd, then $x^n \to \infty$ as $x \to \infty$ but $x^n \to -\infty$ as $x \to -\infty$.
> If $a_n < 0$, then the polynomial takes the opposite sign to $x^n$ as $x$ becomes very large.

When sketching the graph of a polynomial it is useful to know any turning points of the curve.

> **Definition 6.2 — Turning Points of a Polynomial**
> In three of the graphs shown in Figure 6.2 there are places where the slope of the curve changes sign. The $(x, y)$ coordinates where this happens are known as turning points, and this idea will be formalised when we study calculus.

We also need to know where it crosses the $x-$axis.

> **Definition 6.3 — Polynomial Roots**
> If the graph of a polynomial $f(x)$ crosses the $x-$axis at $x_0$ then $f(x_0) = 0$. We call $x_0$ a *root* of the polynomial. A polynomial of degree $n$ can have up to $n$ different, real roots.

Of course, knowing the location of any roots of the polynomial will help us in determining the approximate locations of turning points (if there are any), as we shall see in the example below.

---

**Tip**

Remember that the root of a polynomial can be a turning point of the graph as well. If $f(x_0) = 0$, that does not necessarily mean that $f(x)$ changes sign at $x_0$.

---

**Example 6.1**

Consider the polynomial $f(x) = x^3 + 8x^2 + 19x + 12$.
  (a) Complete the following table.

| $x$    | $-5$ | $-4$ | $-3$ | $-2$ | $-1$ | $0$ | $1$  |
|--------|------|------|------|------|------|-----|------|
| $f(x)$ |      | $0$  |      | $-2$ |      |     | $40$ |

  (b) Given that all roots of the polynomial are shown in this table sketch $y = f(x)$.

**Solution:**

  (a) Substituting the values of $x$ given into $f(x)$ we can complete the table as follows.

| $x$    | $-5$ | $-4$ | $-3$ | $-2$ | $-1$ | $0$  | $1$  |
|--------|------|------|------|------|------|------|------|
| $f(x)$ | $-8$ | $0$  | $0$  | $-2$ | $0$  | $12$ | $40$ |

  (b) From the table we see that the roots of $f(x)$ occur when $x = -4$, $x = -3$ and $x = -1$. The $y$-intercept of $f(x)$ is the point $(0, 12)$.
  Since $f(-5)$ is negative we note that the slope of the curve between $x = -5$ and $x = -4$ must be positive. This leads us to determine that a turning point, with positive $y$ coordinate must exist between $x = -4$ and $x = -3$. Similarly we can deduce that there exists a turning point with negative $y$ coordinate between $x = -3$ and $x = -2$. With this information we can sketch the graph pf $y = f(x)$.

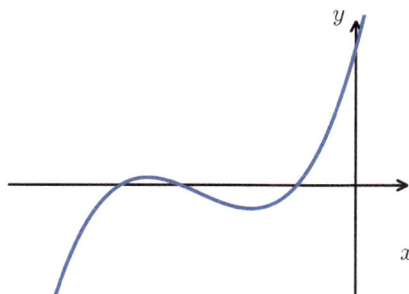

---

**Tip**

If asked to sketch the graph of a polynomial in the exam, ensure that any requested features (such as roots, interception points) are shown on the sketch.

**Exercise 6.1**

Q1. Consider the polynomial expression $f(x) = x^3 - 4x^2 - x + 4$.
   (a) Show that the equation $f(x) = 0$ is satisfied by $x = -1$, $x = 1$ and $x = 4$.
   (b) Given that these are the only solutions of $f(x) = 0$, sketch a graph of $y = f(x)$.
       Label the points where the curve crosses an axis.

Q2. Explain why the graph $y = -3x^5 + 2x^4 - 3x^2 + 10x + 12$ must cross the $x$−axis at least once.

Q3. Consider the polynomial $f(x) = x^3 + ax^2 + bx + 4$ where $a$ and $b$ are constant values.
   (a) Given that $f(-2) = -16$ and $f(2) = 0$, calculate the values of $a$ and $b$.
   (b) Complete the following table.

| $x$ | $-2$ | $-1$ | 0 | 1 | 2 | 3 | 4 |
|---|---|---|---|---|---|---|---|
| $f(x)$ | $-16$ | | | | 0 | | 20 |

   (c) Given that all of the roots of the polynomial are included in the table, sketch $y = f(x)$. Label the roots of $f(x)$.

Q4. A rectangular lawn is twice as long as it is wide. The gardener wants to buy some lawn feed to sprinkle over the lawn, and some edge fencing to go all the way around.
   (a) As we consider larger and larger lawns, which will eventually cost more: the feed or the fencing?
   (b) The feed costs £1 per square metre, and the fencing costs £2 per metre. Write down a polynomial expression for the total price of the feed and fencing.
   (c) The gardener finds that the feed and the fencing will cost the same amount. What is the area of the lawn?

## 6.2  Multiplying Polynomials

If asked to multiply a small number by a larger number in our head, say $4 \times 38$, we might multiply 4 by 30 and add this to 4 multiplied by 8, as follows.

$$4 \times 38 = 4 \times [(3 \times 10) + 8] = [4 \times (3 \times 10)] + [4 \times 8] = 120 + 32 = 152.$$

We can rewrite this calculation, replacing the number 10 with a variable quantity $x$. Instead of writing $(3 \times 10)$ we write $3x$, and so the term 38 becomes $3x + 8$. The same calculation as before now becomes

$$4 \times (3x + 8) = [4 \times 3x] + [4 \times 8] = 12x + 32.$$

Notice that we multiply both terms in the polynomial by 4, just as we multiplied 4 by both 30 and 8. Now consider a long multiplication problem. To multiply 43 by 38, we calculate $43 \times 30$ and we add the result to $43 \times 8$. This can be set out as follows.

$$
\begin{array}{r}
4\,3 \\
\times \quad 3\,8 \\
\hline
1\,2\,9\,0 \\
3\,4\,4 \\
\hline
1\,6\,3\,4 \\
\hline
\end{array}
$$

Multiplying polynomials uses the same method. Instead of multiplying a number such as $43 = (4 \times 10) + 3$, we again replace the 10 with a variable quantity $x$ to give an expression such as $4x + 3$.

Multiplying $(4x + 3)$ by $(3x + 8)$ can be written out in the same way as for $43 \times 38$.

$$
\begin{array}{r}
4x + \phantom{0}3 \\
\times \quad 3x + \phantom{0}8 \\
\hline
12x^2 + \phantom{0}9x \phantom{ + 24} \\
32x + 24 \\
\hline
12x^2 + 41x + 24 \\
\end{array}
$$

Notice that every term in the first expression is multiplied by every term in the second. If the first bracket contains $r$ terms and the second contains $s$ terms, then we will have $rs$ terms in the result. Some of these terms will be of the same order, such as $9x$ and $32x$ in our example. We collect all of the terms of the same order, and add them together to give just one term for each power of $x$.

---

**Tip**

When multiplying two polynomial expressions, make sure that every term in the first expression is multiplied by every term in the second expression.

If both expressions contain two terms, then there will be four terms in the expansion.

$$(a + b)(c + d) = ac + ad + bc + bd.$$

The $ac$ term comes from multiplying the **F**irst term in each bracket. The $ad$ term is the product of the **O**uter two terms. The $bc$ term comes from the **I**nner two terms and the final product, $bd$, comes from the **L**ast two terms in each bracket.

We can remember this technique with the acronym **FOIL**.

If there are more than two terms in either of the expressions, we take each term in one of the expressions and multiply it by every term in the other.

$$
\begin{aligned}
(a + b + c)(d + e + f) &= a(d + e + f) + b(d + e + f) + c(d + e + f) \\
&= ad + ae + af + bd + be + bf + cd + ce + cf.
\end{aligned}
$$

Following a systematic procedure like this will ensure that every pair of terms is multiplied exactly once.

---

When we multiply two polynomials, the highest power of $x$ in the result comes from the product of the two highest powers in each expression. For example,

$$(3x^5 - 8x^3 + 2x^2 + x - 3)(2x^2 + 1) = 6x^7 + \text{ lower order terms.}$$

This means that, if the polynomial $f(x)$ has degree $n$ and the polynomial $g(x)$ has degree $m$, and neither polynomial is identically zero, then the polynomial $h(x) = f(x)g(x)$ has degree $n + m$.

Expanding binomials and trinomials will be familiar from study at GCSE; the following two examples should serve as a reminder.

---

**Example 6.2**
Expand $(2x + 1)(x + 3)$.

**Solution:**
Working systematically we ensure that we multiply each term in the first bracket by every term in the second bracket, before collecting like terms.

$$(2x + 1)(x + 3) = (2x \times x) + (2x \times 3) + (1 \times x) + (1 \times 3),$$
$$= 2x^2 + 6x + x + 3,$$
$$= 2x^2 + 7x + 3.$$

---

**Example 6.3**
Expand $(x - 2)(2x + 3)(3x + 4)$.

**Solution:**
First we consider the last two brackets.

$$(2x + 3)(3x + 4) = (2x \times 3x) + (2x \times 4) + (3 \times 3x) + (3 \times 4),$$
$$= 6x^2 + 8x + 9x + 12,$$
$$= 6x^2 + 17x + 12.$$

We then multiply every term in the first bracket by every term in the above quadratic,

$$(x - 2)(6x^2 + 17x + 12) = (x \times 6x^2) + (x \times 17x) + (x \times 12) + (-2 \times 6x^2)$$
$$+ (-2 \times 17x) + (-2 \times 12)$$
$$= 6x^3 + 17x^2 + 12x - 12x^2 - 34x - 24$$
$$= 6x^3 + 5x^2 - 22x - 24$$

---

When we have linear combinations of products of polynomials we can consider the products individually and then combine them by the properties of addition and subtraction.

---

**Example 6.4**
Expand $(x - 2)(x + 1) - 2x(x + 5)$ and collect terms of the same order.

**Solution:**
First, we expand the brackets.

$$(x - 2)(x + 1) - 2x(x + 5) = x^2 + x - 2x - 2 - 2x^2 - 10x.$$

Then we collect terms with the same power of $x$.

$$x^2 + x - 2x - 2 - 2x^2 - 10x = -x^2 - 11x - 2.$$

---

We can apply the same methods when there are more than two terms per bracket. In this case, working systematically ensures that no pairwise products are missed.

**Example 6.5**
Expand $(x^2 + 2x + 1)(x^3 - x^2 + 4x + 5)$, collecting any like terms.

**Solution:**
Working from the left in the first bracket we proceed, systematically, to consider all pairwise products of terms.

$$\begin{aligned}
(x^2 + 2x + 1)(x^3 - x^2 + 4x + 5) &= (x^2 \times x^3) + (x^2 \times (-x^2)) + (x^2 \times 4x) \\
&\quad + (x^2 \times 5) + (2x \times x^3) + (2x \times (-x^2)) \\
&\quad + (2x \times 4x) + (2x \times 5) + (1 \times x^3) \\
&\quad + (1 \times (-x^2)) + (1 \times 4x) + (1 \times 5), \\
&= x^5 - x^4 + 4x^3 + 5x^2 + 2x^4 - 2x^3 + 8x^2 + 10x \\
&\quad + x^3 - x^2 + 4x + 5, \\
&= x^5 + x^4 + 3x^3 + 12x^2 + 14x + 5.
\end{aligned}$$

**Interactive Activity 6.1**
Using the interactive matching pairs game in the digital book, match the correct expanded polynomial with its simplified counterpart.

**Exercise 6.2**
Q1. Expand the following polynomial expressions and collect terms of the same order in $x$.
  (a) $(x^2 - 3 + 4x) - (7 - x^3 + 2x)$.
  (b) $x(3ax - 2x^2) + 2ax^2 - 3(x + 2)$.
  (c) $(x - 2)(x + 2)$.
  (d) $(x - 2)(2x^2 + 3x - 1) + (x - 2)(x + 3)$.
  (e) $(2x^2 - x + 1)(x^2 - 4x - 1) - x^2(2x^2 + 1)$.
Q2. (a) Write the polynomial $f(x) = (x - 3)(x + c) + 2$ in the form $a_2 x^2 + a_1 x + a_0$.
  (b) Given that $f(2) = 1$, find the value of $c$. Hence find the value of $f(0)$.

## 6.3 Division and Factorisation

Consider how long division is performed. To calculate $578 \div 17$, the calculation can be written out as follows.

```
          34
    17 | 578
         51
        ‾‾‾
         68
         68
        ‾‾‾
          0
```

We now apply the same method, but instead of using numbers such as $578 = (5 \times 10^2) + (7 \times 10) + 8$, we again replace each 10 with a variable quantity $x$, so we have polynomials such as $5x^2 + 7x + 8$.

For example, suppose we want to divide the cubic expression $x^3 - 9x^2 + 21x - 13$ by $(x-1)$. We write it out in the same way as we do for long division.

$$(x-1) \overline{\left) x^3 \quad -9x^2 \quad +21x \quad -13 \right.}$$

Now we divide the highest order term in the polynomial by the highest order term in the divisor. In this case, we divide $x^3$ by $x$, to give $x^2$. This is the first term in the result.

$$\begin{array}{r} x^2 \\ (x-1) \overline{\left) x^3 \quad -9x^2 \quad +21x \quad -13 \right.} \end{array}$$

Following the standard long division method, we multiply $x^2$ by the divisor $(x-1)$, subtract this from the polynomial, and bring down the next term.

$$\begin{array}{r} x^2 \\ (x-1) \overline{\left) x^3 \quad -9x^2 \quad +21x \quad -13 \right.} \\ \underline{x^3 \quad - x^2} \\ -8x^2 \quad +21x \end{array}$$

We now divide the highest term in the expression $-8x^2 + 21x$ by the highest term in the divisor. This gives us $-8x$ which becomes the next term in our result.

$$\begin{array}{r} x^2 \quad - 8x \\ (x-1) \overline{\left) x^3 \quad -9x^2 \quad +21x \quad -13 \right.} \\ \underline{x^3 \quad - x^2} \\ -8x^2 \quad +21x \\ \underline{-8x^2 \quad + 8x} \\ +13x \quad -13 \end{array}$$

Finally, we divide $-13x$ by $x$ to give $-13$, which is the constant term in the result.

$$\begin{array}{r} x^2 \quad - 8x \quad - 13 \\ (x-1) \overline{\left) x^3 \quad -9x^2 \quad +21x \quad -13 \right.} \\ \underline{x^3 \quad - x^2} \\ -8x^2 \quad +21x \\ \underline{-8x^2 \quad + 8x} \\ +13x \quad -13 \\ \underline{+13x \quad -13} \\ 0 \end{array}$$

We have a zero remainder, and so $(x-1)(x^2 - 8x - 13) = x^3 - 9x^2 + 21x - 13$. We can confirm this by expanding the brackets.

---

**Example 6.6**
Divide $f(x) = 4x^3 + 2x^3 - 2x^2 + 90$ by $(x+3)$. Hence, rewrite the polynomial in the form $f(x) = (x-2)g(x)$, where $g(x)$ is to be determined.

**Solution:**

We set out the calculation in the same way as before, noting that we must include a $0x$ term in our layout to avoid errors.

$$(x+3) \, \overline{\left| \, 4x^3 \;\; + 2x^2 \;\; + 0x \;\; + \;\; 3 \right.}$$

We divide the highest term in the polynomial, $4x^3$ by the highest term in the divisor, $x$. This gives us the first term in our result, $4x^2$.

$$
\begin{array}{r}
4x^2 \qquad\qquad\qquad\qquad \\
(x+3) \, \overline{\left| \, 4x^3 \;\; + \;\; 2x^2 \;\; + 0x \;\; + \;\; 3 \right.} \\
\underline{4x^3 \;\; + \;\; 12x^2 \qquad\qquad\qquad} \\
- 10x^2 \;\; + 0x \qquad\qquad
\end{array}
$$

We repeat the process, dividing $-10x^2$ by $x$ to give $-10x$ as the next term and continue to give.

$$
\begin{array}{r}
4x^2 \quad\; - 10x \;\; + 30 \\
(x+3) \, \overline{\left| \, 4x^3 \;\; + \;\; 2x^2 \;\; + \;\; 0x \;\; + \;\; 3 \right.} \\
\underline{4x^3 \;\; + \;\; 12x^2 \qquad\qquad\qquad\qquad} \\
- 10x^2 \;\; + \;\; 0x \qquad\qquad \\
\underline{- 10x^2 \;\; - \;\; 30x \qquad\qquad} \\
30x \;\; + 90 \\
\underline{30x \;\; + 90} \\
0
\end{array}
$$

Hence,

$$4x^3 + 2x^3 - 2x^2 + 90 = (x+3)(4x^2 - 10x + 30).$$

---

**Example 6.7**

Divide $f(x) = x^4 + 2x^3 - 2x^2 + x + 3$ by $(x - 2)$.

Rewrite the polynomial in the form $f(x) = (x - 2)g(x) + c$, where $c$ is a constant value to be determined.

**Solution:**

We set out the calculation in the same way as before.

$$(x-2) \, \overline{\left| \, x^4 \;\; + 2x^3 \;\; - \;\; 2x^2 \;\; + x \;\; + \;\; 3 \right.}$$

We divide the highest term in the polynomial, $x^4$ by the highest term in the divisor, $x$.

This gives us the first term in our result, $x^3$.

$$
\begin{array}{r}
x^3 \phantom{+2x^3 - 2x^2 + x + 3} \\
(x-2)\overline{\big)\, x^4 \;+\; 2x^3 \;-\; 2x^2 \;+\; x \;+\; 3} \\
\underline{x^4 \;-\; 2x^3 \phantom{-2x^2 + x + 3}} \\
4x^3 \;-\; 2x^2 \phantom{+ x + 3}
\end{array}
$$

We repeat the process, dividing $4x^3$ by $x$ to give $4x^2$ as the next term and continue to give.

$$
\begin{array}{r}
x^3 \;+\; 4x^2 \;+\; 6x \;+\; 13 \\
(x-2)\overline{\big)\, x^4 \;+\; 2x^3 \;-\; 2x^2 \;+\; x \;+\; 3} \\
\underline{x^4 \;-\; 2x^3} \\
4x^3 \;-\; 2x^2 \\
\underline{4x^3 \;-\; 8x^2} \\
6x^2 \;+\; x \\
\underline{6x^2 \;-\; 12x} \\
+\,13x \;+\; 3 \\
\underline{+\,26x \;-\; 26} \\
29
\end{array}
$$

Therefore, the remainder of the division is 29 and we can express the polynomial as follows.

$$x^4 + 2x^3 - 2x^2 + x + 3 = (x-2)(x^3 + 4x^2 + 6x + 13) + 29.$$

---

**Interactive Activity 6.2**

Using the interactive matching pairs game in the digital book, match the polynomial division problem with the correct result.

---

This method of division described above suggests that we can divide any polynomial by a linear expression with either no remainder, or a constant remainder.

---

**Theorem 6.4**

Given a linear expression $(x - b)$ and a polynomial

$$f(x) = a_n x^n + a_{n-1} x^{n-1} + \cdots + a_2 x^2 + a_1 x + a_0,$$

there exists a polynomial $g(x)$ of degree $n - 1$ and a constant $c$, such that

$$f(x) = (x - b)g(x) + c.$$

---

We can prove this result by induction on the degree of the polynomial. For a polynomial of degree one, $f(x) = a_1 x + a_0$, we can write

$$f(x) = (x - b)(a_1) + (ba_1 + a_0).$$

Therefore, the result holds for polynomials with degree one, with $g(x) = a_1$ and $c = ba_1 + a_0$. The polynomial $g$ is of zero degree and $c$ is a constant, as required.
Now suppose that the result is true for all polynomials up to degree $n - 1$. We prove that the result is true for all polynomials of degree $n$.
Consider a polynomial of degree $n > 1$,

$$f(x) = a_n x^n + a_{n-1} x^{n-1} + \cdots + a_2 x^2 + a_1 x + a_0,$$

with $a_n \neq 1$. We can write

$$f(x) = (x - b)a_n x^{n-1} + \underbrace{ba_n x^{n-1} + a_{n-1} x^{n-1} + \cdots + a_2 x^2 + a_1 x + a_0}_{p_{n-1}(x)}.$$

The expression $p_{n-1}(x)$ is a polynomial of degree $n-1$, and so, by the inductive assumption, it can be written in the form $(x - b)g(x) + c$ where $g$ is a polynomial of degree $n - 2$ and $c$ is a constant value. Therefore

$$f(x) = (x - b)a_n x^{n-1} + (x - b)g(x) + c = (x - b)\left(a_n x^{n-1} + g(x)\right) + c.$$

Now, $a_n x^{n-1} + g(x)$ is a polynomial of degree $n - 1$ and $c$ is a constant value, and so we have found an expression for $f(x)$ in the required form. Therefore, the result is true for polynomials of any degree.

**Remark**
This proof is not required for the exam. Indeed, proof by induction is contained within the Further Mathematics syllabus.

**Theorem 6.5 — The Remainder Theorem**
When a polynomial $f(x)$ is divided by a linear expression $(x - b)$, the remainder is a constant term equal to $f(b)$.

This result follows directly from the previous theorem. Given a polynomial $f(x)$, we can write it in the form

$$f(x) = (x - b)g(x) + c,$$

for some polynomial $g$ and constant remainder term $c$. We can find the value of $c$ by evaluating $f(b)$.

$$f(b) = (b - b)g(x) + c = c.$$

Therefore, the remainder term when $f(x)$ is divided by $(x - b)$, is equal to $f(b)$.

**Theorem 6.6 — The Factor Theorem**
The linear expression $(x - b)$ is a factor of the polynomial $f(x)$ if and only if $f(b) = 0$.

The factor theorem is a special case of the remainder theorem. A polynomial $f(x)$ is divisible by $(x - b)$ if there is a polynomial $g(x)$ such that $f(x) = (x - b)g(x)$, with no remainder term. We must prove that this is equivalent to the condition $f(b) = 0$.

First, suppose that $(x - b)$ is a factor of $f(x)$. Then there is a polynomial $g(x)$ such that $f(x) = (x - b)g(x)$. We evaluate this expression at $x = b$.

$$f(b) = (b - b)g(x) = 0.$$

Conversely, suppose that $f(b) = 0$. The remainder theorem tells us that we can write $f(x) = (x - b)g(x) + f(b)$ for some polynomial $g(x)$. Since $f(b) = 0$, this becomes $f(x) = (x - b)g(x)$, and so $(x - b)$ is a factor of $f(x)$.

The factor theorem is a quick way of finding out whether linear factors are factors of a polynomial. If $f(b) = 0$, then $(x - b)$ is a factor. Otherwise, $(x - b)$ is not a factor.

---

**Tip**

The remainder theorem gives us a faster way to calculate the remainder of a polynomial division. It can also be useful for detecting errors in the calculation. If the remainder term when $f(x)$ is divided by $(x - b)$ is not equal to $f(b)$, then the calculation contains an error.

For the exam, it is not necessary to be able to apply the remainder theorem in general. However, the factor theorem must be known.

---

Suppose that the polynomial $f(x)$ has a root at $x = b$. The factor theorem tells us that $f(x) = (x - b)g(x)$ for some polynomial $g$. Now suppose that $g$ *also* has a root at $x = b$. This means that $(x - b)$ is a factor of $g(x)$, and so

$$f(x) = (x - b)^2 h(x),$$

for some polynomial $h(x)$. In this case, we say that $f(x)$ has a repeated root at $x = b$.

---

**Exercise 6.3**

Q1. Divide the following polynomials by the given linear expression.
    (a) Divide $2x^2 + 2x - 12$ by $(x + 3)$.
    (b) Divide $x^3 + x^2 - 5x + 3$ by $(x - 3)$.
    (c) Divide $2x^4 - x^3 - 6x^2 + 8x + 12$ by $(2x + 3)$.
    (d) Divide $-2x^4 + 7x^3 - 14x^2 + 19x - 6$ by $(-x + 2)$.

Q2. Divide the polynomial $f(x) = x^3 + 11x^2 + 4x - 60$ by $(x + 10)$.
    Hence, factorise $f(x)$ completely.

Q3. Consider the polynomial

$$f(x) = 6x^3 - 5x^2 - 3x + 10.$$

    (a) What is the remainder when $f(x)$ is divided by $(x - 1)$.
    (b) Express $f(x)$ in the form $(x - 1)g(x) + c$ where $g(x)$ is a polynomial and $c$ is a constant value.

Q4. (a) The linear expression $(x + 1)$ is a factor of $x^4 - 5x^3 + px^2 + 5x - 6$. Find the value of $p$.
    (b) Find the remainder when the polynomial is divided by $(x - 4)$.

Q5. Consider the polynomial

$$f(x) = ax^3 - x^2 - bx + 2,$$

where $a$ and $b$ are constant values.
  (a) When $f(x)$ is divided by $(x-1)$ the remainder is 2. When it is divided by $(x+2)$ the remainder is $-40$. Calculate $a$ and $b$.
  (b) Find the remainder when $f(x)$ is divided by $(x-3)$.

Q6. (a) When the polynomial $4x^2 - ax + 2$ is divided by $(x-1)$, the remainder is $-3$. Find the value of $a$.
  (b) Find the roots of $4x^2 - ax + 2$.
  (c) Given

$$f(t) = 4^{2t+1} - 9.4^t + 2,$$

  find the values of $t$ for which $f(t) = 0$.

Q7. For some integer $|k| \le 3$, the polynomial $f(x)$ is given by

$$f(x) = (x-2)(x+4k) - kx.$$

  (a) Express $f(x)$ in the form $x^2 + a_1 x + a_0$ where $a_1$ and $a_0$ are expressions in terms of $k$.
  (b) Hence, or otherwise, express $f(x)$ in the form $(x-p)^2 - q$, where $p$ and $q$ are expressions in terms of $k$.
  (c) The polynomial $f(x)$ has a repeated root. Use your expression of $f(x)$ to find a quadratic equation such that $g(k) = 0$.
  (d) Use the factor theorem to factorise $g(k)$ and hence find the repeated root of $f(x)$.

---

**Key Facts — Polynomials**

- A polynomial is an expression of the form

$$a_0 + a_1 x + a_2 x^2 + \cdots + a_n x^n,$$

  where $n$ is some positive integer, called the *degree* of the polynomial, and $a_0, a_1, \ldots, a_n$ are constants, with $a_n \ne 0$.
- The asymptotic behaviour of any polynomial is:
    - If $n$ is even, then $x^n \to \infty$ both as $x \to \infty$ and $x \to -\infty$.
    - If $n$ is odd, then $x^n \to \infty$ as $x \to \infty$ but $x^n \to -\infty$ as $x \to -\infty$.
    - If $a_n < 0$, then the polynomial takes the opposite sign to $x^n$ as $x$ becomes very large.
- The $(x, y)$ coordinates where slope of the graph of a polynomial changes sign are known as turning points.
- Roots of a polynomial $f(x)$ are those values of $x$ such that $f(x) = 0$. A polynomial of degree $n$ can have up to $n$ different, real roots.
- When multiplying two polynomial expressions, make sure that every term in the first expression is multiplied by every term in the second expression.

- When a polynomial $f(x)$ is divided by a linear expression $(x - b)$, the remainder is a constant term equal to $f(b)$. This is called the Remainder Theorem.
- The linear expression $(x - b)$ is a factor of the polynomial $f(x)$ if and only if $f(b) = 0$. This is called the factor Theorem.

**Chapter Assessment — Polynomials**

Download and sit the 30 minute assessment for this chapter from the digital book.

$$\sin(x)$$

$$P(A|B) = \frac{P(A \cap B)}{P(B)}$$

$$2ab \le a^2 + b^2$$

$$+ \frac{f(x_2)}{2} + \ldots + \frac{f(x}{2}$$

$$d \sin$$

# 7. Direct Proof

$$\cos(x) \qquad \sin^2$$

Proof is one of the cornerstones of mathematics. This is what sets it apart from the other sciences. Once a mathematical result has been proved it holds forevermore; it is not subject to revision in the light of new experimental evidence. In 1637, the French mathematician Pierre de Fermat conjectured that there were no positive integer solutions of the equation $a^n + b^n = c^n$, for $n > 2$.

(a)

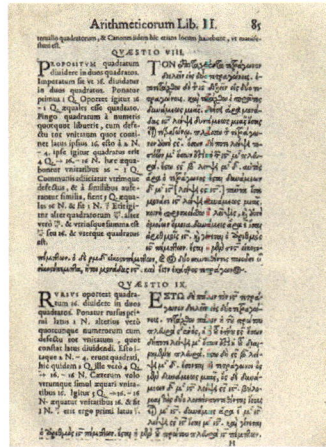
(b)

Figure 7.1: Pierre de Fermat (a) and the original problem against which he wrote his conjecture.

He made this statement in the margin of his copy of Diophantus' Arithmetica, next to Problem II.8 and claimed "It is impossible to separate a cube into two cubes, or a fourth power into two fourth powers, or in general, any power higher than the second, into two like powers. I have discovered a truly marvellous proof of this, which this margin is too narrow to contain."

The conjecture that no such solutions exist for $n > 2$ remained unproven until 1994, when the English mathematician Andrew Wiles released a proof.

> **Remark**
> For $n = 2$ we know that there exist infinitely many positive integer solutions to the equation $a^2 + b^2 = c^2$; these are Pythagorean triples.

A *mathematical argument* is a process whereby we start with one or more basic premises, then follow a series of logical steps to reach a conclusion. The conclusion we reach may then be used as a premise in a further argument that leads to yet another conclusion. Starting with fundamental mathematical concepts, such as counting and straight-line geometry, repeated use of this process allows us to derive all of the mathematics that is known today. This is often referred to as *proof by deduction*, and is a method of *direct proof*.

> **Activity 7.1 — Exploring Methods of Proof**
> The *arithmetic mean* of two numbers $a$ and $b$ is given by
>
> $$m_a = \frac{a+b}{2}.$$
>
> Notice that the numbers $a, m_a$ and $b$ form an arithmetic sequence. This is the same as the commonly used definition of the *mean* applied to two numbers.
> The *geometric mean* of two numbers $a$ and $b$ is given by
>
> $$m_g = \sqrt{ab}.$$
>
> In this case the name arises due to the numbers $a, m_g$ and $b$ forming a geometric sequence.
> The *arithmetic-geometric mean inequality* states that for two non-negative numbers, $a$ and $b$, then
>
> $$\frac{1}{2}(a+b) \geq \sqrt{ab}. \qquad (7.1)$$
>
> In this activity we explore possible approaches to proving (7.1).
>   Q1. **Critiquing a Proof**
>       A student attempting to prove (7.1) produces the following:
>
> $$\frac{1}{2}(a+b) \geq \sqrt{ab}$$
> $$\frac{1}{4}(a+b)^2 \geq ab$$
> $$(a+b)^2 \geq 4ab$$
> $$a^2 + 2ab + b^2 \geq 4ab$$
> $$a^2 - 2ab + b^2 \geq 0$$
> $$(a-b)^2 \geq 0$$

What is wrong with this "proof"?

**Q2. Graphical Proofs**

(a) Explain how the image below can be used to prove (7.1).

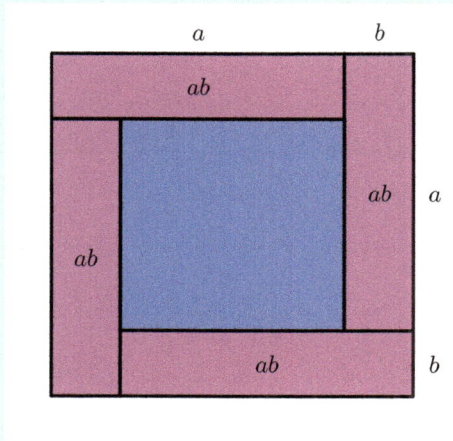

(b) The triangle shown below is right-angled with angle $RPQ = 90°$. Use this to derive an alternative graphical proof of (7.1).

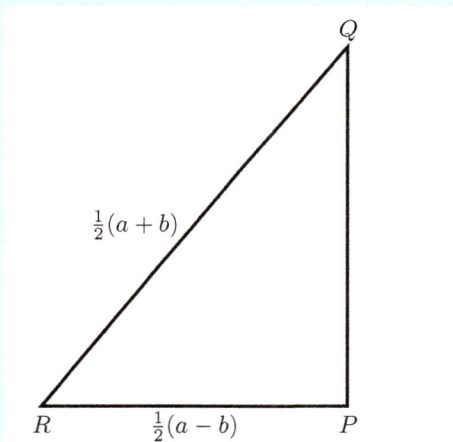

**Q3. Deductive Proofs**

Starting with the statement $(a + b)^2 \geq 0$, by making a sequence of deductions, prove (7.1). Why is this initial statement valid?

**Q4. Proof by Contradiction**

Proof by contradiction does not appear in the A-Level syllabus until the Second Year and will be discussed further in the next book. However, the concept can be illustrated nicely here. Assume that the arithmetic-geometric mean inequality does not hold, *i.e.* $\frac{1}{2}(a + b) < \sqrt{ab}$. Starting with this statement, arrive at a contradiction. This contradiction means that your initial assumption must have been false, hence (7.1) must hold.

**Remark**

The arithmetic-geometric mean inequality can be generalised for a list of $n$ non-negative numbers $x_1, x_2, \ldots, x_n$,

$$\frac{x_1 + x_2 + \cdots + x_n}{n} \geq \sqrt[n]{x_1 \cdot x_2 \cdot \ldots \cdot x_n}.$$

This inequality has many uses; for example, it is used to prove the convergence properties of some numerical methods.

We consider some motivating examples that should be familiar from studies at GCSE level.

**Example 7.1**

Prove that $(2x - 1)(x + 3) = 2x^2 + 5x - 3$ for all real $x$.

**Solution:**

Expanding the brackets on the left-hand side, we obtain

$$(2x - 1)(x + 3) = 2x^2 + 6x - x - 3 = 2x^2 + 5x - 3.$$

Since we have shown that the left and right-hand sides of the identity are equivalent, we have proven directly that the identity is true for all real $x$.

**Example 7.2**

Given that the interior angles of a convex quadrilateral sum to $360°$, use the diagram below to prove that corresponding angles are of equal size.

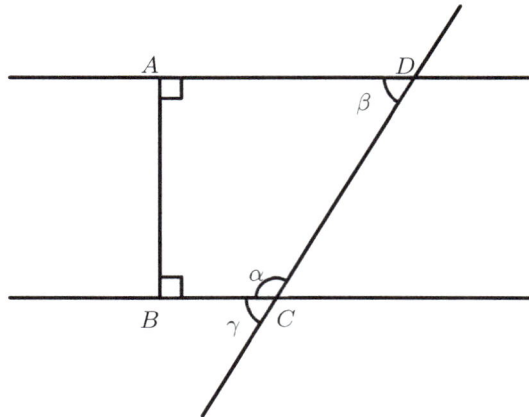

**Solution:**

Clearly, $ABCD$ is a quadrilateral, so its interior angles sum to $360°$. Since angles $DAB$ and $ABC$ are right-angles, $\alpha + \beta = 360° - 180° = 180°$. Furthermore, angles $\alpha$ and $\gamma$ lie along a straight line, so $\alpha + \gamma = 180°$. Thus, $\alpha + \beta = \alpha + \gamma = 180° \Rightarrow \beta = \gamma$, so the corresponding angles $a$ and $c$ are equal.

**Example 7.3**

Prove that the sum of any four consecutive integers is even.

**Solution:**

We let the first number be $n$, and so the second, third and fourth numbers are $n + 1$, $n + 2$, $n + 3$ respectively. Adding these,

$$n + (n + 1) + (n + 2) + (n + 3) = 4n + 6,$$
$$= 2(2n + 3).$$

Since this is a multiple of two, the sum of any four consecutive integers is even.

---

**Example 7.4**

Consider the diagram below.

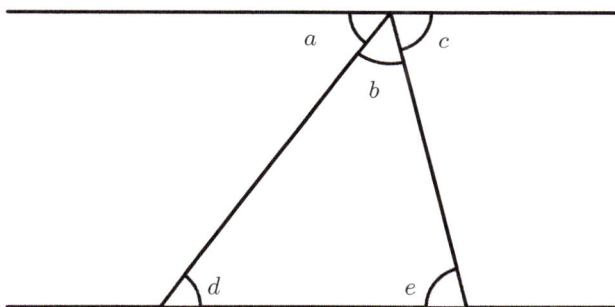

Use properties of parallel lines to prove that the interior angles of a triangle sum to 180°.

**Solution:**

Firstly, angles $a$ and $d$ are alternate angles, so $a = d$. Similarly, $c$ and $e$ are alternate, so $c = e$. Secondly, $a$, $b$ and $c$ together form a straight line, so $a + b + c = 180°$. Combining these facts, we obtain $b + d + e = 180°$. Since none of the angles were specified, this equation holds for all possible angles and therefore all possible triangles. Hence, the interior angles of all triangles sum to 180°.

---

**Tip**

Initial assumptions and known facts should be stated clearly before starting the proof. Once this has been done, ensure that a written proof is concise and easy to follow.

**Exercise 7.1**

Q1. Prove that $(x + y)(x - y) = x^2 - y^2$ for all real $x$ and $y$.

Q2. Prove that if $(x - p)$ is a factor of the polynomial expression $f(x)$ then $f(p) = 0$.

Q3. Prove that $x^3 + x^2 - 2x - 8$ only has one real factor.

Q4. The equation $kx^2 + 2kx - 3 = 0$, where $k$ is a constant, has no real roots. Prove that $k$ satisfies the inequality $-3 < k \leq 0$.

Q5. For which values of the constant $k$ does the equation $kx^2 - 3x + k = 0$ have two distinct real roots?

Q6. (a) Prove that the interior angles of a quadrilateral sum to $360°$.
    (b) Prove that the interior angles of an $n$-sided polygon sum to $180(n - 2)°$.

Q7. Prove that $x^2 - 6x + 10 \geq 1$ for all real $x$.

Q8. The line with equation $y = mx + 2$ intersects the circle with equation $(x + 1)^2 + (y - 2)^2 = r$, $r > 0$, at two distinct points. Prove that $r > \dfrac{m^2}{1 + m^2}$.

Q9. Consider the diagram below, which shows a square of side-length $c$ inscribed in a larger square of side-length $a + b$.

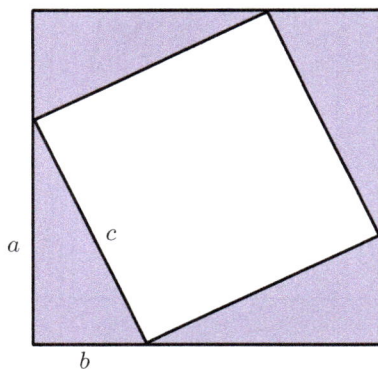

Use the diagram to prove Pythagoras' theorem, $a^2 + b^2 = c^2$.

## 7.1 Mathematical Language

Mathematical arguments are precise and unambiguous. We must learn how to use mathematical language correctly in order to produce sensible mathematical arguments. In this section, we introduce and practise using mathematical language correctly by providing *direct proofs* of well-known mathematical facts and identities.

### 7.1.1 One-way Implication

When making mathematical arguments, we construct them using objects called *statements*, which we denote with letters such as $A$, $B$, etc. Statements are connected using *implication arrows* '$\Rightarrow$', '$\Leftarrow$' and '$\Leftrightarrow$' to form *compound statements*.

The following statement (for real $x$) often causes confusion:

$$x < 0 \quad \Rightarrow \quad x \leq 0. \tag{7.2}$$

This says that "if $x < 0$ then $x \leq 0$". Is (7.2) true? We can make two contradictory arguments:

- If $x < 0$ then $x$ cannot be greater than 0. So $x \leq 0$ and so (7.2) is true.
- If we say $x \leq 0$ then this allows for the possibility that $x = 0$. This cannot be true if $x < 0$, therefore (7.2) must be false.

To determine which, if any, of these is correct we must be rigorous about what we mean by statements such as $A \Rightarrow B$.

---

**Definition 7.1 — Implication**

If we write $A \Rightarrow B$ then this is the compound statement "$A$ implies $B$", where $A$ and $B$ are statements. This is equivalent to the compound statement "if $A$ then $B$". If $A$ holds, then $B$ must hold also.

---

**Example 7.5**

Applying the above definition we see that $x < 0 \quad \Rightarrow \quad x \leq 0$ is a true statement.

---

**Example 7.6**

With the above definition, the statement

$$x = 1 \quad \Rightarrow \quad x^2 = x,$$

for $x$ a real number, is true.

---

**Remark**

The statement $A \Rightarrow B$ can also be written as "$A$ only if $B$", *i.e.* $A$ can only hold if $B$ holds.

---

It is very important to note that "if $A$ then $B$" is *not* the same as "$A$ if $B$". When we write "$A$ if $B$", this means "if $B$ then $A$", *i.e.* $B \Rightarrow A$. This is known as the *converse*.

---

**Remark**

The statement $B \Rightarrow A$ is equivalent to $A \Leftarrow B$.

---

If the statement $A \Rightarrow B$ is true, the converse $(B \Rightarrow A)$ is not necessarily true also.

---

**Example 7.7**

The statement

$$x = 1 \quad \Rightarrow \quad x^2 = x,$$

is true. Its converse is

$$x^2 = x \quad \Rightarrow \quad x = 1.$$

This statement is false, since $x = 0$ also satisfies $x^2 = x$.

### 7.1.2  Two-way Implication

If the statements $A \Rightarrow B$ and $B \Rightarrow A$ are both true, then we have a two-way implication.

---

**Definition 7.2 — Two-way implication**

If $A \Rightarrow B$ and $A \Leftarrow B$, we write

$$A \Leftrightarrow B.$$

This is often expressed as "$A$ if and only if $B$".

Recall that $A \Leftarrow B$ is "$A$ if $B$" and $A \Rightarrow B$ is "$A$ only if $B$".

---

**Remark**

For the statement $A \Leftrightarrow B$ we often write $A$ iff $B$, where iff represents *if and only if*.

---

**Example 7.8**

The statement

$$x + 5 = 3 \quad \Leftrightarrow \quad x = -2,$$

is a two-way implication, since the statements

$$x + 5 = 3 \quad \Rightarrow \quad x = -2,$$

and

$$x = -2 \quad \Rightarrow \quad x + 5 = 3,$$

are both true.

---

### 7.1.3  Strength of Statements

For statements $A$ and $B$, if $A$ implies $B$, but $B$ does *not* imply $B$, then statement $A$ is *stronger* than statement $B$. The strongest possible compound statement we can write is then $A \Rightarrow B$. However, if $A$ implies $B$ *and* $B$ implies $A$, then the strongest possible compound statement is $A \Leftrightarrow B$.

---

**Example 7.9**

Let the statement $A$ be $x > 2$ and the statement $B$ be $x > 0$ for $x \in \mathbb{R}$.

We have $A \Rightarrow B$, since all real numbers greater than 2 are also greater than 0. However, $B \not\Rightarrow A$, since not all real numbers greater than 0 are also greater than 2, *e.g.* $x = 1$. Hence, $A$ is a stronger statement than $B$ and the strongest compound statement we can write is $A \Rightarrow B$.

---

### 7.1.4 Statements Involving "and" and "or"

We can produce more detailed statements by combining simple statements using "and" and "or".

- The statement "if $A$ and $B$ then $C$" means that $C$ holds if **both** $A$ and $B$ hold.
- The statement "if $A$ or $B$ then $C$" means that $C$ holds if **at least one** of $A$ or $B$ holds.

---

**Example 7.10**

Let $A$ be the statement $x \geq 5$ and $B$ be the statement $x < 12$. Then the statement "$A$ and $B$" is equivalent to

$$5 \leq x < 12.$$

---

**Example 7.11**

Let $A$ be the statement $x < 0$ and $B$ be the statement $x = 0$. Then the statement "$A$ or $B$" is equivalent to

$$x \leq 0.$$

---

**Exercise 7.2**

Q1. Insert one of the symbols $\Rightarrow$, $\Leftarrow$, $\Leftrightarrow$ to make the strongest possible compound statement from each of the following, where $n$ is an integer and $x$ is a real number in each case.

(a) $S$ is a rhombus _____ $S$ is a square.
(b) $X$ is a rhombus _____ $X$ is a parallelogram.
(c) $n^2 > 8$ and $n > 0$ _____ $n \geq 3$.
(d) $x^2$ is rational _____ $x$ is rational.

Q2. This question is about constructing sensible mathematical arguments. In each case, does the conclusion Q follow logically from the premises P1 and P2? Here, $x$, $y$, $z$, where they occur, are real numbers.

(a)  P1:  All cats have wings.
     P2:  All winged creatures have four legs.
     Q:   All cats have four legs.
(b)  P1:  If $x^2 + y^2 \leq 1$ then $-1 \leq x \leq 1$.
     P2:  $4x = 1$.
     Q:   $x^2 + y^2 \leq 1$.
(c)  P1:  $y \leq 4$ only if $x \geq 3$.
     P2:  If $y \leq 4$ then $z^2 > x$.
     Q:   If $z = 0$ then $y > 0$.
(d)  P1:  If $x > 0$ then $y > 5$.
     P2:  If $x \leq 0$ then $z < 7$.
     Q:   $y > 5$ only if $z \geq 7$.
(e)  P1:  $y > 3$ only if $x^2 > 4$.
     P2:  If $y > 3$ then $z \geq 5$.
     Q:   If $y > 3$ then $z > x^2$.

---

**Activity 7.2 — Welcome to Mathtopia**

Citizens living in the faraway republic of Mathtopia organise their hobbies according to a very strict set of rules. In Mathtopia, it is forbidden to write or speak in ambiguous terms, so the rules are written using mathematical language to guarantee no ambiguity! The rules are as follows:

1) No Mathtopian that eats bagels cannot swim.
2) No Mathtopian without a calculator plays poker.
3) Mathtopians who have an abacus all eat bagels.
4) No Mathtopian who can swim likes fencing.
5) No Mathtopian has a calculator unless they have an abacus.

Using these 5 rules, what conclusion can we legitimately draw about citizens of Mathtopia?

*Hint:* Label the various properties (e.g. $B$ for "eats bagels") and try to find a chain of implications.

Use $\neg A$ to indicate "not $A$".

Note that "$A$ unless $B$" means $\neg B \Rightarrow A$.

---

## 7.2    Proof by Exhaustion

We often consider statements for which there are a finite number of cases. We can then perform a direct calculation for each case to deduce that the statement holds in all possible cases.

---

**Example 7.12**

Prove that for every positive integer $n$, where $3 \leq n \leq 8$ that the positive integer $n^2 + 3n$ is even.

**Solution:**

We consider each possible $n$.

$$n = 3: \quad 3^2 + 3 \times 3 = 18,$$
$$n = 4: \quad 4^2 + 3 \times 4 = 28,$$
$$n = 5: \quad 5^2 + 3 \times 5 = 40,$$
$$n = 6: \quad 6^2 + 3 \times 6 = 54.$$
$$n = 7: \quad 7^2 + 3 \times 7 = 70.$$
$$n = 8: \quad 8^2 + 3 \times 8 = 86.$$

These are all even, and so the statement is true as it has been verified for every value of $n$ directly.

---

**Remark**

The above statement is in fact true for all integer $n$.

**Example 7.13**

Prove that for every positive integer $n$, where $7 \leq n \leq 14$, the positive integer $n^2 - n - 1$ is either prime or the product of two primes.

**Solution:**

Considering each possible $n$,

$$n = 7: \qquad 7^2 - 7 - 1 = 41 \quad \text{(prime)},$$
$$n = 8: \quad 8^2 - 8 - 1 = 55 = 5 \times 11 \quad \text{(product of two primes)},$$
$$n = 9: \qquad 9^2 - 9 - 1 = 71 \quad \text{(prime)},$$
$$n = 10: \qquad 10^2 - 10 - 1 = 89 \quad \text{(prime)},$$
$$n = 11: \qquad 11^2 - 11 - 1 = 109 \quad \text{(prime)},$$
$$n = 12: \qquad 12^2 - 12 - 1 = 131 \quad \text{(prime)},$$
$$n = 13: \qquad 13^2 - 13 - 1 = 155 = 5 \times 31 \quad \text{(product of two primes)},$$
$$n = 14: \qquad 14^2 - 14 - 1 = 181 \quad \text{(prime)}.$$

We have verified the statement for all possible values of $n$ directly. Hence, the statement is true.

---

**Example 7.14**

Prove that all cube numbers are either a multiple of 9, or 1 more, or 1 less than a multiple of 9.

**Solution:**

All positive integers are either a multiple of 3, or one more or one less than a multiple of 3. We consider each case individually.

- $n$ is a multiple of 3, *i.e.* there exists positive integer $k$ such that $n = 3k$. Then $n^3 = 27k^3 = 9 \times 3k^3$, so $n^3$ is a multiple of 9.
- $n$ is one more than a multiple of 3, *i.e.* there exists positive integer $k$ such that $n = 3k + 1$. Then $n^3 = 27k^3 + 27k^2 + 9k + 1 = 9k(3k^2 + 3k + 1) + 1$, so $n^3$ is one more than a multiple of 9.
- $n$ is one less than a multiple of 3, *i.e.* there exists positive integer $k$ such that $n = 3k - 1$. Then $n^3 = 27k^3 - 27k^2 + 9k - 1 = 9k(3k^2 - 3k + 1) - 1$, so $n^3$ is one less than a multiple of 9.

Since we have found the statement consistent in all possible cases, we have proven that all cube numbers are either a multiple of 9 or 1 more or 1 less than a multiple of 9.

**Example 7.15**

Prove that $x = n^3 - n$ is divisible by 12 for all odd integers $n > 2$.

**Solution:**

Factorising gives

$$n^3 - n = n(n^2 - 1) = (n-1)n(n+1),$$

so $x$ is the product of 3 consecutive positive integers.

Since $n$ is odd, $(n-1)$ and $(n+1)$ are even, which means that $x$ is divisible by 4 for all $n$. We then have two cases:

- $n$ is a multiple of 3, so $x$ is divisible by $3 \times 4 = 12$.
- $n$ is not a multiple of 3. However, exactly one of any 3 consecutive integers is a multiple of 3, so one of $(n-1)$ and $(n+1)$ is a multiple of 3, so $x$ is divisible by $3 \times 4 = 12$.

Since we have found that the statement is true for all possible cases, it is true for all odd $n > 2$.

## 7.2.1 Proofs of Statements Involving the Positive Integers

Many proofs involving positive integers require us to consider odd and even numbers separately, or the statement itself may only concern odd (or even) numbers. Because of this, we can use some basic algebra to write:

- The positive integer $m$ is *even* if there exists a positive integer $p$ such that $m = 2p$,
- The positive integer $n$ is *odd* if there exists a positive integer $q$ such that $n = 2q - 1$.

**Example 7.16**

Prove directly that

Q1. the sum of two even positive integers is even,

Q2. the sum of two odd positive integers is even.

**Solution:**

Q1. Let $m$ and $n$ be even numbers, *i.e.* there exist positive integers $p$ and $q$ such that $m = 2p$ and $n = 2q$. Then

$$m + n = 2p + 2q = 2(p + q).$$

Since $p + q$ is a positive integer, $m + n$ has the form of an even number. Hence, the sum of two even numbers is even.

Q2. Let $m$ and $n$ be odd numbers, *i.e.* there exist positive integers $p$ and $q$ such that $m = 2p - 1$ and $n = 2q - 1$. Then

$$m + n = 2p - 1 + 2q - 1 = 2p + 2q - 2 = 2(p + q - 1).$$

Since $p + q - 1$ is an integer, $m + n$ has the form of an even number. Hence, the sum of two odd numbers is even.

---

**Example 7.17**

Prove that the product of two consecutive odd numbers is 1 less than a multiple of 4.

**Solution**

We can express two consecutive odd numbers $m$ and $n$ as $m = 2k - 1$ and $n = 2k + 1$, where $k$ is a positive integer. Multiplying,

$$mn = (2k - 1)(2k + 1) = 4k^2 - 1.$$

Hence, the product of two consecutive odd numbers is one less than a multiple of 4.

---

## 7.3  Disproof by Counter Example

To prove that a statement is false, it is sufficient to find just one counter example.

---

**Example 7.18**

A student is trying to prove the following inequality for all integers $x$ and $y$,

$$(x + y)^4 \leq x^4 + y^4.$$

Explain why they will not be able to prove this to be true.

**Solution:** If we consider $x = 0, y = 1$, then,

$$(x + y)^4 = (0 + 1)^4,$$
$$= 1,$$
$$x^4 + y^4 = 0^4 + 1^4,$$
$$= 1.$$

However, considering $x = 1, y = 2$, we have

$$(x + y)^4 = (1 + 2)^4,$$
$$= 81,$$
$$x^4 + y^4 = 1^4 + 2^4,$$
$$= 17.$$

Hence, we have found a counter example, disproving the statement. A student cannot possibly prove a statement to be true if there exists a counter example.

---

**Example 7.19**

A Mersenne prime is a prime number that can be written in the form $2^n - 1$ for some positive integer $n > 1$. Prove that the statement

"For every prime number $p$, the number $2^p - 1$ is a Mersenne prime"

is false.

**Solution:**

If we consider the first few prime numbers, we obtain

$$2^2 - 1 = 3, \quad 2^3 - 1 = 7, \quad 2^5 - 1 = 31.$$

It appears that the statement is true so far! However, we have only tried three cases, where there are in fact infinitely many choices of $p$. We only need to show that the statement fails for **one** choice of $p$ to disprove the statement. Considering $p = 11$, we obtain

$$2^{11} - 1 = 2047 = (23)(89).$$

Hence, we have found a counterexample, disproving the statement.

---

**Exercise 7.3**

Q1. Find a counter example to disprove the following statements.
   (a) For all positive integers $n \leq 10$, $n! < 3^n$.
   (b) All non-prime integers have an even number of distinct prime factors.
   (c) $n^2 + 2n - 2$ is even for all positive integer values of $n$.
   (d) All straight lines of the form $y = mx + c$ intersect the $x$-axis.

Q2. (a) Prove that for all positive integer values of $a$ and $b$,

$$\frac{a}{b} + \frac{b}{a} \geq 2.$$

   (b) Find a counter example to disprove this statement when at least one of $a$ or $b$ is negative.

Q3. Prove the following statements about positive integers directly.
   (a) If $n$ is odd, then $n^2$ is odd.
   (b) If $m$ and $n$ are odd, then $mn$ is also odd.
   (c) If $m$ and $n$ are even, then $mn$ is also even.
   (d) If $m$ is odd and $n$ is even, then $mn$ is even.
   (e) If $n$ is even, then $7(n + 4)$ is even.

Q4. For each of the statements in Q3, write down the converse and either prove that the new statement is true or provide a counter example to show that it is false.

Q5. Prove that the sum of the squares of two positive integers is less than the square of the sum of the numbers.

Q6. For the following statements, either prove the statement directly or find a counter example to disprove it.
   (a) The difference of the squares of two consecutive even numbers is a multiple of 4.
   (b) The difference of the squares of two consecutive odd numbers is a multiple of 4.
   (c) The difference of the squares of two consecutive numbers is a multiple of 4.

Q7. Find a counter example to disprove the statement that for positive integer $n \geq 3$, $n! - 1$ is prime.

Q8. **(a)** Let $a$, $b$ and $c$ be positive integers. Prove that if $b$ and $c$ are divisible by $a$, then $b + c$ is also divisible by $a$.

   **(b)** Let $a$, $b$ and $c$ be positive integers. Prove that if $b$ is divisible by $a$ and $c$ is divisible by $b$, then $c$ is divisible by $a$.

Q9. Do there exist integers $a$, $b$ and $c$ such that

$$4a + 3 = b^2 + c^2?$$

**Activity 7.3 — Shaping Up**

Find all possible rectangles, having sides of positive integer length, which have the property that the perimeter is equal (numerically) to the area. It is not enough to identify some examples with this property; you need to prove that there are no more solutions that you have overlooked.

*Hint:* The side-lengths can be labelled as $x$ and $y$ in such a way that $x < y$. Obtain a relationship between $x$ and $y$ and think about $1/x$ and $1/y$. How large can $x$ be?

**Extension:** Find all possible cuboids, having sides of positive integer length $p$, $q$ and $r$, for which the total surface area (numerically) is twice the volume.

*Hint:* Similar to the first problem, assume $p \leq q \leq r$. Consider $1/p$, $1/q$, $1/r$ and how large $p$ can be.

**Learning Resource 7.1 — Structuring Proofs**

From the digital book an activity can be downloaded where a framework for structuring proofs is explained.

**Learning Resource 7.2 — Implication Arrows**

In the digital book there is a learning resource concerning the use of implication arrows. It is important when writing a proof that implication arrows and other connectives are used for the proof to read well. The questions in this activity provide practise on when, and how, they should be used.

**Key Facts — Direct Proof**

- The arithmetic-geometric mean inequality states that $\frac{1}{2}(a+b) \geq \sqrt{ab}$.
- Initial assumptions and known facts should always be stated at the beginning of a proof.
- $A \Rightarrow B$ means "$A$ implies $B$" and is an example of a compound statement. It can also be written as "$A$ only if $B$".
- $A \Rightarrow B$ is equivalent to $A \Leftarrow B$.
- If $A \Rightarrow B$ and $A \Leftarrow B$ then we can write $A \Leftrightarrow$.
- The statement "if $A$ and $B$ then $C$" means that $C$ holds if **both** $A$ and $B$ hold.
- The statement "if $A$ or $B$ then $C$" means that $C$ holds if **at least one** of $A$ or $B$ holds.
- When we are wishing to prove a conjecture for a finite number of cases a possible approach is to prove each case separately.
- The positive integer $m$ is *even* if there exists a positive integer $p$ such that $m = 2p$,
- The positive integer $n$ is *odd* if there exists a positive integer $q$ such that $n = 2q-1$.
- To prove that a statement is false it is sufficient to find just one counter-example.

**Chapter Assessment — Direct Proof**
Download and sit the 30 minute assessment for this chapter from the digital book.

# 8. The Binomial Theorem

We are often presented with the problem of performing expansions of the form

$$(a + bx)^n, \tag{8.1}$$

where $a, b \in \mathbb{R}$ and $n \in \mathbb{N}$.

As we shall see in Chapter 30 expanding functions of this form can lead to an important statistical distribution known as the Binomial distribution. Historically the binomial expansion proved useful when calculating high powers of numbers close to 1, since mechanical calculators, such as those shown in Figure 8.1, can perform operations such as addition easily but not exponentiation.

Figure 8.1: A mechanical calculator.

If we wish to compute $1.021^{10}$, we first note that $1.021 = 1 + 0.021$ and then consider the expansion below.

$$(1 + x)^{10} = 1 + 10x + 45x^2 + 120x^3 + 210x^4 + 252x^5 + 210x^6 + 120x^7$$
$$+ 45x^8 + 10x^9 + x^{10}.$$

From Chapter 6 we know that we can obtain a good approximation by neglecting the higher order powers when $x$ is small. Hence, we consider the approximation

$$(1+x)^10 \approx 1 + 10x + 45x^2 + 120x^3. \tag{8.2}$$

Letting $x = 0.021$ we can compute $1.021^{10}$ via the approximation

$$(1 + 0.021)^10 \approx 1 + 10 \times 0.021 + 45 \times 0.021^2 + 120 \times 0.021^3,$$
$$= 1.23095632.$$

In making this approximation we commit an error of 0.003% which for many purposes is acceptable. This value could be the multiplier when calculating the maximum repayment of a loan at a rate of 2.1% compounded annually for 10 years.

## 8.1   Expanding Binomials

> **Definition 8.1 — Binomial**
> Let $a$ and $b$ be real numbers, $x$ be an indeterminate variable, and $p$ and $q$ be distinct nonnegative integers. Then a binomial is
>
> $$ax^p + bx^q,$$
>
> *i.e.* a polynomial that is the sum of two terms. We refer to each of the terms in the sum as *monomials*.

The term inside the brackets in (8.1) is a specific case of a binomial, where $p = 0$, $q = 1$. Naïvely, we could approach the expansion of (8.1) by multiplying out the brackets term by term.

---

**Example 8.1**
Let $n = 2$.

$$(a + bx)^2 = (a + bx)(a + bx)$$
$$= a^2 + 2abx + b^2x^2.$$

---

**Example 8.2**
Let $n = 3$.

$$(a + bx)^3 = (a + bx)(a + bx)(a + bx)$$
$$= (a + bx)(a^2 + 2abx + b^2x^2)$$
$$= a^3 + 3a^2bx + 3ab^2x^2 + b^3x^3.$$

---

In Example 8.2, we can see that the calculations are already becoming rather tedious for $n = 3$. Clearly this method is not practical for large $n$, such as $n = 20$. We can approach the expansion in a systematic manner using the *Binomial Theorem*.

## 8.2   Derivation

To find a general formula for the expansion of (8.1), we begin by inspecting the expansions for the first few values of $n$. We write the expansions in ascending powers of $x$ to try and find a pattern in the coefficients.

$(a + bx)^0 = \mathbf{1},$

$(a + bx)^1 = \mathbf{1}a + \mathbf{1}bx,$

$(a + bx)^2 = \mathbf{1}a^2 + \mathbf{2}abx + \mathbf{1}b^2x^2,$

$(a + bx)^3 = \mathbf{1}a^3 + \mathbf{3}a^2bx + \mathbf{3}ab^2x^2 + \mathbf{1}b^3x^3,$

$(a + bx)^4 = \mathbf{1}a^4 + \mathbf{4}a^3bx + \mathbf{6}a^2b^2x^2 + \mathbf{4}ab^3x^3 + \mathbf{1}b^4x^4,$

$(a + bx)^5 = \mathbf{1}a^5 + \mathbf{5}a^4bx + \mathbf{10}a^3b^2x^2 + \mathbf{10}a^2b^3x^3 + \mathbf{5}ab^4x^4 + \mathbf{1}b^5x^5.$

We note that the coefficients in the $n$th expansion can be read from the $n$th row of Pascal's Triangle.

$$
\begin{array}{ccccccccccc}
&&&&& 1 &&&&& \\
&&&& 1 && 1 &&&& \\
&&& 1 && 2 && 1 &&& \\
&& 1 && 3 && 3 && 1 && \\
& 1 && 4 && 6 && 4 && 1 & \\
1 && 5 && 10 && 10 && 5 && 1 \\
\end{array}
$$

Figure 8.2: The first 6 rows of Pascal's Triangle. Each row begins and ends with 1. The remaining entries are calculated by summing the two entries directly above.

**Remark**

In the following, we always count from zero when referring to $n$th rows of Pascal's Triangle. For example, note that the coefficients in the expansion of $(a + bx)^2$ are given by the 3rd physical row of Pascal's Triangle, which is row 2 when counting from 0.

Hence, the coefficient of $a^r b^{n-r} x^{n-r}$ in the expansion of $(a + bx)^n$ is given by the $r$th entry in row $n$ of Pascal's Triangle.

Using Pascal's Triangle is still inefficient for expansions with large $n$. For example, if we wanted to compute $(a + bx)^{20}$, we would need to compute the 21st row of Pascal's Triangle. However, we can compute the $r$th entry in the $n$th row directly using the following definitions.

**Definition 8.2 — Factorial**

Consider a positive integer $n$. The factorial of $n$ is denoted $n!$ We compute $n!$ by multiplying $n$ by every positive integer less than $n$, *i.e.*

$$n! = n(n-1)(n-2)\dots(2)(1).$$

**Special case:** we define $0!$ to be 1.

**Definition 8.3 — Binomial Coefficient**

The $r$th entry in the $n$th row of Pascal's Triangle (counting from zero), which is equivalent to the coefficient of $a^r b^{n-r} x^{n-r}$ in the expansion of $(a + bx)^n$, is given by

$$\binom{n}{r} = n\,\mathrm{C}\,r = \frac{n!}{r!(n-r)!}, \quad r \leq n.$$

We use $\binom{n}{r}$ to denote binomial coefficients in written work.

We pronounce $\binom{n}{r}$ "$n$ choose $r$". This refers to counting the number of ways of choosing $r$ objects from $n$ possible objects. We use Binomial Coefficients in probability theory to formulate the *Binomial Distribution*.

**Formula 8.1 — Symmetry of the Binomial Coefficient**

Using Definition 8.3, we note that

$$\binom{n}{r} \equiv \binom{n}{n-r}.$$

**Use of Technology 8.1**

We can use the '$n\,\mathrm{C}\,r$' button on a calculator to compute binomial coefficients directly without manually applying the definition above.

**Remark**

To explain the emergence of the Binomial Coefficient $\binom{n}{r}$, a good approach is to consider writing out the first two expansions as follows:

$$(x+y)^2 = xx + (xy + yx) + yy,$$
$$(x+y)^3 = xxx + (xxy + xyx + yxx) + (xyy + yxy + yyx) + yyy,$$

i.e. treat each term as a string of $r$ $x$s and $(n-r)$ $y$s. Since the product is commutative, the order of the $x$s and $y$s in each term is irrelevant. By the definition of $\binom{n}{r}$, there are $\binom{n}{r}$ individual terms containing $r$ $x$s and $(n-r)$ $y$s when the order is discounted. Applying this to each term for $0 \leq r \leq n$, we deduce binomial coefficients are equivalent to $\binom{n}{r}$.

**Example 8.3**

Prove Pascal's identity (or Pascal's rule) which states that

$$\binom{n}{r-1} + \binom{n}{r} = \binom{n+1}{r} \tag{8.3}$$

**Solution:**

We can prove this identity by applying the definition of the binomial coefficient. Consider the left hand side,

$$\begin{aligned}
\binom{n}{r-1} + \binom{n}{r} &= \frac{n!}{(r-1)!(n-(r-1))!} + \frac{n!}{r!(n-r)!}, \\
&= \frac{n!}{(r-1)!(n-r+1)!} + \frac{n!}{r!(n-r)!}, \\
&= n!\left[\frac{1}{(r-1)!(n-r+1)!} + \frac{1}{r!(n-r)!}\right], \\
&= n!\left[\frac{r+(n-r+1)}{r!(n-r+1)!}\right], \\
&= n!\left[\frac{n+1}{r!(n-r+1)!}\right], \\
&= \frac{(n+1)!}{r!(n+1-r)!}, \\
&= \binom{n+1}{r}.
\end{aligned}$$

**Remark**

To deduce the formula

$$\binom{n}{r} = \frac{n!}{r!(n-r)!},$$

we use the following example.

Suppose we need to count how many ways we can choose $r$ people from a group of $n$ people (clearly $r \leq n$). The first person in the lineup can be chosen in $n$ ways. For each one of these ways, the second person can be chosen in $(n-1)$ ways. Repeat this process $r$ times. It follows that the number of ways of choosing the lineup, $N$, is

$$N = n(n-1)(n-2)\cdots(n-r+1).$$

In terms of factorials, $N$ can be expressed as

$$N = \frac{n!}{(n-r)!}.$$

By counting the lineups in this way, we repeatedly count groups of the same $k$ people that have been selected in different orders. There are $k!$ different orderings of the same $k$ people. We are not interested in the order that the people were chosen. In the context of expanding binomials, we consider $x^2y$ to be equivalent to $yx^2$. Hence, we divide $N$ by $k!$ to account for the different orderings, to obtain

$$\binom{n}{r} = \frac{N}{r!} = \frac{n!}{r!(n-r)!}.$$

**Exercise 8.1**

Q1. Evaluate the following:

    (a) $5!$.

    (b) $6!$.

    (c) $7!$.

    (d) $\dfrac{7!}{5!}$.

Q2. Compute, without a calculator,

    (a) $\binom{5}{2}$;

    (b) $\binom{2}{1}$;

    (c) $\binom{4}{3}$;

    (d) $\binom{7}{2}$;

    (e) $\binom{9}{4}$.

Q3. Evaluate the following, where $n \in \mathbb{N}$:

    (a) $\binom{4}{1}$.

    (b) $\binom{10}{0}$.

    (c) $\binom{5}{3}$.

    (d) $\binom{6}{3}$.

    (e) $\binom{n}{0}$.

    (f) $\binom{n}{1}$.

    (g) $\binom{n}{n}$.

Q4. Write the following rows from Pascal's triangle using the binomial coefficient notation.

    (a) 2nd row;

    (b) 5th row.

Q5. Calculate, using a calculator, or other technology,

    (a) $\binom{90}{5}$;

    (b) $\binom{26}{13}$;

    (c) $\binom{47}{6}$;

    (d) $\binom{50}{23}$;

    (e) $\binom{120}{20}$.

Using the definition of the binomial coefficient, we finally arrive at the Binomial Theorem for positive integer $n$.

---

**Theorem 8.4 — Binomial Theorem for Positive Integer $n$**

Let $a, b \in \mathbb{R}$, and $n \in \mathbb{N}$. Then the expansion of $(a + bx)^n$ is given by

$$(a + bx)^n = \binom{n}{0}a^n + \binom{n}{1}a^{n-1}bx + \binom{n}{2}a^{n-2}b^2x^2 + \ldots + \binom{n}{n-1}ab^{n-1}x^{n-1}$$

$$+ \binom{n}{n}b^nx^n,$$

$$= \sum_{r=0}^{n} \binom{n}{r}a^{n-r}b^rx^r.$$

---

**Tip**

The powers of $a$ and $b$ always add up to $n$ in the binomial formula. This fact can be used to verify that they have applied the formula correctly in their calculations.

---

**Example 8.4**

Find the following expansions using the Binomial Theorem:
(a) $(2 + 3x)^4$;
(b) $(2 - 3x)^4$.

**Solution:**

(a) Here, $a = 2$, $b = 3$, $n = 4$. Using the formula in Theorem 8.4, we obtain

$$(2 + 3x)^5 = \binom{4}{0}(2^4) + \binom{4}{1}(2^3)(3^1)x + \binom{4}{2}(2^2)(3^2)x^2 + \binom{4}{3}(2^1)(3^3)x^3$$

$$+ \binom{4}{4}(2^0)(3^4)x^4$$

$$= 16 + 96x + 216x^2 + 216x^3 + 81x^4.$$

(b) In this case, $a = 2$, $b = -3$, $n = 4$. Using the Binomial Theorem again,

$$(2 - 3x)^5 = \binom{4}{0}(2^4) + \binom{4}{1}(2^3)(-3)^1x + \binom{4}{2}(2^2)(-3)^2x^2$$

$$+ \binom{4}{3}(2^1)(-3)^3x^3 + \binom{4}{4}(2^0)(-3)^4x^4$$

$$= 16 - 96x + 216x^2 - 216x^3 + 81x^4.$$

**Remark**

Note that the coefficients in part (b) are numerically the same as those in part (a). The presence of a minus sign in the second term of the binomial makes terms in odd powers of $x$ negative in the binomial expansion.

**Tip**

When beginning to use the binomial expansion it can be helpful to write the expansion out vertically in order to make the pattern more apparent. For example,

$$
\begin{aligned}
(x+5)^4 = \quad & \binom{4}{0} x^4 5^0 \\
+ & \binom{4}{1} x^3 5^1 \\
+ & \binom{4}{2} x^2 5^2 \\
+ & \binom{4}{3} x^1 5^3 \\
+ & \binom{4}{4} x^0 5^4, \\
= \; & x^4 + 20x^3 + 150x^2 + 500^x + 625.
\end{aligned}
$$

**Remark**

The binomial expansion can be seen as an application of the distributive property.

$$
\begin{aligned}
(a+b)^3 &= (a+b)(a+b)(a+b), \\
&= a(a+b)(a+b) + b(a+b)(a+b), \\
&= aa(a+b) + ab(a+b) + ba(a+b) + bb(a+b), \\
&= aaa + aab + aba + abb + baa + bab + bba + bbb, \\
&= a^3 + 3a^2 b + 3ab^2 + 3b^3.
\end{aligned}
$$

**Example 8.5**

Find the coefficient of $x^5$ in the expansion of $(1 + 3x)^{10}$.

**Solution:**

Since we are only required to find one coefficient, we do not need to find the whole expansion of $(1 + 3x)^{10}$. Instead, we just pick out the $x^5$ term in the general expansion according to Theorem 8.4, with $a = 1$, $b = 3$, $n = 10$, $r = 5$. Hence, the coefficient is given by

$$
\binom{10}{5}(1^5)(3^5) = 61236.
$$

**Example 8.6**
The coefficient of $x^2$ in the expansion of $(1 + ax)^3$ is 108. Find the value of $a$.

**Solution:** Considering only the coefficient of the $x^2$ term, using Theorem 8.4, we have

$$\binom{3}{2}(1^1)(a^2) = 108,$$
$$\Rightarrow \quad 3a^2 = 108,$$
$$\Rightarrow \quad a^2 = 36,$$
$$\Rightarrow \quad a = 6.$$

**Example 8.7**
Find the first 3 terms in the expansion of $(1 - 2x)^{25}$.

**Solution:**
We use the formula in Theorem 8.4 again, giving

$$(1 - 2x)^{25} = \binom{25}{0}(1^{25}) + \binom{25}{1}(1^{24})(-2)^1 x + \binom{25}{2}(1^{23})(-2)^2 x^2 + \ldots$$
$$= 1 - 50x + 1200x^2 + \ldots.$$

**Example 8.8**
Find the coefficient of $x^2$ in the expansion of $(1 + 2x)^3(1 - 4x)^5$.

**Solution:**
We use Theorem 8.4 to expand $(1 + 2x)^3$ and $(1 - 4x)^5$ separately. We only need to expand up to and including terms in $x^2$, since terms in higher powers of $x$ will not contribute to the $x^2$ term in the expansion of $(1 + 2x)^3(1 - 4x)^5$.

$$(1 + 2x)^3 = \binom{3}{0}(1^3) + \binom{3}{1}(1^2)(2^1)x + \binom{3}{2}(1^1)(2^2)x^2 + \ldots,$$
$$= 1 + 6x + 12x^2 + \ldots,$$
$$(1 - 4x)^5 = \binom{5}{0}(1^5) + \binom{5}{1}(1^4)(-4)^1 x + \binom{5}{2}(1^3)(-4)^2 x^2 + \ldots,$$
$$= 1 - 20x + 160x^2 + \ldots.$$

Hence,

$$(1 + 2x)^3(1 - 4x)^5 = (1 + 6x + 12x^2 + \ldots)(1 - 20x + 160x^2 + \ldots).$$

Considering only terms that contribute to the $x^2$ term in the product of expansions, we obtain

$$(1 + 2x)^3(1 - 4x)^5 = x^2(1 \times 160 + 6 \times (-20) + 12 \times 1) + \ldots = 52x^2 + \ldots,$$

so the coefficient of $x^2$ in the expansion of $(1 + 2x)^3(1 - 4x)^5$ is 52.

**Exercise 8.2**

Q1. Find the first four terms of the binomial expansions, in ascending powers of $x$, of the following expressions.
  (a) $(1 + x)^{11}$.
  (b) $(1 - 2x)^9$.
  (c) $(2 + x)^5$.
  (d) $(2 + 3x)^{-2}$.
  (e) $(1 - x)^{1/2}$.
  (f) $(5 - 2x)^{-3/2}$.

Q2. Use the Binomial Theorem to expand the following fully.

  $(a)$  $(1 - 3x)^4$;  $(b)$  $(2 - x)^5$;  $(c)$  $\left(1 + \dfrac{x}{4}\right)^4$;  $(d)$  $\left(2 + \dfrac{1}{x}\right)^5$;

  $(e)$  $(x + y)^3$;  $(f)$  $(2x - y)^5$;  $(g)$  $\left(x + \dfrac{1}{x}\right)^4$;  $(h)$  $\left(x - \dfrac{1}{x}\right)^5$.

Q3. Find the first five terms of each of the following expansions

  $(a)$  $(1 - 3x)^{12}$;  $(b)$  $\left(1 + \dfrac{3}{2}x\right)^7$;  $(c)$  $(x + 2y)^{15}$;  $(d)$  $\left(2x - \dfrac{2}{x}\right)^{11}$.

Q4. Find the first three terms of the binomial expansions, in ascending powers of $x$, of the following expressions.
  (a) $x^2(1 + x)^4$.
  (b) $x(2 - 4x)^6$.
  (c) $(2 - x^2)\sqrt{1 + 4x}$.
  (d) $(2 - x^2)\sqrt{1 - 4x^2}$.
  (e) $\dfrac{x}{(1 - x)^3}$.
  (f) $\dfrac{x^2}{\sqrt{1 + x}}$.
  (g) $\dfrac{1 - x^3}{2 + 3x}$.
  (h) $\dfrac{(1 + 3x)^6}{(x - 2)^2}$.

Q5. Find the coefficient of $x^3$ in the following expansions

  $(a)$  $\left(1 - \dfrac{x}{2}\right)^8$;          $(b)$  $(1 + x)(2 + 3x)^4$;

  $(c)$  $(2 + x^2)\left(1 - \dfrac{x}{5}\right)^{10}$;  $(d)$  $(1 - x)^3(2 + x)^6$.

Q6. Expand the following fully.

(a) $\left(x + \dfrac{1}{x}\right)^{10}$.

(b) $\left(x - \dfrac{1}{x^2}\right)^{10}$.

Q7. Find the coefficient of the specified term in the binomial expansion of the following expressions.

(a) $(1+x)^{20}$     Term in $x^5$.

(b) $(1-2x)^{12}$     Term in $x^4$.

(c) $(2+3x)^{-5}$     Term in $x^3$.

(d) $(2-x)^{1/2}$     Term in $x^4$.

(e) $(4+2x)^{-3/2}$     Term in $x^3$.

(f) $\left(2+\frac{3}{2}x\right)^8$     Term in $x^5$.

(g) $(x+2y)^5$     Term in $y^3$.

(h) $\left(2x+\frac{1}{2}y\right)^6$     Term in $x^3$.

Q8. Write out the first five terms in the binomial expansion, in ascending powers of $x$, of each of the following expressions, where $n \in \mathbb{N}$, $n > 4$:

(a) $(1-2x)^n$.

(b) $(2+3x)^n$.

Q9. Use the binomial theorem to expand and fully simplify the following expressions.

(a) $\left(1+\sqrt{2}\right)^3$.

(b) $\left(2-3\sqrt{3}\right)^4$.

(c) $\left(\sqrt{2}+2\sqrt{3}\right)^4$.

(d) $\left(\frac{\sqrt{2}}{3}+\frac{\sqrt{3}}{2}\right)^3$.

(e) $\left(1+\sqrt{6}\right)^4 + \left(2-2\sqrt{3}\right)^3$.

(f) $\left(1-\sqrt{5}\right)^6 - \left(2-3\sqrt{5}\right)^4$.

Q10. (a) Find the binomial expansion of $(1+2x)^6$, in ascending powers of $x$, up to and including the term in $x^3$.

(b) Use the expansion in (a) to approximate $(1.02)^6$.

(c) Use the expansion in (a) to approximate $(0.96)^6$.

Q11. (a) Find the binomial expansion of $(3-4x)^5$, in ascending powers of $x$, up to and including the term in $x^3$.

(b) Use the expansion in (a) to approximate $(2.96)^5$.

(c) Use the expansion in (a) to approximate $(3.08)^5$.

Q12. of the $x^3$ term in the expansion of $(1+x)^p$ is 10. Find $p$.

Q13. The coefficient of the $x^3$ term in the expansion of $(1+ax)^4$ is 108. Find $a$.

Q14. For sufficiently small $x$ we can ignore terms with power higher than 3, and

$$(3+x)(2+x)^5 = a + bx + cx^2,$$

find $a$, $b$ and $c$.

Q15. Approximate the value of $4.04^6$ using a suitable binomial approximation including terms up to $x^4$. Comment on the accuracy of this approximation.

**Key Facts — The Binomial Theorem**

- A binomial is a polynomial that is the sum of two terms, for example $ax_b^p x^q$ for $a, b \in \mathbb{R}$ and $p, q \in \mathbb{Z}$.
- The factorial of $n$ is denoted $n!$, where $n! = n(n-1)(n-1)\ldots(2)(1)$.
- We define $0!$ to be $1$.
- 

$$\binom{n}{r} = n\,\mathrm{C}\,r = \frac{n!}{r!(n-r)!}, \quad r \le n.$$

- 

$$\binom{n}{r} \equiv \binom{n}{n-1}$$

- 

$$\binom{n}{r-1} + \binom{n}{r} = \binom{n+1}{r}$$

- Let $a, b \in \mathbb{R}$, and $n \in \mathbb{N}$. Then the binomial expansion of $(a+bx)^n$ is given by

$$(a+bx)^n = \binom{n}{0}a^n + \binom{n}{1}a^{n-1}bx + \binom{n}{2}a^{n-2}b^2x^2 + \ldots$$

$$+ \binom{n}{n-1}ab^{n-1}x^{n-1} + \binom{n}{n}b^n x^n,$$

$$= \sum_{r=0}^{n} \binom{n}{r}a^{n-r}b^r x^r.$$

**Chapter Assessment — The Binomial Theorem**
Download and sit the 30 minute assessment for this chapter from the digital book.

# 9. Graph Sketching

Following the modelling of a physical situation, we often end up with a mathematical equation that needs to be interpreted in terms of the physical parameters of the situation. A sketch of this equation is often very illuminating, especially if we are interested in the long term behaviour of a quantity over some time frame.

> **Remark**
> This long term behaviour of a function is referred to as the *asymptotic behaviour* of the function and is an important branch of mathematics in its own right.

Suppose that a clothing manufacturer has been releasing a pollutant into a lake, and that a mathematical model of the situation suggests that the quantity of the pollutant at time $t$ satisfies the following equation

$$q(t) = \frac{10}{449}\left[-1445\exp\left(\frac{-7t}{10}\right) + 280\sin(2t) + 98\cos(2t) + 1347\right],$$

$$= -\frac{14450}{449}\exp\left(\frac{-7t}{10}\right) + \frac{2800}{449}\sin(2t) + \frac{980}{449}\cos(2t) + 30. \tag{9.1}$$

At a glance it is hard to describe the long term behaviour of this function, however a quick sketch reveals that after a few years the quantity of pollutant present in the river begins to oscillate around $q = 30$.

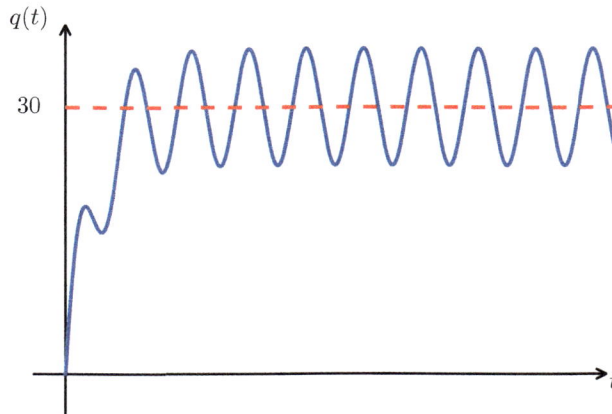

**Remark**

This function for the quantity $q(t)$ is the solution of a differential equation modelling the lake and concentration of the pollutant entering the lake.

Suppose that the lake contains $10^7$ litres of fresh water, and that this capacity remains constant over time. Every year $7 \times 10^6$ litres of contaminated water flows into and out of the lake. The concentration of the pollutant in the incoming water is modelled with the function,

$$\gamma(t) = 3 + 2\cos(2t) \text{ g}\,\text{l}^{-1},$$

where $t$ is measured in years. Letting $Q(t)$ represent the quantity of pollutant in grams at time $t$, then the rate of change of $Q$ is given by,

$$\frac{\mathrm{d}Q}{\mathrm{d}t} = \text{rate in} - \text{rate out}.$$

The rate in to the lake of the pollutant can be found by multiplying the concentration of pollutant by the flow rate of water entering the lake, and so is given by

$$\text{rate in} = (7 \times 10^6)(3 + 2\cos(2t)) \text{ g}\,\text{yr}^{-1}.$$

Since the volume of water in the lake remains constant,

$$\text{rate out} = (7 \times 10^6)\frac{Q(t)}{10^7} \text{ g}\,\text{yr}^{-1},$$
$$= \frac{7Q(t)}{10} \text{ g}\,\text{yr}^{-1}.$$

Hence, we have a differential equation to model how the quantity of pollutant changes over time,

$$\frac{\mathrm{d}Q}{\mathrm{d}t} = (7 \times 10^6)(3 + 2\cos(2t) - \frac{7Q(t)}{10}.$$

> To simplify working, we perform a rescaling by letting $q(t) = \frac{Q(t)}{10^6}$, and then a simple rearrangement. We wish to solve
>
> $$\frac{dq}{dt} + \frac{7q}{10} = 21 + 14\cos(2t),$$
>
> subject to the initial condition $q(0) = 0$. This initial condition states that at time $t = 0$ years there is no pollutant in the lake.
> In Year two we will study the forming of differential equations such as these and learn analytical techniques for their solution.

In this chapter we begin to develop the skills required to sketch a complex, multi term function such as that shown in (9.1). Knowing the shape of a graph can help us to deduce the number of solutions to equations as well as other properties such as approximate values and signs (+ or -) of solutions. We will discuss the limiting behaviour of curves to gain insight into their global properties as well as asymptotic behaviour.
Curve sketching is an important skill which is highly valued by university mathematics departments and can be used to gain insight in many areas of mathematics.

## 9.1 Sketching Graphs

So far we have studied graphs of linear functions and quadratics. A natural extension of these concepts is to consider functions of the form $f(x) = x^n$, starting with positive integers $n$.

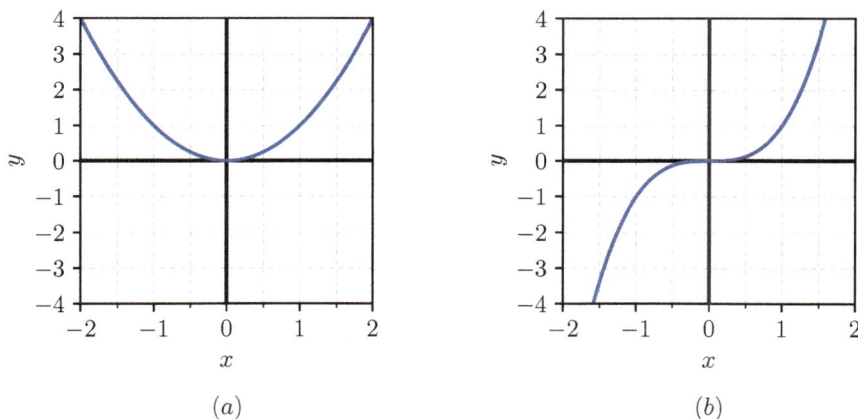

Figure 9.1: (a) $f(x) = x^2$ (b) $f(x) = x^3$

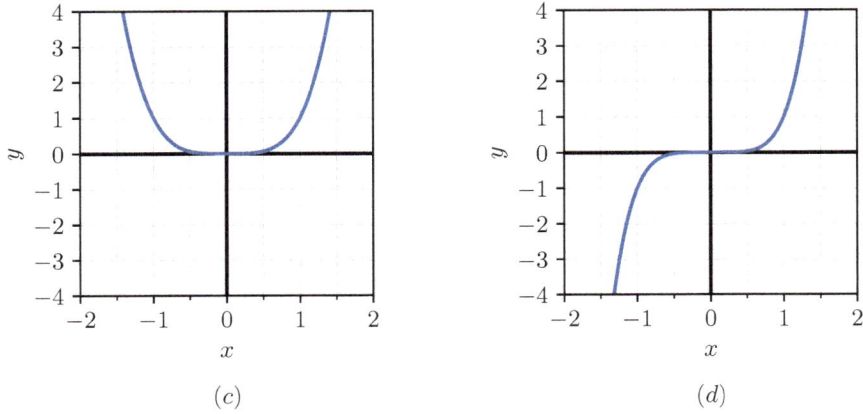

$(c)$                                             $(d)$

Figure 9.2: (a) $f(x) = x^4$  (b) $f(x) = x^5$.

| $x$ | $x^2$ | $x^3$ | $x^4$ | $x^5$ |
|-----|-------|-------|-------|-------|
| -2 | 4 | -8 | 16 | -32 |
| -1.5 | 2.25 | -3.375 | 5.0625 | -7.59375 |
| -1 | 1 | -1 | 1 | -1 |
| -0.5 | 0.25 | -0.125 | 0.0625 | -0.03125 |
| 0 | 0 | 0 | 0 | 0 |
| 0.5 | 0.25 | 0.125 | 0.0625 | 0.03125 |
| 1 | 1 | 1 | 1 | 1 |
| 1.5 | 2.25 | 3.375 | 5.0625 | 7.59375 |
| 2 | 4 | 8 | 16 | 32 |

Table 9.1: Table of values for (a), (b), (c) and (d) with $x \in [-2, 2]$.

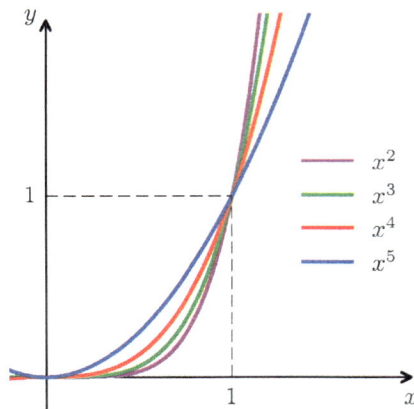

Figure 9.3: Graphs of $x^n$ always pass through (0,0) and (1,1).

We know that each of the graphs above will cross the axes at $(0,0)$. We also know that as $x \to \infty$, $f(x) \to \infty$ and as $x \to -\infty$, $f(x) \to \infty$ for even powers of $x$ or $f(x) \to -\infty$ for odd powers of $x$. Looking closer at the shape in the upper right quadrant we see some interesting behaviour near $x = 1$ as shown in Figure 9.3.

For $0 < x < 1$, we have $x^n > x^{n+1}$ and for $x > 1$ we have $x^n < x^{n+1}$ as shown in Table 9.1.

We now consider fractional powers of $x$ and negative powers of $x$, some of which are shown below.

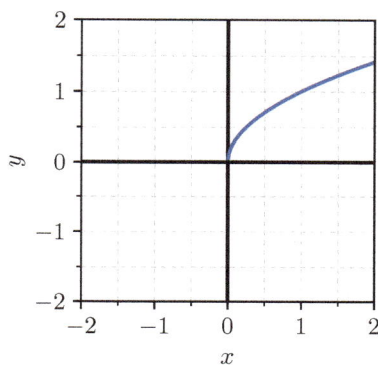

(a)

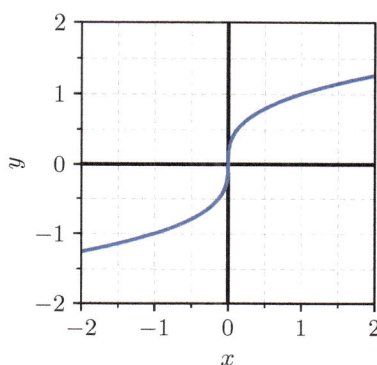

(b)

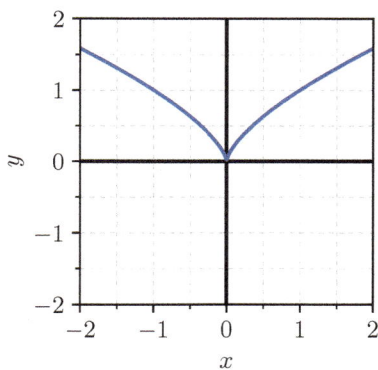

(c)

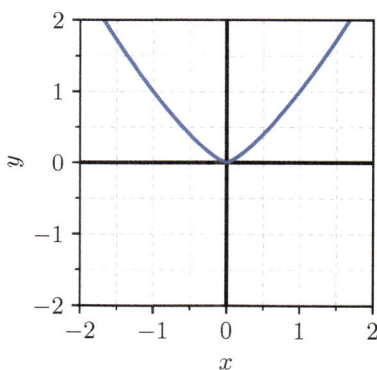

(d)

Figure 9.4: (a) $f(x) = x^{\frac{1}{2}}$ (b) $f(x) = x^{\frac{1}{3}}$ (c) $f(x) = x^{\frac{2}{3}}$ (d) $f(x) = x^{\frac{4}{3}}$

(e)                                                (f)

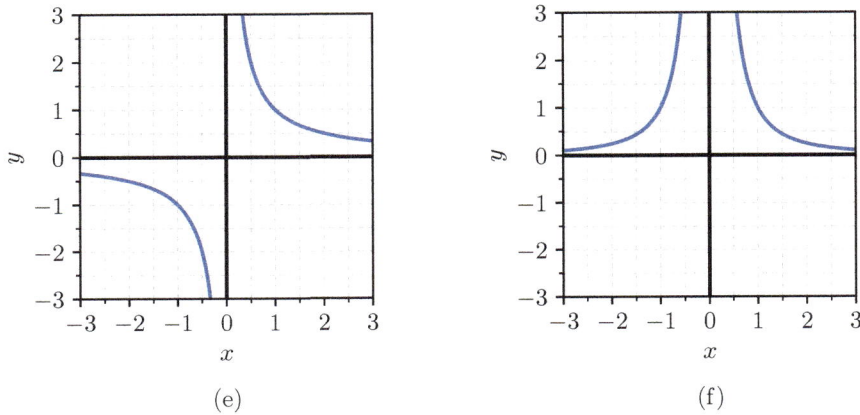

Figure 9.5: (e) $f(x) = \frac{1}{x}$  (f) $f(x) = \frac{1}{x^2}$.

For $f(x) = \frac{1}{x}$ and $f(x) = \frac{1}{x^2}$ we see there are 'breaks', known as discontinuities, in the curves as these functions are undefined at $x = 0$. These functions each have *asymptotes* on the $x$-axis and $y$-axis meaning they get closer and closer to the axes, but never intersect them.

Now that we have seen a number of basic graph shapes, we discuss how to sketch more complicated curves. If we know key features of a graph such as the location of any axis intercepts, turning points and asymptotes along with knowing the general shape of the curve, we start to build up a sketch in a logical fashion.

---

**Example 9.1**

Sketch the graph of $y = (x - 2)(x + 1)(x + 4)$.

**Solution:**

First we find where the curve intersects the axes. When $x = 0$, $y = -8$ and we have three values of $x$ which give $y = 0$ there are $x = 2$, $x = -1$ and $x = -4$. We see this is a cubic equation and since the highest power of $x$ will dominate when $x$ is large, we know that as $x \to \infty$, $y \to \infty$ and as $x \to -\infty$, $y \to -\infty$.

| $x$ | $x < -4$ | $-4 < x < -1$ | $-1 < x < 2$ | $x > 2$ |
|---|---|---|---|---|
| $y$ | $-$ve | $+$ve | $-$ve | $+$ve |

A quick inspection of $y$-values near each root, as shown in the table above, helps us complete the sketch.

---

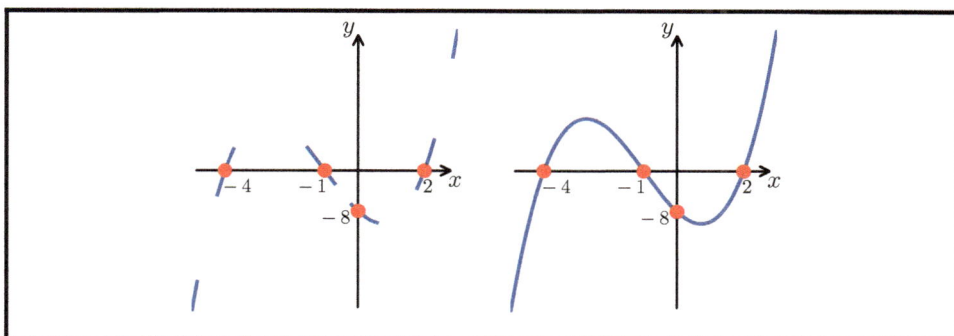

**Example 9.2**

Sketch the graph of $y = x^4 - x^3 - x^2 + x$.

**Solution:**

Similar to the previous example, we find where the curve crosses the axes. By the factor theorem, $x$, $(x-1)$ and $(x+1)$ are factors of he polynomial, so the curve can be expressed as $y = x(x-1)(x+1)(Ax+B)$. Comparing coefficients , it is clear that $A = 1$ and $B = -1$, hence $y = x(x-1)^2(x+1)$. The curve intersects the axes at $(-1,0)$, $(0,0)$ and $(1,0)$. To consider the overall shape of the curve, observe that the highest power of $x$ is 4, so we expect the curve to behave like $y = x^4$ for large values of $x$.

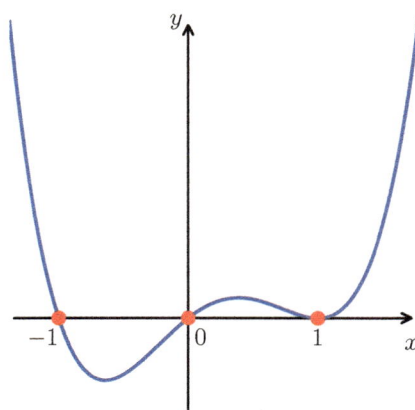

Notice that there is a repeated root at $x = 1$ so the curve touches the $x$-axis at that point.

**Example 9.3**

Sketch the graph of

$$y = \frac{1-x}{2x-3}.$$

**Solution:**

To find the axis intercepts, we substitute $x = 0$ which gives $y = -\frac{1}{3}$ and substitute $y = 0$ which gives $x = 1$. There will be a vertical asymptote at the $x$-value for which the function is undefined, *i.e.* when $2x - 3 = 0$ so $x = \frac{3}{2}$. To find the horizontal asymptote we consider the large values of $x$, as $x \to \pm\infty$, $y \to -\frac{1}{2}$. A table of values will help us to identify behaviour near the asymptotes.

| $x$ | -100 | 1.499 | 1.501 | 100 |
|---|---|---|---|---|
| $y$ | -0.4975 | 249.5 | -250.5 | -0.5025 |

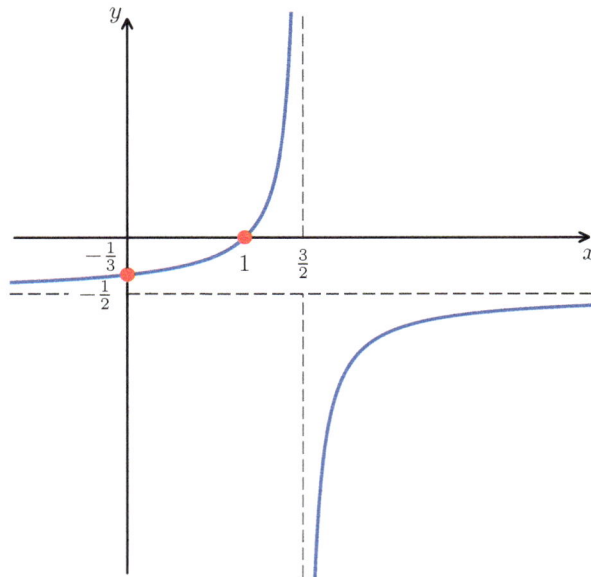

**Exercise 9.1**

Q1. Sketch each of the following curves showing where they intersect the coordinate axes.

(a) $y = x^7$;

(b) $y = x^{\frac{1}{5}}$;

(c) $y = x^{\frac{5}{4}}$;

(d) $y = x^2 - 3x$;

(e) $y = (x-2)(x+7)$;

(f) $y = x^2 + 6x + 9$;

(g) $y = 2x^2 + 7x - 4$;

(h) $y = x(x+2)(x-2)$;

(i) $y = (1-x)(x+2)^2$;

(j) $y = (5-x)^2(x+3)$;

(k) $y = (x+1)(x+2)^2(x+3)$;

(l) $y = x(x-2)(x-4)^2$;

(m) $y = (x+1)(x-1)^3(x-2)$.

Q2. Sketch each of the following curves showing where they intersect the coordinate axes and state the equations of any asymptotes.

(a) $y = \frac{1}{x-1}$;

(b) $y = \frac{x+1}{x}$;

(c) $y = \frac{2+x}{2-x}$;

(d) $y = \frac{1}{(x-4)^2} + 1$.

## 9.2 Graphical Solutions to Equations

Our understanding of the shapes of graphs can be exploited to gain insight into solutions of equations.

**Example 9.4**

Display graphically the solution to the equation $\frac{1}{x} = x^2 + 1$, hence determine the number of real roots of the equation $0 = x^3 + x - 1$.

**Solution:**

First we sketch graphs of $y = \frac{1}{x}$ and $y = x^2 + 1$ on the same axes as shown below.

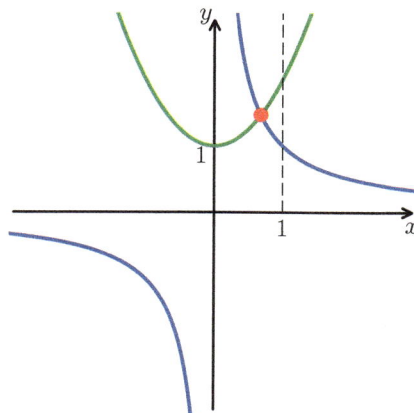

It is clear from the sketch that there is one point at which these curves intersect and it is somewhere in the range $0 < x < 1$. If we rearrange the first equation we can obtain the second equation as follows,

$$\frac{1}{x} = x^2 + 1,$$

$$\Rightarrow \quad 1 = x^3 + x, \quad \text{for } x \neq 0,$$

$$\Rightarrow \quad 0 = x^3 + x - 1.$$

Since these equations are simply a rearrangement of each other, their solutions will be identical, hence $0 = x^3 + x - 1$ also has one solution in the range $0 < x < 1$.

### Activity 9.1 — Graphical Solutions

This activity concerns the graphical solution of various equations that can be interpreted as the intersection of a given curve with other curves or lines.

Q1. On an accurate copy of the graph $y = 3x^2 - 2x - 2$ draw the line $y = 2$ and hence find approximate solutions to the equation $0 = 3x^2 - 2x - 4$.

Q2. On the same graph draw an appropriate line or curve to show graphically the solutions to the following equations.
  (a) $0 = 3x^2 - 3x - 3$
  (b) $0 = 3x^2 - \frac{3}{2}x - 5$
  (c) $0 = 4x^2 - 2$

### Learning Resource 9.1 — Axes for Activity 9.1

The pair of axes available for download below can be used when completing Activity 9.1.

### Exercise 9.2

Q1. Show that the equation $0 = 3x^3 + x^2 - 2$ can be rearranged to the form $\frac{2}{x^2} = 3x + 1$. Hence, by drawing an appropriate sketch, determine the number of real roots of the equation $0 = 3x^3 + x^2 - 2$.

Q2. Show on a sketch that there are exactly two values of $k$ for which $y = -2x + k$ is tangent to $y = \frac{1}{x+1}$.

Q3. Suggest possible equations of the form $y = (x - a)(x - b)$, where $a$ and $b$ are integers, such that the curves intersect the circle with equation $(x - 4)^2 + y^2 = 16$,
  (a) Twice,
  (b) Three times,
  (c) Four times.

Q4. Below is the graph of the hyperbola with equation $x^2 - 3y^2 = 1$. Suggest possible cubic equations that have the following number of intersections with the hyperbola.

  (a) Zero intersections,
  (b) Four intersections,

(c) Six intersections.

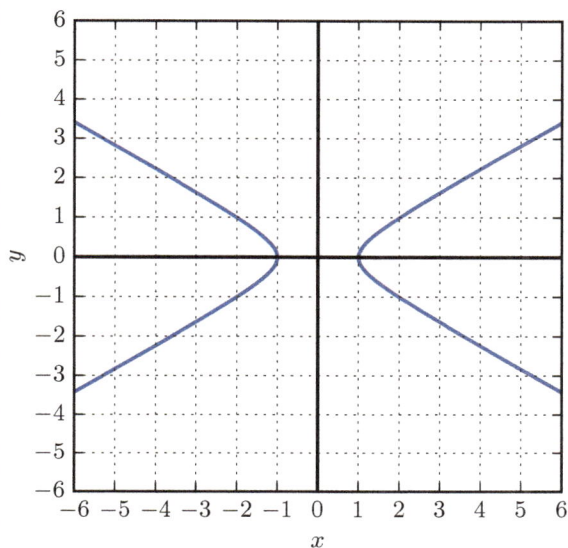

## 9.3 Proportional Relationships

In this section we shall consider graphs of quantities that vary in proportion. If a quantity $y$ is *proportional* to a quantity $x$, we write $y \propto x$ meaning there is some non-zero constant $k$ such that $y = kx$. Similarly, if $y$ is *inversely proportional* to $x$, we write $y \propto \frac{1}{x}$, meaning there is a non-zero constant $k$ such that $y = \frac{k}{x}$.

---

**Example 9.5**

Samples of copper, of varying volumes, are weighed and the mass is recorded in the table below. Show that the mass of copper is proportional to its volume and determine the constant of proportionality. What does the constant of proportionality represent in this case? Sketch a graph of this relationship.

| Volume (cm$^3$) | Mass (g) |
|---|---|
| 5 | 44.8 |
| 15 | 134.4 |
| 20 | 179.2 |
| 50 | 448.0 |
| 250 | 2240.0 |

**Solution:**

If mass is proportional to volume then the ratio will be constant.

| Volume (cm$^3$) | Mass (g) | Mass/Volume (g/cm$^3$) |
|---|---|---|
| 5 | 44.8 | 8.96 |
| 15 | 134.4 | 8.96 |
| 20 | 179.2 | 8.96 |
| 50 | 448.0 | 8.96 |
| 250 | 2240.0 | 8.96 |

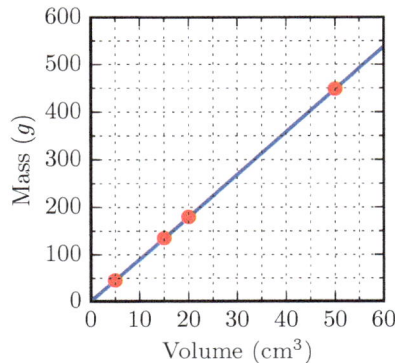

The constant of proportionality is 8.96 g/cm$^3$ and this is the *density* of copper. Notice that the gradient of the straight line is also 8.96.

**Tip**

It is important to always state the units of a constant of proportionality. In Example 9.5 the density is given as 8.96 g/cm$^3$, without these units it would be impossible to understand what the number represents physically.

**Example 9.6**

A builders merchant is intending to sell cement powder in square faced cuboid boxes which hold a fixed volume of 0.25 m$^3$. If the square face has width and height $x$ and the length of the cuboid is $y$, show that $y$ is inversely proportional to the square of $x$ and sketch a graph of this relation.

**Solution:**

The volume of the box can be expressed as $V = x^2 y$. Since the volume is fixed at $V = 0.25$ we have $0.25 = x^2 y$ which can be rearranged to $y = \frac{0.25}{x^2}$ hence $y \propto \frac{1}{x^2}$.

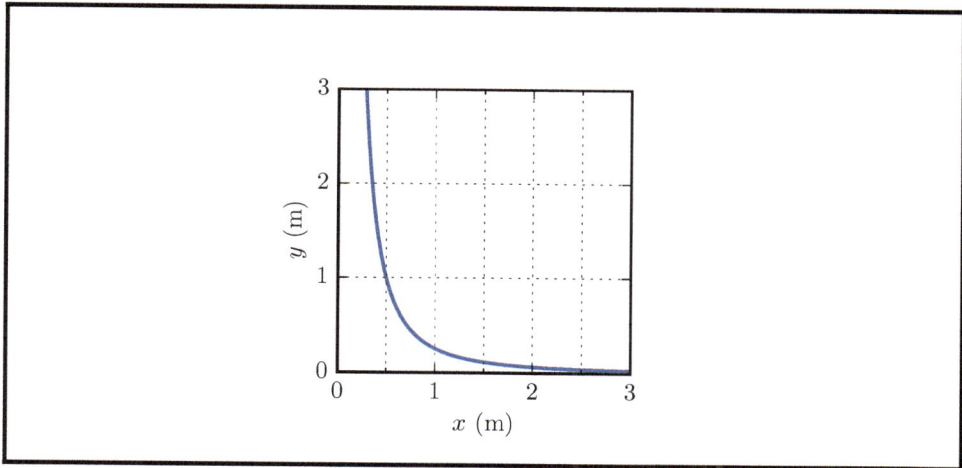

**Exercise 9.3**

Q1. A rectangle with a fixed area of $15\,\text{cm}^3$ has length $x$ and width $y$. Show that $y$ is inversely proportional to $x$ and sketch the graph of this relation.

Q2. The volume of a hemisphere is proportional to the cube of its radius. State the constant of proportionality and sketch the graph of this relation.

Q3. Below is a conversion table for Celsius $(C)$ to Fahrenheit $(F)$. Show that $C$ is *not* proportional to F.

| Celsius $(C)$ | Fahrenheit $(F)$ |
| --- | --- |
| $-40$ | $-40$ |
| $-20$ | $-4$ |
| 0 | 32 |
| 20 | 68 |
| 40 | 104 |

Q4. Four cylinders of Aluminium, each $5\,\text{cm}$ tall, have the following masses and radii. Show that mass is proportional to the square of the radius and determine the density of Aluminium.

| Radius (cm) | Mass (g) |
| --- | --- |
| 5 | 1060.3 |
| 10 | 4241.2 |
| 20 | 16 964.6 |
| 35 | 51 954.1 |

Q5. A fixed amount of gas is held in a cylinder with a piston at one end. The radius of the cylinder is $3\,\text{cm}$. As the piston is slowly pushed into the cylinder, the pressure of the gas is measured. The table below shows pressure for various heights of the piston from the bottom of the cylinder. Here, pressure is measured in Pascals (Pa). $1\,\text{Pa} = 1\,\text{N/m}^2$.

| Length (cm) | Pressure (Pa) |
|:-----------:|:-------------:|
| 9 | 101 325 |
| 7 | 130 275 |
| 5 | 182 385 |
| 3 | 303 975 |

Show that pressure is inversely proportional to volume and state the constant of proportionality.

---

**Key Facts — Graph Sketching**

- When sketching graphs label
  - Points of intersection with the axes.
  - Points of intersection with other functions (if sketching more than one function on the same axes).
  - The equations of any asymptotes.
  - Turning points if required.
- If a quantity $y$ is *proportional* to a quantity $x$, we write $y \propto x$ meaning there is some non-zero constant $k$ such that $y = kx$
- If $y$ is *inversely proportional* to $x$, we write $y \propto \frac{1}{x}$, meaning there is a non-zero constant $k$ such that $y = \frac{k}{x}$
- The graphs of directly proportional relationships pass through the origin.
- The units of a constant of proportionality should always be stated as this provides information on the physical meaning of the constant.

---

**Chapter Assessment — Graph Sketching**

Download and sit the 30 minute assessment for this chapter from the digital book.

$$f(x) = \frac{1}{\sqrt{2\pi\sigma^2}} e^{-\frac{(x-\mu)^2}{2\sigma^2}}.$$

# 10. Transformation of Graphs

Often, in real life situations, the graphs of data we observe or predict are merely transformations of graphs we already have information about. For example, a normal distribution may be fitted to some experimental data, with mean $\mu$ and standard deviation $\sigma$. The equation for this normal distribution probability density function is

$$f(x) = \frac{1}{\sqrt{2\pi\sigma^2}} e^{-\frac{(x-\mu)^2}{2\sigma^2}}.$$

In itself, this equation is not particularly enlightening. Instead, we can understand how the graph of this function will look by *transforming* the graph of the *standard* normal distribution probability density, where $\mu = 0$ and $\sigma = 1$.

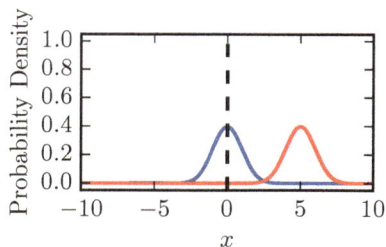
(a)

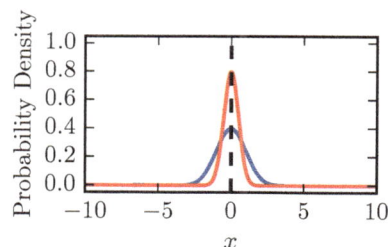
(b)

Figure 10.1: Transformation of probability density function for the Normal Distribution: change in mean $\mu$ (a) and change in standard deviation $\sigma$ (b).

In Figure 10.1(a), we can see the effect of changing the mean so that $\mu = 5$ and, in Figure 10.1(b), we can see the effect of changing the standard deviation so that $\sigma = 0.5$. In both figures, the blue curve is the standard normal distribution, which is centred about the origin. We see that changing $\mu$ has shifted the curve to right, while changing $\sigma$ has resulted in the curve becoming more 'spiky'. Explaining what causes these graph transformations is the aim of this chapter.

> **Remark**
> The normal distribution is a very important distribution in probability theory, we will meet it again in Year 2. At this stage, we need not concern ourselves with how the normal distribution probability density function is derived, or with what it means.

> **Learning Resource 10.1 — Introduction to Graph Transformations**
> Before studying this chapter, investigate graph transformations by attempting the activity below.

> **Use of Technology 10.1 — Using Software to Investigate Transformations**
> Use suitable software to complete the same problems in Teaching Resource 10.1.

> **Interactive Activity 10.1 — Single Transformations**
> The Geogebra applet below (in the digital book) can be used to examine the transformations in this activity, and throughout this chapter.

## 10.1  Translations

Given the function $y = f(x)$, the transformation described by $f(x) \pm a$ causes a translation of the graph of $f(x)$, $\pm a$ units parallel to the $y$-axis.

> **Remark**
> In earlier work on translations we have seen the use of column vectors to indicate how much a given shape or coordinate has moved by. The column vector $\begin{pmatrix} a \\ b \end{pmatrix}$ indicates that the shape moves $a$ units horizontally and $b$ units vertically. For $a > 0$ the horizontal movement is to the right, but $a < 0$ indicates a movement to the left. For $b > 0$ the vertical movement is upwards, but for $b < 0$ this movement is downwards.

**Example 10.1**

The function $y = f(x)$ is shown below, where $f(x) = x^3 - 2x$. Sketch the graph of:

(a) $y = f(x) + 3$;

(b) $y = f(x) - 1$.

For (a) and (b), state the column vector of translation and give the coordinates of the translated point.

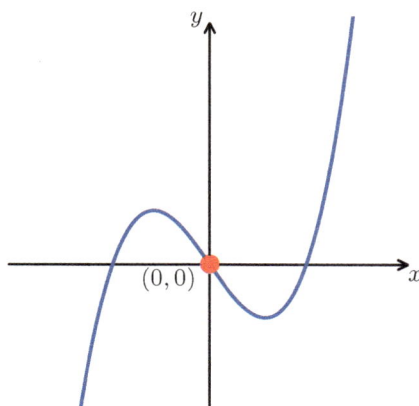

**Solution:**

(a) This is a translation parallel to the $y$-axis by 3 units. Therefore, associated column vector of translation is $\begin{pmatrix} 0 \\ 3 \end{pmatrix}$.

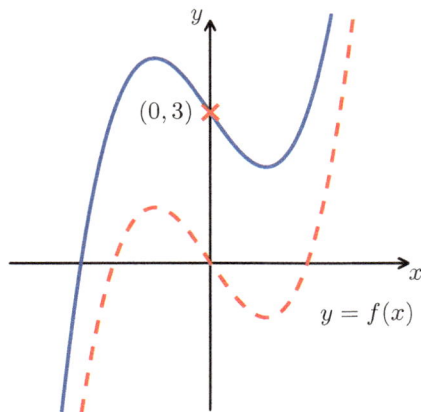

(b) This is a translation parallel to the $y$-axis by $-1$ unit. The associated column vector of translation is $\begin{pmatrix} 0 \\ -1 \end{pmatrix}$.

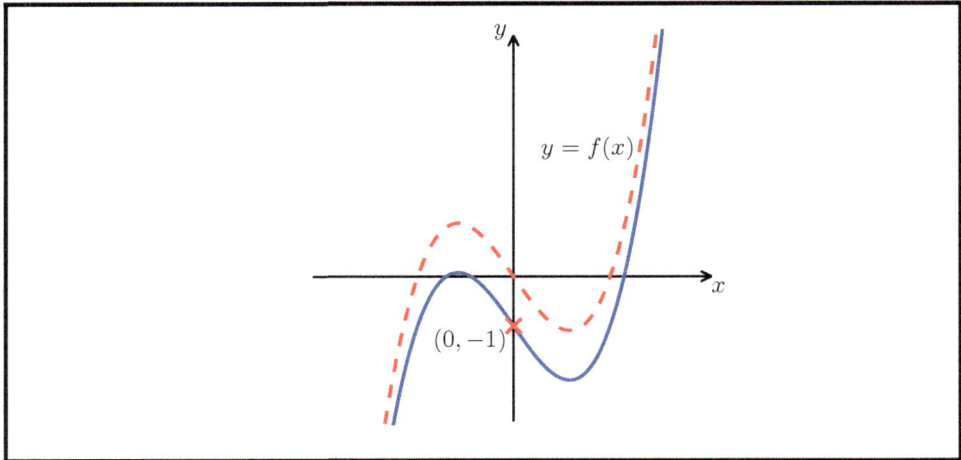

Given the function $y = f(x)$, the transformation described by $f(x \pm a)$ causes a translation of the graph of $f(x)$, $\mp a$ units parallel to the $x$-axis. To see why this is, consider $y_1 = f(x_1)$. To obtain the same value $y_1$ using $f(x+a)$, we would need $x+a = x_1$, hence $x = x_1 - a$, *i.e.* the point which gives $y_1$ is shifted to the left by $a$. A similar argument holds for $f(x-a)$.

**Example 10.2**
The function $y = f(x)$ is shown below, where $f(x) = x^2 - 3x + 2$.

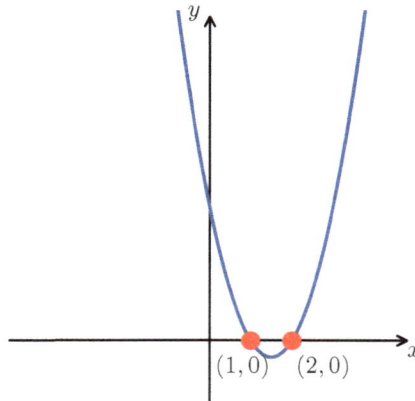

Sketch the graph of:
  (a) $y = f(x+4)$;
  (b) $y = f(x-2)$.
For (a) and (b), state the column vector of translation and give the new coordinates of intersection.

**Solution:**

(a) This is a translation parallel to the $x$-axis by $-4$ units. Therefore, the associated column vector of translation is $\begin{pmatrix} -4 \\ 0 \end{pmatrix}$.

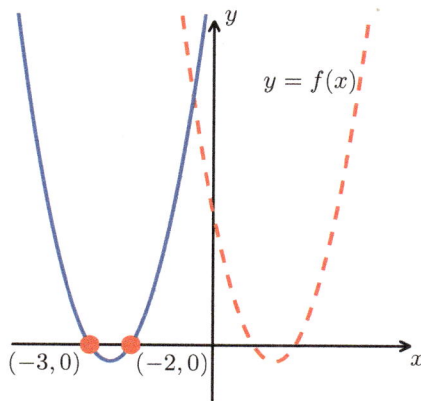

(b) This is a translation parallel to the $y$-axis by 2 units. The associated column vector of translation is $\begin{pmatrix} 2 \\ 0 \end{pmatrix}$.

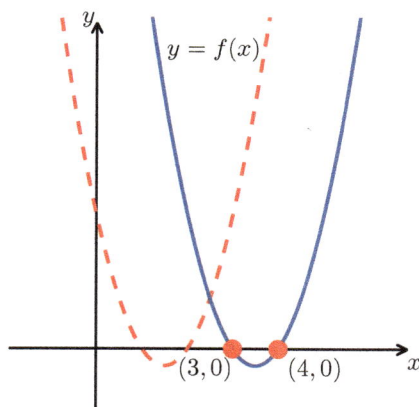

**Example 10.3**

Sketch the graph of $f(x) = \frac{1}{x}$, and label the asymptotes.

(a) Examine what happens to the asymptotes for a transformation defined by $f(x) + a$.

(b) Examine what happens to the asymptotes for a transformation defined by $f(x + a)$.

**Solution:**

The graph of $f(x) = \frac{1}{x}$ is shown below.

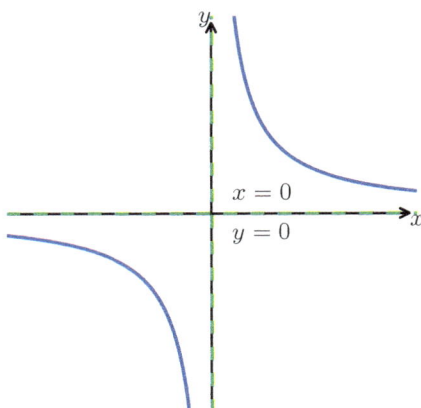

The asymptotes are $x = 0$ and $y = 0$.

(a) For a transformation defined by $f(x) + a$, since the graph will be translated by the column vector $\begin{pmatrix} 0 \\ a \end{pmatrix}$, the equation of the asymptote $x = 0$ will remain unchanged. The asymptote $y = 0$ will be translated to $y = a$.

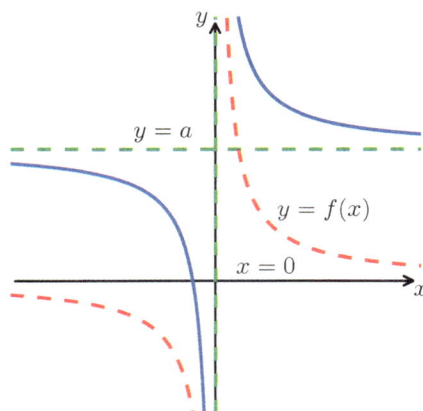

(b) For a transformation defined by $f(x + a)$ the graph will be translated by the column vector $\begin{pmatrix} -a \\ 0 \end{pmatrix}$. Therefore, the equation of the asymptote $y = 0$ will remain unchanged, but the asymptote $x = 0$ will be translated to $x = -a$.

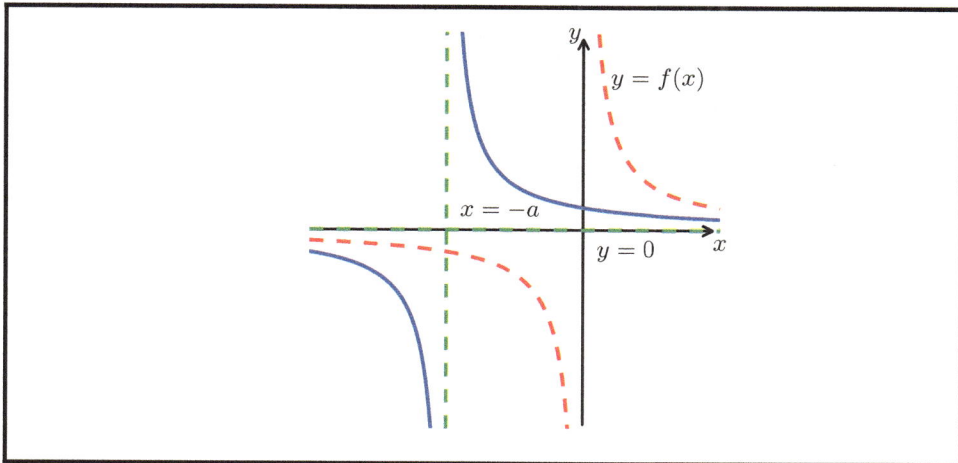

**Exercise 10.1**

Q1. Sketch $f(x) = (x + 2)(x - 3)(x + 1)$ and also
    (a) $f(x - 4)$;
    (b) $f(x + 3)$;
    (c) $f(x) + 2$.

Q2. A sketch of $f(x)$ is given below.

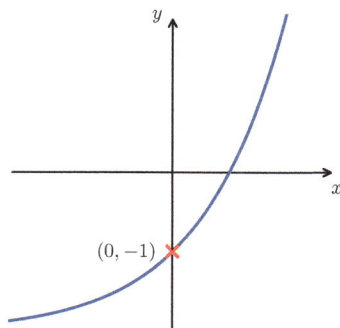

Write the following in terms of $f(x)$, along with a sketch of the transformation.
    (a) Translation by $\begin{pmatrix} 0 \\ -5 \end{pmatrix}$;
    (b) Translation by $\begin{pmatrix} 0 \\ 2 \end{pmatrix}$;
    (c) Translation by $\begin{pmatrix} 3 \\ 0 \end{pmatrix}$;
    (d) Translation by $\begin{pmatrix} -4 \\ 0 \end{pmatrix}$.

Q3. The diagram below shows the sketch of $g(x)$.

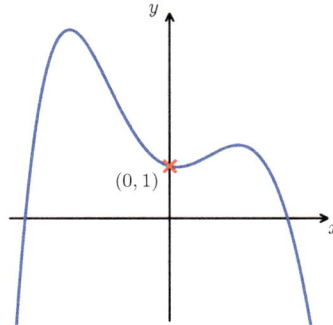

Which of the graphs below could be a sketch of $f(x-1)$? Select all possibilities and justify these selections.

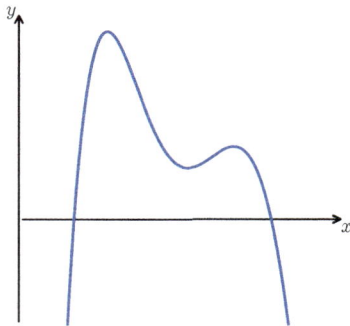

(a)

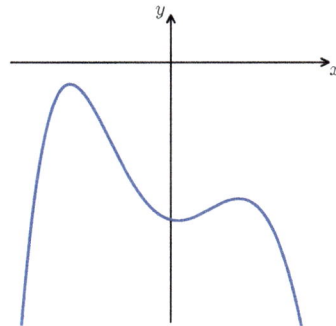

(b)

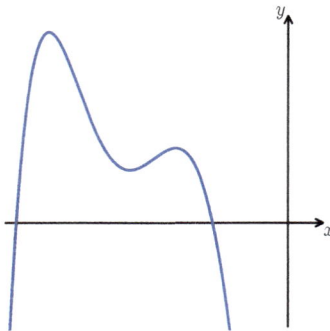

(c)

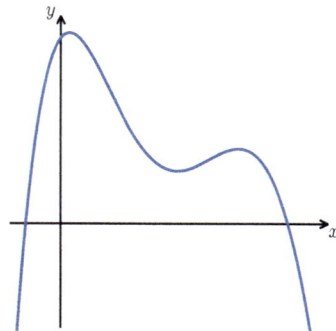

(d)

Q4. (a) Sketch the graph of $f(x) = 3 - x^2$;

(b) If we translate $f(x)$ 3 units down, give the associated column vector, and write this in terms of $f(x)$.

    (c) Write the equation of the resulting function and sketch the graph on the same set of axes.

Q5. Sketch the graph of $f(x) = -\frac{1}{x}$. For each of the following questions, sketch the transformation indicating clearly the equations of the asymptotes.

    (a) $f(x+3)$;

    (b) $f(x) - 2$.

Q6. A sketch of $f(x)$ is given below.

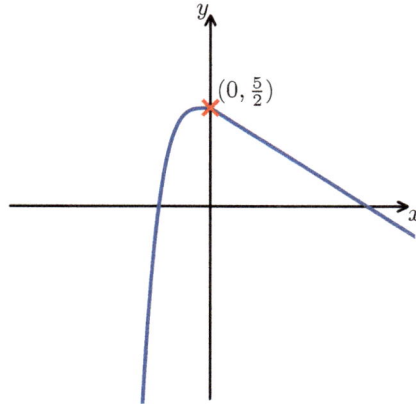

Sketch the graphs of the following translations, giving the coordinates of the translated point,

    (a) $f(x) + 3$;

    (b) $f(x-4)$;

    (c) translation by the column vector $\begin{pmatrix} -1 \\ -2 \end{pmatrix}$, and indicate how this can be represented in terms of $f(x)$.

## 10.2  Reflections

Given the function $y = f(x)$, the transformation described by $-f(x)$ causes a reflection of the graph of $f(x)$ in the $x$-axis. To see why, for a given $x_1$, if $y_1 = f(x_1)$, then $-f(x_1) = -y_1$.

---

**Example 10.4**

The function $y = f(x)$ is shown below.

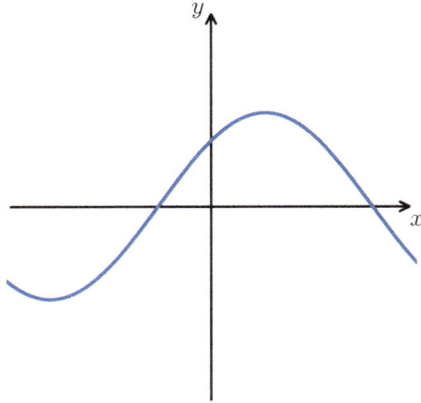

Sketch the graph of $y = -f(x)$.

**Solution:**

As stated above, $y = -f(x)$ results in a reflection in the $x$-axis. We obtain the sketch shown below.

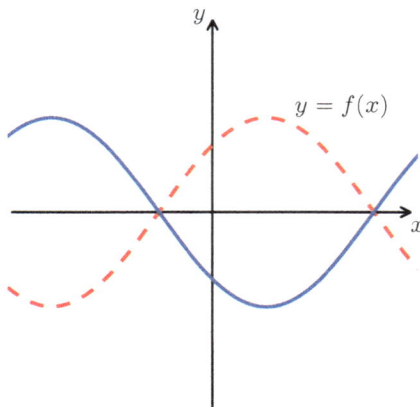

Given the function $y = f(x)$, the transformation described by $f(-x)$ causes a reflection of the graph of $f$ in the $y$ axis. To show this, let $y_1 = f(x_1)$, then $y_1 = f(-(-x_1))$.

**Example 10.5**

Sketch $f(x) = x^2 - 4x - 12$ and show $f(-x)$ on the same graph.

**Solution:**

First, we sketch the graph of $f(x) = x^2 - 4x - 12$.

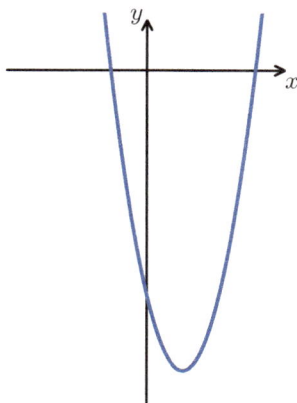

We are then asked to sketch $f(-x)$ which, as stated above, is a reflection of $f(x)$ in the $y$-axis. We obtain the sketch shown below.

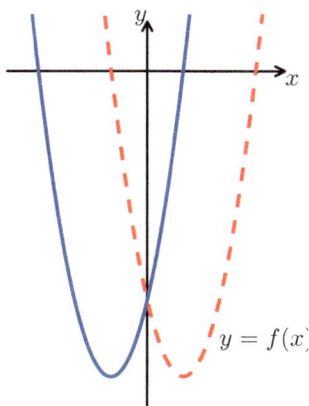

## 10.3   Stretches

Given the function $y = f(x)$, the transformation described by $af(x)$ causes a stretch of the graph of $f(x)$, parallel to the $y$-axis by a scale factor of $a$.

---

**Example 10.6**

The diagram below shows the graph of $y = f(x)$.

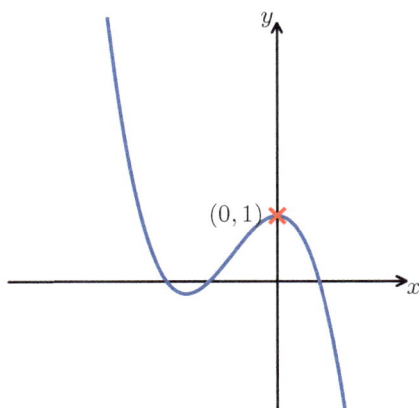

Sketch the graph of
  (a)  $y = 3f(x)$;
  (b)  $y = \frac{1}{2}f(x)$.

**Solution:**
  (a)  The transformation $y = 3f(x)$ is a stretch of the graph by scale factor 3 parallel to the $y$-axis.

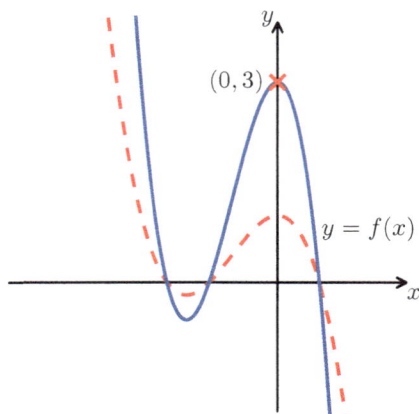

---

(b) The transformation $y = \frac{1}{2}f(x)$ is a stretch of the graph by scale factor $\frac{1}{2}$ parallel to the $y$-axis.

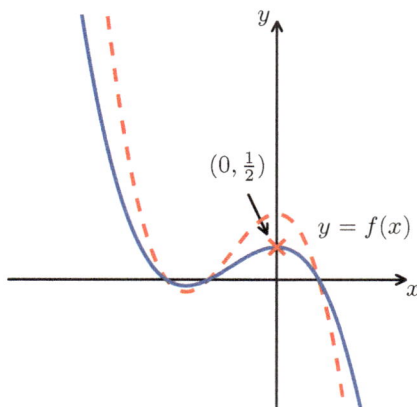

Given the function $y = f(x)$, the transformation described by $f(ax)$ causes a stretch of the graph of $f(x)$, parallel to the $x$-axis by a scale factor of $\frac{1}{a}$. To see this, we notice that if $y_1 = f(x_1)$, then, if $y_1 = f(ax)$, then $ax = x_1$ and hence, $x = \frac{x_1}{a}$.

**Example 10.7**

Sketch
  (a) $f(x) = 16 - x^2$.
  (b) Show on the same graph as part (a) $f(2x)$.
  (c) On a new set of axes sketch $f(x)$ and $f\left(\frac{1}{4}x\right)$.
Label the coordinates of intersection with the $x$-axis.

**Solution:**
  (a) First, we sketch the graph of $f(x) = 16 - x^2$.

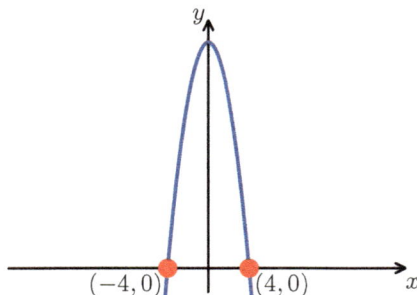

(b) The transformation $f(2x)$ causes a stretch parallel to the $x$-axis by scale factor $\frac{1}{2}$. This results in the graph shown below.

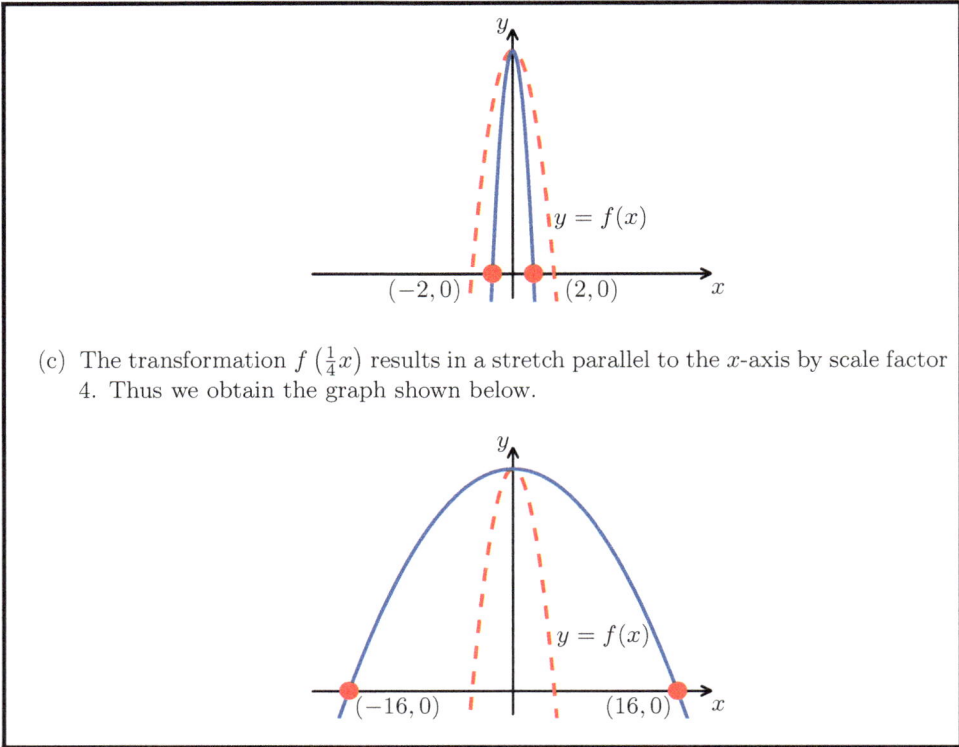

(c) The transformation $f\left(\frac{1}{4}x\right)$ results in a stretch parallel to the $x$-axis by scale factor 4. Thus we obtain the graph shown below.

**Exercise 10.2**

Q1. The diagram below shows the graph of $f(x)$ and $g(x)$.

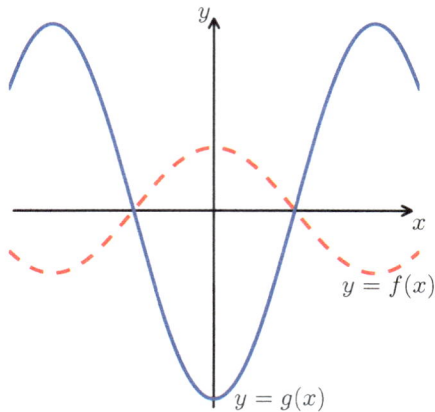

   (a) Which of the following equations could give the transformation of $f(x)$ to $g(x)$.

      i. $y = 3f(x)$;

      ii. $y = -3f(x)$;

      iii. $y = f(3x)$;

      iv. $y = f(-3x)$.

   (b) Hence, write the equation of the transformation of $g(x)$ to $f(x)$.

**Q2.** Given that $f(x) = x^2 - 7x - 18$, sketch the following functions indicating where they intersect with the $x$ axis.

   (a) $y = 2f(x)$;

   (b) $y = \frac{1}{2}f(x)$;

   (c) $y = f\left(\frac{1}{2}x\right)$;

   (d) $y = f(2x)$.

**Q3.** Describe clearly the transformation which maps the graph of $y = 4x^2 - 2x + 12$ to the graph of $y = -2x^2 + x - 6$.

**Q4.** The graph of $y = f(x)$ is shown below.

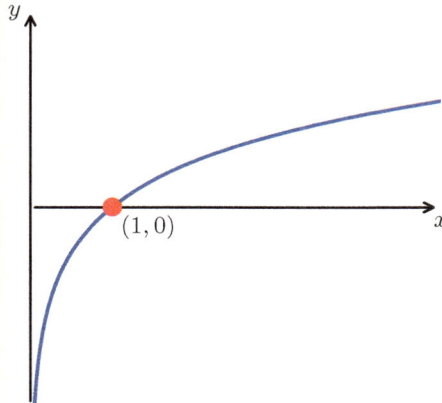

Sketch clearly the the graphs of the following transformations, indicating clearly where the labelled point is now located.

   (a) $y = f(4x)$;

   (b) $y = f(-2x)$;

   (c) $y = 3f(x)$;

   (d) $y = -f(x)$.

**Q5.** Sketch the graph of $f(x) = (x - 1)(x - 2)(x - 3)$.

   (a) Write down the equation of the graph obtained by the transformation which stretches the graph vertically by scale factor -2.

   (b) Sketch the graph of the obtained equation indicating coordinate axes intersections.

Q6. The graph of $y = f(x)$ is shown below.

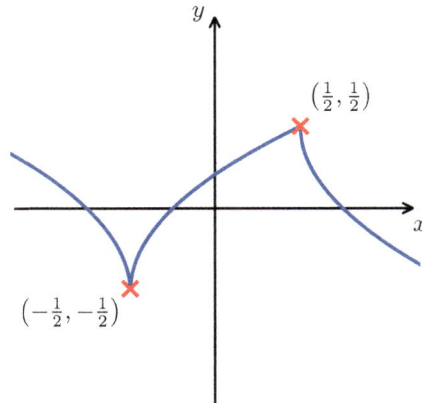

Sketch the graphs of the following transformations, indicating clearly where the labelled points are now located.
(a) $y = f\left(\frac{1}{2}x\right)$;
(b) $y = 2f(x)$;
(c) $y = -f(x)$.

**Interactive Activity 10.2 — Introduction to Combinations of Transformations**
Using the interactive file in your digital textbook, examine the effects on a function $f(x)$ when transformed by $af(bx + c) + d$. The variables $a$, $b$, $c$ or $d$ can take any value including $0$.

**Exercise 10.3**
Q1. (a) Write the equation of transformation of $y = f(x)$ when transformed by
    i. a stretch parallel to the $y$ axis by scale factor $\frac{1}{4}$;
    ii. a translation parallel to the $x$ axis by column vector $\begin{pmatrix} 4 \\ 0 \end{pmatrix}$;
    iii. a reflection in the $x$ axis;
    iv. a translation parallel to the $y$ axis by column vector $\begin{pmatrix} 0 \\ -3 \end{pmatrix}$;
    v. a stretch parallel to the $x$ axis by scale factor $-2$.
  (b) Describe fully the transformation of $y = f(x)$ given by
    i. $f(-x)$.
    ii. $f(x - 1)$.
    iii. $f(x) + 2$.
    iv. $\frac{1}{2}f(x)$.

v. $f\left(\frac{1}{2}x\right)$.

Q2. The function $y = f(x)$ is shown below.

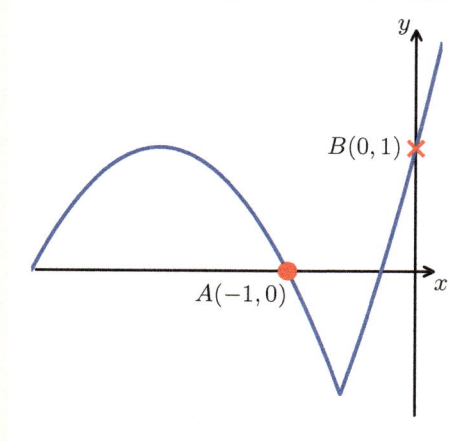

For each of the transformations below, determine which of the labelled $x$ and $y$ intercepts are transformed, stating their new coordinates if this is the case.

(a) $f(-2x)$;

(b) $f(x-2)$;

(c) $3f(x)$;

(d) $f(x) + 1$.

Q3. (a) Sketch the function $y = f(x)$, given that $f(x) = 3x - 4$.

(b) Given that $y = f(x)$ is reflected in the $x$ axis, state the equation of the image obtained.

(c) Following the reflection from (b), the function is translated horizontally $-2$ units. State the equation for this function in terms of $f(x)$.

Q4. The point $A(3, -1)$ lies on the curve with equation $y = f(x)$. State the coordinates of its image when the curve is transformed to

(a) $y = f\left(\frac{1}{2}x\right)$;

(b) $3y = f(x)$;

(c) $y = f(x+3)$.

Q5. The diagram below shows the graph of $y = f(x)$.

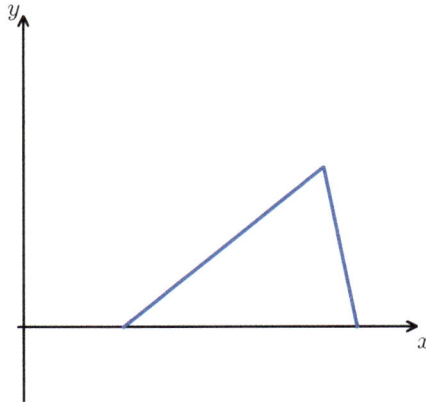

   (a) Copy a sketch of the function and, on the same diagram, draw the graph of $y = f(x + 4)$.

   (b) Describe fully the transformation which transforms the graph of $y = f(x)$ to the graph of $y = 3f(x)$.

Q6. The graph of $y = f(x)$ is shown below.

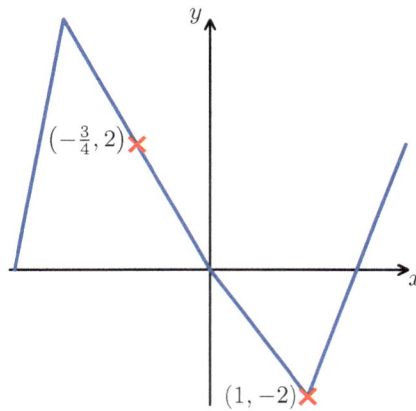

(a) Find the equation for the transformation shown in terms of $f(x)$.

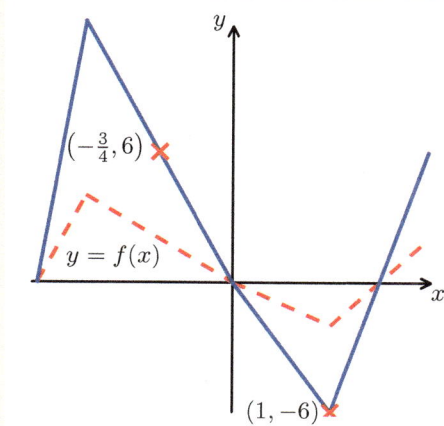

$\left(-\frac{3}{4}, 6\right)$

$y = f(x)$

$(1, -6)$

(b) Find the equation for the transformation shown in terms of $f(x)$.

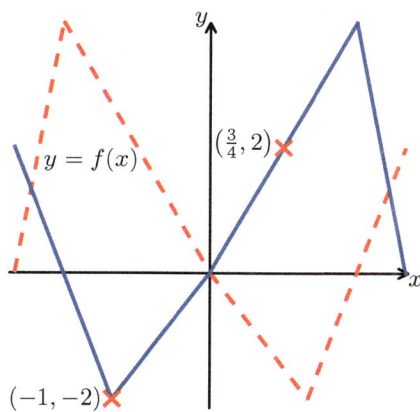

$y = f(x)$

$\left(\frac{3}{4}, 2\right)$

$(-1, -2)$

(c) Find the equation for the transformation shown in terms of $f(x)$.

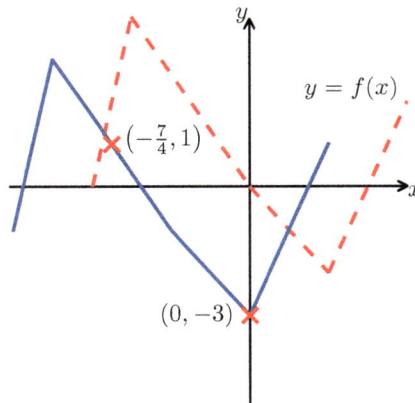

## Key Facts — Transformation of Graphs

- $y = f(x) + a$ translates the graph of $y = f(x)$ $a$ units in the $y$-direction;
- $y = f(x + a)$ translates the graph of $y = f(x)$ $-a$ units in the $x$-direction;
- $y = -f(x)$ reflects the graph of $y = f(x)$ about the $x$-axis;
- $y = f(-x)$ reflects the graph of $y = f(x)$ about the $y$-axis;
- $y = af(x)$ stretches the graph of $y = f(x)$ by a factor $a$ parallel to the $y$-axis;
- $y = f(ax)$ stretches the graph of $y = f(x)$ by a factor $\frac{1}{a}$ parallel to the $x$-axis.
- Points of intersections should be shown whenever sketching graph transformations.

## Chapter Assessment — Transformation of Graphs

Download and sit the 30 minute assessment for this chapter from the digital book.

# 11. Simultaneous Equations

Simultaneous equations arise in any situation when we are trying to find the values of two (or more) unknown variables. Such situations arise throughout mathematics and its applications in statistics and mechanics. Consider, as an example, a power station, where a steam turbine is used to generate electricity. The turbine comprises a number of rotors and stators (Figure 11.1(a)), consisting of a set of blades (Figure 11.1(b)).

(a)                                                                 (b)

Figure 11.1: A steam turbine before installation (a) and the turbine blades on a set of rotors/stators (b).

The turbine blades should be designed to allow an optimal flow of steam through the turbine to maximise energy production. Modelling the flow of steam through the turbine leads to a set of simultaneous *differential* equations (the famous Navier-Stokes equations) involving the velocity of the steam in three space directions, its pressure and its energy. Figures 11.2(c) and (d) show a prediction of the steam velocity in the $x$ and $y$ direction, respectively, through a single rotor. This prediction requires sophisticated mathematics that converts the differential set of equations to a set of a million or more algebraic simultaneous equations.

(a)                                                    (b)

Figure 11.2: Fluid velocity through a cascade of turbine blades (a) $x$-velocity and (b) $y$-velocity.

In this chapter, we shall investigate the rather more simple case where we have just two or three simultaneous equations. We shall also investigate the important connection between algebra and graphing: solving simultaneous equations corresponds to finding where graphs of more than one function intersect.

**Remark**
Differential equations will be studied further in the second year of A-Level.

**Activity 11.1**
Answer the following questions.
  Q1. What do we mean when we say "Solve $2x + 7 = 5$"?
  Q2. Write down the number of solutions for each of the equations below.
      (a) $2x - 2 = 0$.
      (b) $x^2 - 2 = 0$.
      (c) $x^2 + 2 = 0$.
      (d) $\sqrt{x} - 2 = 0$.
  Q3. Find a value of $x$ that satisfies *both* of these equations:

$$\begin{cases} x^2 - 3x + 2 = 0, & ① \\ x^2 + x - 6 = 0. & ② \end{cases}$$

      Are there any other ways of solving this problem? How do these compare with the first method chosen?
  Q4. Find values of $x$ and $y$ that satisfy this equation:

$$x + 2y = 9.$$

      How many solutions are there? Remember that $x$ and $y$ are not necessarily integers. What would be an effective way to represent *all of the solutions*?

Q5. Find values of $x$ and $y$ that satisfy *both* of these equations:

$$\begin{cases} x + 2y = 9, & \textcircled{1} \\ 3x + 2y = 31. & \textcircled{2} \end{cases}$$

Are there any alternative methods of solving this problem?

We will study two analytical techniques for solving simultaneous equations:

**Elimination:** Uses mathematical operations, such as addition or subtraction, to produce a single equation involving just one of the variables.

**Substitution:** Involves substituting an expression from one equation into the other. This will again produce a single equation involving just one of the variables.

## 11.1  Solving by Elimination

The five examples shown below can each be efficiently solved by elimination:

$$\begin{cases} 3x - 4y = -8, \\ 2x + 4y = 28. \end{cases} \quad \begin{cases} x + 2y = 9, \\ 3x + 2y = 31. \end{cases} \quad \begin{cases} 3x + 2y = 13, \\ 5x + 11y = 37. \end{cases} \quad \begin{cases} xy = -12, \\ \dfrac{x}{y} = -3. \end{cases} \quad \begin{cases} xy^2 = 20, \\ xy^4 = 80. \end{cases}$$

**Tip**

It is always good practice to *number* the equations before we start manipulating them. That way we can add meaningful labels to our working so that it is easier for someone else to follow our process.

**Example 11.1**
Solve

$$\begin{cases} 3x - 4y = -8, & \textcircled{1} \\ 2x + 4y = 28. & \textcircled{2} \end{cases}$$

**Solution:**
Adding the two equations together:

$$\textcircled{1} + \textcircled{2} \Rightarrow 5x = 20,$$
$$\Rightarrow x = 4.$$

We now find the corresponding value of $y$ by substituting $x = 4$ into either equation ① or ②. Equation ② has no negative signs and thus we can reduce the risk of making mistakes by choosing that one.

$$\text{Substituting } x = 4 \text{ into } ② \Rightarrow 8 + 4y = 28,$$
$$\Rightarrow 4y = 20,$$
$$\Rightarrow y = 5.$$

Then, to summarise clearly, our solution is:

$$(x, y) = (4, 5).$$

As we know from Chapter 4, the equations ① and ② can be represented as straight lines on a graph. We plot these in the figure below and immediately notice that the point at which the lines cross, *i.e.*, the *point of interection*, has coordinate $(x, y) = (4, 5)$, which corresponds to the solution of our set of simultaneous equations.

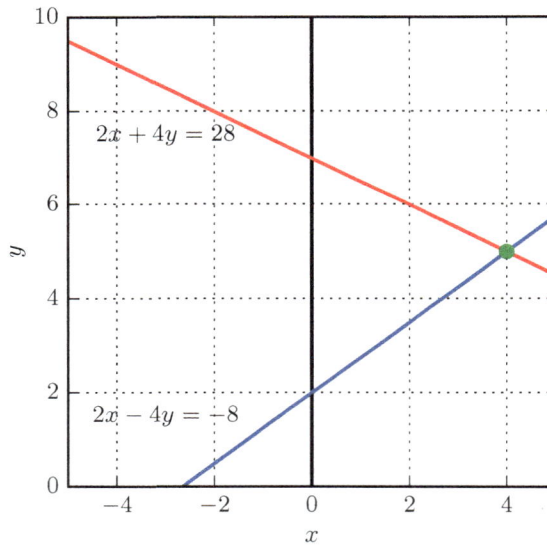

---

**Remark**

Why is adding (or subtracting, multiplying or dividing) two equations mathematically valid? Consider the following simplified situation: $A = B$ and $C = D$. Beginning from $A = B$, we can add $C$ to both sides giving $A + C = B + C$. Now, as $C = D$, we know that $B + C = B + D$. Therefore, $A + C = B + D$. This argument can easily be adapted for subtraction, multiplication and division. As an extension, consider if it is also valid for *inequalities*.

**Example 11.2**

Solve:

$$\begin{cases} x + 2y = 9, & ① \\ 3x + 2y = 31. & ② \end{cases}$$

**Solution:**

Subtracting:

$$② - ① \Rightarrow 2x = 22,$$
$$\Rightarrow x = 11.$$

Now, to find the corresponding value of $y$, equation ① is simpler than ②, so it would be sensible to choose that one.

$$\text{Substituting } x = 11 \text{ into } ① \Rightarrow 11 + 2y = 9,$$
$$\Rightarrow 2y = -2,$$
$$\Rightarrow y = -1.$$

Thus our solution is

$$(x, y) = (11, -1).$$

Again, equations ① and ② can be represented as straight lines on a graph. Plotting these below, we see that the point of intersection $(x, y) = (11, -1)$ coincides with the solution of the set of simultaneous equations.

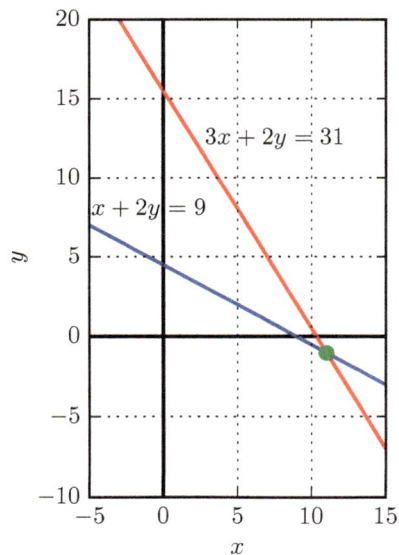

In some situations, if we want to eliminate a variable, both equations may need multiplying by a constant to facilitate this.

---

### Example 11.3

Find the solution pair of the simultaneous equations given below.

$$\begin{cases} 3x + 2y = 13, & \text{①} \\ 5x + 11y = 37. & \text{②} \end{cases}$$

**Solution:**

To eliminate one of the variables here, we must first multiply each equation by different values. These values are chosen to make the coefficients of one of the variables equal. Suppose we wish to eliminate the $x$ variable, then multiplying ① by 5 and ② by 3 will result in the coefficient of $x$ being 15 in each case, thus facilitating elimination. So we consider the following pair of equations,

$$\text{① } \times 5 \Rightarrow \quad 15x + 10y = 65, \quad \text{③}$$
$$\text{② } \times 3 \Rightarrow \quad 15x + 33y = 111. \quad \text{④}$$

Subtracting ③ from ④,

$$23y = 46,$$
$$\Rightarrow \quad y = 2.$$

Substituting this value for $y$ into ① (which we have chosen as it is the simplest equation to work with) gives

$$3x + 2y = 13,$$
$$\Rightarrow \quad 3x + 4 = 13,$$
$$\Rightarrow \quad 3x = 9,$$
$$\Rightarrow \quad x = 3.$$

Thus our solution is

$$(x, y) = (3, 2).$$

As before, equations ① and ② can be represented as straight lines on a graph. Plotting these below, we see again that the point of intersection $(x, y) = (3, 2)$ coincides with the solution of the set of simultaneous equations.

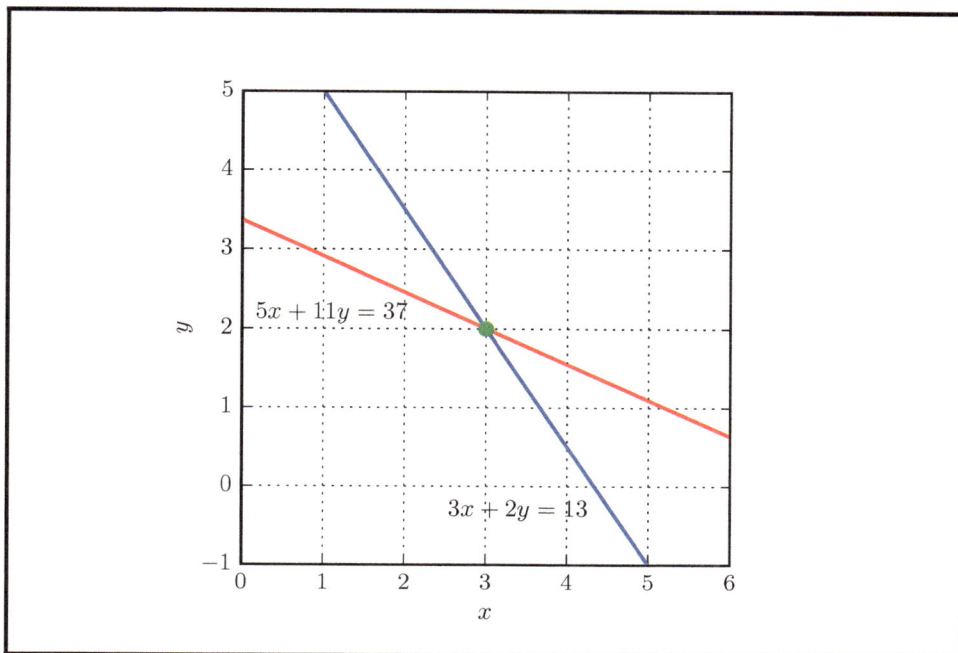

**Example 11.4**

Find the solutions of

$$\begin{cases} xy = -12 & \text{①} \\ \dfrac{x}{y} = -3. & \text{②} \end{cases}$$

**Solution:**

Eliminating a variable can also be achieved by multiplication or division.

$$\text{Multiplying ① and ②} \Rightarrow \quad xy \cdot \frac{x}{y} = (-12)(-3),$$

$$\Rightarrow \quad x^2 = 36,$$

$$\Rightarrow \quad x = \pm 6.$$

Here we will need to find the $y$ value corresponding to $x = 6$ and another $y$ value corresponding to $x = -6$.

$$\text{Substituting } x = 6 \text{ into ①} \Rightarrow \quad 6y = -12,$$

$$\Rightarrow \quad y = -2.$$

$$\text{Substituting } x = -6 \text{ into ①} \Rightarrow \quad -6y = -12,$$

$$\Rightarrow \quad y = 2.$$

Thus, our solutions are:

$$(x, y) = (6, -2) \qquad \text{and} \qquad (x, y) = (-6, 2).$$

In this situation, equation ① can not be represented by a straight line graph, in fact it has the characteristic *hyperbolic* shape. By rearranging equation ② so that $y = -x/3$, it is possible to see that this can be represented by a straight line graph. The equations are plotted below. Notice that, because the original equation ② is not well-defined when $y = 0$, the point $(x, y) = (0, 0)$ must be removed from the line. This is done by drawing an unfilled circle at the point that must be excluded. We see that there are now two points of intersection at $(x, y) = (6, -2)$ and at $(x, y) = (-6, 2)$, which again correspond to the solutions of the simultaneous equations.

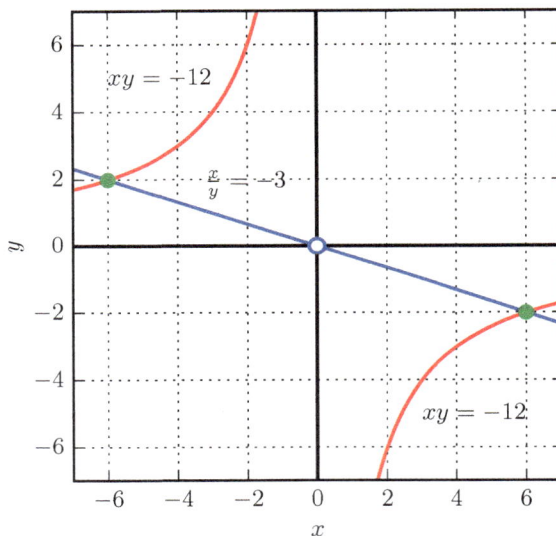

---

**Example 11.5**

Find the solutions of the following equations

$$\begin{cases} xy^2 = 20, & ① \\ xy^4 = 80. & ② \end{cases}$$

**Solution:**
An efficient technique here is to use division to eliminate $x$:

$$② \div ① \Rightarrow \quad \frac{xy^4}{xy^2} = \frac{80}{20},$$
$$\Rightarrow \quad y^2 = 4,$$
$$\Rightarrow \quad y = \pm 2.$$

Now to find the $x$ values:

$$\text{Substituting } y = 2 \text{ into } ① \Rightarrow \quad x(2)^2 = 20,$$
$$\Rightarrow \quad 4x = 20,$$
$$\Rightarrow \quad x = 5.$$

$$\text{Substituting } y = -2 \text{ into } ① \Rightarrow \quad x(-2)^2 = 20,$$
$$\Rightarrow \quad 4x = 20,$$
$$\Rightarrow \quad x = 5.$$

Our solutions are:

$$(x, y) = (5, 2) \qquad \text{and} \qquad (x, y) = (5, -2).$$

A graph showing the equations is below. Notice that, for each equation there are two branches, due to the quadratic and quartic terms in the equations. As we now expect, the solutions of the two equations are the points of intersection of the graphs.

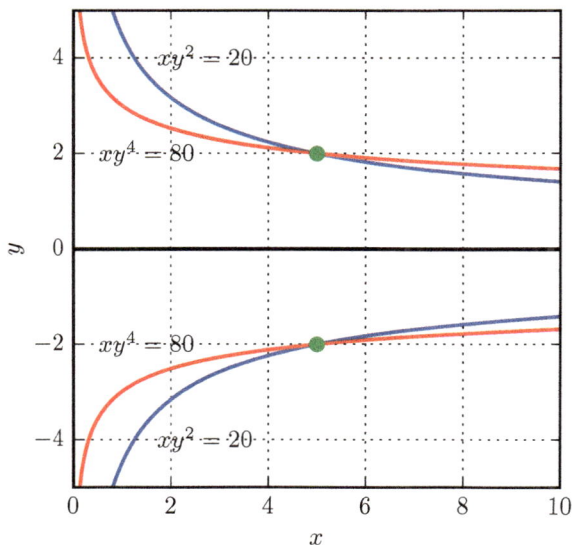

---

**Interactive Activity 11.1 — Deciding on an Elimination Approach**

The decision of how to eliminate a variable is important. Practise this aspect of the solution process using the interactive activity below.

---

### 11.1.1  Solving by Substitution

It is not always easy to see how to directly eliminate one of the variables from simultaneous equations. The alternative approach is to use substitution. We will look at these two examples:

$$\begin{cases} y = x + 1, \\ x^2 + y^2 = 5, \end{cases} \qquad \begin{cases} x - y = 2, \\ xy = 4. \end{cases}$$

---

**Tip**

It is always a good idea to sketch graphs whose equations have been given. We can then use these sketches to check if our final answers are sensible.

---

**Example 11.6**

Find the solutions of the following equations

$$\begin{cases} y = x + 1, & \text{①} \\ x^2 + y^2 = 5. & \text{②} \end{cases}$$

**Solution:**

Equation ① gives us an expression for $y$ in terms of $x$. We can *substitute* this expression into the second equation, replacing $y$ with the expression $x + 1$:

$$\begin{aligned} \text{Substituting ① into ②} \Rightarrow \quad & x^2 + (x+1)^2 = 5, \\ \Rightarrow \quad & x^2 + x^2 + 2x + 1 = 5, \\ \Rightarrow \quad & 2x^2 + 2x - 4 = 0, \\ \Rightarrow \quad & x^2 + x - 2 = 0, \\ \Rightarrow \quad & (x+2)(x-1) = 0, \end{aligned}$$

from which we deduce

$$x = -2 \text{ or } x = 1.$$

Now to find the $y$ values:

$$\begin{aligned} \text{Substituting } x = -2 \text{ into ①} \Rightarrow y = -2 + 1, \\ \Rightarrow y = -1. \end{aligned}$$

Substituting $x = 1$ into ① $\Rightarrow y = 1 + 1$,
$$\Rightarrow y = 2.$$

Therefore, the solutions are

$$(x, y) = (-2, -1) \qquad \text{and} \qquad (x, y) = (1, 2).$$

The plot below shows the graphs of the equation. Equation ① represents a straight line, while equation ② represents a circle of radius $\sqrt{5}$ centred at the origin. In this case, the straight line intersects the circle at two points, which we found above. There is a maximum of two intersections between a circle and straight line, but it is possible there could be no intersections or just one.

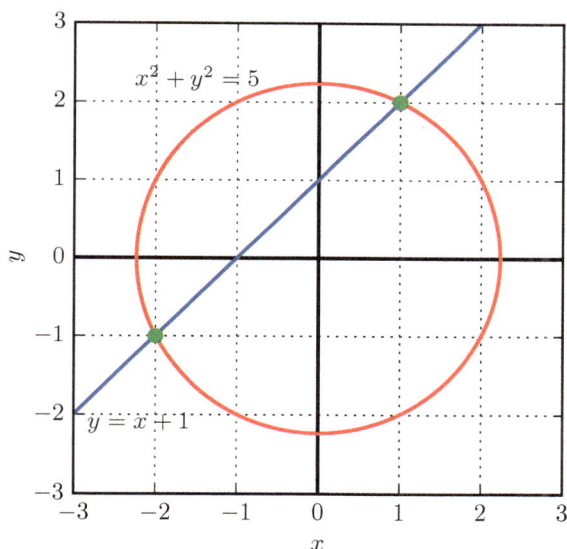

**Example 11.7**

Find the solutions of the following equations

$$\begin{cases} x - y = 2, & \text{①} \\ xy = 4. & \text{②} \end{cases}$$

**Solution:**

To begin the process of substitution here, we need to rearrange one of the equations so that either $x$ or $y$ is the subject. This is a point in the process where we face a *choice* and we should choose on the basis of making our lives easier. Rearranging equation ① to make $x$ the subject avoids any negative signs and fractions:

$$① \Rightarrow x = y + 2$$

We can now proceed as before by substituting this expression:

$$\text{Substituting into } ② \Rightarrow \quad (y+2)y = 4,$$
$$\Rightarrow \quad y^2 + 2y = 4,$$
$$\Rightarrow y^2 + 2y - 4 = 0.$$

As this quadratic cannot be factorised, we could use the quadratic formula or complete the square:

$$y^2 + 2y - 4 = 0$$
$$\Rightarrow (y+1)^2 - 5 = 0$$
$$\Rightarrow \qquad y + 1 = \pm\sqrt{5}$$
$$\Rightarrow \qquad y = -1 \pm \sqrt{5}.$$

Returning to the rearranged form of equation ①, $x = y+2$, we can find the corresponding values of $x$:

$$y = -1 + \sqrt{5} \quad \Rightarrow \quad x = 1 + \sqrt{5}$$
$$y = -1 - \sqrt{5} \quad \Rightarrow \quad x = 1 - \sqrt{5}.$$

A plot showing the graphs of the two equations and the two intesection points is below.

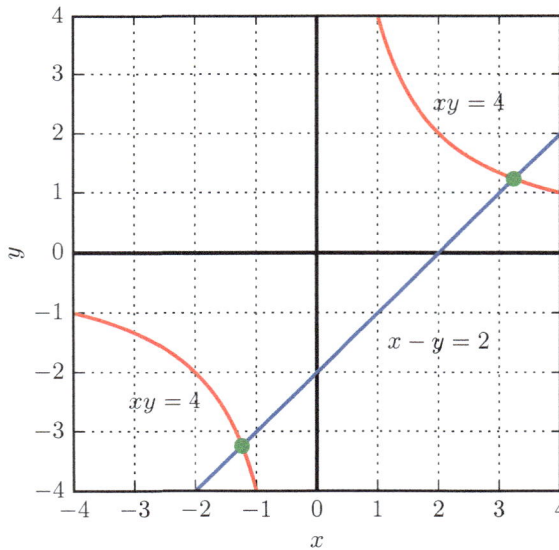

To summarise, we have the solutions:

$$(x, y) = (1 + \sqrt{5}, -1 + \sqrt{5}) \quad \text{and} \quad (x, y) = (1 - \sqrt{5}, -1 - \sqrt{5}).$$

**Exercise 11.1**

Q1. Solve the following pairs of simultaneous equations by elimination.

(a)
$$2x + 2y = 14,$$
$$2x + 3y = 18.$$

(b)
$$3x + 4y = 21,$$
$$7x + 4y = 10.$$

(c)
$$7x + 3y = 21,$$
$$7x + 6y = 30.$$

(d)
$$3t + 4p = 56,$$
$$t + 4p = 32.$$

Q2. Solve the following pairs of simultaneous equations by elimination.

(a)
$$2x + y = 11,$$
$$4x - y = 19.$$

(b)
$$-x + 5y = 12,$$
$$x + 3y = 20.$$

(c)
$$3x - 2y = 15,$$
$$5x + 2y = 22.$$

(d)
$$4x - 5y = 33,$$
$$3x + 5y = 56.$$

Q3. Solve the following pairs of simultaneous equations by elimination.

(a)
$$xy = 16,$$
$$\frac{x}{y} = 4.$$

(b)
$$xy = 20,$$
$$\frac{x}{y} = 5.$$

(c)
$$2xy = 30,$$
$$\frac{x}{y} = 8.$$

(d)
$$xy = \frac{10}{4},$$
$$\frac{x}{y} = 4.$$

Q4. Solve the following pairs of simultaneous equations by elimination.

(a)
$$xy^2 = 12,$$
$$xy^4 = 48.$$

(b)
$$xy^2 = 45,$$
$$xy^4 = 405.$$

(c)
$$xy^2 = 12,$$
$$xy^4 = 54.$$

(d)
$$y^3x = 8,$$
$$y^5x = 32.$$

Q5. Solve the following pairs of simultaneous equations by elimination.

(a)
$$x + 2y = 8,$$
$$3x + 3y = 18.$$

(b)
$$3p - q = 12,$$
$$4p + 3q = 29.$$

(c)
$$4x + 2y = 21,$$
$$5x - y = 18.$$

(d)
$$7x - 2y = 47,$$
$$8x + y = 68.$$

Q6. A brake calliper costs three times as much as a brake cable. Given the cost of buying one of each is £30, find the cost of each item.

Q7. Solve the following pairs of simultaneous equations.

(a)
$$y = x,$$
$$x^2 + y^2 = 9.$$

(b)
$$y = 3x + 2,$$
$$x^2 + y^2 = 16.$$

(c)
$$y = x + 1,$$
$$x^2 + y^2 = 9.$$

(d)
$$x^2 + y^2 = 4,$$
$$y = x - 1.$$

Q8. In a department store, side plates cost half as much as a dinner plate. If an order for 12 side plates and 6 dinner plates comes to £170.40, how much does each item cost?

Q9. Given that the triangle shown below has a perimeter of 30 units and is isosceles, find the values of $x$ and $y$.

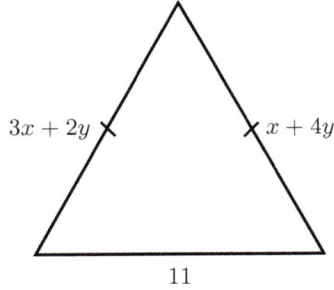

## 11.2 Tangents

This is an important section conceptually, as it connects simultaneous equations to the theory of the discriminant and repeated roots from Chapter 5.

**Activity 11.2**

Solve each of these sets of simultaneous equations.

$$\begin{cases} y = x + 7, \\ y = x^2 + 3x + 4, \end{cases} \quad \begin{cases} y = x + 3, \\ y = x^2 + 3x + 4, \end{cases} \quad \begin{cases} y = x - 1, \\ y = x^2 + 3x + 4. \end{cases}$$

What can you deduce about the graphs of these functions in each case?

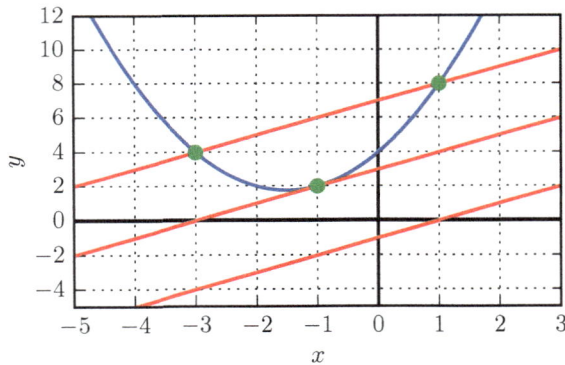

Figure 11.3: The three possible cases when finding intersections between a parabola and a straight line: two distinct points, a single point of tangency, or no intersections.

Figure 11.3 shows the parabola and the three corresponding straight lines. How does this diagram reflect the algebra encountered in Activity 11.2?

As we have seen from Activity 11.2, different situations can arise when finding intersections between a parabola and a straight line:

- *Distinct solutions* of simultaneous solutions correspond to distinct points of intersection of the graphs.
- A *repeated solution* from simultaneous equations corresponds to a point of tangency between the two graphs.
- If the simultaneous equations have *no real solutions*, then those graphs do not intersect.

**Activity 11.3**

By sketching appropriate curves, answer the following questions.

Q1. Show that exactly the same three possibilities from Activity 11.2 arise when finding intersections between a circle and a straight line.

Q2. Show that there are potentially 0, 1, 2, 3, or 4 points of intersection between a circle and a parabola.

Q3. Determine the possible numbers of points of intersection between a circle and a general cubic curve.

**Interactive Activity 11.2 — A Circle and a Parabola**

Investigate further the intersections in Question 2 of the above activity using the Geogebra applet below.

**Interactive Activity 11.3 — A Circle and a Cubic**

Investigate further the intersections in Question 3 of the above activity using the Geogebra applet below.

**Exercise 11.2**

Q1. Solve the simultaneous equations

$$\begin{cases} y = x^2 + 1, \\ y = -(x - 2)^2 + 3. \end{cases}$$

Hence, draw a sketch of the two curves, indicating any points of intersection.

Q2. The diagram below shows the following two graphs:

$$y = 12x^2 - 15x + 11 \quad \text{and} \quad y = 10x - 2.$$

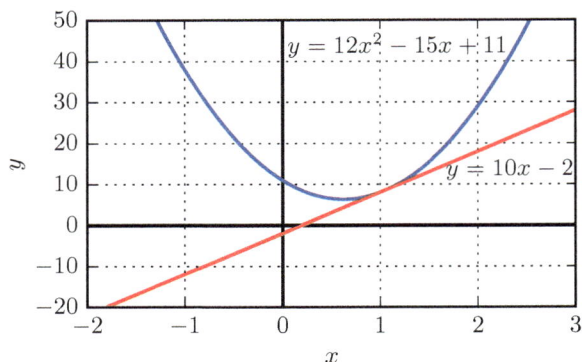

How many points of intersection do there appear to be between the two graphs? Solve the simultaneous equations to find the exact coordinates of any points of intersection.

Q3. Solve the simultaneous equations

$$\begin{cases} \dfrac{1}{x} + \dfrac{1}{y} = 1, \\ x^2 + y^2 = 8. \end{cases}$$

Q4. The perimeter of a rectangle is $12\,\mathrm{cm}$ and its area is $(6 - \sqrt{8})\,\mathrm{cm}^2$. Find its length and width.

Q5. In mechanics, the quantities $u$ and $a$ can represent the *initial velocity* and the *acceleration* of an object. Suppose we know the velocity of an object at two given times: after 2 seconds it is travelling at $10\,\mathrm{m\,s}^{-1}$ and after 3.5 seconds it is travelling at $16\,\mathrm{m\,s}^{-1}$. Solve the following pair of simultaneous equations to find the object's initial velocity and its acceleration:

$$\begin{cases} 10 = u + 2a, \\ 16 = u + 3.5a. \end{cases}$$

Q6. In statistics, the variables $\mu$ (mu) and $\sigma$ (sigma) represent the *mean* and *standard deviation* of a population. Here is a typical pair of simultaneous equations to solve:

$$\begin{cases} 1.2 = \dfrac{150 - \mu}{\sigma}, \\ -1.1 = \dfrac{120 - \mu}{\sigma}. \end{cases}$$

Find the values of $\mu$ and $\sigma$ correct to three significant figures. (We will see how these equations arise when we study the normal distribution.)

Q7. The diagram below shows three objects connected together by ropes. When released, the objects will begin to accelerate: $P$ will move downwards, $Q$ will move to the left, and $R$ will move upwards.

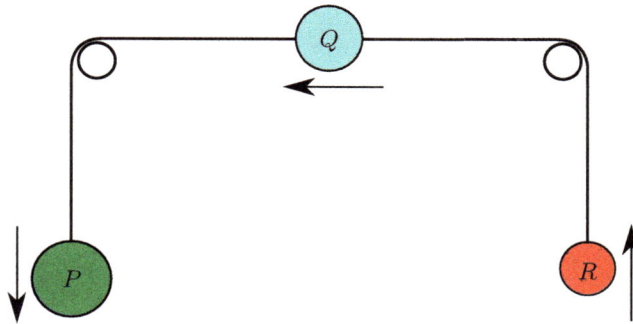

Below is a system of three simultaneous equations. The quantities $T_1$ and $T_2$ represent tension forces in the two different pieces of rope, and $a$ represents the acceleration of the objects.

$$98 - T_1 = 10a$$
$$T_1 - T_2 = 2a$$
$$T_2 - 49 = 5a.$$

Solve the equations to find the values of all the variables. (You will study the situations that produce these equations in Chapter 28.)

Q8.

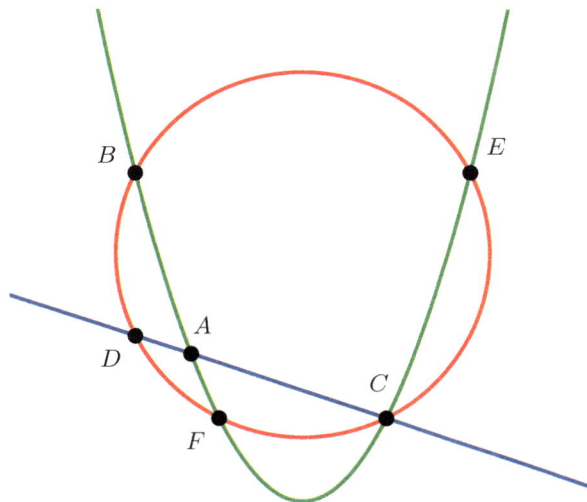

The diagram above shows, without axes, a configuration that comprises a circle, a parabola and a straight line. The equations for these graphs are

$$x + 3y + 1 = 0, \qquad (x-3)^2 + (y-2)^2 = 20, \qquad y = \frac{1}{2}(x-3)^2 - 4.$$

Find the exact coordinates of each of the points of intersection, labelled from A to F.

Q9. Solve this pair of simultaneous equations, giving $x$ and $y$ in terms of the constants $a$, $b$, $c$, $d$, $e$ and $f$.

$$\begin{cases} ax + by = e \\ cx + dy = f. \end{cases}$$

State any conditions on these constants for which solutions do not exist.

Q10. Following the notation of Exercise Q9., show that if $a$, $b$, $e$, $c$, $d$ and $f$ form an arithmetic sequence, then the solution is $(x, y) = (-1, 2)$.

Q11. Following the notation of Exercise Q9., show that if $a$, $b$, $e$, $c$, $d$ and $f$ are any six consecutive terms from the Fibonacci sequence, then the solution is $(x, y) = (1, 1)$.
*Hint:* The Fibonacci sequence begins 1, 1, 2, 3, 5, 8 ..., where each term is found by summing the two previous terms.

Q12. Show that finding the roots of the quadratic equation

$$ax^2 + bx + c = 0$$

is equivalent to finding the solutions of the simultaneous equations

$$\begin{cases} ax_1 + ax_2 = -b \\ ax_1 - ax_2 = -\Delta, \end{cases}$$

where $\Delta$ is the discriminant of the quadratic, $\Delta = b^2 - 4ac$.

Q13. Given that

$$x + y = 7 \qquad \text{and} \qquad xy = 2,$$

find the value of

$$\frac{1}{x} + \frac{1}{y}.$$

Q14. Find the coordinates of the points of intersection $A$ to $F$ shown in the diagram below.

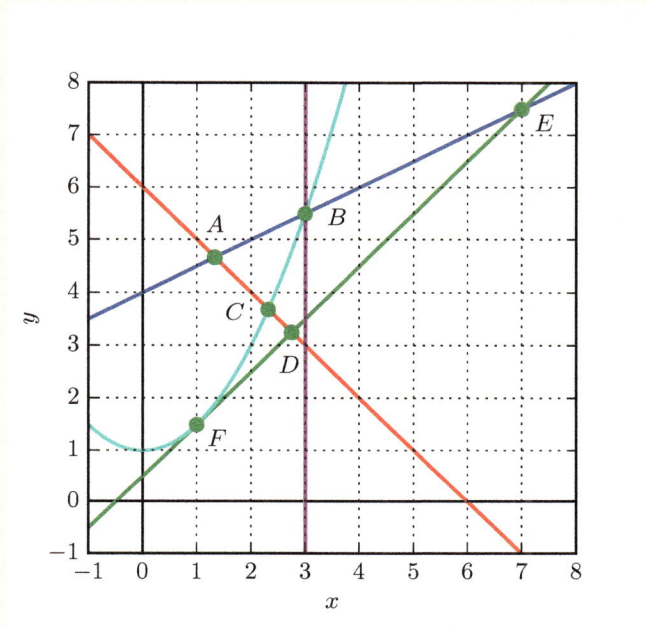

The green line has equation $2x - 2y + 1 = 0$ and the quadratic has equation $y = \frac{1}{2}x^2 + 1$.

Q15. The Highway Code lists the following stopping distances for different speeds, in miles per hour.

| Speed (mph) | Stopping Distance (m) |
|:---:|:---:|
| 20 | 12 |
| 30 | 23 |
| 40 | 36 |
| 50 | 53 |
| 60 | 73 |
| 70 | 96 |

It is suggested that it is suitable to model this data with a quadratic function of the form $y = ax^2 + bx + c$.

(a) Suggest why it could be argued that a quadratic model is suitable here.

(b) Find the values of the coefficients $a$, $b$ and $c$.

(c) Explain why the value for $c$ is unphysical.

**Key Facts — Simultaneous Equations**

- Simultaneous equations can be solved by either elimination or substitution.
- It is recommended to sketch the graphs of the equations we are solving so that we can check if our final answers are sensible.
- Solving a pair of simultaneous linear equations can result in 0 or 1 point of intersection.
- A quadratic function and a linear function may intersect twice, once or not at all.
- A quadratic function and a circle may intersect twice, once or not all.
- A cubic function and a circle may intersect 0, 1 , 2, 3, 4, 5, or 6 times.

**Chapter Assessment — Simultaneous Equations**

Download and sit the 30 minute assessment for this chapter from the digital book.

$$\frac{d \sin(x)}{dx}.$$

$$\Rightarrow$$

$$\Rightarrow$$

$$4(x-2) \leq 3(x+1),$$
$$4x - 8 \leq 3x + 3,$$
$$x \leq 11.$$

$$+ \frac{f(x)}{2}$$

$$\cos(x)$$

$$\sin^2(x)$$

# 12. Inequalities

Inequalities are used in a number of different areas, one such area is linear programming. This is a method of optimisation which may, for example, allow us to find the maximum profit a company can make from sales, or the minimum costs incurred for a company to produce a certain product.

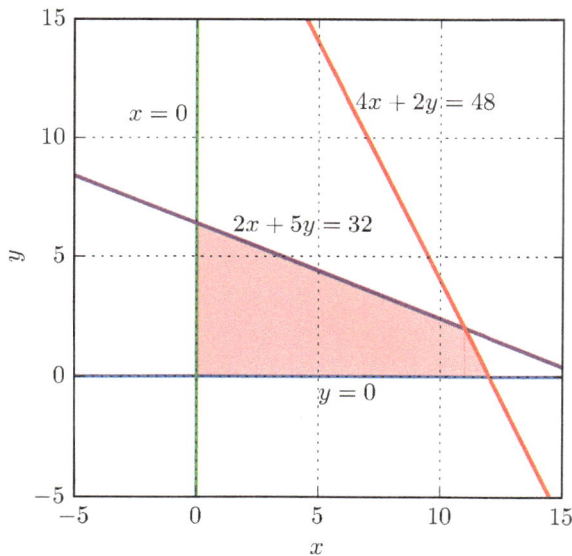

Figure 12.1: The linear programming problem for shoe type A and B.

Let us consider a small company that make two different types of shoe. Shoe type A takes 2 hours of machine time and 4 hours of manual labour, whilst shoe type B take 5 hours of machine time and 2 hours of manual labour. Each day there are 32 hours of machine time available and 48 hours of manual labour. The profit on each sale of shoe type A is £30 and £22 for each sale of shoe type B. This means we want to maximise the objective function $P = 30x + 22y$, where $x$ and $y$ are the number of sales of shoe type A and B, respectively. This is subject to the inequalities $2x + 5y \leq 32$, $4x + 2y \leq 48$, that arise from the machine time and manual labour available. For the problem to make sense, we also require $x \geq 0$ and $y \geq 0$. This results in the region seen in Figure 12.1.

The well known puzzle Sudoku can also be formulated using a system of inequalities as a linear programming problem.

In this chapter we will focus on solving inequalities and how they can be represented graphically.

## 12.1 Solving Inequalities Involving a Single Variable

In this section we begin by solving simple linear inequalities in one variable, before extending the technique to quadratic inequalities and simultaneous inequalities.

### 12.1.1 Linear Inequalities

A linear inequality is one where the power of any term involving the unknown variable is no greater than one. The solutions to a linear inequality are a set of values, and these can be represented using a number line. Figure 12.2 shows how the solutions to the inequality $x > 3$ can be represented on a number line. Note that an 'unfilled' circle is used to indicate that the value 3 is not included in the solutions.

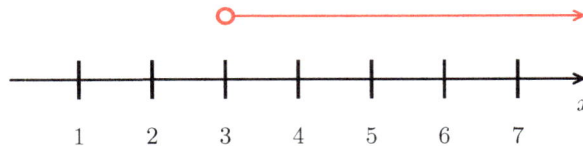

Figure 12.2: The solutions $x > 3$ represented on the number line.

If instead we have $x \geq 3$ we use a solid circle to indicate that the value 3 is included, as shown in Figure 12.3.

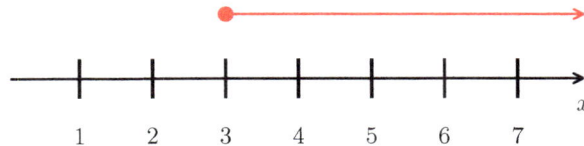

Figure 12.3: The solutions $x \geq 3$ represented on the number line.

**Example 12.1**

Solve the inequality $3x - 5 > 7$.

**Solution:**

We rearrange the inequality to make $x$ the subject, in a similar fashion to when solving linear equations. Hence,

$$3x - 5 > 7,$$
$$\Rightarrow \quad 3x > 12,$$
$$\Rightarrow \quad x > 4.$$

The set of solutions which satisfy the inequality is therefore $\{x : x > 4\}$.

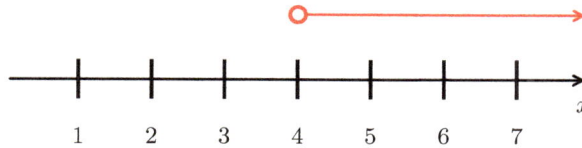

**Remark**

Recall that the solution $\{x : x > 4\}$ is read as '$x$ such that $x$ is greater than 4'.

**Example 12.2**

Solve the inequality $4(x - 2) \leq 3(x + 1)$.

**Solution:**

We rearrange the inequality as follows.

$$4(x - 2) \leq 3(x + 1),$$
$$\Rightarrow \quad 4x - 8 \leq 3x + 3,$$
$$\Rightarrow \quad x \leq 11.$$

The set of solutions which satisfy the inequality is therefore $\{x : x \leq 11\}$.

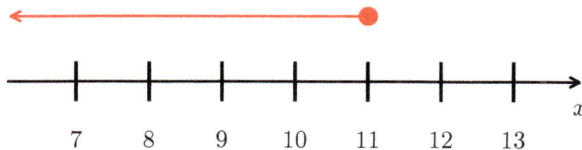

**Example 12.3**

Find the set of values of $x$ which satisfy $-1 < 2x + 7 \leq 8$.

**Solution:**

In this question we have two inequalities to solve. These can be solved together by applying the same operations to all expressions in the inequality. Hence,

$$-1 < 2x + 7 \leq 8,$$
$$\Rightarrow \quad -8 < \quad 2x \quad \leq 1,$$
$$\Rightarrow \quad -4 < \quad x \quad \leq \frac{1}{2}.$$

The set of solutions which satisfy the inequality is therefore $\{x : 4 < x \leq \frac{1}{2}\}$.

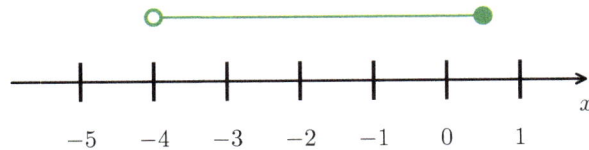

**Remark**

When working with inequalities it is important to be careful when multiplying or dividing by negative numbers.

Considering the true statement $3 < 7$, if we multiply both sides by $-1$ this results in the incorrect statement $-3 < -7$. The implication when multiplying by $-1$ is in fact that $-3 > -7$.

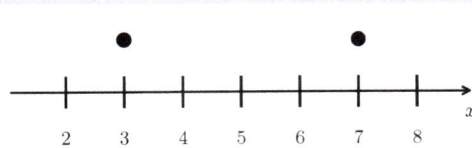

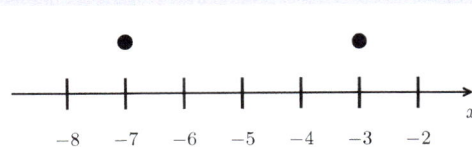

Therefore, when multiplying or dividing by negative numbers we must be careful to change the direction of the inequality.

**Exercise 12.1**

Q1. Find the set of values for $x$ which satisfy:
  (a) $4x - 5 \geq 15$;
  (b) $x + 5 < 3x + 2$;
  (c) $7 - x \leq 2x - 5$;
  (d) $5 - 2x \geq 3 - x$.

Q2. Find the solutions to the following inequalities:
  (a) $3(x - 4) > 8$;
  (b) $5(2 - x) + 2(3x + 2) > 13$;
  (c) $3x - 6(2x - 3) \leq 2(x - 2)$;
  (d) $2(x - 3) + 4(1 - x) < 3(2x + 8) + 2(x - 3)$.

Q3. Solve the inequalities given below:
  (a) $\frac{3}{4}x - 1 > 5$;
  (b) $\frac{2x+1}{2} \geq \frac{3x-1}{2}$;
  (c) $\frac{1}{3}(2x + 1) > \frac{2}{5}(1 - x)$;
  (d) $\frac{x+2}{4} < \frac{1}{3}(3x - 2)$.

Q4. Find the set of $x$ values for which:
  (a) $-2 < \frac{1}{3}x - 2 \leq 6$;
  (b) $4 \leq \frac{1}{2}(3x + 1) < 11$;
  (c) $0 < \frac{x+9}{2} + 2(x + 1) < 7$.

## 12.1.2 Quadratic Inequalities

A quadratic inequality is one where the power of one of the variables is two, but no higher. The set of solutions to a quadratic inequality can be found by finding the roots of the associated quadratic equation. When solving quadratic inequalities it is helpful to draw a sketch of the quadratic to help identify the set, or sets, of solutions.

**Example 12.4**

Solve the inequality $x^2 - 11x + 33 > 9$.

**Solution:**

To solve this inequality, we first rearrange. Subtracting 9 from both sides of the inequality gives

$$x^2 - 11x + 24 > 0.$$

This is a factorisable quadratic which gives

$$(x - 8)(x - 3) > 0.$$

We can sketch this quadratic using the methods described in Chapter 5, as shown below.

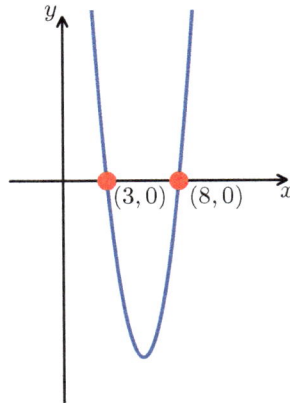

The set of solutions to the inequality are given when this quadratic is greater than 0. Therefore the set of values for $x$ which satisfies this inequality is the union of two disjoint sets; $\{x : x < 3\} \cup \{x : x > 8\}$.

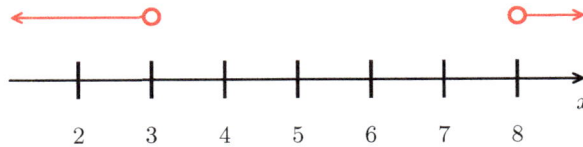

---

**Remark**

Note that, when multiplying by an unknown value, we must take into consideration that the unknown could be negative. In Example 12.5, we will be multiplying by $x^2$, which will be non-negative for all values of $x$.

---

**Example 12.5**

Solve the inequality $3 < \frac{7x+6}{x^2}$.

**Solution:**

To solve this inequality, we must first consider what will happen when we multiply throughout by the unknown value $x^2$. For all values of $x$ this will remain non-negative, therefore,

$$3 < \frac{7x+6}{x^2},$$
$$\Rightarrow \qquad 3x^2 < 7x + 6,$$
$$\Rightarrow \quad 3x^2 - 7x - 6 < 0.$$

This quadratic factorises to give,

$$(3x + 2)(x - 3) < 0.$$

We can sketch this quadratic:

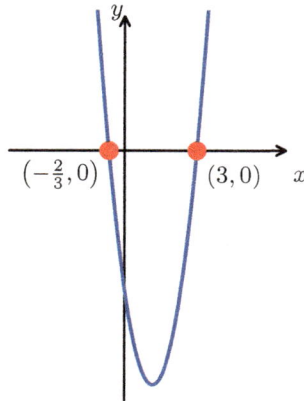

The set of solutions to the inequality are given when this quadratic is less than 0. Therefore the set of values for $x$ which satisfies this inequality is $\{x : -\frac{2}{3} < x < 3\}$.

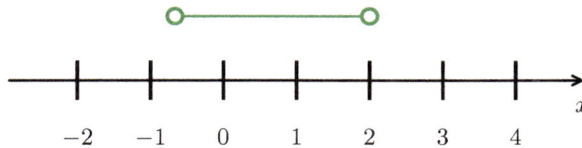

---

**Exercise 12.2**

Q1. Find the solutions to the following inequalities:
 (a) $x^2 - 4 \leq 21$;
 (b) $x^2 + 8 > 17$;
 (c) $10 - x^2 > -6$;
 (d) $23 - 2x^2 \geq 5$.

Q2. Solve the inequalities given below:
 (a) $x^2 - 7x - 18 < 0$;
 (b) $x^2 + 3x - 4 \leq 6$;
 (c) $x(x + 9) \geq 10$;
 (d) $\frac{7x-3}{x^2} < 4$.

### 12.1.3 Simultaneous Inequalities

Sometimes we need to find a set of values which satisfies more than one inequality. In order to do this, we find the solutions to each inequality individually and then consider the set of values which satisfy all of the inequalities.

---

**Example 12.6**

Find the set of values for which:

$$\frac{1}{3}(2x+4) < 2(x-2) \tag{12.1}$$

and

$$4(1+2x) > 7x. \tag{12.2}$$

**Solution:**

First solve (12.1). By algebraic manipulation,

$$\frac{1}{3}(2x+4) < 2(x-2),$$
$$\Rightarrow \quad 2x+4 < 6(x-2),$$
$$\Rightarrow \quad 2+4 < 6x-12,$$
$$\Rightarrow \quad 16 < 4x,$$
$$\Rightarrow \quad x > 4.$$

Therefore, the set of solutions which satisfies (12.1) will be $\{x : x > 4\}$.

Next, we look for the solutions to (12.2). By a similar approach,

$$4(1+2x) > 7x,$$
$$\Rightarrow \quad 4+8x > 7x,$$
$$\Rightarrow \quad x > -4.$$

Therefore, the set of solutions which satisfies (12.2) will be $\{x : x > -4\}$.

In the number line below, the solutions to (12.1) and (12.2) are represented by the red and green arrows, respectively. We can see from this graphical representation that the set of values which will satisfy both inequalities is for $x > 4$. Therefore, the set of values for which (12.1) and (12.2) are true is $\{x : x > 4\}$.

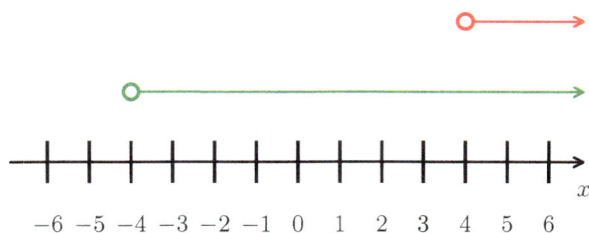

**Example 12.7**

Find the set of values which satisfy simultaneously both of the following inequalities:

$$x(2x - 5) \leq 3 \tag{12.3}$$

and

$$\frac{5}{2}x \geq 5. \tag{12.4}$$

**Solution:**

Using algebraic manipulation to find the solutions to (12.3) we have

$$x(2x - 5) \leq 3,$$
$$\Rightarrow \quad 2x^2 - 5x \leq 3,$$
$$\Rightarrow \quad 2x^2 - 5x - 3 \leq 0.$$

This factorises to give

$$(2x + 1)(x - 3) \leq 0.$$

We sketch the quadratic:

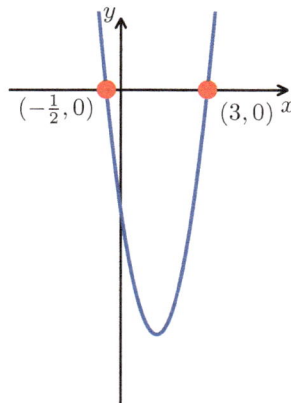

The set of solutions to this inequality is when the quadratic is less than or equal to 0. Therefore the set of values for $x$ which satisfies the inequality (12.3) is $-\frac{1}{2} \leq x \leq 3$. Next we must consider the set of values which satisfies (12.4). By manipulation,

$$\frac{5}{2}x \geq 5,$$
$$\Rightarrow \quad 5x \geq 10,$$
$$\Rightarrow \quad x \geq 2.$$

Therefore, the set of solutions for the inequality (12.4) is $x \geq 2$ .

In the number line below, the solutions to (12.3) and (12.4) are represented by the green and red lines, respectively.

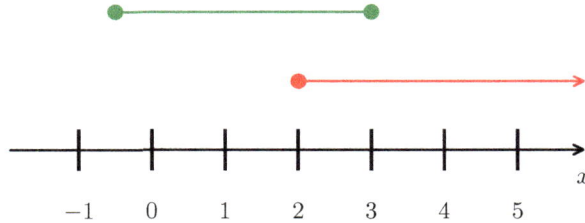

We can see from this graphical representation that the set of values which will satisfy both inequalities will be $\{x : 2 \leq x \leq 3\}$.

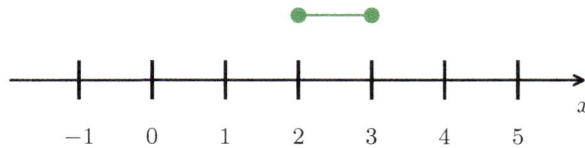

**Exercise 12.3**

Q1. Find the set of values which simultaneously satisfy the following pairs of inequalities:

(a) $1 - x \leq 2x + 4$ and $2(2x - 3) < 2$.

(b) $4(x - 1) > \frac{1}{2}x$ and $x \geq \frac{2}{3}(5 - x)$;

(c) $\frac{5}{3}x > 10$ and $3(1 - x) \leq 2(3x + 1)$.

Q2. Find the set of values for which:

(a) $x^2 - 3x > 4$ and $\frac{2}{3}x < 2$;

(b) $x(x - 5) + 9 < 3$ and $x^2 > 2(4 - x)$;

(c) $\frac{1}{2}x(x + 4) < \frac{2}{3}(4x + 1)$ and $x^2 + 4x > x$.

## 12.2 Multi-variable Inequalities

When we are presented with inequalities involving two variables, these can be represent by a region on a graph. If we consider the inequality $y \geq x + 1$ (as seen in Example 12.8), all coordinates $(x, y)$ in the shaded region will satisfy the inequality, and therefore represents the possible solutions.

**Example 12.8**

Show the region which is represented by the inequality $y \geq x + 1$.

**Solution:**

To find the region of values which satisfy this inequality, we must first draw the boundary, which is indicated as $y = x + 1$. We plot this as a solid line, since the '$\geq$' symbol indicates points on this boundary are included.

We should then consider whether points in the region above or below this boundary satisfy the given inequality. Let us consider the point $(1, 1)$ (which in this case lies below the boundary). When substituting these values into the inequality we obtain the invalid result $1 \geq 2$. Therefore, the point $(1, 1)$ does not lie in the region represented by this inequality. Thus, the region which requires shading is above the boundary. We should check a point above the boundary for completeness. For the point $(1, 3)$, we have $3 > 2$, which satisfies the inequality.

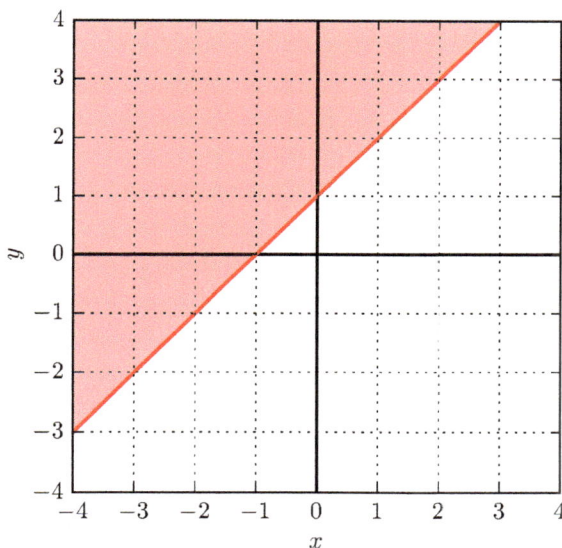

**Remark**

When we have a strict inequality, we use a dashed line to indicate that the points on the boundary are not included in the region. If in Example 12.8, we instead had $y > x + 1$, the plot below would have been obtained.

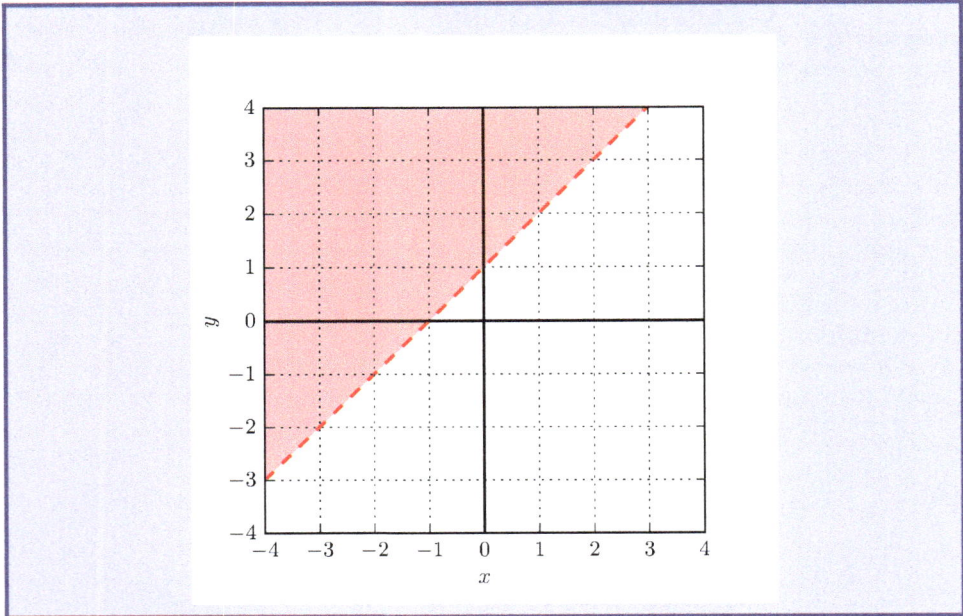

**Example 12.9**

Indicate, using a graphical representation, the region such that $y < x^2 + x - 2$.

**Solution:**

As in Example 12.8, we first draw the boundary, which in this case is $y = x^2 + x - 2$.

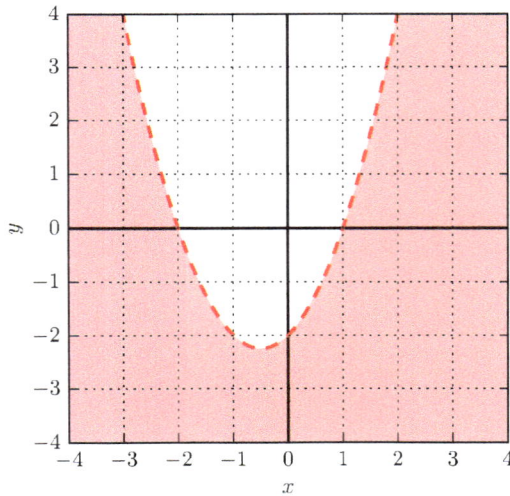

We then consider any point which does not lie on the boundary. Let us consider the point $(1, 1)$ (which in this case lies above the boundary). When substituting these values into the equality we obtain $1 < 0$, which is invalid. Therefore, $(1, 1)$ does not lie in the region represented by the inequality $y < x^2 + x - 2$. Thus, the region which requires shading is below the boundary as shown above.

**Example 12.10**

Show the region of solutions to the inequalities

$$y \leq 2x + 1,$$
$$y \geq x^2.$$

**Solution:**

When representing regions where more than one inequality is concerned, we must ensure that all boundaries are drawn before we begin to consider regions. The boundaries in this case are $y = 2x + 1$ and $y = x^2$.

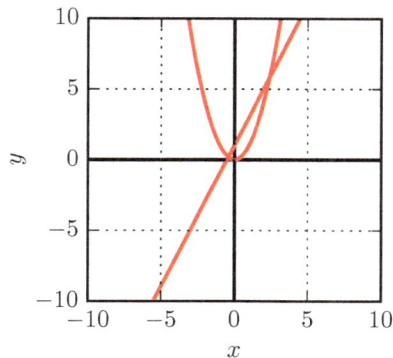

From our plot, we can see there are five possible regions which could satisfy the two inequalities. Whilst we could consider substituting points from each region into each inequality, this would be a long and laborious process. Instead, we consider what each inequality tells us about their solutions. The inequality $y \leq 2x + 1$ indicates that we are looking for all values of $y$ which are less than or equal to $2x + 1$. In other words we are considering the region of values *below* the line $y = 2x + 1$.

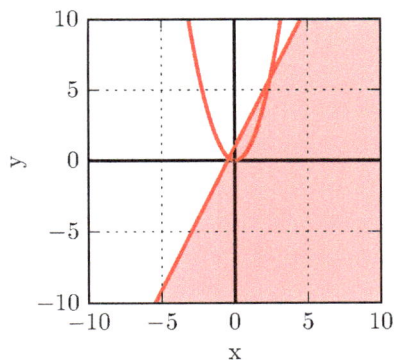

Additionally, we want the values which satisfy the inequality $y \geq x^2$, so all values of $y$ which are greater than or equal to $x^2$. In other words we want the region of values *above* the curve $x^2$, which also lie in the shaded region.

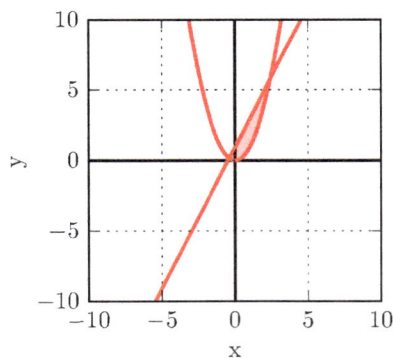

**Interactive Activity 12.1 — Testing the Regions**

The Geogebra interactive below (in the digital book) can be used to test which region should be shaded when solving inequalities in two variables. Up to three inequalities can be entered, and the point then moved using the sliders.

**Exercise 12.4**

Q1. Represent, graphically, the regions which satisfy the inequalities below:
   - (a) $y > 3x - 6$;
   - (b) $y \leq 3 - x$;
   - (c) $y \geq 1 - 2x$.

Q2. Show the region of solutions to the inequality:
   - (a) $y \geq x^2 - 10x + 24$;
   - (b) $y < x(3 - x)$;
   - (c) $y \leq x^2 - 3x + 7$.

Q3. Plot the region which represents the solutions to the pair of inequalities below:
   - (a) $y \leq 3x$ and $y > -x - 2$;
   - (b) $y > x^2 - 3x - 5$ and $y < 1 - x$
   - (c) $y \geq x^2 + 3$ and $y < 5 + 2x - x^2$.

**Learning Resource 12.1 — Solving Inequalities Card Sort**

Download the attached card sorting activity from the digital book to practise solving inequalities.

**Exercise 12.5**

Q1. Suggest a combination of inequalities which will have a set of solutions represented by the region below.

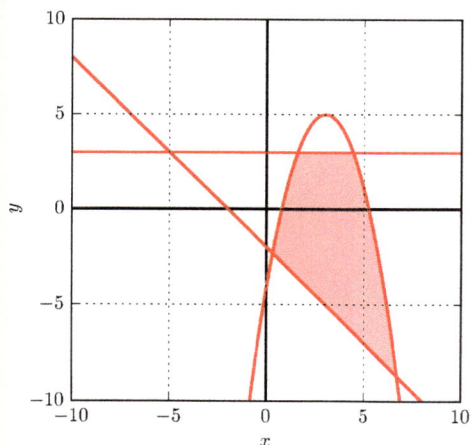

Q2. Indicate which set of solutions below would satisfy the inequalities $3(1 - x) < 2$
   and $x^2 - 4x - 5 \geq 0$.
   (a) $x > \frac{1}{3}$;
   (b) $-1 \leq x \leq 5$;
   (c) $x < -1$;
   (d) $x > 8$.

Q3. A publisher is trying to decide on the size of a book. The book can be any height
   where the width is 6cm less than it's height. Given that the perimeter of the front
   cover should not exceed 60cm,
   (a) determine and solve an inequality involving $x$.
   Given also that the book has a thickness of 2cm and the surface area of the front
   cover must be at least 160cm,
   (b) determine and solve second inequality involving $x$.
   (c) Suggest the set of values for $x$ which will satisfy both inequalities.

---

**Key Facts — Inequalities**

- When representing the solutions of inequalities on a number line we use an unfilled
  circle if the inequality is strict to indicate that the value is not included in the
  solution. Conversely, if the inequality is not strict we use a filled circle to indicate
  that the value is included in the solution.
- To solve inequalities we can proceed similarly to solving equations.
- If solving a set of simultaneous inequalities, solve each inequality separately and
  represent their solutions on a single number line. The solution of the simultaneous
  inequalities is then the region on the number line where both solutions exist.
- For multivariable inequalities we plot the boundary of the inequality and then test
  a point either side of this boundary to decide which region should be shaded.

---

**Chapter Assessment — Inequalities**

Download and sit the 30 minute assessment for this chapter from the digital book.

$\sin(x)$

$$x^2 + y^2 = r^2.$$

$$\underbrace{\phantom{xxxx}}_{} = 0$$

$d\sin$

$+ \dfrac{f(x}{2}$

$\cos(x)$

$\sin^2$

# 13. Geometry of the Circle

Circles have fascinated humankind, especially mathematicians, for millennia. Perhaps this is because the sun and moon appear as circles in the sky, even though we now know they are not. Perhaps it is because wheels are circular and allow for smooth transport. Perhaps it is because of the relationship between the area/circumference of the circle and $\pi$ and in turn to the links with trigonometry. Whatever the reason, understanding the geometry of circles is as important today as it always has been.

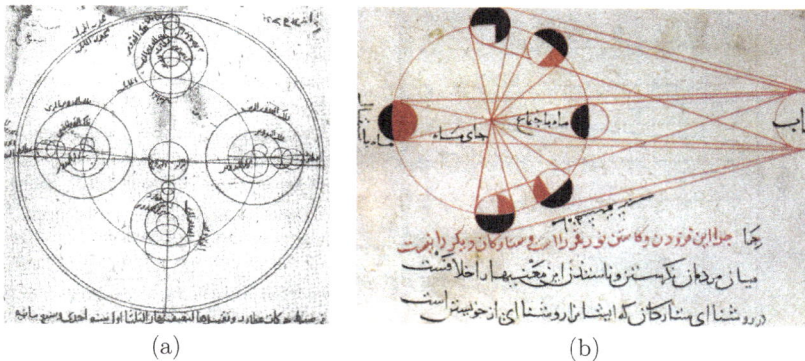

(a)                                   (b)

Figure 13.1: Circular geometry in use in ancient Arabic astronomy to predict the appearance of Mercury (a) and the phases of the moon (b).

Circles have played an important role in astronomy and understanding their geometry has led to further understanding of how the planets move. Figures 13.1(a) and (b) show circles being used to predict the appearance of Mercury and the phases of the moon, respectively, by ancient Arabic scholars. Until the 17th century, planets were believed to travel in circular orbits around the Sun, but Johannes Kepler showed that they actually follow elliptical paths. Circles are examples of *conic sections*, which also include ellipses, parabolae and hyperbolae, these curves can be found by cutting a (double) cone with a plane.

> **Remark**
> Even though circles are primarily used for transportation, other shapes can be used. These are the so-called *shapes of constant width*. We are familiar with examples of these shape from our everyday life, as both 20 pence and 50 pence coins have constant width. What can be observed when two rulers are laid on a table top with two fifty pence pieces between them and pushed in opposite directions?

## 13.1  Cartesian Description

Let us consider a circle of radius $r$ which is centred at the origin $O$, as in Figure 13.2. Can we formulate an equation to describe this circle?

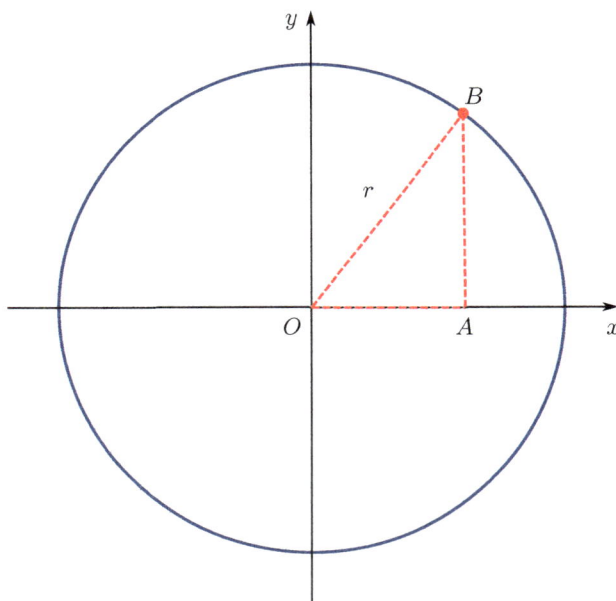

Figure 13.2: The equation of a circle centred at the origin can be found using a right angled triangle.

Suppose that the point $B(x, y)$ lies on the circle, then as long as neither $x = 0$ nor $y = 0$, we can construct a right angled triangle with sides $OA$, $AB$ and $BO$, where $A(x, 0)$, see Figure 13.2. We know that $|OA| = |x|$, $|AB| = |y|$ and $|BO| = r$. Pythagoras' theorem then reveals that $|x|^2 + |y|^2 = r^2$. Now, $|x|^2 = x^2$ and $|y|^2 = y^2$, so we have

$$x^2 + y^2 = r^2.$$

Now suppose $x = 0$, then we cannot form a right-angled triangle. However, for $B(0, y)$ to lie on the circle, $y = \pm r$. In this case, we again must have

$$\underbrace{x^2}_{=0} + y^2 = r^2.$$

Similarly, if $y = 0$, then, for $B(x, 0)$ to lie on the circle, $x = \pm r$ and once again

$$x^2 + \underbrace{y^2}_{=0} = r^2.$$

These results are summarised in the formula below.

> **Formula 13.1 — Cartesian Equation of a Circle Centred at the Origin**
> A circle of radius $r > 0$, whose centre is at the origin, has the following equation.
>
> $$x^2 + y^2 = r^2. \tag{13.1}$$

---

**Example 13.1**
Write down the equation of the circle centred at the origin with radius 2.

**Solution:**
We have $r = 2$, hence the equation of the circle is:

$$x^2 + y^2 = 4.$$

---

In general, a circle may not be centred at the origin, but could have centre at the point $C(a, b)$, as in Figure 13.3. In a similar manner to before, we can use Pythagoras' theorem to construct a Cartesian equation for a general circle.

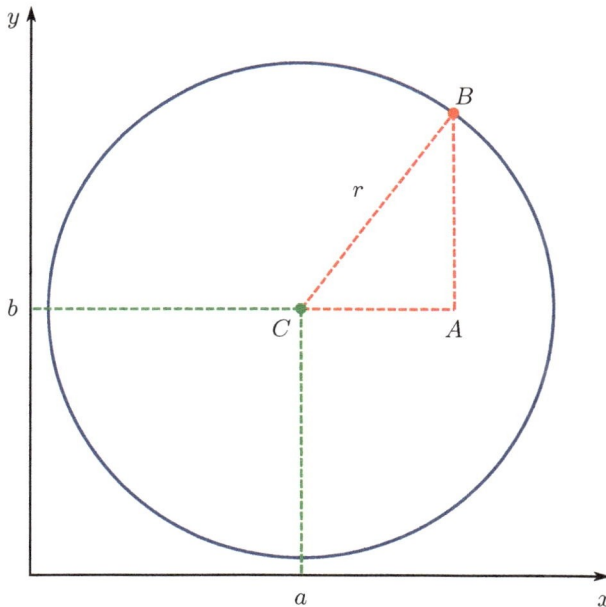

Figure 13.3: The equation of a general circle can be found using a right angled triangle and the centre of the circle.

Once again, suppose the point $B(x, y)$ lies on the circle, this time with $x \neq a$ and $y \neq b$. Then a right angled triangle can be constructed with sides $CA$, $AB$ and $BC$, where $A = (x, b)$ and $C = (a, b)$, see Figure 13.3. Now, $|CA| = |x - a|$, $|AB| = |y - b|$ and $|AB| = r$. Pythagoras' theorem then yields that $|x - a|^2 + |y - b|^2 = r^2$, or

$$(x - a)^2 + (y - b)^2 = r^2.$$

In the special case when $x = a$, then, for $B$ to lie on the circle, $|y - b| = r$, hence the same equation applies

$$\underbrace{(x - a)^2}_{=0} + (y - b)^2 = r^2.$$

Similarly, if $y = b$, then $|x - a| = r$ and once more

$$(x - a)^2 + \underbrace{(y - b)^2}_{=0} = r^2.$$

The general equation for a circle is summarised in the following formula.

---

**Formula 13.2 — General Cartesian Equation of a Circle**
A circle of radius $r > 0$ whose centre is at the point $(a, b)$ has the following equation.

$$(x - a)^2 + (y - b)^2 = r^2. \tag{13.2}$$

---

**Remark**
Above, we used the notation $A = (x, b)$. Throughout this section we shall use the notation $A(x_A, y_A)$ and $A = (x_A, y_A)$ interchangeably for clarity and conciseness.

---

**Example 13.2**
Write down the equation of the circle centred at the origin with radius 3 whose centre has coordinates $(-3, 2)$.

**Solution:**
We have $r = 3$ and hence the equation of the circle is:

$$(x + 3)^2 + (y - 2)^2 = 9.$$

---

**Exercise 13.1**
Q1. Write down the equations of the circle with:
    (a) radius $r = 2$, centre $(0, 0)$;
    (b) radius $r = 4$, centre $(0, 0)$;
    (c) radius $r = 1$, centre $(3, 3)$;
    (d) radius $r = 6$, centre $(-1, 2)$;
    (e) radius $r = 3$, centre $(-4, -5)$;
    (f) radius $r = 5$, centre $(6, -4)$.
Q2. Identify which of the following equations represent a circle and, for the ones that do, find the radius and centres of the circles.

(a) $x^2 + y^2 = 8$;
(b) $x^2 = 4 - y^2$;
(c) $0 = -(x^2 + y^2 - 12)$;
(d) $0 = x^2 + y^3 + 3$;
(e) $(x-4)^2 + (y-2)^2 = 4$;
(f) $(y-3)^2 - (3-x)^2 = 9$;
(g) $x^2 + y^2 + 2x + 1 = 5$;
(h) $x^2 + y^2 - 4y + 4 = 36$;

As seen in parts 2(g) and 2(h) of Exercise 13.1, in some instances it may not be obvious that an equation is actually that of a circle, By completing the square the general form from Formula 13.2 can be recovered.

**Example 13.3**
Show that the equation

$$x^2 + 6x + y^2 = 7$$

is an equation of a circle and hence find the centre of the circle and its radius. Finally, sketch the circle.

**Solution:**
It is not immediately obvious that this is the equation of a circle, but by completing the square on the terms involving $x$ (see Chapter 5), we find that

$$x^2 + 6x = (x+3)^2 - 9.$$

Inserting this back into the original equation, we see

$$
\begin{aligned}
x^2 + 6x + y^2 &= 7, \\
\Rightarrow \quad (x+3)^2 - 9 + y^2 &= 7, \\
\Rightarrow \quad (x+3)^2 + y^2 &= 16.
\end{aligned}
$$

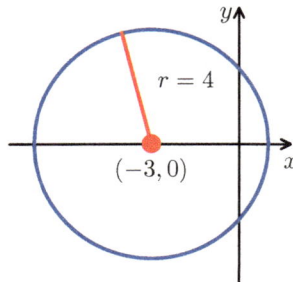

This is now in the usual form for an equation of a circle, whose radius $r = 4$ and centre has coordinates $(-3, 0)$. A sketch of this circle is shown above.

In some cases, we need to complete the square on both the terms involving $x$ and terms involving $y$, as the next example shows.

**Example 13.4**

Show that the equation

$$x^2 + 2x + y^2 + 4y = 4$$

is an equation of a circle and hence find the centre of the circle and its radius. Finally, sketch the circle.

**Solution:**

We complete the square on both the terms involving $x$ and the terms involving $y$. We have that $x^2 + 2x = (x+1)^2 - 1$ and $y^2 + 4y = (y+2)^2 - 4$. Hence,

$$
\begin{aligned}
& x^2 + 2x + y^2 + 4y = 4, \\
\Rightarrow \quad & (x+1)^2 - 1 + (y+2)^2 - 4 = 4, \\
\Rightarrow \quad & (x+1)^2 + (y+2)^2 = 9, \\
\Rightarrow \quad & (x+1)^2 + (y+2)^2 = 3^2.
\end{aligned}
$$

This is exactly in the form of Formula (13.2), with $a = -1$, $b = -2$ and $r = 3$. In other words, this equation is that of a circle centred at the point $(-1, -2)$ with a radius of 3. The circle is sketched below.

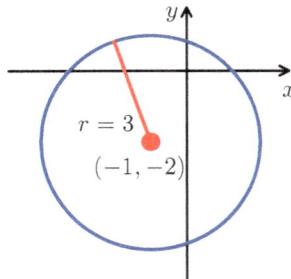

**Activity 13.1**

It may be tempting to think that every equation that involves both $x$ and $y$ to the power of 2, but no higher, is an equation for a circle. This is not the case. Sketch the curve (if such a curve exists) represented by the equations

(a) $x^2 - 2x + y^2 + 8x = 6$;

(b) $x^2 - y^2 = 4$.

**Exercise 13.2**

Q1. By completing the square, determine if the following equations represent circles and hence determine their centres and radii.

(a) $x^2 + 8x + y^2 = 20$;

(b) $x^2 + y^2 - 4y = 12$;

(c) $x^2 + 10x + y^2 - 12y = 20$;

(d) $x^2 - 2x + y^2 - 2y + 4 = 0$;

(e) $2x^2 - 4x + 2y^2 + 4x + 2 = 0$;

(f) $x^2 - 16x + 4y^2 - 16y = 4$;

(g) $x^2 + y^2 = 8x + 4x - 16$;

(h) $9x^2 - 24x + 9y^2 = 12y - 16$;

(i) $4x^2 - 24x = -4y^2 - 16y - 32$

(j) $5x^2 - 40x + 9y^2 - 126y + 521 = 0$;

Q2. Find the equations of the circles where $A$ and $B$ are the endpoints of a diameter of the circle:

(a) $A(5, 2)$, $B(10, -4)$;

(b) $A(11, 2)$, $B(16, 4)$;

(c) $A(-5, 4)$, $B(-1, -6)$;

(d) $A(1, 6)$, $B(10, 3)$.

Q3. Consider the circle with equation $x^2 - 14x + y^2 - 6y + 58 = 16$.

(a) Express the equation in completed square form;

(b) Does the point $(4, 3)$ lie inside or outside of the circle?

(c) Show that the circle does not intersect the $y$-axis;

(d) Let $P$ and $Q$ be the points of intersection of the circle with the $x$-axis. Find $|PQ|$.

Q4. Do the circles with equations $(x+2)^2 + (y+1)^2 = 16$ and $x^2 - 10x + y^2 - 8y + 41 = 16$ intersect?

Q5. For the circle $c : x^2 - 10x + y^2 - 8y + 16$

(a) Find the centre of the circle $C$;

(b) Show that the circle touches the $y$-axis.

(c) Let $l$ be the line through the origin and the centre of the circle. This line meets the circle at point $P$. Find the distance $|OP|$;

(d) Let $A$ and $B$ be the points of intersection of the circle with the $x$-axis, with $A$ closest to the origin. Find the the area of the triangle $OAC$.

## 13.2  Angle in a Semicircle

> **Interactive Activity 13.1 — Angle in a Semicircle**
> Use the Geogebra applet below in the digital book to investigate the angles in a triangle
> with one side a diameter and all three vertices on the circle.

From GCSE, we are familiar with the fact that if a triangle is constructed with one of its
sides being the diameter of a circle and the opposite point lying on the circle, then the
triangle is right angled, see Figure 13.4.

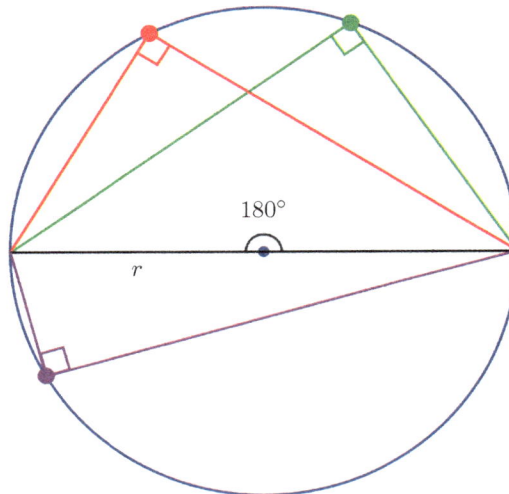

Figure 13.4: The triangle formed from the endpoints of the diameter of a circle and another
point on the circumference of the circle is right angled.

Geometric reasoning offers one means of proving this fact. However, we make use of the
equation of a circle in our proof. As it is easy to see that rotation and translation of the
circle will make no difference to the result, we consider a circle centred at the origin for
simplicity and without loss of generality. Similarly, which diameter is picked is unimportant,
hence we consider a triangle with vertices at $A = (-r, 0)$, $B = (r, 0)$ and $C = (x, y)$, where
$(x, y)$ satisfies $x^2 + y^2 = r^2$. We now calculate the length of each of the sides:

$$
\begin{aligned}
|AB|^2 &= (2r)^2, \\
|BC|^2 &= |(x, y) - (r, 0)|^2 = (x - r)^2 + y^2 = x^2 - 2rx + r^2 + y^2 = 2r^2 - 2rx, \\
|CA|^2 &= |(-r, 0) - (x, y)|^2 = (-r - x)^2 + (-y)^2 = x^2 + 2rx + r^2 + y^2 = 2r^2 + 2rx.
\end{aligned}
$$

Hence, $|BC|^2 + |CA|^2 = |AB|^2$ and, by Pythagoras' theorem, we have a right angled triangle.

---

**Example 13.5**

Consider the triangle $ABC$, where $A = (-4, 1)$, $B = (-2, -5)$ and $C = (0, -1)$. Suppose that these points all lie on a circle.

Show that the line $AB$ is a diameter of the circle and hence find an equation of the circle.

**Solution:**

In order to show that $AB$ is a diameter, it suffices to show that the triangle is right-angled, with $\angle BCA$ being the right angle. To do this, we make use of Pythagoras' theorem and show that

$$|AB|^2 = |BC|^2 + |CA|^2.$$

We find the squares of the lengths:

$$|AB|^2 = (-2 + 4)^2 + (-5 - 1)^2 = 4 + 36 = 40,$$
$$|BC|^2 = (0 + 2)^2 + (-1 + 5)^2 = 4 + 16 = 20,$$
$$|CA|^2 = (-4 + 0)^2 + (1 + 1)^2 = 16 + 4 = 20.$$

Hence, $|AB|^2 = |BC|^2 + |CA|^2$ and we have a right angled triangle with $AB$ the hypotenuse. In fact, $ABC$ is an isosceles, right angled triangle. By properties of circles and triangles, $AB$ is a diameter of the circle.

Now, the centre of the circle will be the midpoint $M$ of $AB$, which we can be found using techniques from Chapter 4. The midpoint is given by

$$M = \frac{A + B}{2} = \left( \frac{-4 - 2}{2}, \frac{1 - 5}{2} \right) = (-3, -2).$$

The radius, $r$, is half the diameter, so

$$r = \frac{|AB|}{2} = \frac{\sqrt{40}}{2} = \frac{2\sqrt{10}}{2}.$$

Then, an equation of the circle is

$$(x + 3)^2 + (y + 2)^2 = 10.$$

A sketch of this situation is shown below.

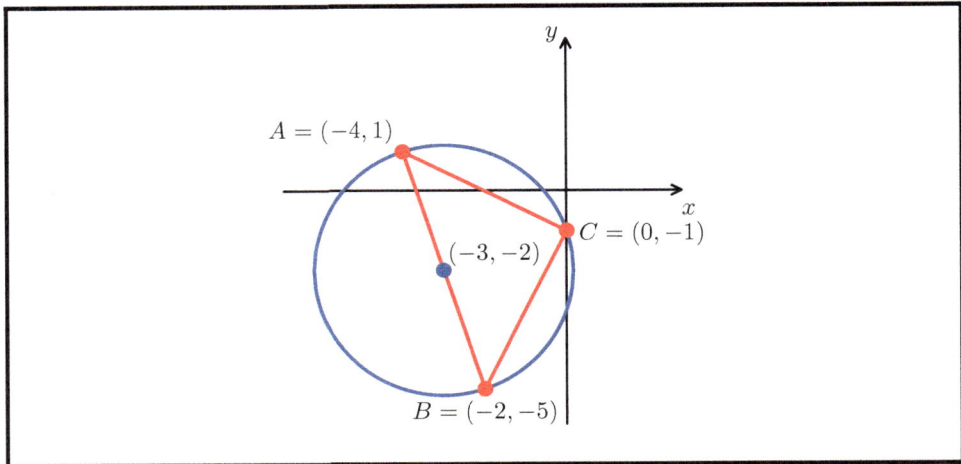

## 13.3  Perpendicular Bisector of a Chord

**Interactive Activity 13.2 — Perpendicular Bisector of a Chord**
Use the Geogebra applet below in the digital book to investigate the perpendicular bisector of a chord of a circle.

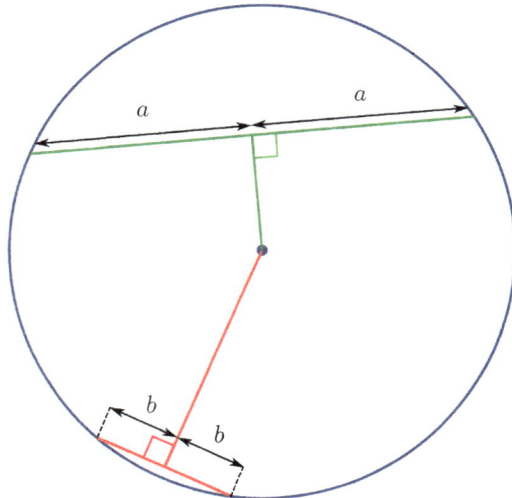

Figure 13.5: The perpendicular bisector of a chord of a circle passes through the centre of the circle.

Suppose that any chord of a circle is drawn and a line is constructed from the midpoint of the chord to the centre of the circle. As can be seen in Figure 13.5, this line bisects the chord at an angle of $90°$, *i.e.* it is a perpendicular bisector of the chord..

In order to prove this result, without loss of generality, we consider a circle centred at the origin $O$ and a chord passing through the points $A$ and $B$ that lie on the circle. This situation can be seen in Figure 13.6. Letting $C$ be the midpoint of $\overrightarrow{AB}$, we will show that the triangle $AOC$ is always right angled and hence the line $OC$ is a perpendicular bisector of $AB$.

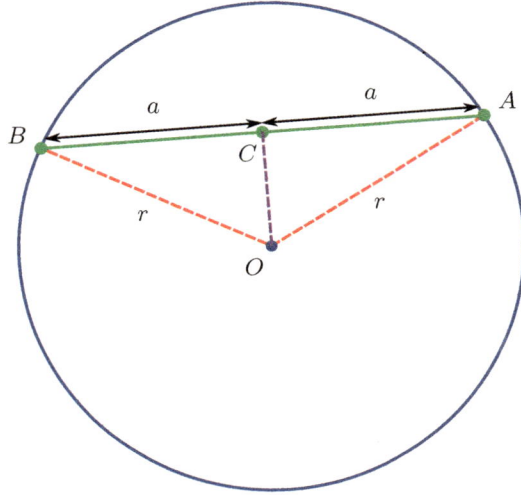

Figure 13.6: Two isosceles triangles formed by a perpendicular of a chord.

Let $A = (x_1, y_1)$ and $B = (x_2, y_2)$, then $C$ has coordinates $((x_1 + x_2)/2, (y_1 + y_2)/2)$. We now find the square of the magnitude of the lines $AC$ and $OC$:

$$|AC|^2 = \left((x_1 - x_2)^2 + (y_1 - y_2)^2\right)/4,$$
$$|OC|^2 = \left((x_1 + x_2)^2 + (y_1 + y_2)^2\right)/4.$$

Adding these together, we find

$$\begin{aligned}
|AC|^2 + |OC|^2 &= \left((x_1 - x_2)^2 + (y_1 - y_2)^2\right)/4 + \left((x_1 + x_2)^2 + (y_1 + y_2)^2\right)/4 \\
&= \left(2(x_1^2 + y_1^2) + 2(x_2^2 + y_2^2)\right)/4 \\
&= \frac{4r^2}{4} \\
&= r^2.
\end{aligned}$$

Now, as $|OA| = r$, Pythagoras' theorem reveals that $AOC$ must be a right angled triangle, as required.

The steps above can be reversed, to show that a perpendicular bisector to a chord must pass through the centre of the circle.

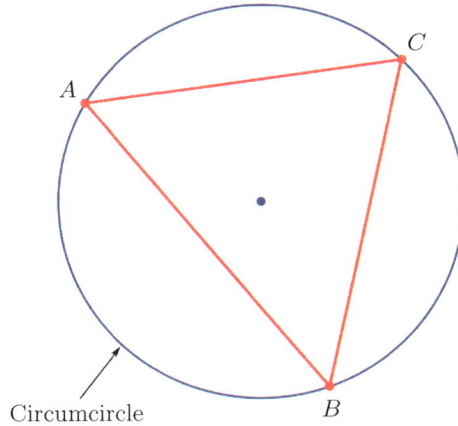

Figure 13.7: A triangle $ABC$ and its circumcircle.

A circumcircle of a triangle $ABC$ is the circle which encloses the triangle with points $A$, $B$ and $C$ lying on the circle. An example of this is given in Figure 13.7.

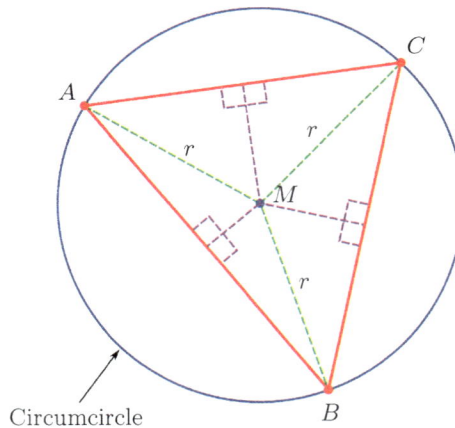

Figure 13.8: Construction of a triangle's circumcircle

How might we construct such a circle? The result above provides an answer to this question. If $A$, $B$ and $C$ lie on the circle, then $AC$, $BC$ and $CA$ are all chords of the circle, hence their perpendicular bisectors must all pass through the centre of the circle, let us call this point $M$. Where any two of these bisectors meet will be $M$. It is then a simple matter to find the radius of the circle by computing the distances $AM$, $BM$ and $CM$, which will all be the same. An example of this construction is shown in Figure 13.8.

---

**Example 13.6**

Let $A = (14, 2)$, $B = (13, 7)$ and $C = (-4, 14)$ be the vertices of a triangle.
  (a) Find the equations of the perpendicular bisectors of $AB$ and $BC$.
  (b) Find the coordinate of the centre of the circumcircle of $ABC$.
  (c) Find the equation of the circumcircle of $ABC$.

**Solution:**
  (a) To find the perpendicular bisector of $AB$, we first find $m_1$, the gradient of $AB$:

$$m_1 = \frac{7-2}{14-13} = -5.$$

Immediately we have that the perpendicular has equation

$$y = -\frac{1}{m_1}x + b = \frac{1}{5}x + b,$$

where $b$ is found by making sure the line passes through the midpoint $M_1$ of $A$ and $B$. We find

$$M_1 = \frac{A+B}{2} = \left(\frac{27}{2}, \frac{9}{2}\right).$$

Hence,

$$\frac{9}{2} = \frac{27}{2} \cdot \frac{1}{5} + b,$$
$$\Rightarrow \quad b = \frac{45}{10} - \frac{27}{10} = \frac{9}{5}.$$

So, the equation of the perpendicular bisector of $AB$ is $y = \frac{1}{5}(x+9)$.
We repeat the process to find the equation of the perpendicular bisector of $BC$, the gradient of $BC$ is $-\frac{17}{7}$ and the midpoint of $BC$ has coordinates $\left(\frac{9}{2}, \frac{21}{2}\right)$. The equation of this perpendicular bisector is

$$y = \frac{1}{7}(17x - 3).$$

  (b) The center of the circumcircle of $ABC$ is where the two perpendicular bisectors, found above, intersect. We solve the following set of simultaneous equations

$$y = \frac{1}{5}(x+9), \quad ①$$
$$y = \frac{1}{7}(17x - 3), \quad ②$$

Equating $y$s above, we have

$$\frac{1}{5}(x+9) = \frac{1}{7}(17x-3),$$
$$\Rightarrow \quad 7(x+9) = 5(17x-3),$$
$$\Rightarrow \quad 7x+63 = 85x-15,$$
$$\Rightarrow \quad 78x = 78,$$
$$\Rightarrow \quad x = 1.$$

(c) Inserting this value of $x$ back into ①, we find that $y = 2$. Hence, the circumcircle has centre $P = (1, 2)$.

(d) To find the equation of the circumcircle we must find the radius. This can be found by computing any of the lengths $|PA|$, $|PB|$ or $|PC|$. We choose $|PA|$ and find

$$r = |PA| = |(14-1, 2-2)| = 13.$$

The equation of the circle is

$$(x-1)^2 + (y-2)^2 = 13^2.$$

Note, to be certain that the centre point has been found correctly, we should check that $|PB|$ and $|PC|$ are both 13.

The circumcircle is sketched below.

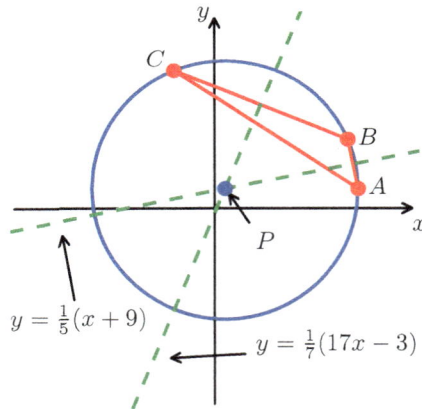

## 13.4  Angles at the Centre of a Circle

Consider the circles shown in Figure 13.9. In both cases, two chords of the circle $AB$ and $AC$ are shown. The angle $\angle BOC$ is always twice that of the $\angle BAC$. We say that "the angle at the centre is twice the angle at the circumference" and have already come across this at GCSE.

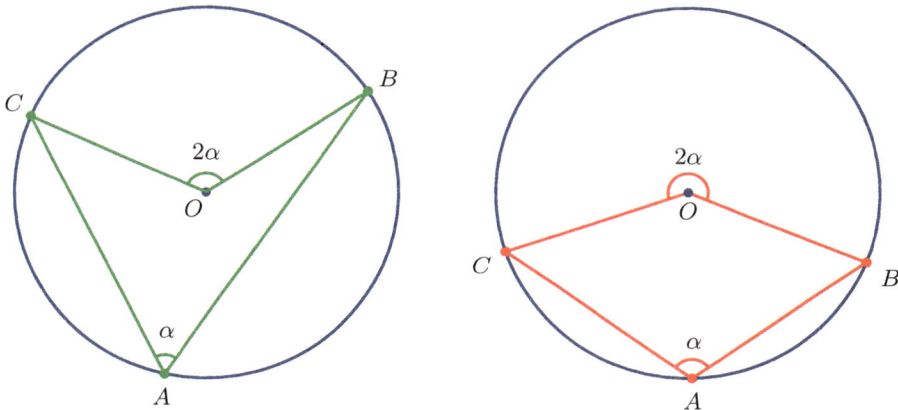

Figure 13.9: The angle at the centre of a circle is twice the angle at the circumference.

In order to prove this result, we first draw a line from $O$ to $A$ and construct two isosceles triangles, $OAB$ and $OCA$, where, in both the cases, the sides of common length have length $r$, see Figure 13.10.

As in Figure 13.10, let $\gamma = \angle ABC = \angle OAB$. Then we know that $\sigma = 180 - 2\gamma$. Similarly, let $\beta = \angle OCA = \angle CAO$, then $\phi = 180 - 2\beta$. We also know that $\alpha = \beta + \gamma$. Thus,

$$
\begin{aligned}
\theta &= 360 - \phi - \sigma \\
&= 360 - (180 - 2\beta) - (180 - 2\gamma) \\
&= 2(\beta + \gamma) \\
&= 2\alpha.
\end{aligned}
$$

Hence, the result is proved.

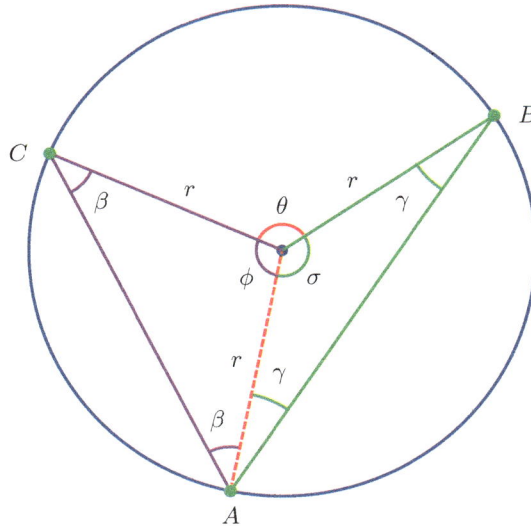

Figure 13.10: Construction of two isosceles triangles when proving the angle at the centre of a circle is twice the angle at the circumference.

---

**Example 13.7**

Find the area of triangle $ABM$ in the figure below, in terms of the radius $r$ of the circle. $M$ is the centre of the circle.

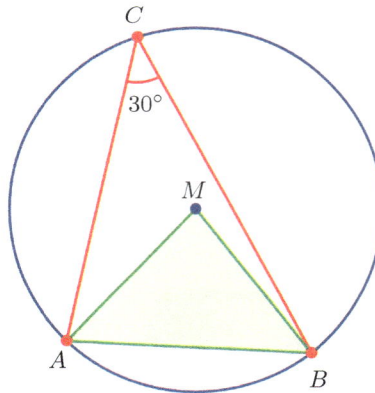

**Solution:**

The angle $\angle BCA$ is $30°$. As $M$ is the centre of the circle, we know that angle $\angle BMA$ is $60°$. Now, $ABM$ is an isosceles triangle and hence, the remaining two angles must also be 60°, *i.e.* we $ABM$ is equilateral with side length $r$. It can be divided into two congruent right angled triangles with hypotenuse of length $r$ and angles of size 30° and 60°. Hence, the lengths of the remaining sides are $r\sin(30) = \frac{r}{2}$ and $r\cos(30) = r\frac{\sqrt{3}}{2}$. The area of the right angled triangle is then $\frac{1}{2} \cdot \frac{r}{2} \cdot r\frac{\sqrt{3}}{2}$ and hence the area of $ABM$ is

$$\text{Area} = \frac{r^2\sqrt{3}}{4}.$$

An alternative way to compute the area of $ABM$ would be to use Heron's formula, which states that, for a general triangle with sides of length $a$, $b$ and $c$,

$$\text{Area} = \sqrt{s(s-a)(s-b)(s-c)},$$

where

$$s = \frac{a+b+c}{2}.$$

As $ABM$ is equilateral, with $a = b = c = r$, then $s = \frac{3r}{2}$ and $s - a = s - b = s - c = \frac{r}{2}$ the area is

$$\begin{aligned}\text{Area} &= \sqrt{\frac{3r}{2} \cdot \frac{r}{2} \cdot \frac{r}{2} \cdot \frac{r}{2}} \\ &= \sqrt{\frac{3r^4}{16}} \\ &= \frac{\sqrt{3}r^2}{4},\end{aligned}$$

as before.

## 13.5 Tangent to a Circle is Perpendicular to the Radius

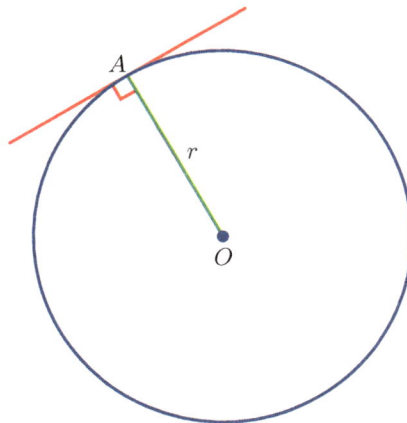

Figure 13.11: A tangent to a circle is perpendicular to the radius.

Another important property of a circle is that, if a tangent is drawn which touches the circle only at a point $A$, then the tangent is perpendicular to the radius drawn from the centre of the circle to point $A$. This is shown in Figure 13.11.

In order to prove this, without loss of generality, we assume the circle is centred at the origin and therefore has equation $x^2 + y^2 = r^2$. Let point $A = (x_1, y_1)$ lie on the circle. We consider the case $x_1, y_1 \neq 0$ first.

In this case, the straight line which passes through the origin $O$ and $A$ has equation $y = m_1 x$, where $m_1 = y_1/x_1$. We now try to find the equation of the tangent to the circle at point $A$. This equation has general form $y = m_2 x + b$ and must intersect the circle *only* once at the point $(x_1, y_1)$. Using the fact that the line intersects the circle at $(x_1, y_1)$, we find that

$$y_1 = m_2 x_1 + b,$$
$$\Rightarrow \quad b = y_1 - m_2 x_1. \tag{13.3}$$

We also know, from Chapter 11, that to find an intersection point of two curves/lines involves solving a set of simultaneous equations. Here, the two equations are

$$y = m_2 x + b, \quad ①$$
$$r^2 = x^2 + y^2. \quad ②$$

Substituting the expression for $y$ from ① directly into ② gives

$$r^2 = x^2 + (m_2 x + b)^2.$$

This is a quadratic equation for $x$ which can be rearranged into the usual form:

$$(1 + m_2^2)x^2 + 2bm_2 x + (b^2 + r^2) = 0.$$

As this is a quadratic equation, we expect there to be zero, one or two solutions, which can be found using the quadratic equation formula:

$$x = \frac{-2bm_2 \pm \sqrt{4b^2 m_2^2 - 4(1 + m_2^2)(b^2 + r^2)}}{2(1 + m_2^2)}.$$

Specifically, we are looking for the case where there is only one solution and hence, we know that the discriminant $4b^2 m_2^2 - 4(1 + m_2^2)(b^2 + r^2) = 0$ and

$$x_1 = -\frac{2bm_2}{2(1 + m_2^2)} = -\frac{bm_2}{(1 + m_2^2)}.$$

Now, using Equation 13.3, we have

$$x_1 = -\frac{(y_1 - m_2 x_1)m_2}{(1 + m_2^2)},$$
$$\Rightarrow \quad x_1(1 + m_2^2) = -(y_1 - m_2 x_1)m_2,$$
$$\Rightarrow \quad \quad x_1 = -y_1 m_2,$$
$$\Rightarrow \quad \quad m_2 = -\frac{x_1}{y_1}.$$

Now, the slope of the line passing through $O$ and $A$ is $m_1 = y_1/x_1$ and the slope of the tangent is $m_2 = -x_1/y_1$. Hence, these two lines are perpendicular, by the result in Equation 4.12.

**Activity 13.2**

Prove that, in the cases when either $x_1 = 0$, or $y_1 = 0$, again the tangent is perpendicular to the radius.

**Remark**

The method shown above, for proving that the tangent and radius are perpendicular, is very long and algebraically heavy. The theories of calculus, specifically of differentiation, which we will meet in Chapter 15, enables us find the equations of tangents much more simply.

**Example 13.8**

A circle of radius $r = \sqrt{2}$ is centred at the origin. Show that the point $P = (-1, 1)$ lies on the circle and find the equation of the tangent that passes through this point.

**Solution:**

As the circle has radius $\sqrt{2}$ and is centred at the origin, we can immediately write down the circle's equation:

$$x^2 + y^2 = 2.$$

The point $P$ lies on the circle because its coordinates satisfy the equation, *i.e* $(-1)^2 + 1^2 = 2$.

We know that the slope of the line joining the centre of the circle to the point $P$ is $m_1 = \frac{1}{-1} = -1$. Hence, we can find the slope of the tangent $m_2 = -\frac{1}{m_1} = -\frac{1}{-1} = 1$. Thus, the equation of the tangent is

$$y = x + b.$$

To find $b$, we ensure that $P$ lies on the line, so that

$$1 = -1 + b$$
$$\Rightarrow \quad b = 2.$$

Thus, the equation of the tangent is

$$y = x + 2.$$

**Example 13.9**

A circle is centred at point $C = (1, 2)$. We are told that the point $A = (1 + \sqrt{3}, 3)$ lies on the circle.
  (a) Find the equation of the circle;
  (b) Find the equation of the tangent to the circle at point $A$.

**Solution:**

(a) As the circle is centred at $C = (1, 2)$, we know that the circle has equation

$$(x - 1)^2 + (y - 2)^2 = r^2,$$

where the radius $r$ is to be determined. Inserting the point $A = (1 + \sqrt{3}, 3)$ into the above equation gives

$$(1 + \sqrt{3} - 1)^2 + (3 - 2)^2 = r^2,$$
$$\Rightarrow \qquad\qquad 3^2 + 1^2 = r^2,$$
$$\Rightarrow \qquad\qquad\qquad 4 = r^2.$$

The radius is therefore 2 and the equation of the circle is

$$(x - 1)^2 + (y - 2)^2 = 4.$$

(b) The line joining the centre of the circle to the point $A$ has slope $m_1$ given by

$$m_1 = \frac{3 - 2}{1 + \sqrt{3} - 1} = \frac{1}{\sqrt{3}}.$$

Hence, the slope of the tangent is $m_2$, where

$$m_2 = -\frac{1}{m_1} = -\sqrt{3}$$

and hence, the equation of the tangent is

$$y = \sqrt{3}x + b,$$

where $b$ is to be determined. By inserting point $A$ into the equation of the tangent, we find

$$2 = -\sqrt{3} + b,$$
$$\Rightarrow \quad b = \sqrt{3} + 2.$$

Hence, the equation of the tangent, after simplification, is

$$y = -\sqrt{3}(x - 1) + 2.$$

---

**Example 13.10**

A circle has radius 2 and we are told that the line with equation $y = \sqrt{3}x - 2$ is tangent to the circle at point $A = (\sqrt{3}, 1)$. Find two possible equations for the circle and sketch the circles and the tangent line.

**Solution:**

**Method 1:**
As the slope of the tangent is $\sqrt{3}$, we immediately know that the slope of the line from

the centre of the circle to point $A$ is $m_1 = -\frac{1}{\sqrt{3}}$. The vector

$$\mathbf{v} = \begin{pmatrix} -\sqrt{3} \\ 1 \end{pmatrix}$$

is parallel to this line. To find the possible centres of the circle, starting at point $A$, we can move a distance of 2 either in the direction of $\mathbf{v}$ or $-\mathbf{v}$. Now, $\mathbf{v}$ has length $|\mathbf{v}| = \sqrt{(-\sqrt{3})^2 + 1^2} = \sqrt{4} = 2$, which is already of the correct length. Thus, the possible centres are $A + \mathbf{v} = (\sqrt{3} - \sqrt{3}, 2) = (0, 2)$ or $A - \mathbf{v} = (\sqrt{3} + \sqrt{3}, 0) = (2\sqrt{3}, 0)$. Hence, the possible equations of the circles are

$$(x^2 + (y - 2))^2 = 4,$$

and $\qquad (x - 2\sqrt{3})^2 + y^2 = 4.$

**Method 2:**
Again, the slope of the tangent is $\sqrt{3}$ and hence, the slope of the line from the centre of the circle to point $A$ is $m_1 = -\frac{1}{\sqrt{3}}$. The equation of this line is therefore

$$y = -\frac{x}{\sqrt{3}} + b_1,$$

where $b_1$ can be found by ensuring the line passes through point $A$, *i.e.*

$$1 = -\frac{\sqrt{3}}{\sqrt{3}} + b_1$$

$$\Rightarrow \quad b_1 = 2.$$

Thus the equation is $y = -\frac{x}{\sqrt{3}} + 2$. We are now looking for points that lie on this line and are at a distance 2 away from point $A$. In other words, we need to find a point $B = (x, y)$ such that $|AB| = 2$. For simplicity, let us instead rewrite this as $|AB|^2 = (x - \sqrt(3))^2 + (y - 1)^2 = 4$. Hence, we have the set of simultaneous equations:

$$y = -\frac{x}{\sqrt{3}} + 2, \qquad\qquad ①$$

$$4 = (x - \sqrt(3))^2 + (y - 1)^2. \quad ②$$

Note that ② is also the equation of a circle. To solve this, we insert the expression for $y$ from ① directly into ② to obtain a quadratic equation:

$$(x - \sqrt{(3)})^2 + \left(-\frac{x}{\sqrt{3}} + 2 - 1\right)^2 = 4,$$

$$\Rightarrow \qquad (x - \sqrt{(3)})^2 + \left(-\frac{x}{\sqrt{3}} + 1\right)^2 = 4,$$

$$\Rightarrow \qquad x^2 - 2\sqrt{3}x + 3 + \frac{x^2}{3} - 2\frac{x}{\sqrt{3}} + 1 = 4,$$

$$\Rightarrow \qquad \left(1 + \frac{1}{3}\right)x^2 - 2\left(\sqrt{3} + \frac{1}{\sqrt{3}}\right)x = 0,$$

$$\Rightarrow \qquad x\left(\left(\frac{4}{3}\right)x - 2\left(\frac{4}{\sqrt{3}}\right)\right) = 0.$$

Hence, the two solutions are $x_1 = 0$ and $x_2 = 2\sqrt{3}$. Inserting these values into ①, we find that $y_1 = 2$ and $y_2 = 0$. Thus, the possible centres of the circle are $(2\sqrt{3}, 0)$ or $(0, 2)$ and their respective equations are

$$(x^2 + (y - 2))^2 = 4,$$

and $\qquad (x - 2\sqrt{3})^2 + y^2 = 4.$

Sketches of the circles and the tangents are show below.

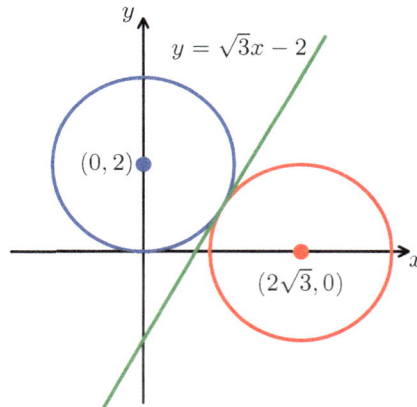

---

**Example 13.11**

Let $A = (4, 6)$ and $B = (-2, -2)$ and the line $AB$ be the diameter of a circle.
  (a) Find the equation of the circle;
  (b) Find the equations of the tangents to the circle at the points $A$ and $B$.

**Solution:**

(a) The centre of the circle is the midpoint of $A$ and $B$, which has coordinates

$$C = \frac{A+B}{2} = \frac{(2,4)}{2} = (1,2).$$

The radius has length

$$\begin{aligned}
\frac{1}{2}|AB| &= \frac{1}{2}|(6,8)| \\
&= \frac{\sqrt{36+64}}{2} \\
&= 5.
\end{aligned}$$

Hence, the equation of the circle is

$$(x-1)^2 + (y-2)^2 = 25.$$

(b) We can find the slope of the radius, $m_1$, by calculating the change in $y$ of $AB$ divided by the change in $x$ of $AB$:

$$m_1 = \frac{8}{6} = \frac{4}{3}.$$

The general equation of the tangents is then

$$y = -\frac{1}{m_1}x + b = -\frac{3}{4}x + b,$$

where $b$ can be found by making sure the lines pass through either $A$ or $B$. To pass through $A$, the equation satisfies

$$\begin{aligned}
6 &= -\frac{3}{4}4 + b, \\
\Rightarrow \quad b &= 6 + \frac{12}{4} \\
&= \frac{36}{4} = 9.
\end{aligned}$$

The equation of the tangent is then

$$y = -\frac{3x}{4} + 9.$$

Similarly, for point $B$, we have

$$\begin{aligned}
-2 &= \frac{3}{4}2 + b, \\
\Rightarrow \quad b &= -2 - \frac{6}{4} \\
&= -\frac{14}{4} = -\frac{7}{2}.
\end{aligned}$$

The equation of the tangent is then

$$y = -\frac{3x}{4} - \frac{7}{2}$$
$$= -\frac{1}{2}\left(\frac{3}{2}x + 7\right).$$

A sketch of the circle and tangents is shown below.

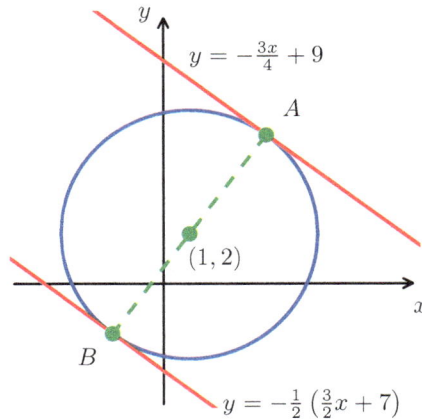

### Example 13.12

Consider the figure below, which shows two circles $A$ and $B$. $A$ has radius 1 and the $x$- and $y$-axes are both tangents to the circle. $B$ has radius $\frac{1}{2}$ and the $x$-axis is tangent to it. $A$ and $B$ intersect at only one point.

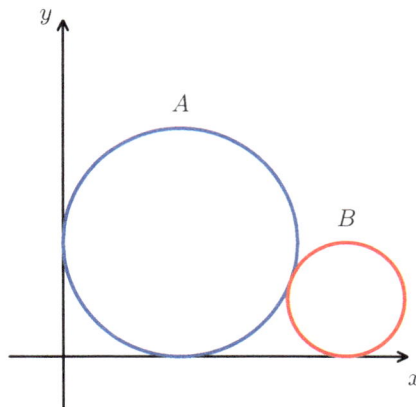

Find the coordinates of the centre of circle $B$ and hence its equation.

**Solution:**

The centre of circle $A$ must be at point $C_1 = (1, 1)$ and the centre of circle $B$ must have $y$-coordinate $\frac{1}{2}$. Now, as $A$ and $B$ intersect at just one point, the circles must be tangent to each other at this point. The two radii passing through this intersection point must, therefore, have the same slope. Hence, the line $C_1C_2$ passes through the intersection points and $|C_1C_2| = 1 + \frac{1}{2} = \frac{3}{2}$. This is shown in the figure below

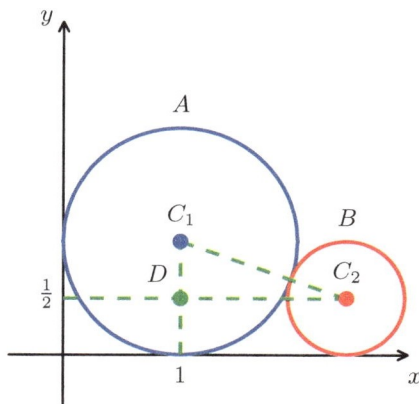

$C_1DC_2$ is a right angled triangle and finding the length $|DC_2|$ will enable us to determine the $x$-coordinate of $C_2$. Using Pythagoras' theroem, we have

$$|C_1C_2|^2 = |DC_1|^2 + |DC_2|^2,$$

$$\Rightarrow \quad \left(\frac{3}{2}\right)^2 = |DC_1|^2 + \frac{1}{4},$$

$$\Rightarrow \quad |DC_1| = \sqrt{\left(\frac{3}{2}\right)^2 - \frac{1}{4}} = \sqrt{2}.$$

The $x$-coordinate of $C_2$ is, therefore, $1 + \sqrt{2}$ and $C_2 = \left(1 + \sqrt{2}, \frac{1}{2}\right)$. The equation of $B$ is then

$$(x - 1 - \sqrt{2})^2 + \left(y - \frac{1}{2}\right)^2 = \frac{1}{4}.$$

**Student Challenge 13.1**

Two straight lines with equations $y = -1$ and $y = -x + 2(1 + \sqrt{2})$ are tangents to a set of circles of varying radius.

(a) Find an equation for the locus of the centres of these circles and sketch the this locus and tangents on a graph.

(b) Find the equation of the circle whose centre has $x$-coordinate 1 and sketch this circle on the same set of axes.

**Exercise 13.3**

Q1. For a circle of radius $r$ and a chord of length $c \le 2d$, what is the shortest distance form the centre of the circle to the chord in terms of $r$ and $c$?

Q2. For a circle of radius $r$, the shortest distance from the centre of the circle to the chord is $a \le r$. What is the length of the chord in terms of $r$ and $a$?

Q3. Consider the figure below.

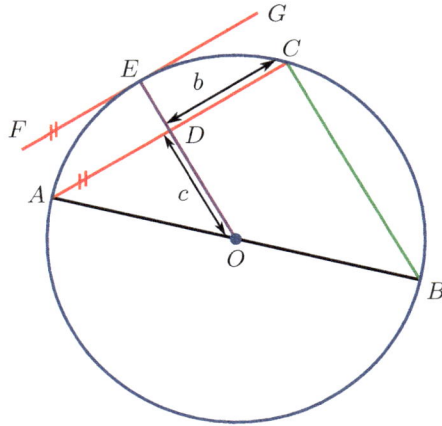

The line $FG$ is a tangent to the circle and passes through point $E$. Suppose the circle has radius $r$ and $\overrightarrow{BC}$ is of length $3r/2$.

   (a) What is the length of $b$?

   (b) What is the length of $c$?

Q4. An equilateral triangle is drawn within a circle of radius $r$ so that the three vertices of the triangle lie on the circle.

   (a) What are the lengths of the sides of the triangle?

   (b) What is the area of the triangle?

Q5. A metal disc with a radius of $5\,\text{cm}$ has a circular hole cut out of the centre. The hole has a radius of $3\,\text{cm}$.

   The points $A$ and $B$ lie on the outer edge of the disc. The line $AB$ lies entirely on the metal ring, and touches the inner edge of the ring at the point $P$.

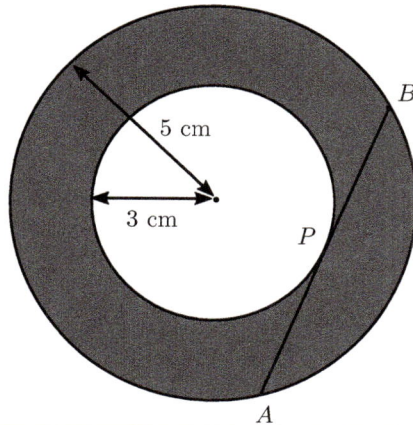

(a) Calculate the length of the line $AB$.

(b) The disc is stood on its edge with the point $A$ resting on a flat surface. Calculate the height of the point $B$ above the surface.

Q6. Consider the equation $x^2 - 4x + y^2 + 2y = k$, where $k$ is a real number.

(a) Calculate the range of values of $k$ for which the equation describes a circle.

(b) Sketch the curve when $k = 11$.

(c) Give a geometrical interpretation of the equation when $k = -5$.

Q7. The points $A = (1, 4)$, $B = (5, -1)$, $C = (-5, -9)$ and $D = (-9, -4)$ form a rectangle $ABCD$. The circle $S_1$ intersects the four corners of the rectangle.

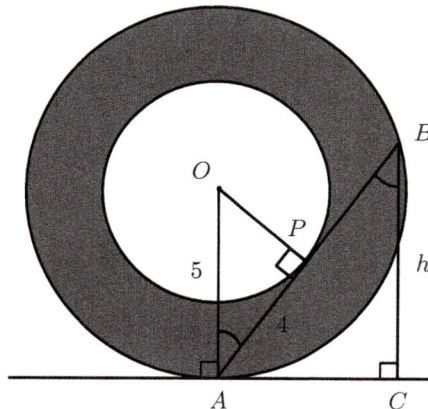

(a) Calculate the equation of the circle $S_1$.

(b) The circle $S_2$ is centred on the point $A$ and passes through $B$. Show that $E = (-5.4, 3.8)$ lies on both circles.

Q8. The cross section of a hollow, regular cone forms an angle of $60°$ at its apex. When a ball of radius $r$ is dropped into the cone, it touches the surface of the cone at a distance of $x$cm from the apex.

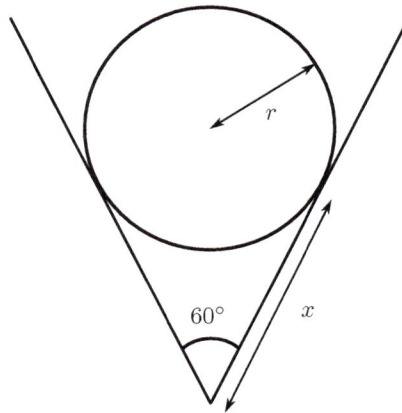

(a) Calculate $x$ in terms of the radius of the ball.
(b) Calculate the distance from the centre of the ball to the apex of the cone in terms of $r$.
(c) If a second ball is dropped into the cone, it will either hit the first ball or it will be stopped by the surface of the cone before reaching the first ball. Calculate, in terms of $r$, the radius of the largest ball which, when dropped into the cone, is small enough to hit the first ball.

**Key Facts — Geometry of the Circle**

- An equation with radius $r$ and centre $C(a,b)$ has equation $(x-a)^2+(y-b)^2 = r^2$;
- Completing the square can be used to convert an equation into the form $(x-a)^2 + (y-b)^2 = r^2$;
- If a triangle is constructed with one of its sides being the diameter of a circle and the opposite point lying on the circle, then the triangle is right angled;
- The perpendicular bisector of a chord of a circle passes through the centre of the circle.
- The angle at the centre of a circle is twice the angle at the circumference.
- A tangent to a circle at point $P$ is perpendicular to the radius of the circle with endpoint $P$.

**Chapter Assessment — Geometry of the Circle**
Download and sit the 30 minute assessment for this chapter from the digital book.

$$\sin(90^\circ - \theta) = \frac{c}{b} = \cos(\theta),$$
$$\cos(90^\circ - \theta) = \frac{a}{b} = \sin(\theta),$$
$$\tan(90^\circ - \theta) = \frac{c}{a} = \frac{1}{\tan(\theta)}.$$

# 14. Introduction to Trigonometry

Trigonometry has been studied since at least the 5th century (although the word trigonometry - from the Greek meaning 'measuring triangles' - was not used until the sixteenth century). It has been used for calculations in astronomy and engineering for hundreds of years. its use is also central to acoustics, architecture, medical imaging, climate studies, crystallography and many more fields. As one example, we can consider the creation of maps for British India in the time of the British Empire. The Great Trigonometrical Survey was begun in 1802 (under the control of the British East India Company) and was first completed in 1872. The aim of the survey was to accurately measure the British territories, as shown in Figure 14.1.

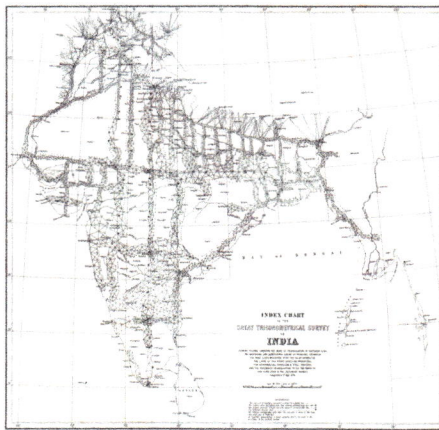

Figure 14.1: Victorian map of India showing the triangulation.

In addition to this, however, was the first accurate measurement of the Himalayan giant mountains, notably Mount Everest.

Figure 14.2: Sunset on Mount Everest

In 1856 Andrew Waugh, after years of work reported that the height of Everest was 8840 m, only 8 metres short of the modern officially accepted height.

We have previously understood the trigonometric functions as the ratios between different pairs of sides of a right angled triangle, however, calculators give an answer for a sine, cosine or tangent of any number that is input, including negative numbers. This chapter explores how the functions can be defined for all input values and examines some applications of the *trigonometric functions*.

> **Remark**
>
> Some of the words used when talking about trigonometry have evolved through translations between multiple languages to arrive at today's common usage.
>
> The word *sine* is from the Latin *sinus*, which means fold in a garment or bend, curve or bosom. The word is believed to have originally been rendered into Medieval Latin in the 12th century where it had been incorrectly translated from the Arabic *jaib* meaning bendle, bosom or fold in a garment rather than the correct *jiba* itself a transliteration of the earlier Sanskrit *jya ardha -half chord* or *bowstring*. This is based on a similar word in an earlier Indian astronomical text. This mistaken translation can be understood due to the majority of vowel sounds being missing from the writing of Semitic languages and understood by usage.
>
> *Cosine* is short for *complementary sine* meaning the sine of the complementary angle - for the angle $\theta$ the complementary angle is $90° - \theta$.
>
> *Tangent* is the Latin for "touching" from the verb *tangere* - to touch.

## 14.1 Trigonometric ratios for acute angles

We have met the trigonometric ratios of acute angles at GCSE - we will recap these definitions first, before exploring the ratios for larger and smaller angles.

> **Remark**
>
> The convention for labelling triangles is to use an uppercase letter for a vertex and the corresponding lowercase letter for the opposite side (see Figure 14.3 below). Unless otherwise indicated in diagrams, this can be assumed in this chapter and throughout the

book. It is also convention that for a triangle $ABC$, the vertices appear in anticlockwise order. The angle at vertex $A$, say, will be denoted by $\angle A$, but when no confusion will occur, we may call this angle simply $A$.

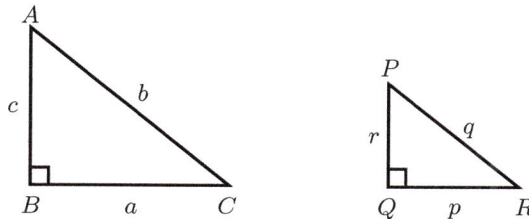

Figure 14.3: Similar right angled triangles $ABC$ and $PQR$.

The definitions of the trigonometric ratios for acute angles use the sides of a right angled triangle. If we start by considering two right angled similar triangles $ABC$ and $PQR$, as in Figure 14.3, the theory of similar triangles tells us that there is a constant ratio between the corresponding sides.

$$\frac{AB}{PQ} = \frac{BC}{QR} = \frac{AC}{PR} \quad \text{or} \quad \frac{c}{r} = \frac{a}{p} = \frac{b}{q},$$

which can be rearranged to give

$$\frac{c}{a} = \frac{r}{p}, \quad \frac{b}{a} = \frac{q}{p}, \quad \frac{c}{b} = \frac{r}{q}.$$

There are an infinite number of such pairs of triangles, but the relationships of the pairs of sides does not change. The ratio of the lengths of the sides of the triangle will determine the *exact* shape of the triangle we have. We can recognise that these fractions are then the definitions for the trigonometric ratios that we saw at GCSE.

The trigonometric ratios allow us to compute unknown angles and lengths of the sides of triangles. To help them with these computations, mathematicians of the past found it very useful to keep tables of the trigonometric ratios, but now our calculators give this information almost instantly.

**Definition 14.1 — Trigonometric ratios of acute angles**
Consider the right angled triangle $ABC$ shown in Figure 14.4, then:

$$\sin(\theta) = \frac{o}{h}, \qquad \cos(\theta) = \frac{a}{h} \qquad \text{and} \qquad \tan(\theta) = \frac{o}{a}. \tag{14.1}$$

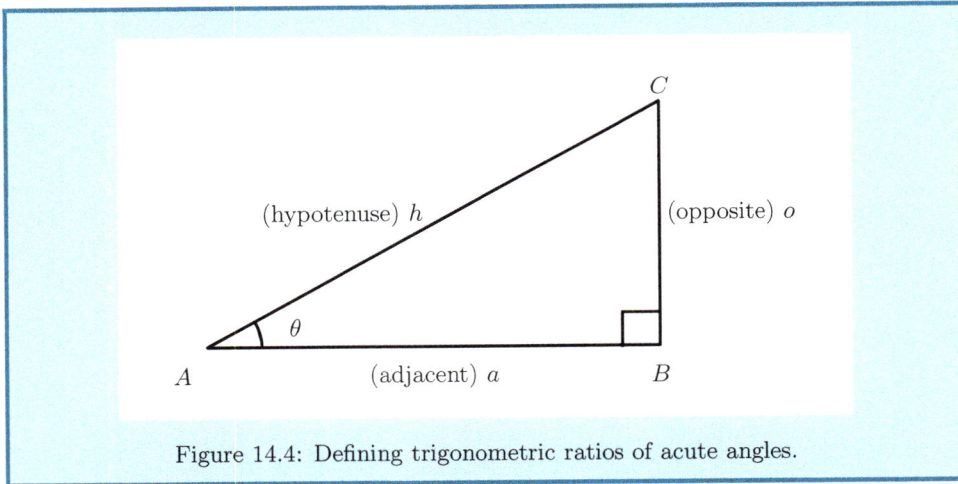

Figure 14.4: Defining trigonometric ratios of acute angles.

**Remark**
Remember that the longest side of a right angled triangle, which is always the side opposite the right angle, is termed the *hypotenuse*. Calling one of the other angles $\theta$, then the side nearest to $\theta$ (not the hypotenuse) is called the *adjacent* side. The remaining side is termed the *opposite* side as it is opposite to $\theta$. Hence, the opposite and adjacent sides of a right angled triangle depend on the angle being considered, while the hypotenuse is fixed.

**Tip**
It can be helpful to invent a mnemonic sentence, the funnier the better, to remember these ratios. Although many different ones can be found on the internet, it is the creation process which plays the most important role in aiding the memory.

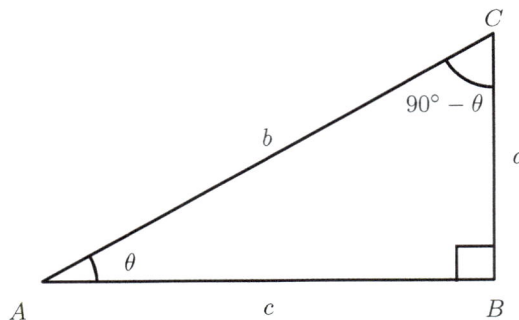

Figure 14.5

By considering the triangle shown in Figure 14.5, noticing that $\angle C = 90° - \theta$ and applying the trigonometric ratios at $C$, we have

$$\sin(90° - \theta) = \frac{c}{b} = \cos(\theta), \tag{14.2}$$

$$\cos(90° - \theta) = \frac{a}{b} = \sin(\theta), \tag{14.3}$$

$$\tan(90° - \theta) = \frac{c}{a} = \frac{1}{\tan(\theta)}. \tag{14.4}$$

When we know values of one of these trigonometric ratios, we can find the angle $\theta$ using the *inverse sine*, *inverse cosine* or *inverse tangent* which are denoted $\arcsin(\theta)$, $\arccos(\theta)$, $\arctan(\theta)$, respectively, or $\sin^{-1}(\theta)$, $\cos^{-1}(\theta)$, $\tan^{-1}(\theta)$.

---

**Example 14.1**

Find the angle $\theta$ shown in the triangle below.

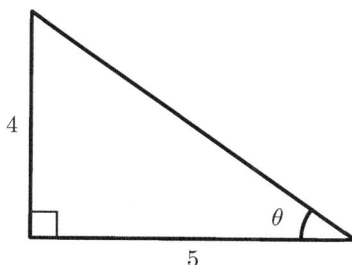

**Solution:**

We can proceed in a couple of different ways.

**Method 1:** We are given the sides adjacent to, and, opposite the angle $\theta$ and so we can use the tangent ratio;

$$\tan(\theta) = \frac{4}{5},$$

$$\Rightarrow \qquad \theta = \arctan\left(\frac{4}{5}\right)$$

$$= 38.65980825$$

$$= 38.66 \quad \text{to 2 d.p.}$$

**Method 2:** Since we have a right angle triangle we can use Pythagoras to find the hypotenuse. Letting $x$ represent the hypotenuse,

$$x = \sqrt{4^2 + 5^2}$$

$$= \sqrt{41}.$$

---

We can then calculate $\theta$ using either the cosine or sine ratios:

$$\cos(\theta) = \frac{5}{\sqrt{41}},$$

$$\Rightarrow \qquad \theta = \arccos\left(\frac{5}{\sqrt{41}}\right)$$

$$= 38.66 \quad \text{to 2 d.p.},$$

or,

$$\sin(\theta) = \frac{4}{\sqrt{41}},$$

$$\Rightarrow \qquad \theta = \arcsin\left(\frac{4}{\sqrt{41}}\right)$$

$$= 38.66 \quad \text{to 2 d.p.}$$

**Tip**

Either of these methods would score full marks in an exam. However, mathematicians value elegance in calculation and Method 1 is certainly more elegant and concise than Method 2. Where possible, aim to produce an elegant and concise correct solution to a problem.

**Example 14.2**

A triangle $ABC$ has side lengths $AB = 5$, and $BC = 13$. Given that the $BC$ is the hypotenuse, find angle $C$.

**Solution:**

Letting $\theta$ represent angle $C$, this angle will be opposite side $AB$. As we have also been given the length of the hypotenuse we use the sine ratio.

$$\sin(\theta) = \frac{5}{13},$$

$$\Rightarrow \qquad = \arcsin\left(\frac{5}{13}\right),$$

$$\Rightarrow \qquad = 22.61986495,$$

$$\Rightarrow \qquad = 22.62 \quad \text{to 2 d.p.}$$

**Example 14.3**

Can right angled trigonometry be used to find the missing angles in the triangle $ABC$, where $AB = 20$, $BC = 21$ and $AC = 27$.

**Solution:**

$|AB|^2 + |BC|^2 = 20^2 + 21^2 = 841 \neq 27^2 = 729$, and so Pythagoras' Theorem does not hold, so the triangle is not right angled.

**Remark**

The $\arcsin(\theta)$, $\arccos(\theta)$, $\arctan(\theta)$ forms are preferred to avoid confusion with the notation $x^{-1}$ and hence with $\frac{1}{\sin(\theta)}$, especially since, when we write $(\sin(\theta))^2$, it is the convention to shorten it to $\sin^2(\theta)$. The $\sin^{-1}$ notation has become the more commonly used version on calculators (presumably because of limited space for printing). In fact, there is a very good geometrical reason to prefer the arcsin notation since the values returned are directly proportional to the arc lengths on a *unit circle* (*i.e.* a circle of radius 1).

**Activity 14.1**

Use the trigonometric ratios in a right angled triangle to label the remaining sides and angles of these triangles.

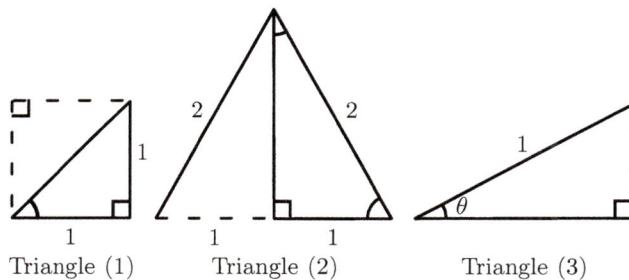

Triangle (1)    Triangle (2)    Triangle (3)

**Remark**

We are often expected to know the ratios for some standard angles; recognising the numbers and surds involved can help us to instantly see an easier way through some problems. An easy way of learning them or checking them is to draw small diagrams of the triangles above. These values should be known:

| Angle | Sine | Cosine | Tangent |
|-------|------|--------|---------|
| 0° | 0 | 1 | 0 |
| 30° | $\frac{1}{2}$ | $\frac{\sqrt{3}}{2}$ | $\frac{1}{\sqrt{3}}$ |
| 45° | $\frac{1}{\sqrt{2}}$ | $\frac{1}{\sqrt{2}}$ | 1 |
| 60° | $\frac{\sqrt{3}}{2}$ | $\frac{1}{2}$ | $\sqrt{3}$ |
| 90° | 1 | 0 | - |

These ratios are required knowledge for GCSE, but that they do not need to be memorised if students are only doing AS.

**Example 14.4**
Using trigonometry find the height of the tree in the figure below.

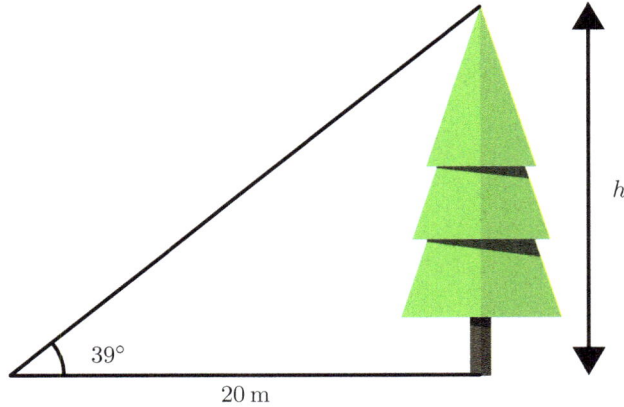

**Solution:**
From the Figure, the height of the tree is opposite the angle of inclination that we know. Since we also have the length of of the adjacent side we use the tangent ratio.

$$\tan(39) = \frac{h}{20},$$
$$\Rightarrow \qquad h = 20\tan(39)$$
$$= 16.19568066.$$

Answers must be given to a suitable degree of accuracy, so rounding to three significant figures gives $h = 16.2\,\text{m}$.

**Exercise 14.1**

Q1. Use the definitions of $\sin(\theta)$, $\cos(\theta)$ and $\tan(\theta)$ as appropriate with the given information to determine the missing sides and ratios in the right-angled triangle $ABC$ with $C = 90°$.
    (a) $\sin(A) = \frac{3}{5}$, $AB = 5$cm;
    (b) $\tan(A) = \frac{4}{3}$, $AB = 0.1$m;
    (c) $\cos(B) = \frac{5}{13}$; $AC = 48$mm.

Q2. It is given that $\theta = 30°$, the side opposite $\theta$ is $x$ and the hypotenuse is $x + 3$. Find the value of $x$.

Q3. Given that $\sin(\theta) = \frac{2}{3}$, find $\cos(\theta)$ and $\tan(\theta)$.

Q4. If $\tan(\theta) = \frac{1}{3}$, find $\sin(\theta)$ and $\cos(\theta)$.

Q5. Find the angle $\theta$ in the triangles below.

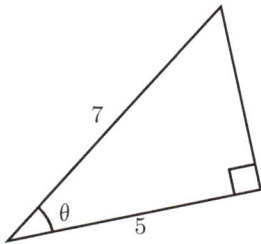

(a)

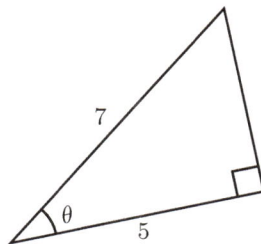

(b)

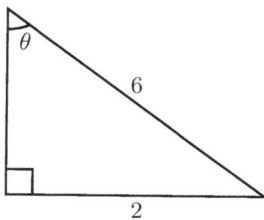

(c)

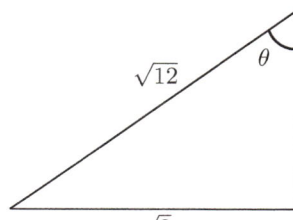

(d)

Q6. Find $x$ and the length of the hypotenuse in the case that $\theta = 30°$.

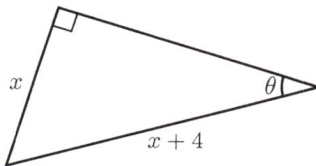

Q7. Find the indicated sides in the triangles below.

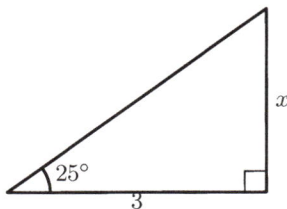

(a)

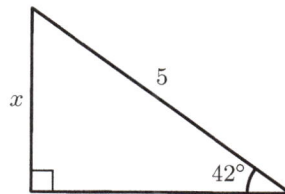

(b)

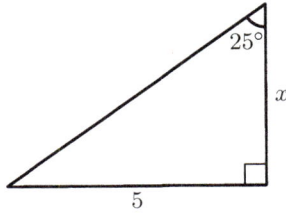

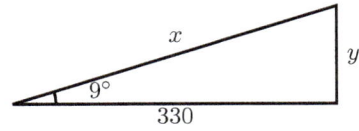

(c)                                                    (d)

Q8. John stands at a point $50\,\mathrm{m}$ from the Seattle Space Needle and measures two angles of inclination, $\theta_1 = 72.6°$ and $\theta_2 = 74.8°$ as shown in the figure below. Given that $\theta_1$ measures the angle of inclination to the base of the restaurant level, calculate the distance between the restaurant and the top of the tower.

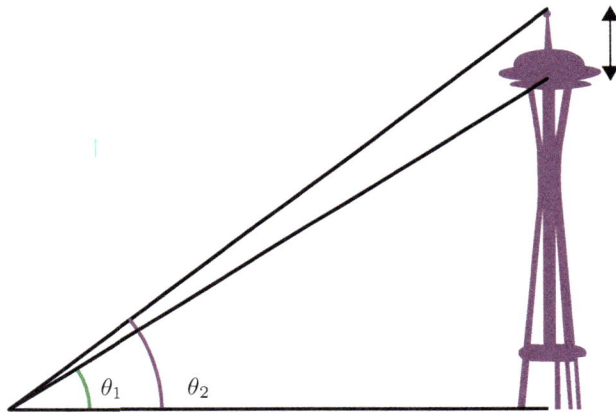

## 14.2  Trigonometric Functions for Angles of any Magnitude

As we know, angles need not only be acute, but can be obtuse or reflex and a glance at sports such as snowboarding tells us we can continue to give meaning to angles even larger than $360°$. For example, a 1080 (said "ten-eight") is a spin making three complete turns. In some instances we may find it useful to refer to negative angles. In all of these cases it is less clear what the trigonometric ratios mean and so, the definitions of the trigonometric functions are extended, as described below.

Let us consider Triangle (3) from Activity 14.1 and place it at the origin on a Cartesian plane, as in Figure 14.6. We can see that point $C$ has coordinates $(\sin(\theta), \cos(\theta))$. We can also see that there are three other points with coordinates which could immediately be written down.

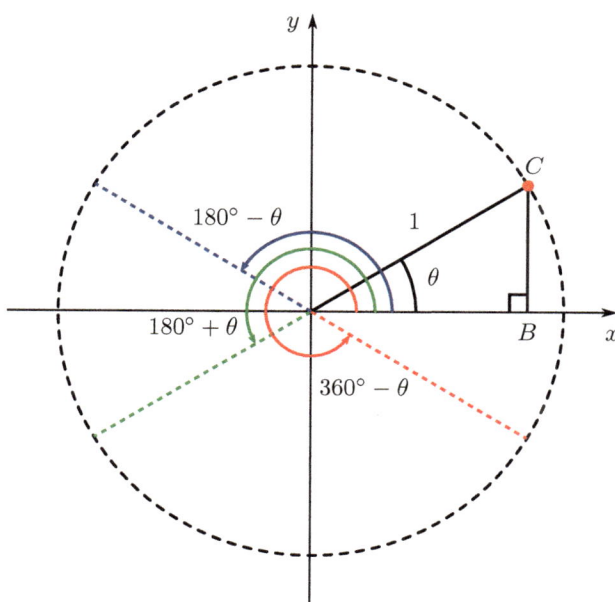

Figure 14.6: Defining trigonometric ratios of any angle

**Definition 14.2 — Trigonometric functions of any angle**

We define trigonometric ratios for any angle as

$$\sin(\theta) = y \quad \text{and} \quad \cos(\theta) = x, \tag{14.5}$$

for a point $C$ with coordinates $(x, y)$ on a unit circle, where $\theta$ is the angle that $OC$ makes with the positive $x$-axis when measured anticlockwise.

**Activity 14.2 — Describing the Tangent Function**

What does $\tan(\theta)$ describe?

**Interactive Activity 14.1**

By using the Geogebra activities below, explore the shape of the curves defined by the trigonometric functions and the relationships to the coordinates of a point which moves around a unit circle.

After a full revolution has been made (with $\theta = 360°$) we arrive back at the starting point on the $x$-axis. Hence, the angle $360°$ is essentially the same as $0°$. If we continue to increase $\theta$, the same path around the circle is traced out and the definitions of sine and cosine will mean their curves are repeated. Similarly, if we begin at $\theta = 0$ and decrease $\theta$, we will revolve clockwise around the circle and the same curves will once again be drawn. Plots of the sine and cosine functions are presented in Figure 14.7 over the interval $-360 \leq \theta \leq 720$. We immediately observe the repeating pattern of both sine and cosine and this repetition occurs every $360°$. We say that sine and cosine are *periodic* with *period* $360°$. In mathematical terms we have

$$\sin(\theta) = \sin(\theta + 360°),$$
$$\cos(\theta) = \cos(\theta + 360°).$$

We also notice, since $(\sin(\theta), \cos(\theta))$ represent coordinates on a unit circle, that $-1 \leq \sin(\theta) \leq 1$ and $-1 \leq \cos(\theta) \leq 1$.

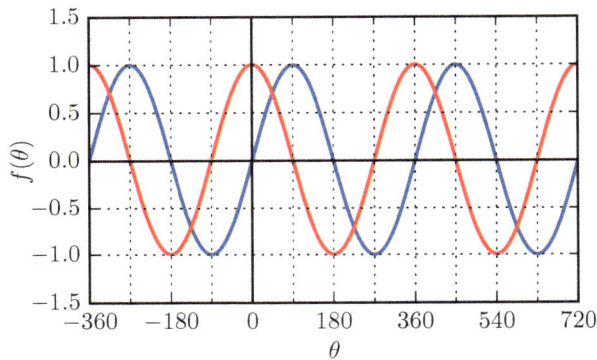

Figure 14.7: Plots of the sine (blue curve) and cosine (red curve) functions for $-360° \leq \theta \leq 720°$.

By considering Figure 14.7 in more detail, we see that the curve of cosine has the same shape as sine, but shifted to the left by $90°$. We also observe that there is extra symmetry in both the sine and cosine curves. In summary, we have the following important properties.

---

**Formulae 14.1 — Sine and Cosine Identities**

$$\sin(\theta) = \sin(\theta + 360°), \tag{14.6}$$
$$\cos(\theta) = \cos(\theta + 360°), \tag{14.7}$$
$$\cos(\theta) = \sin(\theta + 90°), \tag{14.8}$$
$$\sin(\theta) = \sin(180° - \theta) = -\sin(180° + \theta) = -\sin(360° - \theta), \tag{14.9}$$
$$\cos(\theta) = -\cos(180° - \theta) = -\cos(180° + \theta) = \cos(360° - \theta). \tag{14.10}$$

**Example 14.5**

Plot, on the same pair of axes, the functions $y = \cos(x)$, $y = 2\cos(x)$ and $y = \cos(x + 30°)$.

**Solution:**

We use the techniques described in Chapter 10 to transform the graph of $y = \cos(x)$. $y = 2\cos(x)$ is a stretch parallel to the $y$-axis by a scale factor of three and is shown in red. The graph of $y = \cos(x + 30°)$ is a translation of $y = \cos(x)$, 30 degrees to the left, and is shown in green. The original function, $y = \cos(x)$ is shown in blue.

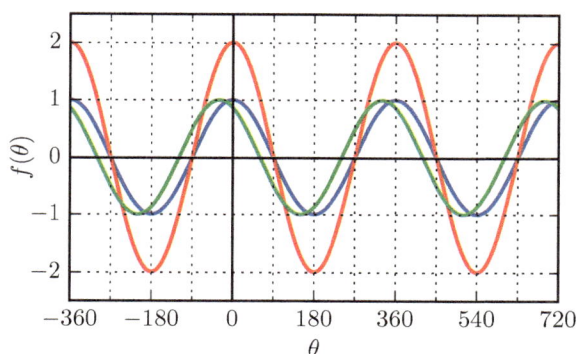

Using the definition $\tan(\theta) = \frac{\sin(\theta)}{\cos(\theta)}$, it is simple to plot the tangent function over the same interval $-360° \leq \theta \leq 720°$, as in Figure 14.8. On this occasion we observe that $\tan(\theta)$ is again a periodic function, but with period 180°. We also notice that $-\infty < \tan(\theta) < \infty$ and the curve has asymptotes at $\theta = -90°$, $\theta = 90°$, *etc.* The function is undefined at these values because $\cos(\theta) = 0$ when $\theta$ is a multiple of 90°. Again, there is extra symmetry to the curve of th                                                                                              ities.

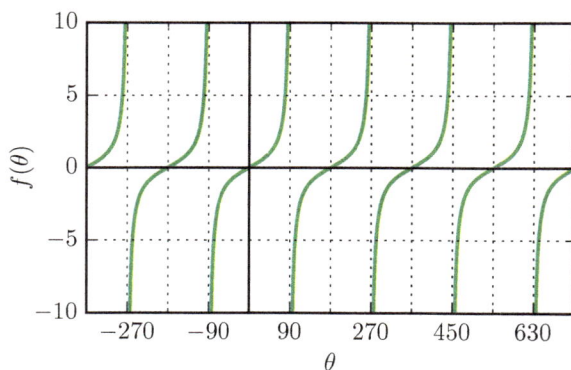

Figure 14.8: Plot of the tangent function for $-360° \leq \theta \leq 720$

**Formula 14.2 — Tangent identities**

$$\tan(\theta) = -\tan(180° - \theta) = \tan(180° + \theta) = -\tan(360° - \theta). \qquad (14.11)$$

**Example 14.6**

Given that $\sin(26°) = 0.438\,(3\text{ s.f.})$, write down $\sin(206°)$ and find $\cos(116°)$.

**Solution:**
Using formula (14.9), we have

$$\sin(206°) = \sin(26° + 180) = -\sin(26°) = -0.438\,(3\text{ s.f.})$$

Similarly, using (14.8), we have

$$\cos(116°) = \sin(116° + 90°) = \sin(206°) = -0.438\,(3\text{ s.f})$$

It is often useful to refer to a point or the size of an angle as being in a specific *quadrant*, which means which quarter of the Cartesian plane it is in (with angles being measured anticlockwise from the positive $x$-axis). We number the quadrants from first to fourth anticlockwise starting with the area where $x$ and $y$ are positive.

**Tip**

Many people find it helpful to use a mnemonic to help them remember in which quadrants the different ratios are positive. Two common ones are CAST (Cosine, All, Sine, Tangent) taking the quadrants anticlockwise starting with the 4th then 1st to 3rd, another is ACTS starting in the first quadrant and progressing clockwise.

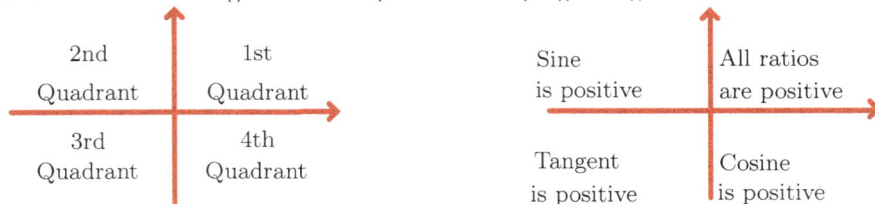

|  |  |
|---|---|
| 2nd Quadrant | 1st Quadrant |
| 3rd Quadrant | 4th Quadrant |

|  |  |
|---|---|
| Sine is positive | All ratios are positive |
| Tangent is positive | Cosine is positive |

Figure 14.9: Positivity of trigonometric functions.

**Example 14.7**

The line segment between point $P$ and the origin has gradient $\sqrt{3}$ and $P$ has $x$-coordinate $-4$. Find the $y$-coordinate and the angle $OP$ makes with the positive $x$-axis.

**Solution:**
We recognise that one possible angle is $\sqrt{3} = \tan(60°)$ (or we could use our calculator to find $\arctan(\sqrt{3})$).
The tangent is positive, so $OP$ will either be in quadrant 1 or 3, however, we discard quadrant 1 as the $x$-coordinate of $P$ is negative and hence the angle cannot be $60°$.

Considering Figure 14.9, $P$ must therefore be in quadrant 3 and (14.11) reveals that $OP$ must be at an angle of $180° + 60° = 240°$ (or $-(180° - 60°) = -120°$) to the $x$-axis.

A sketch of this situation is shown below.

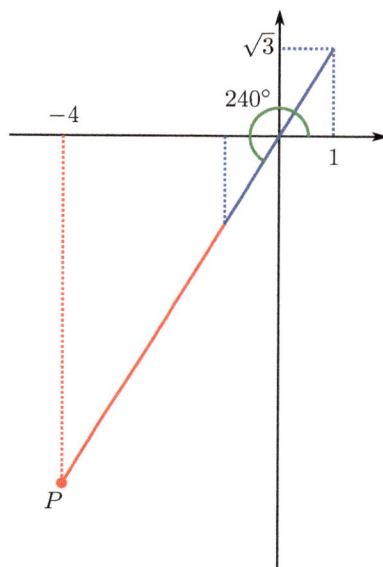

We can see that we have a right angled triangle with base 4 and base angle $60°$. Hence, the magnitude of the $y$-coordinate is $4\sqrt{3}$ (by scaling useful Triangle (3)) and the diagram reminds us that it will take a negative value in the third quadrant.

Thus, point $P$ has coordinates $(-4, -4\sqrt{3})$.

**Exercise 14.2**

Q1. (a) Sketch $y = \sin(x)$;
    (b) Sketch $y = \sin(2x)$;
    (c) Sketch $y = 2\sin(x)$;
    (d) Sketch $y = \sin(x + 30°)$

Q2. (a) Sketch $y = \cos(x)$;
    (b) Sketch $y = \cos(3x)$;
    (c) Sketch $y = 3\cos(x)$;
    (d) Sketch $y = \cos(x - 45°)$

Q3. (a) Sketch $y = \tan(x)$;
    (b) Sketch $y = \tan\left(\frac{1}{2}x\right)$;
    (c) Sketch $y = 4\tan(x)$;
    (d) Sketch $y = \tan(x + 60°)$

Q4. By definition $\sin(60°) = \frac{\sqrt{3}}{2}$. Find the other values $x$, such that $-540° < x < 540°$

and $\sin(x) = \frac{\sqrt{3}}{2}$.

Q5. By definition $\cos(120°) = -0.5$. Find the other values $x$, such that $-540° < x < 540°$ and $\cos(x) = -0.5$.

Q6. By definition $\tan(60°) = \sqrt{3}$. Find the other values $x$, such that $-360° < x < 360°$ and $\tan(x) = \sqrt{3}$.

Q7. By definition $\tan(45°) = 1$. Find the other values $x$, such that $-360° < x < 360°$ and $\tan(3x) = 1$.

Q8. The line segment between point $P$ and the origin has gradient $\frac{1}{\sqrt{3}}$

  (a) $P$ has $x$-coordinate $-3$. Find the $y$-coordinate and the angle $OP$ makes with the positive $x$-axis.

  (b) $Q$ has $x$-coordinate 2. Find the $y$-coordinate and the angle $OQ$ makes with the positive $x$-axis.

## 14.3   The Area of a Triangle

Using these trigonometric functions, it is straightforward to derive the formula for the area of any triangle.

> **Formula 14.3 — Area of a Triangle**
> The area $A_T$ of a triangle $ABC$ with sides of length $a$, $b$, and $c$ is $A_T = \frac{1}{2}ab\sin(C)$.

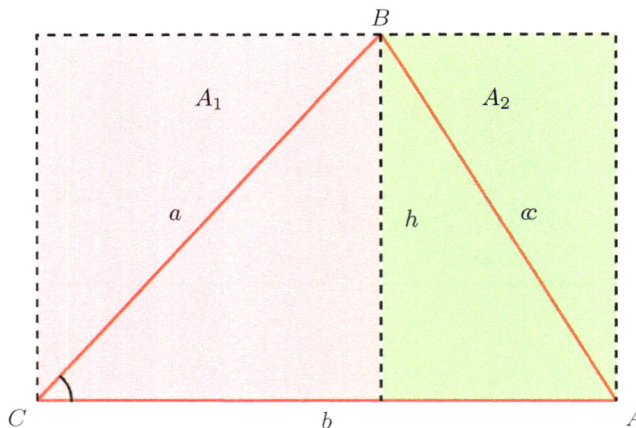

Figure 14.10: Finding the area of a triangle.

Consider the triangle $ABC$ in Figure 14.10 and the rectangle which encloses it. Clearly the triangle can be split into two smaller right angled triangles, with areas $\frac{A_1}{2}$ and $\frac{A_2}{2}$. Hence, $A_T = \frac{A_1+A_2}{2}$. Now, $A_1 + A_2$ is the area of the enclosing rectangle and $A_1 + A_2 = bh$. $h$ is found using trigonometry and is given by $h = a\sin(C)$. Thus, $A_T = \frac{1}{2}ab\sin(C)$.

> **Activity 14.3 — Area of a Triangle With an Obtuse Angle**
> Show that the formula for the area of a triangle holds if the triangle has an obtuse angle at $C$.

**Example 14.8**

Find the area of triangle $ABC$ where $AB = 5.83$, $AC = 7.62$ and $\angle BAC = 54.16°$.

**Solution:**

Considering triangle $ABC$, $\angle BAC = 54.16°$ is between sides $AB$ and $AC$. Thus, by Formula 14.3 we have that the area, A, is

$$A = \frac{1}{2} \times 5.83 \times 7.62 \times \sin(54.16),$$

$$= 18.00651747,$$

$$= 18.01 \quad \text{to 2 d.p.}$$

**Example 14.9**

The shape shown below is the design for a commercial logo. Given that the logo is symmetric along the line $AD$ find the area of the logo.

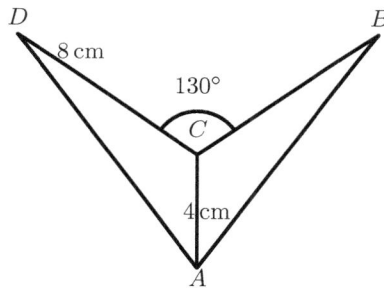

**Solution:**

Since the logo is symmetric angle $ACB$ is the same as angle $ACD$. Hence,

$$\angle ACB = \frac{360 - 130}{2}$$

$$= \frac{230}{2}$$

$$= 115°$$

Consider the triangle ACD; we have been given two sides and we have just found the angle between these sides. Hence, we can use Formula 14.3 to find its area, which for simplicity we call A.

$$A = \frac{1}{2} \times 4 \times 8 \times \sin(115)$$

$$= 14.50092459$$

$$= 14.50 \quad \text{to 2 d.p.}$$

Hence, the area of the whole logo is double this, namely $29\,\text{cm}^2$.

**Exercise 14.3**

Q1. Find the area of the triangles below, giving answers to two decimal places.

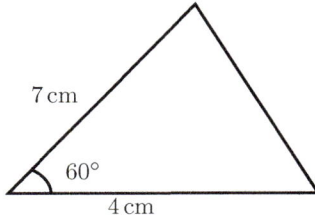

(a)

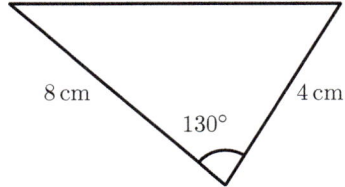

(b)

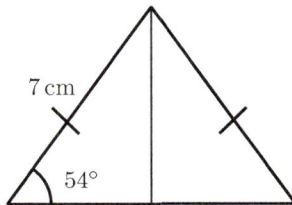

(c)

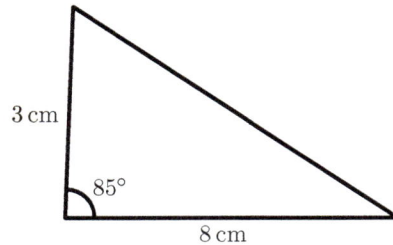

(d)

Q2. For the triangle shown below,
  (a) Find the area using:
    i. the formula for the area of a right angled triangle;
    ii. the general formula for the area of a triangle.
  (b) Which is the most efficient method? Why?

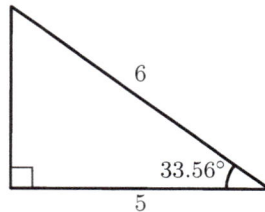

Q3. An equilateral triangle has area $16\sqrt{3}$ cm$^2$. Find the length of each side.

Q4. An isosceles triangle with two side lengths of $x$ cm and one of $y$ cm with base angles $55°$ has area 21 cm$^2$. Find the value of $x$.

Q5. Given that the area of the triangle below is $10\sqrt{3}$ cm$^2$, find $x$.

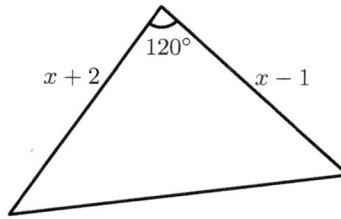

Q6. Given that the area of the triangle below is $35$ cm$^2$, find the possible values for the angle $\theta$.

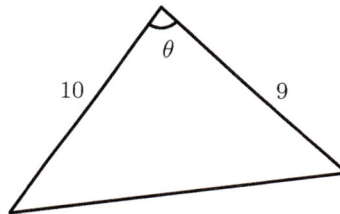

## 14.4 The Sine Rule

So far, we have only used trigonometry to find the lengths of sides/angles of right angled triangles. In many situations this is not sufficient. The *sine rule* enables us to find missing angles/lengths of sides of general triangles in certain circumstances.

> **Formula 14.4 — Sine Rule**
> For any triangle $ABC$
> $$\frac{\sin(A)}{a} = \frac{\sin(B)}{b} = \frac{\sin(C)}{c}. \tag{14.12}$$

### Proof

This rule is a direct extension of the basic trigonometric definitions. It can be proved first by considering a right angled triangle and then moving on to a general triangle.
In a right angled triangle $ABC$ (right-angle at $B$) we can immediately state that:

$$\sin(A) = \frac{a}{b} \quad \text{and} \quad \sin(C) = \frac{c}{b}. \tag{14.13}$$

Rearranging these we find

$$\frac{\sin(A)}{a} = \frac{1}{b} = \frac{\sin(C)}{c}.$$

Since $\sin(90°) = 1$, we already have the sine rule relationship.

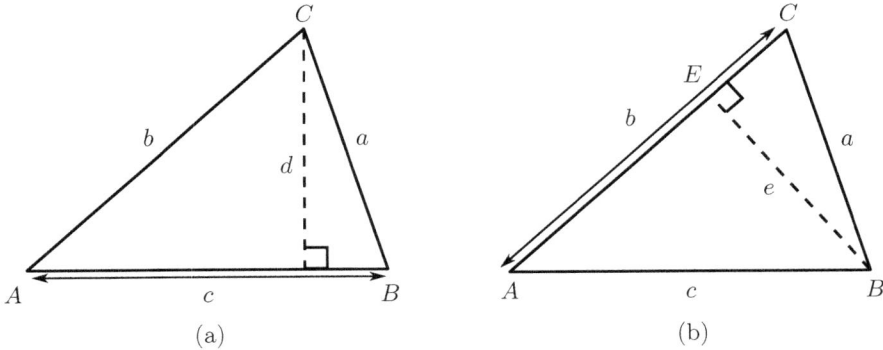

Figure 14.11: A triangle can be partitioned into two right-angled triangles in different ways.

We can now consider a more general triangle $ABC$ shown in Figure 14.11. First it is partitioned into two right-angled triangles by dropping a perpendicular from vertex $C$ to side $AB$, as in Figure 14.11(a). We use the ratios

$$\sin(A) = \frac{d}{b} \quad \text{and} \quad \sin(B) = \frac{d}{a}.$$

Rearranging gives

$$d = b\sin(A) = a\sin(B)$$
$$\Rightarrow \qquad \frac{\sin(A)}{a} = \frac{\sin(B)}{b}. \tag{14.14}$$

Alternatively, splitting into two right angles triangles by dropping a perpendicular from vertex $B$ to side $AC$, as in Figure 14.11(b), leads us to the following equation

$$\frac{\sin(A)}{a} = \frac{\sin(C)}{c}. \tag{14.15}$$

The two equations (14.14) and (14.15) are combined to create the three part equation (14.12) which is the Sine rule.

If the triangle has an obtuse angle, the perpendicular will drop from the vertex to the opposite side *produced* (this means extended in a straight line). In this case we make use of the fact that $\sin(180° - \theta) = \sin(\theta)$.

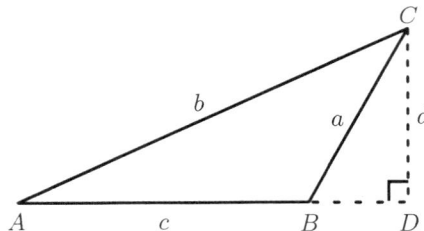

Figure 14.12: For a triangle with an obtuse angle, the line $AB$ is produced to the point $D$.

Consider Figure 14.12, we have,

$$\sin(A) = \frac{d}{b} \quad \text{and} \quad \sin(180° - B) = \sin(B) = \frac{d}{a}.$$

These equations now rearrange as before.

---

**Example 14.10**

Consider the triangle shown below. Find the missing angle and use the sine rule to determine the missing side lengths.

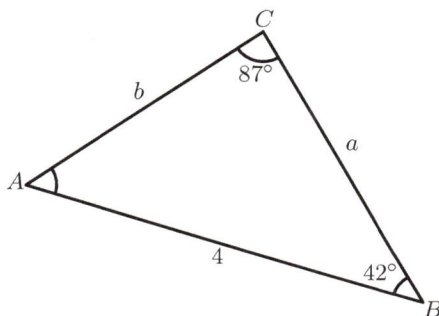

**Solution:**

We have $\angle A = 180° - 42° - 87° = 51°$. Then, using the sine rule, we have that

$$\frac{\sin(87°)}{4} = \frac{\sin(51°)}{a} = \frac{\sin(42°)}{b}.$$

Hence, rearrangement of both parts of these equations gives

$$a = \frac{4\sin(51°)}{\sin(87°)} = 3.11 \text{ to 4 d.p.}$$

and

$$b = \frac{4\sin(42°)}{\sin(87°)} = 2.68 \text{ to 4 d.p.}$$

---

**Tip**

Note that the fractions in the sine rule could equally well have been written as

$$\frac{a}{\sin A} = \frac{b}{\sin B} = \frac{c}{\sin C} \tag{14.16}$$

---

**Remark**

We can use whichever orientation suits our needs. In Example 14.10, the alternative form given in equation 14.16, would have been easier to work with. When in this form the ratios are all equal to the diameter of the triangle's *circumcircle* (*i.e.* the circle which passes through all three vertices).

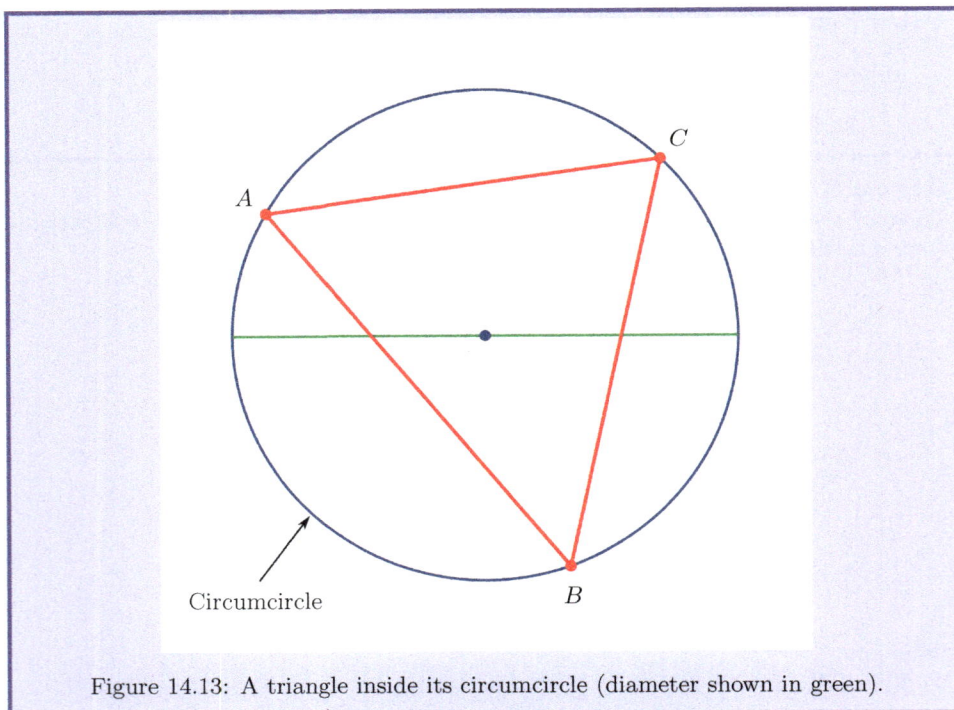

Figure 14.13: A triangle inside its circumcircle (diameter shown in green).

**Activity 14.4 — Exploring the Circumcircle**
In this activity we look in more detail at the property described in the remark above.
  Q1. Prove that $\frac{a}{\sin A} = d$ (where $d$ is the diameter of the circumcircle).
  Q2. In this orientation of the sine rule, each fraction has a sine value as the denominator. Sine is a function which can be zero, which would be an issue in a denominator. Explain why this is not a problem.

## 14.4.1  The Ambiguous Case

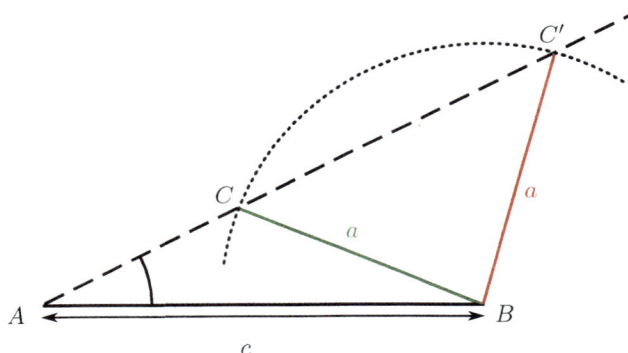

Figure 14.14: The ambiguous case when using the sine rule.

Suppose we are given just one angle and the lengths of two sides of a triangle, one of which is opposite the angle. Then suppose we we try to construct the triangle (using ruler and compasses). As can be seen in Figure 14.14, there are two possible triangles. This is known as *the ambiguous case*.

---

**Remark**

If we think back to GCSE work on congruent triangles, we might remember that the situation SSA (side, side, angle) did not give congruent triangles, whereas, SAS (side, angle, side) did. Here the same SSA situation arises when we are told the side opposite the only given angle. Hence, we are led to the ambiguous case. The geometrical reason is the same in both cases.

---

**Example 14.11**

Find the remaining sides and angles of triangle $ABC$ when $\angle A = 28°$, $a = 5$ and $b = 8$.

**Solution:**

We start by using two parts of the sine rule equation (14.12) to find the angle at $B$.

$$\frac{\sin(A)}{a} = \frac{\sin(B)}{b},$$

$$\Rightarrow \quad \frac{\sin(28°)}{5} = \frac{\sin(B)}{8},$$

$$\Rightarrow \quad \sin(B) = \frac{8\sin(28°)}{5}$$

$$= 1.6 \times 0.46947\ldots.$$

Hence, $\angle B = \arcsin(0.75115...) = 48.7°$ (1 d.p.).

We notice that the sides given include the one opposite to the given angle, so this might be an ambiguous case.

The calculator will give an acute angle and we must find the related obtuse angle which might be valid: $\angle B = 180° - 48.7° = 131.3°$(1 d.p.).

There is one angle and one side left to find.

$\angle C = 180° - 28° - 48.7° = 103.3°$ (1 d.p.) or $\angle C = 180° - 28° - 131.3° = 20.7$ (1 d.p.). Using the sine rule again (this time version (14.16)), we find the length of $c$ in both cases:

$$\frac{a}{\sin(A)} = \frac{c}{\sin(C)},$$

$$\Rightarrow \quad \frac{5}{\sin(28°)} = \frac{c}{\sin(C)},$$

$$\Rightarrow \quad c = \frac{5\sin(C)}{\sin(28°)}.$$

The obtuse angle then gives

$$c = \frac{5\sin(103.3°)}{\sin(28°)} = 10.36 \quad \text{to 2 d.p.}$$

while the acute angle gives

$$c = \frac{5\sin(20.7°)}{\sin(28°)} = 3.76 \quad \text{to 2 d.p.}$$

This is an ambiguous case and there are two possible triangle solutions:

$$c = 3.76, \quad \angle B = 48.7°, \quad \angle C = 103.3°$$

or

$$c = 10.36, \quad \angle B = 131.3°, \quad \angle C = 20.7°$$

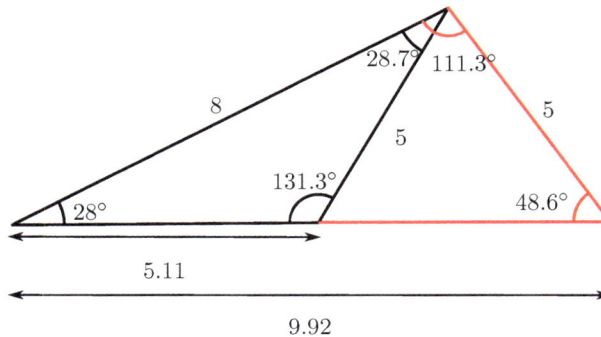

---

**Tip**
Note that in the second part of the example we use the pair $\angle A$ and $a$ (as given in the question) rather than $\angle B$ and $b$. Generally, to reduce the risk of compounding rounding errors, or following on from a mistake we might have made, we choose to use as much given information and as few of our own calculated answers as possible.

Note that we do not always get two possible triangles when we are given this collection of information. Consider the next (very similar) example.

---

**Example 14.12**
Find the remaining sides and angles of triangle $ABC$ where $\angle A = 28°$, $a = 9$ and $b = 8$.

**Solution:**
We start by using two parts of the sine rule equation (14.12) again to find the angle at

$B$ with

$$\frac{\sin(A)}{a} = \frac{\sin(B)}{b}$$

$$\Rightarrow \quad \frac{\sin(28°)}{9} = \frac{\sin(B)}{8}$$

$$\Rightarrow \quad \sin(B) = \frac{8\sin(28°)}{9}$$

Hence,

$$\sin(B) = \frac{8}{9} \times 0.46947\ldots$$

$$\Rightarrow \quad \angle B = \arcsin(0.4173\ldots)$$

$$= 24.7° \text{ (1 d.p.)}$$

We notice that the sides we were given included the one opposite to the only angle given so this might be an ambiguous case.

We find the related obtuse angle: $\angle B = 180° - 24.7° = 155.3°$ (1 d.p.).
But $28° + 155.3° = 183.3° > 180°$, so this cannot be the answer and there is in fact only one possible value for $\angle C$:

$$\angle C = 180° - 28° - 24.7° = 127.3° \text{ (1 d.p.)}.$$

Using the sine rule we find

$$c = \frac{9\sin(127.3)}{\sin(28°)} = 15.2 \text{ (1 d.p.)}.$$

---

**Question**
How can we tell whether the obtuse angle is feasible and, hence, whether we have an ambiguous case?

---

Returning the calculation of the height of a mountain that was discussed in the introduction to this chapter, consider the Example below.

---

**Example 14.13**
A cartographer wishes to calculate the height of Mont Ventoux. To do this he has measured the angle of inclination of the summit of the mountain from a particular point to be $\theta_1 = 7.26°$. He then does the same from a point that is $b = 10\,\text{km}$ further away from the mountain and obtains an angle of inclination of $\theta_2 = 4.37°$.
   (a) Calculate the height of Mont Ventoux.
   (b) What assumptions have been made in this calculation.

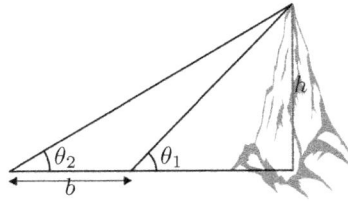

**Solution:**

(a) We first label the diagram of the situation we are considering for clarity.

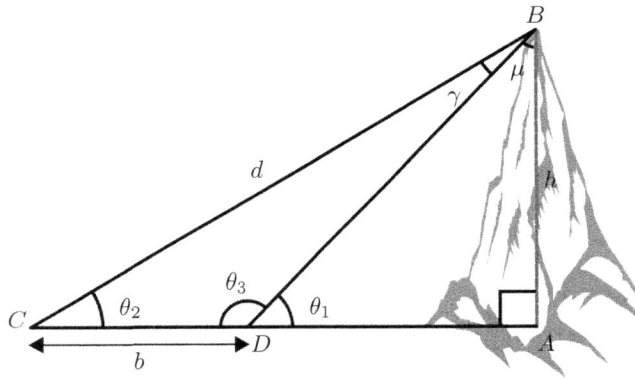

From the information given we know $\theta_1 = 7.26°$, $\theta_2 = 4.37°$ and $d = 10$ km. Since the height of a mountain is usually measured in metres, we convert the distance $d$ to metres, thus $b = 10\,000$ m.

We then calculate some of the missing angles.

- Considering triangle $ABD$,

$$\mu = 180 - 90 - 7.26,$$
$$= 82.74°.$$

- Since $\theta_1$ and $\theta_3$ lie on straight line,

$$\theta_3 = 180 - \theta_1,$$
$$= 172.74°.$$

- Considering the triangle $DBC$ we can calculate $\angle\gamma$ as follows.

$$\gamma = 180 - 172.74 - 4.37,$$
$$= 2.89°.$$

Since we know all angles in triangle $DBC$ and one of the sides we can use the sine rule to find the length $d$.

$$\frac{\sin(2.89)}{10000} = \frac{\sin(172.74)}{d},$$
$$\Rightarrow \qquad d = \frac{10000\sin(172.74)}{\sin(2.89)},$$
$$= 25064.56542.$$

Finally, $d$ is the hypotenuse of the right angled triangle $ABC$, and so we can use right angled trigonometry to determine the height of Mont Ventoux.

$$\sin(4.37) = \frac{h}{d},$$
$$\Rightarrow \qquad h = 25064.5642\sin(4.37),$$
$$= 1909.84374,$$
$$\approx 1909\,\text{m}.$$

(b) The cartographer has assumed that the two locations where he has measured the angle of inclination from are horizontally level. The height of Mont Ventoux is accepted to be $1912\,\text{m}$.

**Remark**

This is a simplified setting. In practise the calculations were more involved as it was hard to keep the location of any measurement points on a horizontal level and this had to be corrected for using other trigonometrical calculations.

**Exercise 14.4**

Q1. Decide if this triangle is physically possible, justifying your answer.

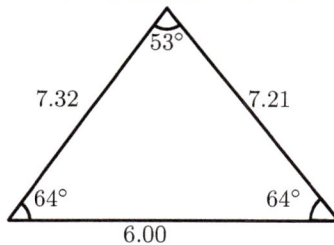

Q2. For the triangle shown below, use the given data in each situation to find $x$.

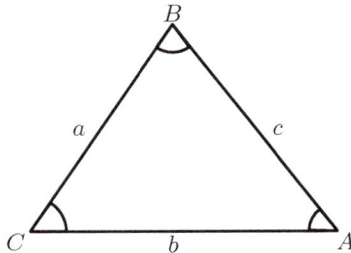

(a) $|AC| = 8$, $\angle ABC = 60°$, $\angle BCA = 45°$, $AB = x$.
(b) $|AB| = 7.3$, $\angle CAB = 30°$, $\angle BCA = 70°$, $BC = x$.
(c) $|AB| = 22.4$, $\angle ABC = 130°$, $\angle BCA = 40°$, $CA = x$.
(d) $|AC| = 9$, $\angle ABC = 50°$, $\angle BCA = 70°$, $AB = x$.
(e) $|BC| = 12$, $\angle CAB = 100°$, $\angle BCA = 30°$, $AB = x$.

Q3. For the triangle shown in Q2., use the given information to find $\theta$ in each case.

(a) $|AB| = 6$, $|BC| = 7$, $\angle CAB = 30°$, $\angle BCA = \theta$.
(b) $|AB| = \sqrt{79}$, $|BC| = 7$, $\angle CAB = 45°$, $\angle BCA = \theta$.
(c) $|AC| = 8$, $|BC| = 7$, $\angle CAB = 50°$, $\angle ABC = \theta$.
(d) $|AB| = 6$, $\angle CAB = 30°$, $\angle BCA = 73$, $\angle ABC = \theta$.
(e) $|CA| = 9$, $|BC| = 11$, $\angle CAB = 72°$, $\angle ABC = \theta$.

Q4. Town B lies 8 miles away from Town A on a bearing of 045°. Town C is located on a bearing of 160° from B and a bearing of 95° from A.
Calculate

(a) The angle $CBA$;
(b) The distance of Town C from Town A.
(c) The distance of Town C from Town B.

Q5. Use the information in the diagram below to calculate the height of Mount Everest. Given that Everest has an accepted height of 8848 m comment on the accuracy of your calculation. What effect would a changing the inclination of 5.05° to 5.75° have?

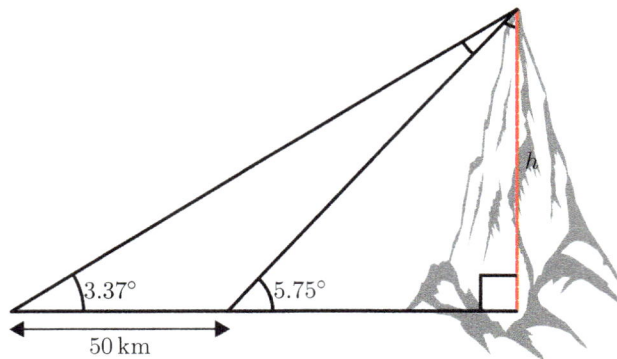

Q6. In triangle $ABC$, the length $AB = x$, the length $BC = x + 2$, angle $CAB = 60°$ and the sine of angle $ACB$ is $\frac{1}{\sqrt{6}}$. Find an expression for $x$.

## 14.5  The Cosine Rule

Consider the sine rule equation in a triangle $ABC$ when the pieces of information given are $a$, $b$ and $\angle C$:

$$\frac{\sin(A)}{\mathbf{a}} = \frac{\sin(B)}{\mathbf{b}} = \frac{\sin(\mathbf{C})}{c}.$$

In this situation, a two part equation containing only one unknown cannot be isolated, hence, we cannot find the missing sides and angles for this triangle. In this case we need an alternative approach: the *cosine rule*.

**Formula 14.5 — Cosine rule**
For a triangle $ABC$

$$c^2 = a^2 + b^2 - 2ab\cos(C). \tag{14.17}$$

**Proof**

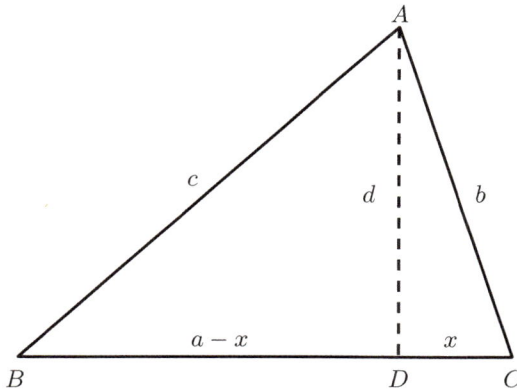

Figure 14.15: Triangle for use in the proof of the cosine rule.

Figure 14.15 shows a triangle $ABC$, split into two right angled triangles so that the line $CD$ is perpendicular to $AB$.

By defining $x = |DC|$, we use Pythagoras' theorem to find $d$ in both triangles $ADC$ and $DBC$

$$d^2 = c^2 - (a - x)^2 \quad \text{and} \quad d^2 = b^2 - x^2.$$

Hence, noting that $x = c\cos(B)$, we have

$$c^2 - (a-x)^2 = b^2 - x^2,$$
$$\Rightarrow \quad c^2 - a^2 + 2ax - x^2 = b^2 - x^2,$$
$$\Rightarrow \qquad\qquad c^2 = a^2 + b^2 - 2ax,$$
$$\Rightarrow \qquad\qquad c^2 = a^2 + b^2 - 2ab\cos(C).$$

We could use any of the angles as the apex of this triangle which leads to two other versions of Cosine Rule:

$$a^2 = b^2 + c^2 - 2bc\cos(A)$$
$$b^2 = a^2 + c^2 - 2ac\cos(B)$$

---

**Tip**
We most commonly see

$$a^2 = b^2 + c^2 - 2bc\cos(A),$$

---

**Activity 14.5 — Proving the Cosine Rule**
Construct a proof for Cosine Rule where angle $C$ is obtuse.

---

**Example 14.14**
Find the length of the side $a$ in the triangle below.

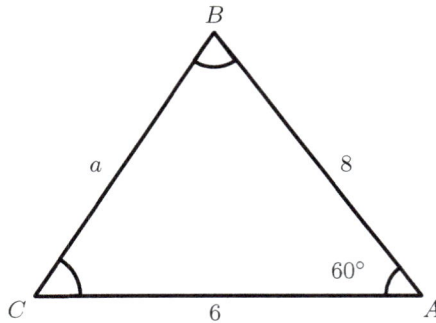

**Solution:**
We use the cosine rule,

$$a^2 = b^2 + c^2 - 2bc\cos(A),$$
$$= 8^2 + 6^2 - 2 \times 8 \times 6 \times \cos(60),$$
$$= 64 + 36 - 96 \times \frac{1}{2},$$
$$= 100 - 48,$$
$$= 52.$$

Hence, $a = \sqrt{52}$.

We can also use the cosine rule to find missing angles, assuming that we know the lengths of all three sides of the triangle.

**Tip**

We can rearrange the cosine rule into a formula to find the cosine of a missing angle.

$$a^2 = b^2 + c^2 - 2bc\cos(A),$$
$$\Rightarrow \quad a^2 + 2bc\cos(A) = b^2 + c^2,$$
$$\Rightarrow \quad 2bc\cos(A) = b^2 + c^2 - a^2,$$
$$\Rightarrow \quad \cos(A) = \frac{b^2 + c^2 - a^2}{2bc}.$$

**Example 14.15**

Find all missing angles in the triangle below.

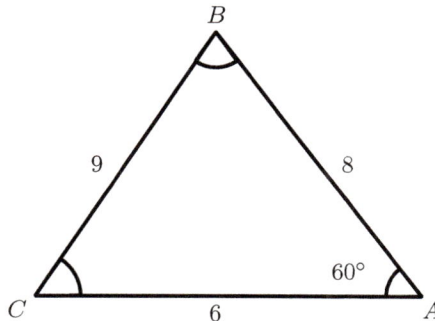

**Solution:**
Let $\theta_1 = \angle CAB$, $\theta_2 = \angle ABC$ and $\theta_3 = \angle BCA$. Using the rearrangement of the cosine

rule,

$$\cos(\theta_1) = \frac{6^2 + 8^2 - 9^2}{2 \times 6 \times 8},$$

$$= \frac{19}{96},$$

$$\cos(\theta_2) = \frac{9^2 + 8^2 - 6^2}{2 \times 9 \times 8},$$

$$= \frac{109}{144},$$

$$\cos(\theta_1) = \frac{9^2 + 6^2 - 8^2}{2 \times 9 \times 6},$$

$$= \frac{53}{108}.$$

Hence, using the inverse tangent function we have,

$$\theta_1 = \arccos\left(\frac{19}{96}\right),$$

$$\approx 78.58°,$$

$$\theta_2 = \arccos\left(\frac{109}{144}\right),$$

$$\approx 40.80°,$$

$$\theta_2 = \arccos\left(\frac{53}{108}\right).$$

$$\approx 60.62°.$$

**Exercise 14.5**

Q1. Find the lengths indicated below.
   (a) For the triangle $ABC$, where $AB = 8.2$cm, $AC = 4.7$cm and angle $\angle CAB = 120°$, find the length of the side $BC$.
   (b) For the triangle $ABC$, where $AB = 4$m, $AC = 2$m and angle $\angle CAB = 60°$, find the length of the side $BC$.
   (c) For the triangle $ABC$, where $AB = 5.6$km, $AC = 3.1$km and angle $\angle CAB = 50°$, find the length of the side $BC$.

Q2. Consider the triangle below

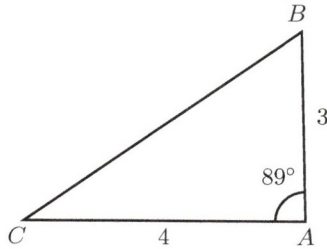

(a) Find the length of the missing side.
(b) If this triangle had been approximated as a right angled triangle what would have been the percentage error made when finding the missing length?

Q3. A lighthouse is located at a point $A$. A ship is located at point $B$ which is 50km away from $A$ on a bearing of $13°$ from $A$. Another ship is located at point $C$ which is on a bearing of $37°$ and at a distance of 70km from the lighthouse. How far apart are the two ships?

Q4. A golfer hits a ball 180m towards a hole that is 290m away. She hits the ball $10°$ off course. How far away from the hole is the ball?

Q5. Calculate the size of the angle labelled in the triangles below.

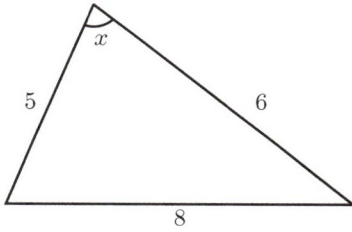

(a)

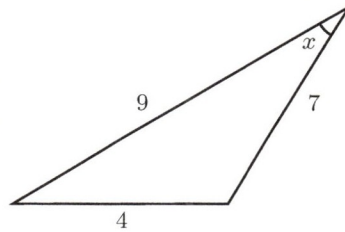

(b)

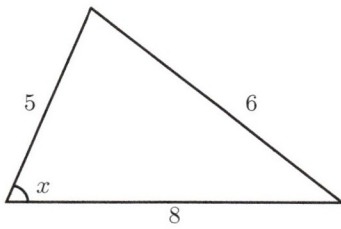

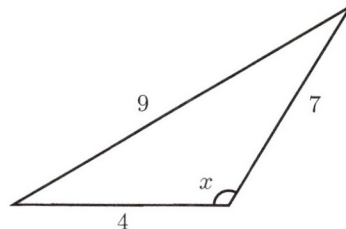

Q6. Consider the triangle below.

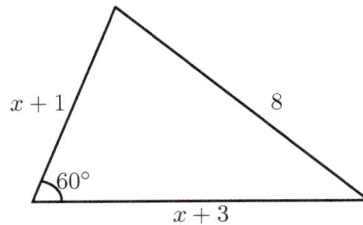

(a) Show that $x^2 + 4x - 57 = 0$.

(b) Hence, or otherwise, find the lengths of the two sides which are given in terms of $x$.

Q7. Show, that for the triangle below, $\cos(A) = \frac{1}{21}$.

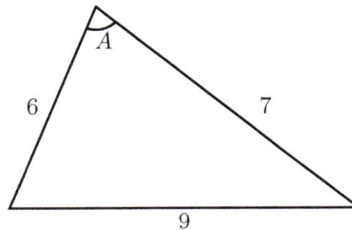

Q8. A plane flies 200km on a bearing of 40° from $A$ to $B$. From $B$ it flies 100km to $C$. Given that the distance $AC$ is 120km, what is the bearing of $C$ from $A$.

**Interactive Activity 14.2 — Using Trigonometry to Model Sunrise and Sunset Times**

Our knowledge of trigonometry and of circle geometry, from this chapter and Chapter 13, respectively, is enough to provide a simple model for sunset/sunrise times at any latitude and at any time of year. The Geogebra applet below (in the digital book) implements one such model. There are two modes in the applet, one is the 'Year View' mode which plots the number of hours of daylight against day of the year, the other is the 'Day View' mode, which plots whether the sun is up or down against hour in the day. Note, that day one is the Winter Solstice in the Northern Hemisphere.

Q1. What assumptions could have been made to derive this model. What improvements could be made?

Q2. What times does the model predict for sunset and sunrise in London on

(a) The Winter Solstice;

(b) The Solstice.

How do these results compare with the actual times for these dates in London? How do the results compare for other cities in the world?

**Key Facts — Introduction to Trigonometry**

- For the right angled triangle $ABC$ shown in the triangle below, then:

$$\sin(\theta) = \frac{o}{h}, \qquad \cos(\theta) = \frac{a}{h} \qquad \text{and} \qquad \tan(\theta) = \frac{o}{a}.$$

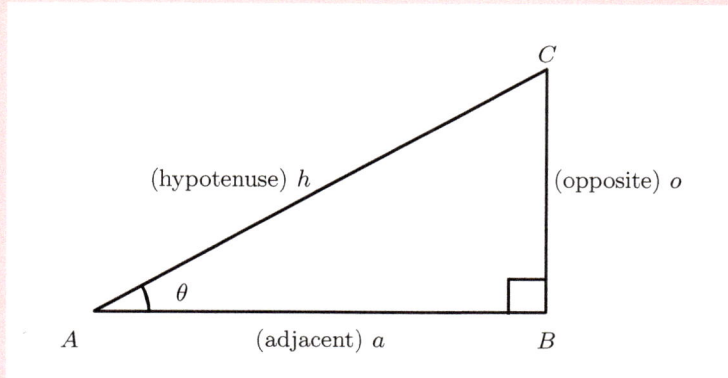

- The following values should be learnt.

| Angle | Sine | Cosine | Tangent |
|-------|------|--------|---------|
| 0° | 0 | 1 | 0 |
| 30° | $\frac{1}{2}$ | $\frac{\sqrt{3}}{2}$ | $\frac{1}{\sqrt{3}}$ |
| 45° | $\frac{1}{\sqrt{2}}$ | $\frac{1}{\sqrt{2}}$ | 1 |
| 60° | $\frac{\sqrt{3}}{2}$ | $\frac{1}{2}$ | $\sqrt{3}$ |
| 90° | 1 | 0 | - |

- The following identities hold for the the trigonometric functions.

$$\sin(\theta) = \sin(\theta + 360°),$$
$$\cos(\theta) = \cos(\theta + 360°),$$
$$\cos(\theta) = \sin(\theta + 90°),$$
$$\sin(\theta) = \sin(180° - \theta) = -\sin(180° + \theta) = -\sin(360° - \theta),$$
$$\cos(\theta) = -\cos(180° - \theta) = -\cos(180° + \theta) = \cos(360° - \theta).$$

$$\tan(\theta) = -\tan(180° - \theta)$$
$$= \tan(180° + \theta)$$
$$= -\tan(360° - \theta).$$

- The area $A_T$ of a triangle $ABC$ with sides of length $a$, $b$, and $c$ is $A_T = \frac{1}{2}ab\sin(C)$.
- The Sine rule states that, for any triangle $ABC$,

$$\frac{\sin(A)}{a} = \frac{\sin(B)}{b} = \frac{\sin(C)}{c},$$

or equivalently

$$\frac{a}{\sin A} = \frac{b}{\sin B} = \frac{c}{\sin C}$$

- Beware of the ambiguous case when using the sine rule.
- The cosine rule states that, for any triangle $ABC$,

$$c^2 = a^2 + b^2 - 2ab\cos(C),$$

or equivalently,

$$\cos(A) = \frac{b^2 + c^2 - a^2}{2bc}.$$

**Chapter Assessment — Introduction to Trigonometry**

Download and sit the 30 minute assessment for this chapter from the digital book.

# 15. Introduction to Differentiation

Differentiation forms one half of "the Calculus". Calculus was first formalised independently by Newton and Leibniz in the 17th Century and is the study of change. Differentiation is used to find the rate of change of a quantity with respect to some variable.

We are often interested in the rate of change of a quantity with respect to time, for example, velocity is the rate of change of displacement with respect to time. This shall be studied further in Chapter 26. As a further example, consider reaction rates of a chemical process. Determining these reaction rates accurately is extremely important, since, amongst others, they enable the effectiveness of catalysts to be ascertained. Reaction rates vary immensely between reactions and depend on many factors, the most important of which are:

- concentration, or pressure of reactants;
- temperature;
- presence of catalysts;
- availability of reactants.

Reaction rates can also vary dramatically during the course of a reaction, meaning that reaction rates can be modelled as functions of time. This can be demonstrated with the classic "marble chip and hydrochloric acid" experiment.

When calcium carbonate (a compound present in marble chips) reacts with dilute hydrochloric acid, carbon dioxide, amongst other products, is released. If we measure the amount of carbon dioxide at fixed time intervals, we can investigate the changing reaction rate of this chemical reaction.

After performing this experiment, the following data was obtained.

| Time (s) | Volume of gas ($cm^3$) |
|:---:|:---:|
| 0 | 0 |
| 15 | 27 |
| 30 | 47 |
| 45 | 61 |
| 60 | 69 |
| 75 | 75 |
| 90 | 80 |
| 105 | 80 |

This data can be plotted on a graph and a smooth curve of best fit drawn. We can then use a tangent to the curve at a given point to estimate the reaction rate at that time. Figure 15.1 shows the estimation of the reaction rate at times $t = 40\,$s and $t = 80\,$s. We find the reaction rates are $0.8\,cm^3s^{-1}$ and $0.36\,cm^3s^{-1}$, respectively.

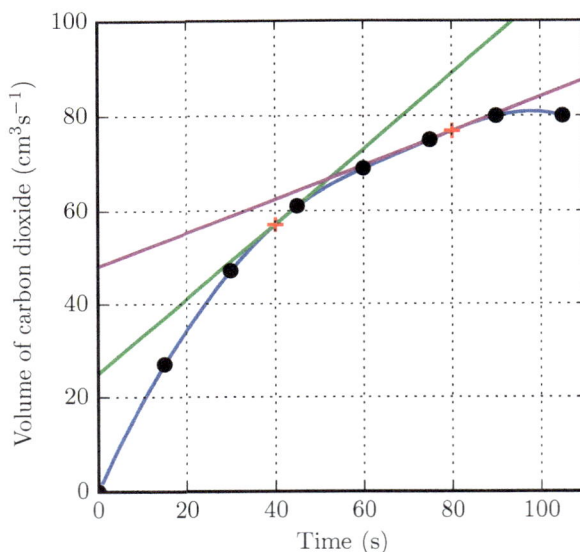

Figure 15.1: The reaction of calcium carbonate with hydrochloric acid, with tangents at $t = 40$ s (green) and at $t = 80$ s (magenta) to determine the rates of reaction.

In this chapter, we introduce differentiation and begin to explore how it can be used to solve problems such as those mentioned above.

## 15.1  The Gradient Function

In Chapter 4, the gradient of a straight line with equation $y = mx + c$ was defined to be the slope of the line, or the change in $y$ per unit change in $x$. For linear functions, such as these, the gradient is constant, and so for any $x$ in the domain of the function, the gradient has the same value.

For a general function, this is not the case and the gradient changes as we move along the curve.

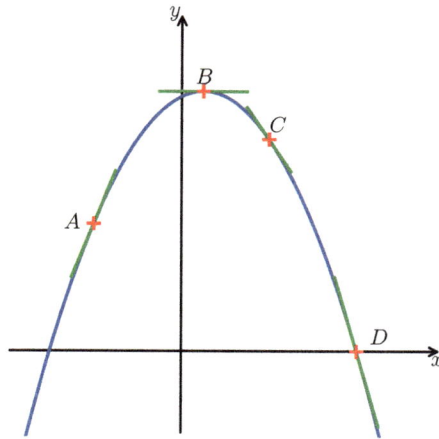

Figure 15.2: A quadratic function (blue) and gradient at different points on the curve (green).

Figure 15.2 shows a quadratic curve where the gradient depends on $x$. At point $A$ the gradient is positive and at point $B$ it is zero. The gradient is negative at both points $C$ and $D$, however it is steeper at $D$ than it is at $C$.

We can formalise the notion of a gradient to a curve as follows.

---

**Interactive Activity 15.1 — A Constantly Changing Gradient**

Use the Geogebra interactive below (in the digital book) to explore the changing value of the gradient for a variety of functions.

---

**Definition 15.1 — Gradient of a Curve**

The gradient of a curve at a given point, $A$, is defined to be the gradient of the tangent at $A$. An example is shown below.

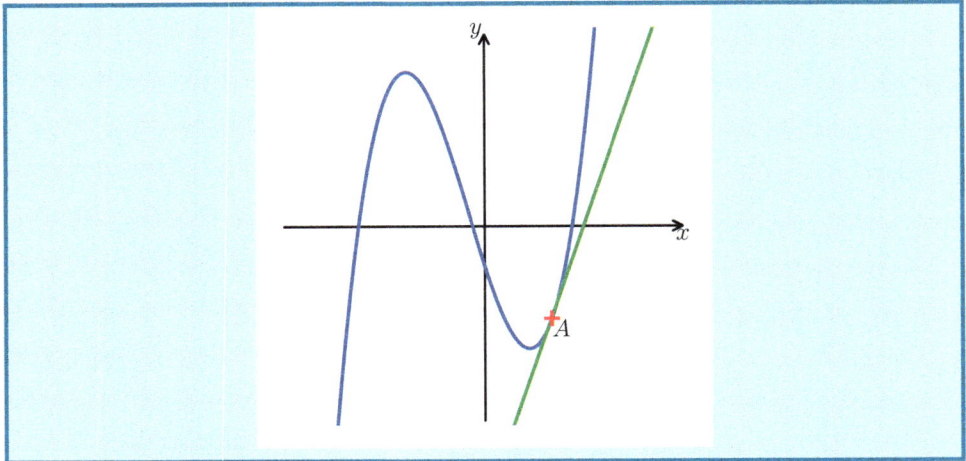

As we move the tangent along the curve of $f(x)$ we gradually build up the gradient function, which we denote $f'(x)$.

**Example 15.1**
Sketch the gradient function for $f(x)$ shown below.

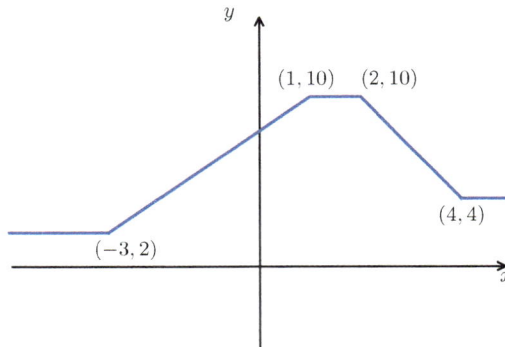

**Solution:**
Any line segments that are horizontal in the graph of $f(x)$ have gradient zero. The gradient of the straight line segments can be easily calculated as follows.

- The gradient of the line joining $(-3, 2)$ to $(1, 10)$ is 2.
- The gradient of the line joining $(2, 10)$ to $(4, 4)$ is $-3$.

With this information, the graph of the function $f'(x)$ can be sketched on the same axes in red, as above.

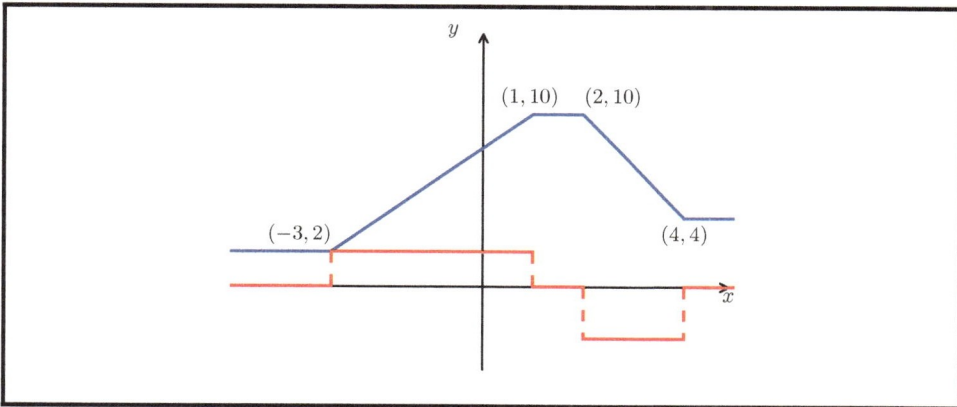

In order to compute the gradient of a function $f(x)$ that is not piecewise linear, we wish to find the gradient of the tangent to the curve at a particular point. Let us call this point $A$, as shown in Figure 15.3. The gradient of the chords $AD$, $AC$ and $AB$ provide increasingly accurate estimates to the gradient of the tangent at $A$. This suggests that, by calculating the gradient of the chord that joins point $A(x_A, f(x_A))$ to point $B(x_B, f(x_B))$ and letting $B$ become closer and closer to $A$, we can find the gradient of $f(x)$ at $A$.

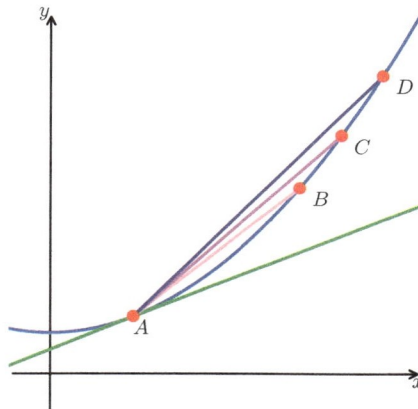

Figure 15.3: Approaching the tangent with chords.

**Interactive Activity 15.2 — Functions and their Gradient Functions**
Using the Tarquin interactive in the digital book match functions with their gradient function. Think about where the slope of the curve changes from positive to negative when completing this activity.

---

**Interactive Activity 15.3 — Gradients of Interesting Functions**

Use the Geogebra applet below (in the digital book) to investigate how the tangent changes for some interesting functions such as $f(x) = \sin\left(\frac{1}{x}\right)$ and $f(x) = x\sin\left(\frac{1}{x}\right)$. Try a couple of other functions.

---

**Example 15.2**

The blue curve in the figure below is the graph of equation $y = \frac{1}{2}x^2$. The green line is the tangent to the curve at the point $A\left(1, \frac{1}{2}\right)$, which has equation $y = x - \frac{1}{2}$.

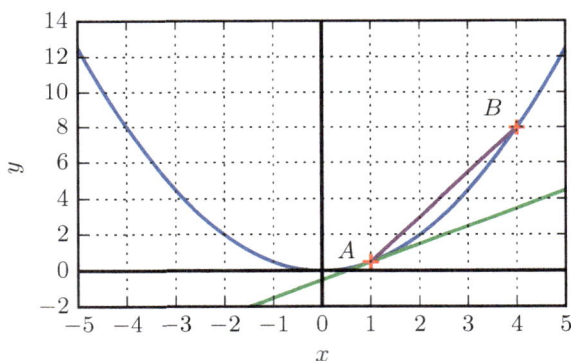

Calculate the gradient of the chord $AB$ where $B$ has the following $x$-coordinates:

(a) $x = 4$.

(b) $x = 3$.

(c) $x = 2$.

(d) $x = 1.5$.

What can be observed?

**Solution:**

Substituting the $x$-coordinate into $y = \frac{1}{2}x^2$ we find the $y$-coordinate of $B$ and we can then use Equation (4.2) from Chapter 4 to work out the gradient of the chord $AB$ in each case.

(a) Let $m_1$ denote the gradient of the chord joining $\left(1, \frac{1}{2}\right)$ to $(4, 8)$, then,

$$
\begin{aligned}
m_1 &= \frac{8 - \frac{1}{2}}{4 - 1} \\
&= \frac{5}{3} \\
&= 2.5.
\end{aligned}
$$

(b) Let $m_2$ denote the gradient of the chord joining $\left(1, \frac{1}{2}\right)$ to $\left(3, \frac{9}{2}\right)$, then,

$$
\begin{aligned}
m_2 &= \frac{\frac{9}{2} - \frac{1}{2}}{3 - 1} \\
&= \frac{4}{2} \\
&= 2.
\end{aligned}
$$

(c) Let $m_3$ denote the gradient of the chord joining $\left(1, \frac{1}{2}\right)$ to $(2, 2)$, then,

$$
\begin{aligned}
m_3 &= \frac{2 - \frac{1}{2}}{2 - 1} \\
&= \frac{\frac{3}{2}}{1} \\
&= 1.5.
\end{aligned}
$$

(d) Let $m_4$ denote the gradient of the chord joining $\left(1, \frac{1}{2}\right)$ to $\left(\frac{3}{2}, \frac{9}{8}\right)$, then,

$$
\begin{aligned}
m_4 &= \frac{\frac{9}{8} - \frac{1}{2}}{\frac{3}{2} - 1} \\
&= \frac{\frac{5}{8}}{\frac{1}{2}} \\
&= 1.25.
\end{aligned}
$$

We observe that the gradients of the chords are getting closer to 1, as point $B$ moves closer to point $A$ and the gradient of the tangent at point $A\left(1, \frac{1}{2}\right)$ is also 1.

The approach in the example above can be formalised in the following way.
For the general function $f(x)$, the gradient of the chord joining the point $A(x, f(x))$ to $B(x + h, f(x + h))$, is given by,

$$
\begin{aligned}
m &= \frac{f(x + h) - f(x)}{x + h - x}, \\
&= \frac{f(x + h) - f(x)}{h}.
\end{aligned}
$$

As the point $B$ approaches the point $A$, then $m$ approaches the gradient of the tangent at $A$, motivating the following definition for the derivative.

**Definition 15.2 — The Derivative**

The derivative of the curve $y = f(x)$ is denoted either as $f'(x)$ or $\frac{dy}{dx}$. We define $f'(x)$ to be

$$f'(x) = \lim_{h \to 0} \frac{f(x+h) - f(x)}{h}. \tag{15.1}$$

**Remark**

In Definition 15.2, $\lim_{h \to 0}$ means "in the limit as h tends to 0". Note that, we cannot evaluate

$$\frac{f(x+h) - f(x)}{h} \tag{15.2}$$

at $h = 0$, as this would involve dividing by zero. However, we can let $h$ become arbitrarily close to 0 and, as this happens, the expression (15.2) approaches a limiting value. It is this value that is defined to be the derivative at $x$.

Limits have a number of useful properties, which we will exploit later. For two functions $f(h)$ and $g(h)$ and $c \in \mathbb{R}$, we have:

$$\lim_{h \to 0} (f(h) + g(h)) = \lim_{h \to 0} f(h) + \lim_{h \to 0} g(h),$$

$$\lim_{h \to 0} cf(h) = c \lim_{h \to 0} f(h).$$

These show that *taking limits* is a linear operation. These two formulae are part of a larger set of rules known as *the algebra of limits*.

The process of finding a derivative is called *differentiation* and we might say "We differentiate $f(x)$", *etc.* A function is called *differentiable at $x$* if it has a derivative at $x$. If a function is differentiable at every $x$, we say the function is differentiable.

**Interactive Activity 15.4 — Differentiation from First Principles**

Explore the use of Definition 15.2 in computing the derivative of linear, quadratic and cubic functions using the Geogebra activity contained within the digital book. The point at which the derivative is being evaluated can be moved and the value of $h$ varied. It is possible to extend the chord joining $(x_A, f(x_A))$ to $(x_A + h, f(x_A + h))$ and observe how this approaches the equation of the tangent at $x_A$.

**Remark**

Equation (15.1) given in Definition 15.2 makes no distinction as to whether $h$ is positive or negative. If we define $h$ to be positive then (15.1) gives the *right derivative*, since the point $(x + h, f(x + h))$ is to the right of the point $(x, f(x))$.

For $h$ still positive, we can also define the *left derivative* by

$$f'(x) = \lim_{h \to 0} \frac{f(x) - f(x-h)}{h}. \tag{15.3}$$

In this case, the point $(x - h, f(x - h))$ lies to the left of $(x, f(x))$.
If instead of considering one point to the right (or left) of $(x, f(x))$ we consider two points, one each side of $(x, f(x))$ we obtain what is known as the *central derivative*

$$f'(x) = \lim_{h \to 0} \frac{f(x+h) - f(x-h))}{2h}. \tag{15.4}$$

The central derivative can be seen to be the mean of the left and right derivatives.

---

**Tip**

The notation $\delta x$ is sometimes used instead of $h$ Definition 15.2. $\delta x$ represents a small increment in $x$. With this notation, Definition 15.2 can be written as

$$f'(x) = \lim_{\delta x \to 0} \frac{f(x + \delta x) - f(x)}{\delta x}.$$

Both versions shall be used within this book.

---

**Interactive Activity 15.5 — The Accuracy of Numerical Approximations to Derivatives**

Use the Geogebra applet in the digital book to examine how the function

$$\frac{f(x+h) - f(x)}{h}$$

approaches the derivative as $h$ tends to zero.

---

Definition 15.2 can be used, along with some algebraic manipulation, to find the derivative of different functions as the examples below show.

---

**Interactive Activity 15.6 — Order the Differentiation Steps**

Using thr Tarquin interactive activity below (in the digital book), correctly order the statements to show how to differentiate the given function from first principles.

**Example 15.3**
Differentiate, from first principles, the function $y = 3x + 6$.

**Solution:**
Let $f(x) = 3x + 6$, then

$$
\begin{aligned}
f'(x) &= \lim_{h \to 0} \frac{f(x+h) - f(x)}{h} \\
&= \lim_{h \to 0} \left( \frac{3(x+h) + 6 - (3x + 6)}{h} \right) \\
&= \lim_{h \to 0} \left( \frac{3x + 3h + 6 - 3x - 6}{h} \right) \\
&= \lim_{h \to 0} (3) \\
&= 3.
\end{aligned}
$$

**Example 15.4**
Differentiate, from first principles, the function $y = x^2$.

**Solution:**
Let $f(x) = x^2$, then

$$
\begin{aligned}
f'(x) &= \lim_{h \to 0} \frac{f(x+h) - f(x)}{h} \\
&= \lim_{h \to 0} \left( \frac{(x+h)^2 - x^2}{h} \right) \\
&= \lim_{h \to 0} \left( \frac{x^2 + 2xh + h^2 - x^2}{h} \right) \\
&= \lim_{h \to 0} (2x + h) \\
&= 2x.
\end{aligned}
$$

The next example demonstrates the use of the central derivative formula in finding derivatives from first principles. Sometimes using the central derivative results in slightly more straightforward working, however, this example does not demonstrate this. We shall see in the Year 2 book, that using the central derivative results in much cleaner proofs for the derivatives of trigonometric functions.

**Example 15.5**
Find the derivative of $f(x) = x^3$ from first principles, using
  (a) Defintion 15.2 for the derivative;
  (b) the definition of the central derivative.

**Solution:**

(a) We use Definition 15.2 to find $f'(x)$:

$$f'(x) = \lim_{h \to 0} \frac{f(x+h) - f(x)}{h}$$

$$= \lim_{h \to 0} \left( \frac{(x+h)^3 - x^3}{h} \right)$$

$$= \lim_{h \to 0} \left( \frac{x^3 + 3x^2 h + 3xh^2 + h^3 - x^3}{h} \right)$$

$$= \lim_{h \to 0} \left( \frac{3x^2 h + 3xh^2 + h^3}{h} \right),$$

$$= \lim_{h \to 0} (3x^2 + 3xh^2 + h^3)$$

$$= 3x^2.$$

(b) Here, instead of Equation (15.1) we use Equation (15.4).

$$f'(x) = \lim_{h \to 0} \frac{f(x+h) - f(x-h)}{2h}$$

$$= \lim_{h \to 0} \left( \frac{(x+h)^3 - (x-h)^3}{2h} \right)$$

$$= \lim_{h \to 0} \left( \frac{x^3 + 3x^2 h + 3xh^2 + h^3 - (x^3 - 3x^2 h + 3xh^2 - h^3)}{2h} \right)$$

$$= \lim_{h \to 0} \left( \frac{6x^2 h + 2h^3}{2h} \right)$$

$$= \lim_{h \to 0} (3x^2 + h^2)$$

$$= 3x^2.$$

**Remark**

When evaluating derivatives numerically (as a scientific calculator does), using the central derivative results in closer approximations.

**Question**

With reference to the above example for $f(x) = x^3$, why does using the central derivative result in more accurate approximations to the value of a derivative at a point than using a right/left derivative?

**Tip**

If we are finding the derivative of $f(x) = x^n$ from first principles, then the Binomial Theorem (see Chapter 8) can be used to efficiently expand $(x+h)^n$.

**Activity 15.1 — Rules of Differentiation**

In this activity we shall discover some rules of differentiation.

Q1. (a) Differentiate $f(x) = x^3$ from first principles.

     (b) Differentiate $f(x) = 4x^3$ from first principles.

Q2. Differentiate $f(x) = 3x^2$ from first principles.

Q3. Differentiate $f(x) = 5x$ from first principles.

Q4. Differentiate $f(x) = 7$ from first principles.

Q5. Differentiate $f(x) = 4x^3 + 3x^2 + 5x + 7$ from first principles.

Q6. What conjectures can we make about the operation of differentiation from the results of the above calculations?

We are now in a position to prove some properties of differentiation for general functions $f(x)$ and $g(x)$.

**Theorem 15.3 — Properties of Differentiation**

Let $f(x)$ and $g(x)$ be continuous, differentiable functions and $c \in \mathbb{R}$, then,

(a) $\frac{d}{dx}[f(x) + g(x)] = \frac{d}{dx}[f(x)] + \frac{d}{dx}[g(x)]$,

(b) $\frac{d}{dx}[f(x) - g(x)] = \frac{d}{dx}[f(x)] - \frac{d}{dx}[g(x)]$,

(c) $\frac{d}{dx}[cf(x)] = c\frac{d}{dx}[f(x)]$.

**Proof:**

(a) Using Definition 15.2,

$$\frac{d}{dx}[f(x) + g(x)] = (f(x) + g(x))'$$

$$= \lim_{h \to 0}\left(\frac{f(x+h) + g(x+h) - (f(x) + g(x))}{h}\right)$$

$$= \lim_{h \to 0}\left(\frac{f(x+h) - f(x) + g(x+h) - g(x)}{h}\right)$$

$$= \lim_{h \to 0}\left(\frac{f(x+h) - f(x)}{h}\right) + \lim_{h \to 0}\left(\frac{g(x+h) - g(x)}{h}\right)$$

$$= \frac{d}{dx}[f(x)] + \frac{d}{dx}[g(x)].$$

In the above we have used the linearity properties of limits.

(b) This is analogous to part (a) with a negative sign instead of the positive sign.

(c) Again, using Definition 15.2,

$$\frac{d}{dx}[cf(x)] = \lim_{h \to 0}\frac{cf(x+h) - cf(x)}{h}$$

$$= \lim_{h \to 0} c\left(\frac{f(x+h) - f(x)}{h}\right)$$

$$= c \lim_{h \to 0}\frac{f(x+h) - f(x)}{h} \qquad \text{(using properties of limits)}$$

$$= c\frac{d}{dx}[f(x)].$$

We will make extensive use of these properties in the next section.

**Exercise 15.1**

Q1. Consider the plot below of the function $y = \frac{1}{2}x^3$.

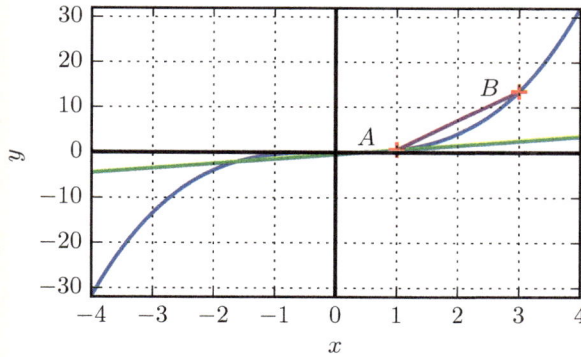

Calculate the gradient of the chord $AB$ when $B$ has the following $x$-coordinates:
(a) $x = 3$;
(b) $x = 2.5$;
(c) $x = 2$;
(d) $x = 1.75$;
(e) $x = 1.5$;
(f) $x = 1.25$;
(g) $x = 1.1$.

What can we conclude about the derivative of $y = \frac{1}{2}x^3$ at $x = 1$.

Q2. Differentiate $y = x^5$ from first principles.

Q3. Show, from first principles, that the derivative of $y = \frac{1}{x}$ is $\frac{dy}{dx} = -\frac{1}{x^2}$.

Q4. (a) Show that the point with coordinate $A(5, 75)$ lies on the curve $y = 3x^2$.
    (b) At point $A$, the gradient has a value $m$, show that $m = \lim_{h \to 0}(h^2 + 15h + 75)$.
    (c) Deduce the value of $m$.

Q5. Show, from first principles that any constant value differentiates to zero.

Q6. Show, from first principles that the derivative of $y = 3x^2 + 5x$ is $y' = 6x + 5$.

Q7. Sketch the gradient functions for the functions shown in the graphs below.

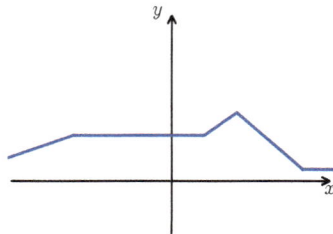

(a)

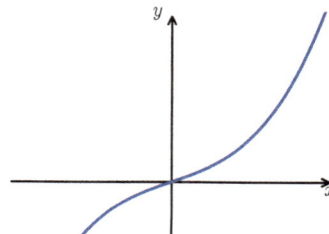

(b)

Q8. Let $f(x) = x^2 + x - 12$
  (a) Sketch $y = f(x)$.
  (b) Sketch, on the same set of axes, the derivative $f'(x)$.
  (c) Why is the $x$-coordinate of the vertex of the quadratic the same as the $x$-coordinate of the point where $f'(x)$ crosses the $x$-axis?

## 15.2  Differentiating Polynomials

Using Definition 15.2, we can find the derivatives of many functions. In this book, we begin by differentiating $y = x^n$. Once we have done this we can differentiate any polynomial.

**Theorem 15.4 — The Derivative of $f(x) = x^n$**
Let $f(x) = x^n$, where $n \in \mathbb{Q}$, then

$$f'(x) = \frac{\mathrm{d}f(x)}{\mathrm{d}x} = nx^{n-1}. \tag{15.5}$$

**Proof:**
We prove this property where $n$ is a positive integer and shall return to complete the proof (for negative and fractional $n$) in Year 2. We first prove the cases $n = 0$ and $n = 1$, before employing the Binomial Theorem from Chapter 8 to expand $(x + h)^n$ for all other positive $n$.

*Case 1: $n = 0$*
In this case $f(x) = x^0 = 1$. By Definition 15.2,

$$\begin{aligned}
\frac{\mathrm{d}f(x)}{\mathrm{d}x} &= \lim_{h \to 0} \frac{f(x+h) - f(x)}{h} \\
&= \lim_{h \to 0} \frac{1 - 1}{h} \\
&= \lim_{h \to 0} \frac{0}{h} \\
&= 0.
\end{aligned}$$

Here, $nx^{n-1} = 0x^{-1} = 0$, and so our assertion is true for $n = 0$. *Case 2: $n = 1$*
For $n = 1$, $f(x) = x^1 = x$. Using Definition 15.2 again

$$\begin{aligned}
\frac{\mathrm{d}f(x)}{\mathrm{d}x} &= \lim_{h \to 0} \frac{f(x+h) - f(x)}{h} \\
&= \lim_{h \to 0} \frac{x + h - x}{h} \\
&= \lim_{h \to 0} \frac{h}{h} \\
&= 1.
\end{aligned}$$

Now, for $n = 1$, $nx^{n-1} = 1x^{1-1} = x^0 = 1$, proving the statement for $n = 1$.

*Case 3: $n > 1$*

We first exapnd $(x + h)^n$ using the Binomial Theorem from Chapter 8 ,

$$(x + h)^n = x^n + nx^{n-1}h + \frac{n(n-1)}{2}x^{n-2}h^2 + \ldots + h^n.$$

Then we use Definition 15.2 to find the derivative of $f(x) = x^n$ from first principles:

$$
\begin{aligned}
\frac{\mathrm{d}f(x)}{\mathrm{d}x} &= \lim_{h\to0} \frac{f(x+h) - f(x)}{h} \\
&= \lim_{h\to0} \frac{(x+h)^n - x^n}{h} \\
&= \lim_{h\to0} \frac{x^n + nx^{n-1}h + \frac{n(n-1)}{2}x^{n-2}h^2 + \ldots + h^n - x^n}{h} \\
&= \lim_{h\to0} \frac{nx^{n-1}h + \frac{n(n-1)}{2}x^{n-2}h^2 + \ldots + h^n}{h} \\
&= \lim_{h\to0} \frac{h\left(nx^{n-1} + \frac{n-1}{2}x^{n-1}h + \ldots h^{n-1}\right)}{h} \\
&= \lim_{h\to0} \left(nx^{n-1} + \frac{n-1}{2}x + \ldots h^{n-1}\right) \\
&= nx^{n-1}.
\end{aligned}
$$

Hence, as we have shown the assertion to be true for $n = 0$, $n = 1$ and integers greater than 1 it is true for all positive integers. Now that this result has been proved, we can use it, in conjunction with Theorem 15.3 to prove a slightly more general result.

---

**Theorem 15.5 — The Derivative of $f(x) = ax^n$**

Let $f(x) = ax^n$, where $n \in \mathbb{Q}$, then

$$f'(x) = \frac{\mathrm{d}f(x)}{\mathrm{d}x} = anx^{n-1} \tag{15.6}$$

---

**Proof:**

It suffices to apply the third part of Theorem 15.3 to the result from Theorem 15.4.

With these two results and Theorem 15.3 from the previous section we can differentiate any polynomial function.

---

**Example 15.6**

For the following functions, $f(x)$, find the derivative, $f'(x)$.
  (a)  $f(x) = x^8$;
  (b)  $f(x) = x^{1/4}$;
  (c)  $f(x) = x^{-4}$;
  (d)  $f(x) = x^3 + 6x$.

---

**Solution:**

(a) Using Theorem 15.4,

$$f'(x) = 8x^7.$$

(b) Using Theorem 15.4,

$$f'(x) = \frac{1}{4}x^{-\frac{3}{4}}.$$

(c) Using Theorem 15.4,

$$f'(x) = -4x^{-5}.$$

(d) We use results from Theorem 15.3 and differentiate term by term to obtain

$$f'(x) = \frac{\mathrm{d}}{\mathrm{d}x}\left[x^3\right] + \frac{\mathrm{d}}{\mathrm{d}x}\left[6x\right]$$
$$= 3x^2 + 6.$$

**Remark**

In the last part of the above example, we used the result from Theorem 15.3 for differentiating the sum of two functions: $\frac{\mathrm{d}}{\mathrm{d}x}\left[f(x) + g(x)\right] = \frac{\mathrm{d}}{\mathrm{d}x}\left[f(x)\right] + \frac{\mathrm{d}}{\mathrm{d}x}\left[g(x)\right]$

We can apply this recursively to differentiate sums of more than two functions.
As an example, suppose that we wish to differentiate $y = x^3 + x^2 + 6x$.

$$\frac{\mathrm{d}y}{\mathrm{d}x} = \frac{\mathrm{d}}{\mathrm{d}x}\left[\underbrace{x^3}_{f_1(x)} + \underbrace{x^2 + 6x}_{g_1(x)}\right]$$

$$= \frac{\mathrm{d}}{\mathrm{d}x}\left[x^3\right] + \frac{\mathrm{d}}{\mathrm{d}x}\left[x^2 + 6x\right]$$

$$= \frac{\mathrm{d}}{\mathrm{d}x}\left[x^3\right] + \frac{\mathrm{d}}{\mathrm{d}x}\left[\underbrace{x^2}_{f_2(x)} + \underbrace{6x}_{g_2(x)}\right]$$

$$= \frac{\mathrm{d}}{\mathrm{d}x}\left[x^3\right] + \frac{\mathrm{d}}{\mathrm{d}x}\left[x^2\right] + \frac{\mathrm{d}}{\mathrm{d}x}\left[6x\right]$$

$$= 3x^2 + 2x + 6.$$

**Example 15.7**

Find the gradient function $f'(x)$ for $f(x) = 4x^5 + 3x^2$.

**Solution:**

We differentiate term-by-term:

$$f'(x) = \frac{\mathrm{d}}{\mathrm{d}x}\left[4x^5 + 3x^2\right]$$
$$= \frac{\mathrm{d}}{\mathrm{d}x}\left[4x^5\right] + \frac{\mathrm{d}}{\mathrm{d}x}\left[3x^2\right]$$
$$= 20x^4 + 6x.$$

It is often necessary to perform some algebraic manipulations before applying the differentiation rules above.

**Example 15.8**

Differentiate $f(x) = (x^2 + 3x)(2x + 4)$ with respect to $x$.

**Solution:**

We have two bracketed terms multiplied together, but we have no simple way of performing the differentiation. Hence, we expand the brackets first, so that we have a sum of powers of $x$, which we can differentiate.

$$(x^2 + 3x)(2x + 4) = 2x^3 + 10x^2 + 12x.$$

Hence,

$$\frac{\mathrm{d}}{\mathrm{d}x}\left[(x^2 + 3x)(2x + 4)\right] = \frac{\mathrm{d}}{\mathrm{d}x}\left[2x^3 + 10x^2 + 12x\right]$$
$$= 6x^2 + 20x + 12.$$

**Example 15.9**

Find the derivative, with respect to $x$, of the function $f(x) = \frac{4x^2 - 7x}{2x}$. Evaluate this derivative at the point $x = 0$.

**Solution:**

First, we note that

$$\frac{4x^2 - 7x}{2x} = \frac{4x^2}{2x} - \frac{7x}{2x},$$
$$= 2x - \frac{7}{2}.$$

We can now differentiate this using the tools we have to obtain

$$f'(x) = 2.$$

When $x = 0$, $f'(x) = 2$.

**Example 15.10**

Differentiate, with respect to $x$, the function

$$f(x) = \frac{3x^3 + 2x^2 + 6x}{x^2}.$$

**Solution:**

We use linearity and then differentiate term-by-term:

$$\begin{aligned}
\frac{\mathrm{d}f(x)}{\mathrm{d}y} &= \frac{\mathrm{d}}{\mathrm{d}x}\left[\frac{3x^3 + 2x^2 + 6x}{x^2}\right] \\
&= \frac{\mathrm{d}}{\mathrm{d}x}\left[\frac{3x^3}{x^2}\right] + \frac{\mathrm{d}}{\mathrm{d}x}\left[\frac{2x^2}{x^2}\right] + \frac{\mathrm{d}}{\mathrm{d}x}\left[\frac{6x}{x^2}\right] \\
&= \frac{\mathrm{d}}{\mathrm{d}x}\left[3x\right] + \frac{\mathrm{d}}{\mathrm{d}x}\left[2\right] + \frac{\mathrm{d}}{\mathrm{d}x}\left[\frac{6}{x}\right] \\
&= 3 - \frac{6}{x^2}.
\end{aligned}$$

---

**Example 15.11**

Find the derivative function of $f(x) = \frac{x^2 + 5x + 6}{(x+3)}$.

**Solution:**

By factorising the quadratic, we obtain $x^2 + 5x + 6 = (x+3)(x+2)$ and so,

$$\begin{aligned}
\frac{x^2 + 5x + 6}{(x+3)} &= \frac{(x+3)(x+2)}{x+3} \\
&= x + 2.
\end{aligned}$$

Hence, differentiating, we obtain,

$$\begin{aligned}
\frac{\mathrm{d}f(x)}{\mathrm{d}x} &= \frac{\mathrm{d}}{\mathrm{d}x}\left[\frac{x^2 + 5x + 6}{(x+3)}\right] \\
&= \frac{\mathrm{d}}{\mathrm{d}x}\left[x + 2\right], \\
&= 1.
\end{aligned}$$

---

**Remark**

At present, we are not able to differentiate functions of the form $f(x) = \frac{g(x)}{h(x)}$ unless $h(x)$ is a factor of $g(x)$. In Year 2, we will learn techniques to perform this differentiation, as well as methods to differentiate products and quotients more efficiently.

If we differentiate the function $f(x)$ twice, we find what is known as the second derivative of $f(x)$, denoted by

$$\frac{\mathrm{d}^2 f(x)}{\mathrm{d}x^2},$$

or by

$$f''(x).$$

The second derivative is understood as the rate of change of the gradient since,

$$\frac{\mathrm{d}^2 f(x)}{\mathrm{d}x^2} = \frac{\mathrm{d}f'(x)}{\mathrm{d}x}.$$

---

**Example 15.12**
For the function $y = 5x^3 + 7x^2 + 2$, find
 (a)

$$\frac{\mathrm{d}y}{\mathrm{d}x};$$

 (b)

$$\frac{\mathrm{d}^2 y}{\mathrm{d}x^2}.$$

**Solution:**
 (a)

$$\frac{\mathrm{d}y}{\mathrm{d}x} = \frac{\mathrm{d}}{\mathrm{d}x} \left[ 5x^3 + 7x^2 + 2 \right]$$
$$= 15x^2 + 14x.$$

 (b)

$$\frac{\mathrm{d}^2 y}{\mathrm{d}x^2} = \frac{\mathrm{d}^2}{\mathrm{d}x^2} \left[ 5x^3 + 7x^2 + 2 \right]$$
$$= \frac{\mathrm{d}}{\mathrm{d}x} \left[ 15x^2 + 14x \right]$$
$$= 30x + 14.$$

---

**Example 15.13**
As we shall see in Chapter 26, given a function $f(t)$ that describes the displacement of a particle, the acceleration of the particle is given by the second derivative of $f(t)$. If $f(t) = t^4 + 3t^2$, what can be said about the acceleration of the particle in question?

**Solution:**
We find the second derivative of $f(t)$.

$$\frac{\mathrm{d}^2 f(t)}{\mathrm{d}t^2} = \frac{\mathrm{d}}{\mathrm{d}t}\left[\frac{\mathrm{d}}{\mathrm{d}t}\left[t^4 + 3t^2\right]\right]$$
$$= \frac{\mathrm{d}}{\mathrm{d}x}\left[4t^3 + 6t\right]$$
$$= 12t^2 + 6.$$

Hence, the acceleration of the particle is always positive.

**Remark**
As velocity is the rate of change of displacement, it naturally follows that differentiation should be performed with respect to the time variable, $t$, and displacement should be a function of $t$ as well. Differentiating with respect to $t$ is the same as for $x$, we just substitute $t$ for $x$ in Definition 15.2.

### 15.2.1  Increasing and Decreasing Functions

By sketching the graph of a function, we have an intuitive understanding of when a function is either increasing or decreasing. This concept can be formalised by the definitions below.

**Definition 15.6 — Increasing Function on the Interval $[a, b]$**
A function, $f(x)$, is said to be increasing on the interval $[a, b]$, if $f(x) \geq 0$ for all $x \in [a, b]$.

**Definition 15.7 — Decreasing Function on the Interval $[a, b]$**
A function, $f(x)$ is said to be decreasing on the interval $[a, b]$, if $f(x) \leq 0$ for all $x \in [a, b]$.

**Example 15.14**
Find the intervals for which $f(x) = x^2 - 4x + 2$ is
  (a) increasing;
  (b) decreasing.

**Solution:**
We differentiate $f(x)$ to find $f'(x)$.

$$f'(x) = \frac{\mathrm{d}}{\mathrm{d}x}\left[x^2 - 4x + 2\right]$$
$$= 2x - 4.$$

(a) The function is increasing when $f'(x) \geq 0$, hence,

$$2x - 4 \geq 0,$$
$$\Rightarrow \quad 2x \geq 4,$$
$$\Rightarrow \quad x \geq 2.$$

(b) The function is decreasing when $f'(x) \leq 0$, hence

$$2x - 4 \geq 0,$$
$$\Rightarrow \quad 2x \leq 4,$$
$$\Rightarrow \quad x \leq 2.$$

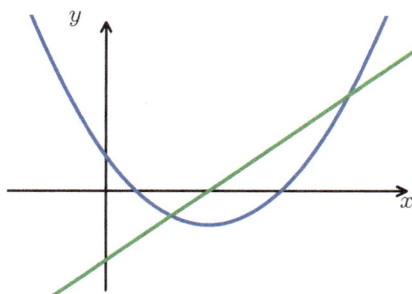

The figure above shows $f(x)$ (blue) and $f'(x)$ (green) on the same axes. Note that, where $f(x)$ changes from being a decreasing to an increasing function, the gradient $f'(x)$ crosses the $x$-axis.

**Example 15.15**

Show that the function $f(x) = \frac{x^3}{3} + 3x^2 + 9x + 6$ is an increasing function for all $x \in \mathbb{R}$.

**Solution:**

We first find the gradient function $f'(x)$:

$$f'(x) = \frac{\mathrm{d}}{\mathrm{d}x}\left[\frac{x^3}{3} + 3x^2 + 9x + 6\right]$$
$$= x^2 + 6x + 9$$
$$= (x + 3)^2.$$

Since $(x + 3)^2 \geq 0$ for all $x \in \mathbb{R}$, $f(x)$ is an increasing function for all real values of $x$. This is shown graphically in the figure below.

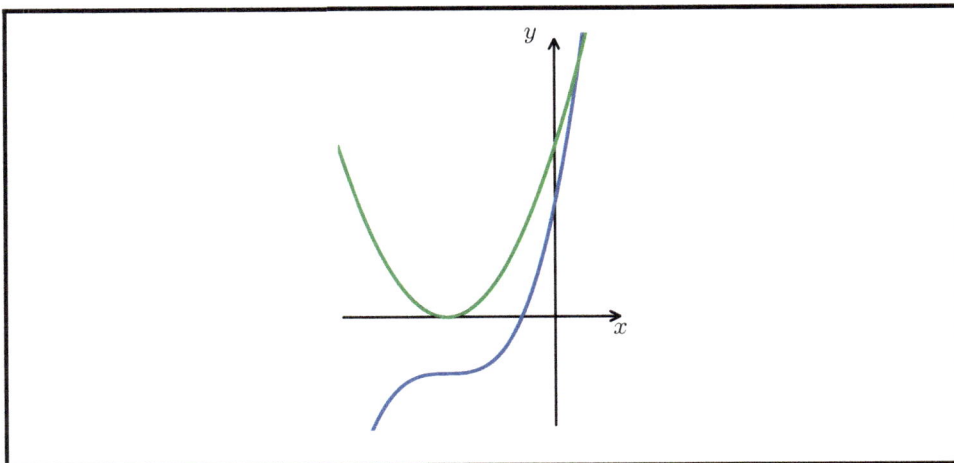

## Exercise 15.2

Q1. Differentiate the following with respect to $x$.

(a) $f(x) = x^6$;

(b) $y = x^8$;

(c) $f(x) = x^9$;

(d) $y = x^{-4}$;

(e) $f(x) = x^{-7}$;

(f) $y = x^{-4}$;

(g) $f(x) = x^{\frac{1}{2}}$;

(h) $y = x^{\frac{1}{5}}$;

(i) $f(x) = x^{\frac{3}{4}}$;

(j) $y = x^3 \times x^4$;

(k) $f(x) = x^6 \times x^{-2}$;

(l) $y = x^4 \times x^{\frac{1}{2}}$;

(m) $f(x) = \frac{x^7}{x^4}$;

(n) $y = \frac{x^8}{x^2}$;

(o) $f(x) = \frac{x^9}{x^{10}}$;

(p) $y = \frac{x^4 \times x^5}{x^3}$;

(q) $f(x) = \frac{x^7 \times x^3}{x}$;

(r) $y = \frac{x^4}{x^3 \times x^7}$;

(s) $f(x) = \frac{x^{\frac{1}{2}} \times x^{\frac{1}{4}}}{x}$;

(t) $y = \frac{x^{\frac{4}{3}} \times x^{\frac{7}{2}}}{x^4}$.

Q2. Differentiate the following with respect to $x$.

(a) $y = 6x$;

(b) $f(x) = 7x^3$;

(c) $y = 4x^7$;

(d) $f(x) = 4x^{-3}$;

(e) $y = 12x^{-4}$;

(f) $f(x) = 7x^{-5}$;

(g) $y = -6x^{-3}$;

(h) $f(x) = -2x$;

(i) $y = -5x^{-1}$;

(j) $f(x) = \frac{1}{2}x^{\frac{1}{4}}$;

(k) $y = -\frac{1}{4}x^{-2}$;

(l) $f(x) = -\frac{1}{8}x^{-\frac{4}{5}}$.

Q3. Evaluate the gradient of the functions below at the indicated point.

(a) $y = 4x^5$ at $x = 4$;

(b) $y = 3x^2 + 3x - 5$ at $x = -2$;

(c) $y = \frac{5x^4 + 2x}{x^3}$ at $x = 1$;

(d) $y = 2x^3 + 4x - 5$ at $x = \frac{13}{27}$.

Q4. For the following functions find the intervals where they are increasing and decreasing:
(a) $f(x) = 2x + 5$;
(b) $f(x) = x^2 + 2x + 1$;
(c) $f(x) = x^3 + 5x + 6$;
(d) $f(x) = x^4 + 6x^2 + 3$.

Q5. Differentiate the following with respect to $x$.

(a) $y = 3x^4 + 2x^3 + 4x + 2$;
(b) $f(x) = 4x^3 + x^2 + 2$;
(c) $y = 6x^6 - 4x^5 + 3x^3 + 4x^2 + x + 10$;
(d) $f(x) = \frac{1}{2}x^3 + \frac{2}{3}x^2 - \frac{3}{4}x + 4$;
(e) $y = \frac{3}{4}x^4 + \frac{2}{3}x^2 - \frac{1}{5}x + \frac{6}{5}$;
(f) $f(x) = \frac{7}{3}x^5 + \frac{2}{3}x^4 + \frac{3}{4}x^3 - \frac{1}{2}x^2 + 2x + 6$;
(g) $y = \frac{3}{2}\sqrt{x} + \frac{4}{5}x^{\frac{5}{2}}$;
(h) $f(x) = \frac{4}{5}x^{\frac{7}{3}} + \frac{3}{5}x^{\frac{4}{5}} - \frac{1}{3}x^{\frac{2}{3}}$;

(i) $y = \frac{3}{5}x^{\frac{6}{5}} - \frac{2}{3}x^{\frac{3}{4}} + \frac{1}{3}\sqrt{x}$;
(j) $f(x) = \frac{4}{x^4} - \frac{3}{x^2} + \frac{2}{3x^3}$;
(k) $y = 3x - \frac{2}{x^{\frac{5}{3}}} + \frac{4}{x^3}$;
(l) $f(x) = 4x^2 + 3x - 6 - \frac{7}{3x} + \frac{10}{7x^2}$;
(m) $y = \frac{2}{3}x^{\frac{5}{4}} + \frac{4}{5}\sqrt{x} - \frac{2}{5\sqrt{x}} - \frac{3}{4x^{\frac{3}{4}}}$;
(n) $f(x) = \frac{4}{3}x^2 - 2\sqrt{x} + \frac{3}{2\sqrt{x}} - \frac{4}{3\sqrt[3]{x}}$;
(o) $y = x^7 - \frac{2}{3}x^{\frac{5}{4}} + \frac{2}{3\sqrt{x}} - \frac{5}{x^{\frac{1}{5}}}$.

Q6. Find the derivatives of the following functions with respect to $x$.

(a) $y = x(x^2 - 3x + 2)$;
(b) $f(x) = 4x(2x^2 + 3x - 4)$;
(c) $y = 3x^2(4x^3 + 2x)$;
(d) $f(x) = (3x - 4)(2x + 1)$;
(e) $y = (2x - 6)(2x + 4)$;
(f) $f(x) = (3x + 4)(4x + 1)$;
(g) $y = (2x^2 + 5)(3x + 1)$;
(h) $f(x) = (3x^2 + 5)(2x - 5)$;
(i) $y = (3x^2 + 4)(3x^2 - 5)$;
(j) $f(x) = \left(\frac{1}{x} + 5\right)(2x - 3)$;
(k) $y = \left(\frac{2}{x} - 5\right)\left(\frac{3}{x} + 1\right)$;
(l) $f(x) = \left(2x + \frac{1}{x}\right)\left(\frac{1}{x^2} + 4x\right)$;
(m) $y = \frac{3x+4}{x}$;
(n) $f(x) = \frac{6x+7}{2x}$;

(o) $y = \frac{8x+9x^2}{2x}$;
(p) $f(x) = \frac{8x+7x^2}{2x^2}$;
(q) $y = \frac{6x^3+3x}{3x^3}$;
(r) $f(x) = \frac{7x^4+3x^2+6x}{2x^2}$;
(s) $y = \frac{2x^3+4x}{\sqrt{x}}$;
(t) $f(x) = \frac{4x^2-3x+2}{\sqrt[3]{x}}$;
(u) $y = \frac{3x^4+2x^2+4x}{3\sqrt{x}}$;
(v) $f(x) = \frac{4x^3+3x+2}{2x^{\frac{2}{3}}}$;
(w) $y = \frac{x^7+6x^2+5x+3}{3x^{\frac{2}{5}}}$;
(x) $f(x) = \frac{2x^3+3x+4}{3x^{\frac{5}{3}}}$.

Q7. Evaluate the second derivative of the following functions at $x = 2$.
(a) $y = x^{\frac{1}{2}}$;
(b) $y = x^2 + 2\sqrt{x}$;
(c) $y = x^3 + 5x + 6$.

## Activity 15.2 — Polynomials and Their Derivatives

Consider the polynomial $f(x) = x^3 + 4x^2 - 17x - 60$.
(a) Find the solutions of $f(x) = 0$.
(b) Differentiate $f(x)$ and find the roots of $f'(x) = 0$.
(c) Calculate the mean of the solutions to $f(x) = 0$ and the mean of the solutions to

$f'(x) = 0$.

(d) What can be said about the answers to (c). Does this property hold for all polynomials and their derivatives?

---

**Key Facts — Introduction to Differentiation**

- The gradient of a curve at a given point, $A$, is defined to be the gradient of the tangent at $A$.
- The derivative of the curve $y = f(x)$ is denoted either as $f'(x)$ or $\frac{dy}{dx}$. We define $f'(x)$ to be

$$f'(x) = \lim_{h \to 0} \frac{f(x+h) - f(x)}{h}.$$

- Let $f(x)$ and $g(x)$ be continuous, differentiable functions and $c \in \mathbb{R}$, then,

$$\frac{d}{dx}\left[f(x) + g(x)\right] = \frac{d}{dx}\left[f(x)\right] + \frac{d}{dx}\left[g(x)\right],$$

$$\frac{d}{dx}\left[f(x) - g(x)\right] = \frac{d}{dx}\left[f(x)\right] - \frac{d}{dx}\left[g(x)\right],$$

$$\frac{d}{dx}\left[cf(x)\right] = c\frac{d}{dx}\left[f(x)\right].$$

- Let $f(x) = ax^n$, where $n \in \mathbb{Q}$, then

$$f'(x) = \frac{df(x)}{dx} = anx^{n-1}$$

- The second derivative is the rate of change of the gradient, since

$$\frac{d^2 f(x)}{dx^2} = \frac{df'(x)}{dx}.$$

- A function, $f(x)$, is said to be increasing on the interval $[a,b]$, if $f(x) \geq 0$ for all $x \in [a,b]$.
- A function, $f(x)$, is said to be decreasing on the interval $[a,b]$, if $f(x) \leq 0$ for all $x \in [a,b]$.

---

**Chapter Assessment — Introduction to Differentiation**

Download and sit the 30 minute assessment for this chapter from the digital book.

# 16. Integration 1 - Indefinite Integration

(a)                                                    (b)

Figure 16.1: Sir Isaac Newton (a) and Gottfried Leibniz (b).

The key concepts underlying integration were independently developed by Isaac Newton and Gottfried Leibniz (Figure 16.1) in the 17th centuries in efforts to calculate the area under a curve.

**Remark**

Throughout their lives, there was a feud between Newton and Leibniz as to who invented *calculus*. There is still no consensus on this issue, but much of the notation we use today can be attributed to Leibniz.

The Choco-Leibniz biscuit is named after Gottfried Leibniz. Its manufacturer, Balhsen,

> are based in Hanover, where Leibniz was resident from 1767-1716. Fig Newtons are not
> named after Newton, they are named after the city of Newton in Massachusetts.

The most important idea is that integration "undoes" differentiation. Integration and
differentiation together make up the mathematical toolkit known as calculus. Calculus
provides a framework for modelling how systems change. Understanding how a system
changes enables engineers and scientists to identify how to control the system and modify
its behaviour. The role of integration in calculus is (i) to find the original function from a
derivative, and (ii) to find the area under a curve.

This chapter will explain how and why we integrate a function, introduce the mathematical
notation for this operation, and introduce techniques of integration.

## 16.1  Finding the area under a curve

We know how to compute the area within 'regular' shapes, such as rectangles, circles
and trapezia. However, we also need a straightforward method for calculating the area
of a shape with curved boundaries. Accurate solutions for the area of these shapes are
important throughout applied mathematics and engineering. This chapter addresses this
area problem through methods for calculating the area $A$ under a general curve $C$ such as
that in Figure 16.2.

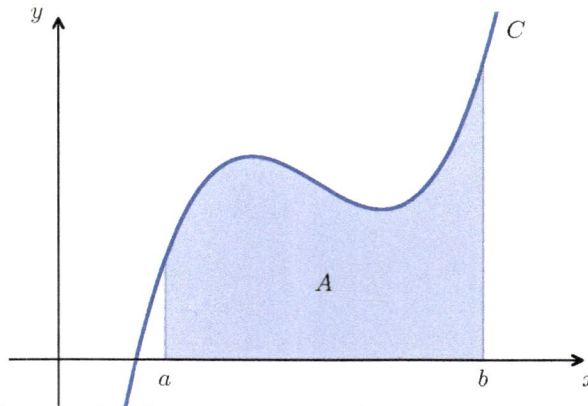

Figure 16.2: The area $A$ under a general curve $C$ between the limits $x = a$ and $x = b$.

> **Remark**
> The methods introduced in this chapter are extremely powerful tools in applied mathe-
> matics. They are used across electronics, mechanics, quantum physics and mathematical
> biology among others. Applications include calculating position from velocity, calculat-
> ing an electric field, finding the rate of flow of a fluid in a pipe, and understanding
> population growth.

### 16.1.1  The Rectangle Method

We first explore whether we can use the area formulae that we already know to help us to
find the area under a curve. Figure 16.3 shows how splitting the area $A$ into a series of

rectangles gives us an approximation to the area $A$.

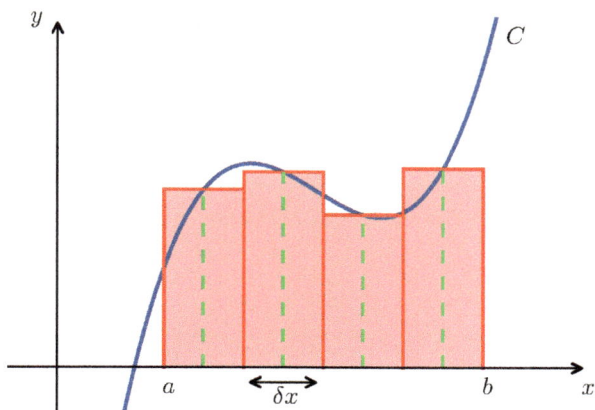

Figure 16.3: An approximation to the area under a general curve $C$ between the limits $x = a$ and $x = b$, using rectangles.

Here the shapes chosen are rectangles with a height equal to curve $C$ evaluated at the midpoint of the rectangle. This is known as the rectangle method for finding $A$, or the midpoint rule. The accuracy of the rectangle method improves the narrower the rectangles become. This is shown in Figure 3 where we see that the total area of the rectangles approaches the area $A$ as $\delta x \to 0$.

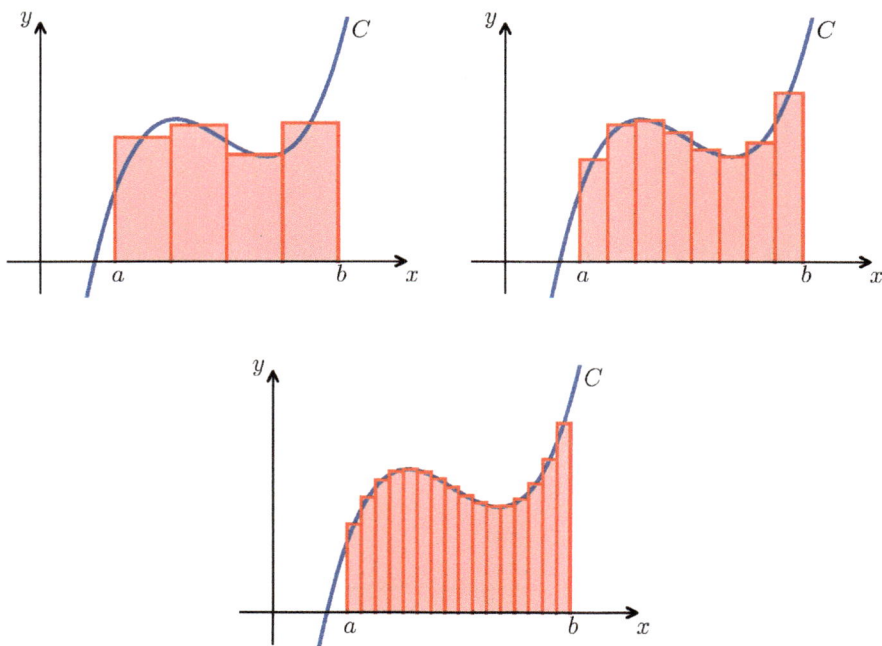

Figure 16.4: The area approximated by the rectangles as the number of rectangles increases and $\delta x \to 0$.

**Interactive Activity 16.1 — Approximating the Area by Rectangles**
Use the Geogebra interactive below to investigate using rectangles to approximate the area under the curve of different functions.

**Activity 16.1 — Accuracy of the Rectangle Method**
Using the example of $y = x^2$, for $a = 1$ and $b = 2$, demonstrate that the rectangle method becomes more accurate as the number of rectangles increases. Assume that the accurate value for the area under $y = x^2$ is $\frac{8}{3}$.

**Remark**
The figures above all use rectangles of height equal to the curve at the midpoint of the rectangle (shown with the green dashed lines in Figure 16.3). We could instead use rectangles of height equal to the curve at one of the edges of the rectangles (e.g. the left edge or the right edge). The figure below shows where the right edge is used for calculating the height of the rectangles.

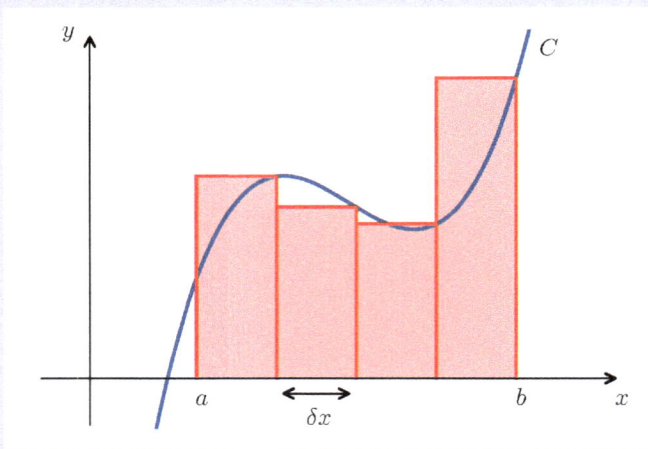

**Example 16.1**
Find an approximation to the area $A$ under $y = x^2$ between $x = 1$ and $x = 2$ by using the rectangle method with 6 rectangles.

**Solution:**

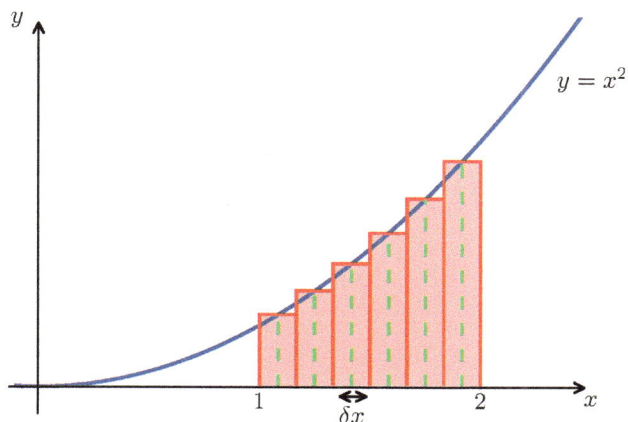

Figure 16.5: Approximating the area under $y = x^2$ between the limits $x = 1$ and $x = 2$.

To divide the area $A$ into 6 rectangles, first split the interval $x = 1$ to $x = 2$ into 6 equal parts each of width $\delta x = \frac{2-1}{6} = \frac{1}{6}$. Construct a rectangle for each with a height equal to $y = x_i^2$ where $x_i$ is the midpoint of the rectangle. The midpoints of the rectangles are given by

$$\frac{13}{12}, \frac{15}{12}, \frac{17}{12}, \frac{19}{12}, \frac{21}{12}, \frac{23}{12}.$$

The heights of the rectangles are then equal to $y = x^2$ evaluated for $x$ at the midpoint of each rectangle. Since the width of each rectangle is $\delta x = \frac{1}{6}$ the area of each rectangle is given by the height of the rectangle multiplied by $\frac{1}{6}$. Therefore, the approximation to the area $A$ is

$$A \approx \frac{1}{6}\left(\left(\frac{13}{12}\right)^2 + \left(\frac{15}{12}\right)^2 + \left(\frac{17}{12}\right)^2 + \left(\frac{19}{12}\right)^2 + \left(\frac{21}{12}\right)^2 + \left(\frac{23}{12}\right)^2\right)$$

$$= \frac{1}{6} \cdot \frac{1}{144}(13^2 + 15^2 + 17^2 + 19^2 + 21^2 + 23^2)$$

$$= \frac{1007}{432}.$$

Therefore, the rectangle method gives $A = 2.3310$ to four decimal places.

### 16.1.2 The Antiderivative

Whilst the rectangle method enables us to approximate the area under a curve, it is often necessary to find a precise solution for the area $A$. Where this is the case we can use the antiderivative method, shown in the following example.

**Example 16.2**
In Activity 16.1 we assumed that the actual area under $y = x^2$ between the limits $x = 0$ and $x = 2$ is $\frac{8}{3}$.
Notice that

$$\frac{\mathrm{d}}{\mathrm{d}x}(x^3) = 3x^2. \tag{16.1}$$

Therefore, if we suppose that $y = x^2$ is the differential, then reversing the differentiation gives us the antiderivative $\frac{1}{3}x^3$.
If we substitute $x = 2$ into this antiderivative we obtain $\frac{8}{3}$. This is the value approached by the approximations in Activity 16.1.

This example suggests that the antiderivative method might give an exact value for the area under a curve. We now proceed to demonstrate that this is the case.
Let us assume that we can define the curve $C$ in Figure 1 by a function $f(x)$ where $f$ is continuous. We also define $A(x)$ to be the area under the curve between the limits $a$ and an arbitrary point $x$ that lies between $a$ and $b$. We now seek a relationship between $A(x)$ and $f(x)$ that will enable us to calculate the area for a given $f$. From Chapter 15 we know that the derivative of $A$ can be calculated as

$$A'(x) = \lim_{h \to 0} \frac{A(x+h) - A(x)}{h}. \tag{16.2}$$

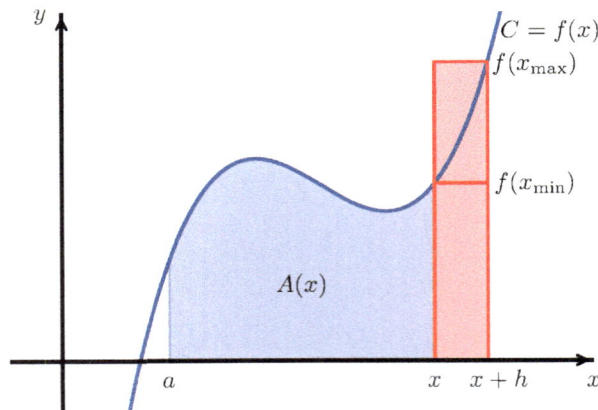

Figure 16.6: The area $A(x)$ under the curve $C = f(x)$ between the limits $x = a$ and an arbitrary point $x$. The rectangles between $x$ and $x + h$ used in the derivation below are also shown.

Now, $A(x+h) - A(x)$ is the area under the curve $f(x)$ between $x$ and $x + h$.
Using the rectangle method we can construct rectangles to approximate this area. Here we consider two possible approximations, each of which uses one rectangle, shown in Figure 16.6. First, consider the lowest possible approximation, which is given when we choose $x_{\min}$ such that $f(x_{\min})$ is the minimum value of $f(x)$ on $[x, x+h]$. The second is the highest possible approximation, which occurs when we choose $x_{\max}$ such that $f(x_{\max})$ is the maximum value of $f(x)$ on $[x, x+h]$.

Calculating the areas of the corresponding rectangles gives

$$f(x_{\min}) \cdot h \le A(x+h) - A(x) \le f(x_{\max}) \cdot h.$$

Therefore,

$$f(x_{\min}) \le \frac{A(x+h) - A(x)}{h} \le f(x_{\max}).$$

We know that as $h \to 0$, $x_{\min} \to x$ and $x_{\max} \to x$. We have assumed that the function $f$ is continuous, so $f(x_{\min}) \to f(x)$ as $x_{\min} \to x$ and $f(x_{\max}) \to f(x)$ as $x_{\max} \to x$. Therefore,

$$\lim_{h \to 0} \frac{A(x+h) - A(x)}{h} = f(x).$$

Substituting into (16.2) we obtain:

$$A'(x) = f(x). \tag{16.3}$$

In other words $f(x)$ is equal to the derivative of the area under $f$ between the limits $a$ and $x$. If we can reverse the differentiation then we can find the area precisely by substituting for $x = b$.

**Remark**

This result is known as the Fundamental Theorem of Calculus. We will state it formally later when we have discussed the notation required.

**Definition 16.1 — Antiderivative**

$F$ is called an antiderivative of a function $f$ if it is differentiable and $F'(x) = f(x)$ for all $x$.

For any $f(x)$ we can find multiple functions $F(x)$ that satisfy the equation in definition 16.1 by adding any constant value. Therefore, each antiderivative must include a general constant in order to capture all possible antiderivatives. It may be possible to find the constant for a specific problem using information that we have about that problem.

**Activity 16.2 — Multiple Antiderivatives**

Show that

$$\frac{1}{3}x^3 + 1, \ \frac{1}{3}x^3 + \sqrt{3}, \text{ and } \frac{1}{3}x^3 - \sin\left(\frac{\pi}{2}\right)$$

are all antiderivatives of $y = x^2$.

**Tip**

In the problems that follow, we will find the area under a curve by "undoing" differentiation. Whenever we do this we should state that we are using the Fundamental Theorem of Calculus.

**Example 16.3**

Use the antiderivative method to find a formula for area $A$ under $y = x^2$ between the limits $x = 1$ and $x = 2$.

**Solution:**

By equation (16.3) we have

$$A'(x) = x^2. \tag{16.4}$$

We know that

$$\frac{\mathrm{d}}{\mathrm{d}x}(x^3) = 3x^2.$$

Therefore, we use the Fundamental Theorem of Calculus to deduce that one solution of (16.4) is

$$A(x) = \frac{1}{3}x^3.$$

In order to capture all possible solutions we need to add a constant giving us the full solution

$$A(x) = \frac{1}{3}x^3 + c.$$

Now, the area under $f(x)$ between the limits $x = 1$ and $x = 2$ must be equal to the area between $x = 0$ and $x = 2$ less the area between $x = 0$ and $x = 1$. That is

$$A = A(2) - A(1)$$
$$= \left[\frac{1}{3}(2)^3 + c\right] - \left[\frac{1}{3}(1)^3 + c\right]$$
$$= \frac{8}{3} + c - \frac{1}{3} - c$$
$$= \frac{7}{3}.$$

Comparing our result with that from Example 16.1, we can examine the error made using the rectangle method with 6 rectangles as a percentage of the exact value:

$$100 \times \left(\frac{\frac{7}{3} - \frac{2014}{864}}{\frac{7}{3}}\right) \approx 0.09\%.$$

**Example 16.4**

Use the antiderivative method to find a formula for area $A$ under $y = \sin(x)$ between the limits $x = 0$ and $x = \pi$.

**Solution:**

By equation (16.3) we have

$$A'(x) = \sin(x). \tag{16.5}$$

In the Year 2 book we shall see that,

$$\frac{\mathrm{d}}{\mathrm{d}x}(\cos(x)) = -\sin(x).$$

Therefore, we use the Fundamental Theorem of Calculus to deduce that one solution of (16.5) is

$$A(x) = -\cos(x).$$

In order to capture all possible solutions we need to add a constant giving us the full solution

$$A(x) = -\cos(x) + c.$$

If we assume that the area at $x = 0$ is 0 then we see that $c = \cos 0 = 1$. Therefore,

$$A(x) = -\cos(x) + 1.$$

Therefore, the area between the limits $x = 0$ and $x = \pi$ is

$$
\begin{aligned}
A &= A(\pi) - A(0) \\
&= [(-\cos \pi + 1) - (-\cos 0 + 1)] \\
&= (1 + 1) - (-1 + 1) \\
&= 2.
\end{aligned}
$$

**Exercise 16.1**

Q1 (a) Use the Fundamental Theorem of Calculus to deduce the antiderivative of $y = x^3$.

    (b) Use the antiderivative method to find the area under the curve $y = x^3$ between the limits of $x = 0$ and $x = 2$.

    (c) Use the rectangle method to find the area under $y = x^3$ between the limits of $x = 0$ and $x = 2$. Use 4 rectangles evaluated at the end-point of each rectangle.

    (d) What is the accuracy of the rectangle method in this case?

Q2 Compare the solutions of the area under $y = 2x + 3$ between $x = 0$ and $x = 2$ derived using

    (a) the midpoint rule with 4 rectangles;

    (b) the antiderivative method; and

> (c) geometric area formulae.
>
> Does the antiderivative method agree with the geometric formulae? What is the error of the rectangle method in this case? Why?

## 16.2   Indefinite Integrals

There are some special symbols that are used to describe antiderivative problems which we introduce now. The equation

$$\int f(x)\,\mathrm{d}x = F(x) + c \tag{16.6}$$

tells us that "The *indefinite integral* of $f(x)$ with respect to $x$ is equal to $F(x) + c$". The combined $\int [\ldots]\,\mathrm{d}\cdot$ symbol is the *indefinite integral sign*. The function $f(x)$ is the *integrand* and $c$ is the *constant of integration*. The process of finding the antiderivative or indefinite integral is called *integration*.

> **Remark**
>
> Although some mathematicians make a distinction between the antiderivative and the indefinite integral, many do not, and for our purposes the antiderivative and the indefinite integral are the same.

### 16.2.1   Integrating Powers of x

Integrating, or finding the antiderivative of, basic functions relies upon our knowledge of differentiation. We work backwards from known derivatives, "undoing" the differentiation. We use these known derivatives to develop a suite of known integrals. Combined with rules that simplify complex integration problems, this provides a methodology for finding integrals of a wide range of functions. We now develop our tools for integrating functions such as $f(x) = x^n$ for $n \neq -1$.

From Section 16.1.2, we already know that

$$\int x\,\mathrm{d}x = \frac{1}{2}x^2 + c,$$

$$\int x^2\,\mathrm{d}x = \frac{1}{3}x^3 + c, \text{ and}$$

$$\int x^3\,\mathrm{d}x = \frac{1}{4}x^4 + c.$$

We have also seen in Chapter 15 how to differentiate $x^{n+1}$. Therefore, it is reasonable to hypothesise that this will generalise to

$$\int x^n\,\mathrm{d}x = \frac{1}{n+1}x^{n+1} + c.$$

Recalling that differentiating an integral takes us back to the original function, we can confirm whether this is correct by differentiating the right hand side and checking that we obtain the original integrand, $x^n$.

$$\frac{\mathrm{d}}{\mathrm{d}x}\left(\frac{1}{n+1}x^{n+1} + c\right) = \frac{1}{n+1}\frac{\mathrm{d}}{\mathrm{d}x}(x^{n+1}) + \frac{\mathrm{d}}{\mathrm{d}x}(c) = \frac{n+1}{n+1}x^n + 0 = x^n.$$

**Formula 16.1 — Integral of $x^n$, $n \neq -1$**

$$\int x^n \, dx = \frac{1}{n+1} x^{n+1} + c$$

**Example 16.5**
Find

$$\int x^5 \, dx.$$

**Solution:**
Substitute $n = 5$ into Formula 16.1 as follows:

$$\int x^5 \, dx = \frac{1}{5+1} x^{5+1} + c$$
$$= \frac{1}{6} x^6 + c.$$

**Activity 16.3**
Why are we not able to use this Formula 16.1 for $n = -1$?

**Activity 16.4 — Integrating fractional powers**
Find

$$\int x^{1/n} \, dx$$

and use differentiation to show that it is correct.

**Tip**
The following is a simple way of thinking about Formula 16.1 that readily applies both to natural number powers and fractional powers.
1. Add one to the power of $x$;
2. Divide by the new power;
3. Add a constant.

### 16.2.2 Integrating Powers of x in simple formulae

Functions describing real-world problems are rarely a simple $x^n$ formula. They typically involve constant multiples $(Kx^n)$, sums $(x^n + x^m)$ and differences $(x^n - x^m)$, or combinations of these three.

The following formulae will help us to reduce complex integration problems into a series of simpler problems that we can solve easily.

**Formulae 16.2**

Any constant multiple of $f(x)$ can be moved outside of the integral sign with the same result:

$$\int Cf(x)\mathrm{d}x = C \int f(x)\mathrm{d}x. \tag{16.7}$$

The integral of a sum of two functions is equal to the sum of the integrals of each function:

$$\int f(x) + g(x)\mathrm{d}x = \int f(x)\mathrm{d}x + \int g(x)\mathrm{d}x. \tag{16.8}$$

The integral of the difference between two functions is equal to the difference between the integrals of each function:

$$\int f(x) - g(x)\mathrm{d}x = \int f(x)\mathrm{d}x - \int g(x)\mathrm{d}x. \tag{16.9}$$

**Activity 16.5**

Prove Formulae 16.2

**Example 16.6**

Find

$$\int 6x^7 \, \mathrm{d}x.$$

**Solution:**

Using equation (16.7) we can simplify the problem as follows:

$$\int 6x^7 \, \mathrm{d}x = 6 \int x^7 \, \mathrm{d}x.$$

From Formula 16.1 we see that

$$\int x^7 \, \mathrm{d}x = \frac{1}{8}x^8 + c.$$

Therefore,

$$\int 6x^7 \, \mathrm{d}x = \frac{6}{8}x^8 + 6c = \frac{3}{4}x^8 + k.$$

**Tip**

The final answer in the example above replaces the $6c$ with $k$. This is correct because both $c$ and $k$ are generic, unknown constants. It is preferred because it is notationally simpler. We will do the same in the following examples.

**Example 16.7**

Find

$$\int x^3 + x^{11} \, \mathrm{d}x.$$

**Solution:**

First, we simplify the problem using equation (16.8):

$$\int x^3 + x^{11} \, \mathrm{d}x = \int x^3 \, \mathrm{d}x + \int x^{11} \, \mathrm{d}x.$$

Then we use Formula 16.1 to solve each of the integration problems on the right hand side:

$$\int x^3 \, \mathrm{d}x = \frac{1}{4}x^4 + c_1, \text{ and}$$

$$\int x^{11} \, \mathrm{d}x = \frac{1}{12}x^{12} + c_2.$$

Therefore,

$$\int x^3 + x^{11} \, \mathrm{d}x = \frac{1}{4}x^4 + c_1 + \frac{1}{12}x^{12} + c_2 = \frac{1}{12}x^4(3 + x^8) + k.$$

**Activity 16.6**

Find

$$F(x) = \int x - x^5 \, \mathrm{d}x$$

given that $F(2) = -8$.

**Example 16.8**

Find the integral of

$$f(x) = (x - 3)(x + 1) + 4x^2(x - x^{1/3}).$$

**Solution:**

Firstly, we expand the brackets so that the function is in a form to which we can apply the integration techniques we have learned.

$$
\begin{aligned}
f(x) &= (x-3)(x+1) + 4x^2(x - x^{1/3}) \\
&= x^2 - 3x + x - 3 + 4x^3 - 4x^{7/3} \\
&= x^2 - 2x - 3 + 4x^3 - 4x^{7/3}.
\end{aligned}
$$

Therefore, we need to find

$$
\int x^2 - 2x - 3 + 4x^3 - 4x^{7/3} \, \mathrm{d}x.
$$

From Formulae 16.2 we know that

$$
\int x^2 - 2x - 3 + 4x^3 - 4x^{7/3} \, \mathrm{d}x = \int x^2 \, \mathrm{d}x - 2 \int x \, \mathrm{d}x - 3 \int x^0 \, \mathrm{d}x
$$
$$
+ 4 \int x^3 \, \mathrm{d}x - 4 \int x^{7/3} \, \mathrm{d}x.
$$

(16.10)

Now we solve each integral problem on the right hand side in turn.

$$
\int x^2 \, \mathrm{d}x = \frac{1}{3}x^3 + c_1,
$$
$$
\int x \, \mathrm{d}x = \frac{1}{2}x^2 + c_2,
$$
$$
\int x^0 \, \mathrm{d}x = x + c_3,
$$
$$
\int x^3 \, \mathrm{d}x = \frac{1}{4}x^4 + c_4,
$$
$$
\int x^{7/3} \, \mathrm{d}x = \frac{3}{10}x^{10/3} + c_5.
$$

Substituting back into (16.10) and grouping all of the constants together into the generic constant $k$ we get

$$
\begin{aligned}
\int f(x) \, \mathrm{d}x &= \frac{1}{3}x^3 + c_1 - 2\left(\frac{1}{2}x^2 + c_2\right) - 3(x + c_3) + 4\left(\frac{1}{4}x^4 + c_4\right) \\
&\quad - 4\left(\frac{3}{10}x^{10/3} + c_5\right) \\
&= \frac{1}{3}x^3 - x^2 - 3x + x^4 - \frac{12}{10}x^{10/3} + k \\
&= x\left(\frac{1}{3}x^2 - x - 3 + x^3 - \frac{6}{5}x^{7/3}\right) + k.
\end{aligned}
$$

**Remark**

In the previous examples we have used $k$ to denote the final constant. Since the $c_i$ and $k$ are all generic constants we could use any letter (except for the variable, $x$, or the exponential, $e$) and we would be correct.

We will, however, need to use more care when dealing with definite integrals as we will see later.

**Interactive Activity 16.2 — Indefinite Integral Match**

In the interactive activity below, match the functions with their indefinite integrals.

**Tip**

The variable following the d· is important and tells us which variable we must integrate with respect to. The problems

$$\int f(x)\,\mathrm{d}x$$

and

$$\int f(x)\,\mathrm{d}t$$

are different. For example, we revisit the problem in example 16.5. In the example, we found that

$$\int x^5\,\mathrm{d}x = \frac{1}{6}x^6 + c.$$

Now consider instead

$$\int x^5\,\mathrm{d}t.$$

In contrast to the example, here we are integrating with respect to the $t$. That means that we are treating $t$ as a variable and all other letters represent constants. Therefore, using (16.7),

$$\int x^5\,\mathrm{d}t = x^5\int 1\,\mathrm{d}t.$$

Using Formula 16.1

$$\int x^5\,\mathrm{d}t = x^5 t + c.$$

**Exercise 16.2**

Q1. In the equation

$$\int 4x^3 \, \mathrm{d}x = x^4 + c,$$

identify the following elements:
  (a) the indefinite integral;
  (b) the integrand;
  (c) the constant of integration; and
  (d) the variable against which we are integrating.

Q2. Find the following:
  (a)

$$\int 3x^2 \, \mathrm{d}x;$$

  (b)

$$\int 6x^{3/2} \, \mathrm{d}x;$$

  (c)

$$\int (5x + 2)(x^2 - 3) \, \mathrm{d}x;$$

  (d) $F(x)$, given that when $x = 1$, $F(x) = 1$, and where

$$F(x) = \int x(x^2 + 2) \, \mathrm{d}x;$$

  (e)

$$\int t^3 \, \mathrm{d}t;$$

  (f)

$$\int \frac{1}{x^3} \, \mathrm{d}x;$$

  (g)

$$\int x^4 \, \mathrm{d}t;$$

  (h) $F(x)$, given that when $x = 3$, $F(x) = 21$, and where

$$F(x) = \int \left(\frac{1}{x^2} + 3\right)(x^2 - 1) \, \mathrm{d}x.$$

**Key Facts — Integration 1 - Indefinite Integrations**

- Integration is motivated by trying to find the area enclosed between a function $f(x)$ and the $x$-axis between limits $x = a$ and $x = b$.

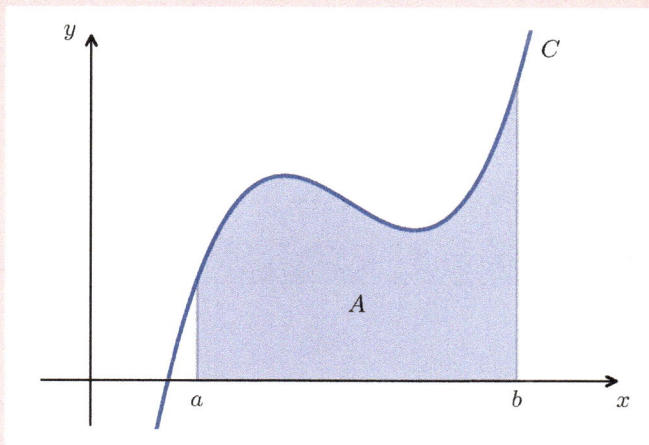

- An *antiderivative* or *indefinite integral* of $f(x)$ is denoted $\int f(x)\,\mathrm{d}x$. By definition, the antiderivative/indefinite integral is a function $F(x)$ such that $\frac{\mathrm{d}F}{\mathrm{d}x}(x) = f(x)$.
- If $F(x)$ is an indefinite integral, then so is $F(x) + c$ for any constant $c \in \mathbb{R}$. $c$ is called the *constant of integration*.
- When finding indefinite integrals, always remember to include the constant of integration.
- For $n \neq -1$, we have

$$\int x^n\,\mathrm{d}x = \frac{1}{n+1}x^{n+1} + c.$$

- For $C \in \mathbb{R}$ and for functions $f(x)$ and $g(x)$, the following formulae hold

$$\int Cf(x)\mathrm{d}x = C \int f(x)\mathrm{d}x,$$

$$\int f(x) + g(x)\mathrm{d}x = \int f(x)\mathrm{d}x + \int g(x)\mathrm{d}x,$$

$$\int f(x) - g(x)\mathrm{d}x = \int f(x)\mathrm{d}x - \int g(x)\mathrm{d}x.$$

**Chapter Assessment — Indefinite Integration**

Download and sit the 30 minute assessment for this chapter from the digital book.

# 17. Exponential Functions

In his book *Liber Abaci* of 1202 Leonardo of Pisa, more commonly known as Leonardo Fibonacci , posed a puzzle concerning the idealised population of rabbits in a field. Fibonacci assumed that:

- At the beginning of month one, a newly born pair of rabbits (one male, one female) are put in a field.
- Rabbits take one month to reach maturity at which point they can mate.
- The gestation period of a rabbit is one month.
- After a gestation period, a female rabbit produces one male and one female kitten.
- A pair of mating rabbits produce a new pair of rabbits, one male and one female every month from the time they are two months old.

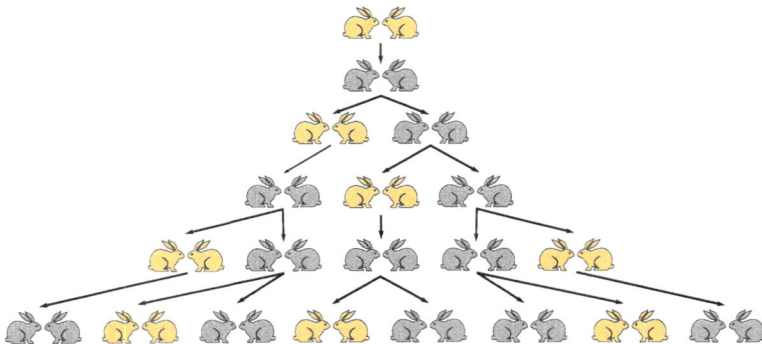

Figure 17.1: Number of pairs of rabbits at the beginning of a month, up to month six, according to Fibonacci's model.

This model is not biologically realistic, not least due to the additional assumption that no rabbits die, however this model exhibits the exponential growth shown by many real-world populations, such as those of bacteria.

Figure 17.1 shows the number of pairs of rabbits present in the field at the beginning of each month. Note that a grey pair of rabbits are mature and can produce offspring, whereas a yellow pair are immature. We can begin to tabulate the numbers of pairs of rabbits present at the end of each month.

| Month | Number of Pairs |
|:-----:|:---------------:|
| 1 | 1 |
| 2 | 1 |
| 3 | 2 |
| 4 | 3 |
| 5 | 5 |
| 6 | 8 |

This sequence is now known as the Fibonacci sequence and exhibits exponential growth. In Figure 17.2 we plot the first 16 Fibonacci numbers and fit an exponential model to them.

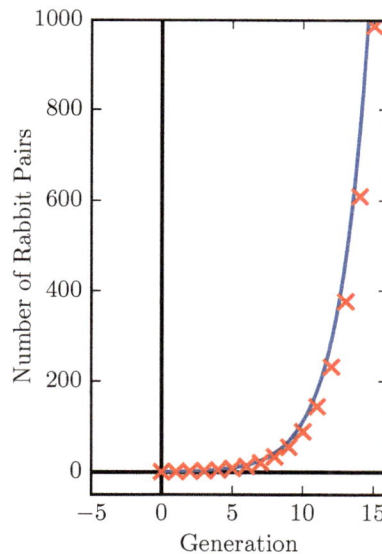

Figure 17.2: The Fibonacci sequence (red crosses) with a fitted exponential model (blue curve).

As is often the case when we model a situation, the exponential model isn't a perfect fit, but it certainly shows the same qualitative behaviour.

Exponential models arise in many contexts, and it is these that are the subject of this chapter.

## 17.1   What is an Exponential Function?

An exponential function is one where the variable appears in the index. The simplest exponential function is a function of the form:

$$f(x) = a^x.$$

One example of an exponential function is $f(x) = 2^x$. To investigate this function, we can substitute values for $x$ into it and tabulate as follows:

| $x$ | $a^x$ | $f(x)$ |
|-----|-------|--------|
| 0   | $2^0$ | 1      |
| 1   | $2^1$ | 2      |
| 2   | $2^2$ | 4      |
| 3   | $2^3$ | 8      |
| 4   | $2^4$ | 16     |

We should also consider negative values for $x$:

| $x$ | $a^x$ | $f(x)$ |
|-----|-------|--------|
| $-1$ | $2^{-1}$ | $\frac{1}{2} = 0.5$ |
| $-2$ | $2^{-2}$ | $\frac{1}{4} = 0.25$ |
| $-3$ | $2^{-3}$ | $\frac{1}{8} = 0.125$ |
| $-4$ | $2^{-4}$ | $\frac{1}{16} = 0.0625$ |

These values are plotted in Figure 17.3 and a smooth curve drawn through them.

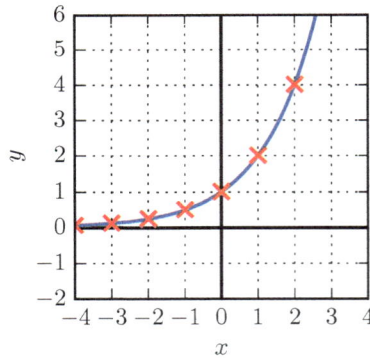

Figure 17.3: Graph of $y = 2^x$ plotted over the interval $[-4, 4]$.

---

**Exercise 17.1**

Q1. Without sketching the graphs, say what the similarities and differences will be between the graph in Figure 17.3 and the graph of $g(x) = 3^x$?

Q2. Sketch the graphs of:

(a) $y = 3^x$, $x \in \mathbb{R}$;

(b) $y = 4^x$, $x \in \mathbb{R}$.

Q3. (a) Sketch the graph of $y = \frac{1}{2}^x$.
   (b) How does this graph differ to the previous two?
   (c) Sketch the graph of $y = 2^{-x}$, for $x \in \mathbb{R}$.
   (d) What do you notice? Why is this the case?
Q4. (a) What are the distinguishing features of the graphs of functions $f(x) = a^x, a > 0, a \neq 1$? Explain your reasoning for each feature.
   (b) Why might the constraints $a > 0$ and $a \neq 1$ be used here?

## 17.2 Exponential Growth and Decay

A quantity that grows or decays by a constant multiplier over regular time intervals is said to display *exponential growth* or *exponential decay*. The constant multiplier is known as the *growth factor*.

Exponential growth is 'faster' than polynomial growth: regardless of the degree of a given polynomial, an exponential function will always 'overtake' the polynomial.

It can be seen in Figure 17.4 below that for values of $x > 4$, the curve for $f(x) = 4^x$ is growing at a faster rate than the curve for $g(x) = x^4$.

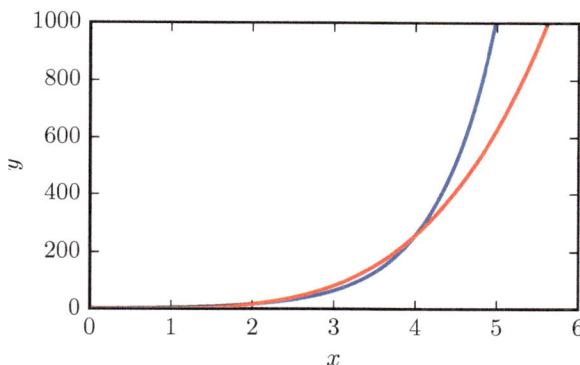

Figure 17.4: $f(x) = 4^x$ (blue) and $g(x) = x^4$ (red) plotted over the interval $[0, 6]$.

### 17.2.1 Modeling Exponential Growth

Functions that describe how things grow over time are useful in many fields of study. Some examples include population growth of bacteria, the spread of viruses and interest payments, to name a few.

$y = a^x$ is one possible model for exponential growth. If the function is changed to $ka^x$, then any data that exhibits exponential growth or decay can be modelled.

---

**Interactive Activity 17.1 — Investigating Exponential Curves**
Use the Geogebra activity below to investigate the graph of $y = ka^x$ using different values for $k$ and $a$.

- How does the graph of $y = k \times 2^x$ differ to the graph in Figure 17.3?
- What would be the significance of $k$ in the context of a real world problem?

## Use of Technology 17.1 — Exploring Exponential Curves With a Graphical Calculator

A graphical calculator can also be used to investigate the effect of $k$ and $a$ on the function $y = ka^x$.

## Definition 17.1 — Growth Function

A general growth function has the form

$$y = ka^x,$$

where $k$ and $a$ are constants.
- The constant $k$ is the initial value (when $x = 0$).
- $a$ is the growth factor (sometimes referred to as percentage increase).

## Interactive Activity 17.2 — Fitting Exponential Data

In this activity we investigate the use of exponential growth functions to model data that exhibits exponential growth or decay (discussed in more detail in Section 17.2.2) For each of the three data sets contained within the Geogebra activity below, use the sliders to find the appropriate values of $k$ and $a$.

## Example 17.1

The table below shows the population of Brazil over a 12 year period.

| Year | Population (millions) |
|------|----------------------|
| 1966 | 86.7 |
| 1970 | 95.5 |
| 1974 | 105.4 |
| 1978 | 116.2 |

Assuming the population growth is approximately exponential, model the population data for the population, $P$ in terms of $t$, the number of years after 1966. Use the model to:

(a) make an estimate for the population in 1972;

(b) project the population in 2012.

**Solution:**

We can check to see if the growth is approximately exponential by comparing the growth factor between each consecutive pieces of data.

| Year | Population (millions) | Growth factor |
|------|----------------------|---------------|
| 1966 | 86.7 | |
| 1970 | 95.5 | 1.101 |
| 1974 | 105.4 | 1.104 |
| 1978 | 116.2 | 1.102 |

Although the growth factors are not identical, they are all 1.10 when rounded to 3 significant figures. Taking this into account, we can say that the growth is approximately exponential. In this case, the initial value $k = 86.7$. However, the annual growth factor is not 1.10, as the table goes up in four year time intervals. Therefore, the annual growth factor can be calculated by

$$\left(\frac{116.2}{86.7}\right)^{\frac{1}{12}} \approx 1.024.$$

The population growth can now be modelled using the equation

$$P = 86.7 \times 1.024^t.$$

We can now use this equation to answer the questions.

(a) Make an estimate for the population in 1972:

Use the equation with $t = 6$

$$P = 86.7 \times 1.024^6,$$
$$P = 100.$$

So an estimate for the population of Brazil in 1972 is 100 million (3 s.f.) The actual population for Brazil in 1972 was 100.8 million, so this model seems appropriate.

(b) Project the population in 2012:

Use the equation with $t = 46$

$$P = 86.7 \times 1.024^{46}$$
$$P = 258$$

So a projection for the population of Brazil in 2012 using the model is 258 million (3 s.f.)

The actual population of Brazil in 2012 was 199 million (3 s.f). This number is not close to the projection of 258 million which we made. This shows that the model cannot be used to predict accurately for large time frames outside the given data set. This is because there are many factors that can alter population trends over time.

As we have seen in Example 17.1, real-world data is often not exactly exponential and we are trying to fit the "best" exponential model to it. By this, we mean the values of $k$ and $a$ that best approximate each of the data points the data points.

---

**Interactive Activity 17.3 — Finding the Best Exponential Model**

As with Interactive 17.2 find the values of $k$ and $a$ which best approximate the data points shown for the three data sets contained within the Geogebra Interactive below.

---

**Exercise 17.2**

Q1. A colony of bacteria initially has 115 bacteria and has a growth factor of 2.7 per hour. Which expression below shows the number of bacteria after t hours?
  (a) $115 \times t^{2.7}$;
  (b) $115 \times 2.7^t$;
  (c) $2.27 \times 115^t$;
  (d) $2.27 \times t^{115}$.

Q2. A woman deposits £400 into a bank account that pays compound interest at 3% per annum. Which expression below shows how much money is in her bank account after 4 years (assuming she has not withdrawn or deposited any money over that time)?
  (a) $400 \times 3 \times 4$;
  (b) $400 \times 3^4$;
  (c) $400 \times 0.03^4$;
  (d) $400 \times 1.03^4$.

Q3. What is the equation for the graph shown below?

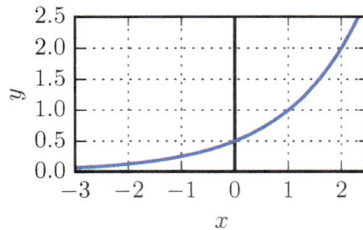

  (a) $2^x$;
  (b) $y = 2 \times \frac{1}{2}^x$;
  (c) $y = \frac{1}{2}^x$;
  (d) $y = \frac{1}{2} \times 2^x$.

Q4. The value $V$ in £s of an investment over $t$ years is given by the formula

$$V = 10\,000 \times 1.024^t. \tag{17.1}$$

    (a) How much money was initially invested?

    (b) What was the interest rate?

    (c) Use the formula to find how much money was in the account after 4 years.

    (d) Use the formula to find out after how many years the value of the initial investment first doubles.

    (e) The owner of the investment withdraws half of the money after 8 years. He leaves the remaining amount in for a further 5 years. How much money is the investment worth at the end of this time?

Q5. In 2000, £2700 was invested into a new company. The value of the investment over the next five years is shown in the table below:

| Year | Value |
|------|-------|
| 2000 | £2700 |
| 2001 | £3240 |
| 2002 | £3890 |
| 2003 | £4670 |
| 2004 | £5600 |
| 2005 | £6720 |

Show that the investment has approximately exponential growth and

    (a) Find the approximate percentage increase in the value of the investment each year.

    (b) Write down an equation for $V$ the value of the investment after $t$ years.

    (c) How much will the investment be worth at the end of 2008, according to this model?

Q6. Write a question that would result in the growth expression $5.4 \times 1.06^t$.

Q7. The graph shows the population of Bangladesh since 2006.

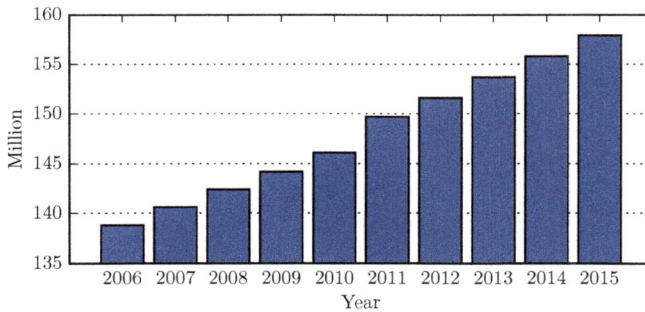

Use this information to create a model to find in which year the population of Bangladesh will first exceed 200 million.

Q8. A bacterial culture contained 7 cells when measured at midday. After 40 minutes it contained 14 cells and after 1 hour 20 minutes it contained 28 cells. I estimate that after 10 hours there will be around 2 million cells. Is this a sensible estimate?

### 17.2.2 Exponential Decay

*Exponential decay* occurs when the growth factor, $a$, lies in the range $0 < a < 1$

---

**Example 17.2**

Sodium-24 is a radioactive isotope. The mass of Sodium-24 remaining in a sample over time was measured as follows:

| Time (hours) | Mass (g) |
|:---:|:---:|
| 0 | 15.00 |
| 1 | 14.93 |
| 2 | 14.26 |
| 3 | 13.62 |
| 4 | 13.01 |
| 5 | 12.42 |

(a) Verify that the Sodium-24 is decaying exponentially.
(b) Write an equation to find the mass, $M$, after $t$ number of hours.
(c) Use the model to find the half-life of Sodium-24 to the nearest hour. (The time it takes for the amount to decay to half the initial amount).

**Solution:**

(a) We can verify that the Sodium-24 is decaying exponentially by comparing the growth factors between each consecutive piece of data:

| Time (hours) | Mass (g) | Growth factor |
|:---:|:---:|:---:|
| 0 | 15.00 | |
| 1 | 14.93 | 0.995 |
| 2 | 14.26 | 0.995 |
| 3 | 13.62 | 0.955 |
| 4 | 13.01 | 0.955 |
| 5 | 12.42 | 0.955 |

As the growth factors are all the same, and are all less than 1, we can verify that the amount has decayed exponentially.

(b) The initial amount is $15\,\text{g}$ and the growth factor is 0.955. Therefore, an equation to find the remaining mass, $M$ after $t$ hours is

$$M = 15 \times 0.955^t$$

(c) We need to find how long it will take until there is $7.5\,\text{g}$ of Sodium-24 left.

We already know that there is $12.42\,\text{g}$ left after 5 hours, so we can use a trial and improvement method to find the answer.
We can start with $t = 10$:

$$M = 15 \times 0.955^{10} = 9.47.$$

Try $t = 13$:

$$M = 15 \times 0.955^{13} = 8.24.$$

Try $t = 15$

$$M = 15 \times 0.955^{15} = 7.52.$$

Try $t = 16$

$$M = 15 \times 0.955^{16} = 7.18.$$

Therefore the half-life of Sodium-24 is 15 hours, to the nearest hour.

**Use of Technology 17.2 — Generating Tables of Values with a Calculator**
Scientific calculators can be used to generate tables of values for user defined functions. Once a table for the function $M = 15 \times 0.955^t$ in Example 17.2 has been created it can be used to find the appropriate value of $t$ for the half-life of Sodium-24.

**Remark**
The Sodium isotope, Sodium-24 is used in medicine to check the flow of sodium in a person's body. Sodium-24 is injected into the body, allowing detectors to trace the radiation, and hence the flow of sodium. An advantage of using sodium-24 in medical tracing is its short half-life. The half-life is relatively short at only 15 hours, therefore there is minimal radiation damage to the body.

**Remark**
There is a far more efficient way of solving equations such as

$$15 \times 0.955^t = 7.5,$$

using *logarithms*. We will study these further in Chapter 18.

**Exercise 17.3**

Q1. In the exponential decay equation $y = 32 \times 0.32^t$, identify
  (a) the initial amount;
  (b) the decay factor;
  (c) the percentage decrease.

Q2. An adult takes a $200\,\text{mg}$ of ibuprofen tablet. The amount of ibuprofen in their system decreases by approximately 30% per hour.

  (a) Write an equation to model the amount of ibuprofen, $I$ in their system after $t$ hours.
  (b) The adult takes another $200\,\text{mg}$ tablet after the recommended 4 hours. How much ibuprofen do they have in their system once they have taken this second tablet?
  (c) How much ibuprofen do they have in their system when they take another tablet four hours later?

Q3. A brand new motor home costs £64,000. The value $V$, of the motor home decreases by 14% every year.

  (a) Write an equation for the value, $V$, of the motor home for a given number of years, $t$.
  (b) Use the model to estimate the value after 2 years.
  (c) Sketch the model.
  (d) Use the graph to estimate when the motor home will have a value of £32,000.

## 17.3  The Natural Base

**Activity 17.1 — Exponential Bases**

We investigate changing the base of exponential functions.
  (a) Write $9^t$ using a different base. Why is it possible do this?
  (b) Rewrite $16^t$ so that it has a different base. Is it possible to write it in more than one way?
  (c) Check the answers to parts (a) and (b) using a graphical calculator or graph plotting package.

It is easy to rewrite powers when one number is a power of another, for example, $27 = 3^3$, so $27^t$ can be rewritten as $3^{3t}$.

It is still possible to change the base when the numbers are not as obviously related as a power of the other. For example, we could check that $5^t$ could be rewritten as $2^{bt}$ by using a graphical calculator or graph plotter as before.

In fact, for any positive value of $b$, except $b = 1$, $a^x$ can be rewritten as $b^{cx}$ for some positive $c$. This means that a universal base could be used for all exponential functions. The base that is used most commonly is e.

**Definition 17.2 — The Natural Base, e**

The *natural base*, e, which is sometimes referred to as the *Euler number* after the Swiss mathematician, Leonhard Euler (1707–1783), is an irrational number, similar to $\pi$.

e lies between 2 and 3 and is written below truncated to 50 decimal places.

$$e = 2.71828182845904523536028747135266249775724709369995\ldots$$

**Remark**

Just like $\pi$, e appears in almost all areas of mathematics. There are many equivalent ways of deriving e. We show one derivation which fits in well with the compound interest ideas we have already encountered in this chapter.

Suppose that we have £1 and we accrue interest at a rate of $\frac{1}{12}$ per month for a year. At the end of the year, we have

$$\left(1 + \frac{1}{12}\right)^{12} = 2.6130\ldots$$

Now suppose we have again £1 and accrue interest at a rate of $\frac{1}{365}$ per day or a year. This time, at the end of the year we have

$$\left(1 + \frac{1}{365}\right)^{365} = £2.7146\ldots$$

This number is already beginning to resemble e. Now suppose interest is accrued hourly over a year at a rate of $\frac{1}{n}$ per hour, where $n = 8760 = 24 \times 365$. In this case, we will have

$$\left(1 + \frac{1}{n}\right)^{n} = £2.7182\ldots,$$

at the end of the year. As $n$ gets larger, our investment is approaching the value of e given above. This motivates the definition

$$e = \lim_{n \to \infty} \left(1 + \frac{1}{n}\right)^{n}.$$

This was the original definition of e given by Jacob Bernoulli in the 17th century. We will meet Bernoulli again when we study probability in Chapter 30. Indeed, $\frac{1}{e}$ appears as the limit of an increasingly large number of successful/unsuccessful Bernoulli trials.

The graph of $f(x) = e^{x}$ has exactly the same features as all of the other exponential functions. The graph of $f(x) = e^{x}$ is shown in Figure 17.5, between the graphs of $g(x) = 2^{x}$ and $h(x) = 3^{x}$.

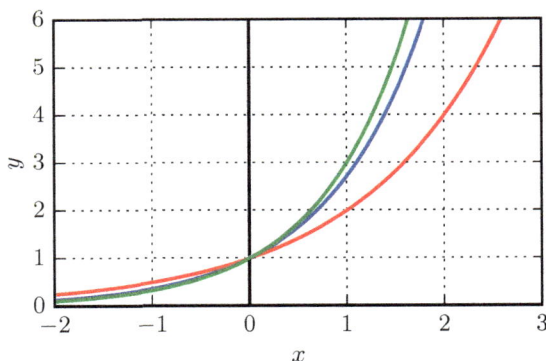

Figure 17.5: $f(x) = e^x$ (blue), $g(x) = 2^x$ (red) and $h(x) = 3^x$ (green).

**Interactive Activity 17.4 — Derivative of $e^x$**

Using the Geogebra applet, investigate the derivative of $y = e^x$ and $y = e^{kx}$ at various values of $x$. Hence, suggest a formula for $\frac{d}{dx}(e^x)$.

**Formulae 17.1 — Derivative of $e^x$**

From the interactive activity above, we should have been able to notice the extremely useful property of $e^x$

$$\frac{d}{dx}(e^x) = e^x.$$

In addition to this, for any $k$,

$$\frac{d}{dx}(e^{kx}) = ke^x.$$

The properties above can be used to explain why exponential functions are a good model for population growth. Let $P(t)$ be the population at a time $t$, then $\frac{dP}{dt}$ is the rate of change of population with time. We can imagine that this rate of change is proportional to the population size. For example, the larger the population, the more births there will be and the more deaths there will be. We will, therefore, have

$$\frac{dP}{dt} \propto P,$$
$$\Rightarrow \quad \frac{dP}{dt} = kP,$$

for some $k$. This is an example of a *differential equation* and $P = e^{kt}$ will satisfy it. In fact, by using properties of differentiation, $P = Ae^{kt}$, for any $A$ will also satisfy the equation.

A more in depth discussion of differential equations will take place in the second year course.

---

**Use of Technology 17.3**

Calculators have an $e^x$ button that can be used to evaluate powers of e, however, for *exact* values, answers should be left in terms of e.

---

**Example 17.3**

A portable electric compressor is rated to inflate car tyres to up to 80psi. An owner is interested to model the time taken to inflate an under-inflated tyre to the rated pressure. He proposes that the tyre pressure, $P$, at time $t$ (measured in minutes) is given by the function $p = 80 - 55e^{\frac{t}{14}}$

(a) What is the pressure of the tyre at time $t = 0$.
(b) Find the pressure of the tyre after 75 seconds.
(c) By tabulating the function for some values of $t$, approximately how long does it take to inflate the tyre to the target pressure of 32psi.
(d) Describe one external factor that could influence the validity of this model.

**Solution:**

(a) When $t = 0$,

$$P = 80 - 55e^{\frac{0}{14}},$$
$$= 80 - 55,$$
$$= 25.$$

So, the initial pressure of the tyre is 25psi.

(b) The function for pressure requires time to be measured in minutes - 75 seconds is 1.25 minutes. At this time the pressure is given by,

$$P = 80 - 55e^{\frac{1.25}{14}},$$
$$= 29.69786754$$

The pressure after 75 seconds is approximately 29.7psi.

(c) We produce a table of values below. From this Table we can say that the pressure is 32psi at approximately 2 minutes.

| Time (minutes) | Pressure (psi) |
|----------------|----------------|
| 0.25 | 25.973 |
| 0.50 | 26.929 |
| 0.75 | 27.868 |
| 1.00 | 28.791 |
| 1.25 | 29.697 |
| 1.50 | 30.588 |
| 1.75 | 31.462 |
| 2.0 | 32.321 |

(d) The external air pressure will also affect the pressure inside the tyre. Air pressure can be dependent on altitude and temperature.

**Exercise 17.4**

Q1. Sketch the graphs of:
   (a) $y = e^{2x}$;
   (b) $y = e^{-3x}$;
   (c) $y = \frac{3}{2}e^{-2x}$;
   (d) $y = e^{\frac{x}{2}}$.

Q2. The value of a car, $V$, of the number of years, $t$ is represented by the formula $V = 24000e^{-0.2t}$.
   (a) Use the formula to find the original value of the car.
   (b) Calculate the value after 3 years.
   (c) Find how many full years it will take for the car to be worth less than half its original value.
   (d) Sketch the graph of $V$ against $t$.

Q3. A species of owl is declining exponentially. The number of owls was modelled using the formula $P = 12\,000e^{-0.4t}$. According to this model, how many years will it take before there are fewer than 1000 owls?

Q4. Scientists are investigating the spread of a pathogen within an isolated community of sheep. They model the number, $I$, of infected sheep by the formula $I = 150 - 120e^{-0.16t}$, where $t$ represents the number of days after the initial infection.
   (a) How many sheep were initially infected with the pathogen.
   (b) How many sheep will be infected 10 days after the initial infection.
   (c) What long term prediction does this model make.
   (d) Sketch the graph of $I$ versus $t$ for $t > 0$.

**Key Facts — Exponential Functions**

- A general growth function has the form $f(x) = ka^x$. $k$ is the initial value and $a$ is the growth factor.
- The growth function can be used to model many physical phenomena.
- If $a > 1$ the growth function exhibits growth, an example of which is shown below.

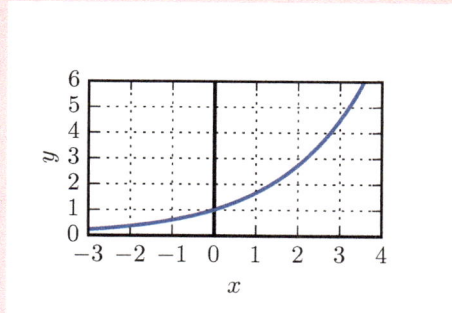

- If $0 < a < 1$, then the growth function exhibits decay, an example of which is shown below.

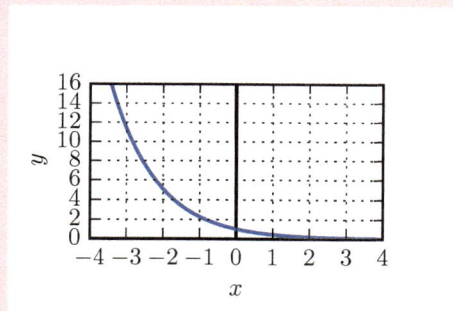

- The natural base is e $= 2.7182818284\ldots$ and the natural exponential function is $\mathrm{e}^x$.
- The natural base e has the following properties:

$$\frac{\mathrm{d}\mathrm{e}^x}{\mathrm{d}t} = \mathrm{e}^x,$$

$$\frac{\mathrm{d}\mathrm{e}^{kx}}{\mathrm{d}t} = k\mathrm{e}^{kx},$$

**Chapter Assessment — Exponential Functions**

Download and sit the 30 minute assessment for this chapter from the digital book.

# 18. Logarithms

In Chapter 17 we saw equations of the form $y = e^{kx}$ and solved them using a trial and error approach. Logarithms were discovered in the 17th century by Scottish mathematician John Napier and they revolutionised mathematics because, as we shall see, they allow equations of this form to be solved very efficiently. Tables of logarithms and slide rules were used for the early computation of logarithms, but now calculators and computers perform this task for us.

Logarithms are common throughout nature as well. One example is the so-called *logarithmic spiral* shown in Figure 18.1(a).

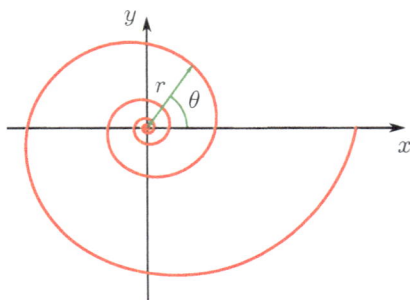

(a)                                              (b)

Figure 18.1: A logarithmic spiral (a) and a nautilus shell exhibiting such a spiral (b).

This spiral has the property that the angle, $\theta$, between the $x$-axis and a point on the spiral satisfies the following equation:

$$\theta = \frac{1}{b} \ln\left(\frac{r}{a}\right),$$

where $r$ is the distance from the centre of the spiral to the points, $a$ and $b$ are some constants and ln is the natural logarithm, which we shall define later in this chapter. Spirals such as this often appear when things grow around a central point, for example, Figure 18.1(b) shows the shell of a nautilus, which exhibits approximately a logarithmic spiral. Hurricanes, spiral galaxies and the Mandelbrot fractal also have logarithmic spiral shapes, as can be seen in Figure 18.2.

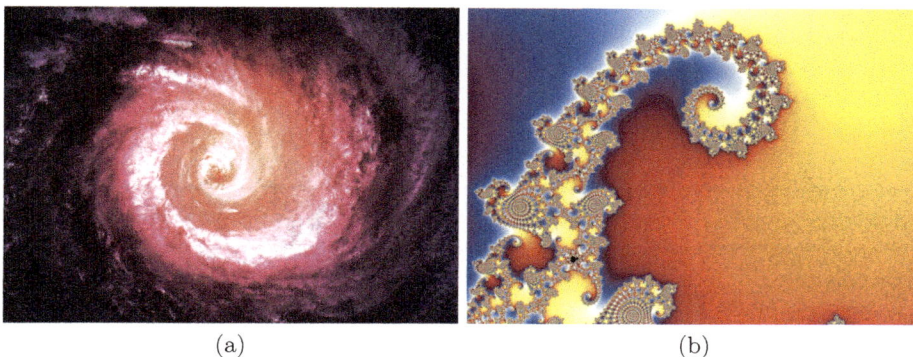

(a)                                                    (b)

Figure 18.2: Examples of logarithmic spirals in a galaxy (a) and a Mandelbrot fractal (b).

## 18.1  Logarithms

In Chapter 17, we were introduced to exponential functions of the form $y = a^x$, that is a function which raises $a$, the base, to the power $x$. A logarithm is a function, which is the inverse of the exponential function. We formalise this notion below.

**Definition 18.1**
Given a value $y > 0$ and a base $a > 0$, $a \neq 1$, the logarithm of $y$ to base $a$, denoted $\log_a(y)$, is the value of $x$ such that

$$y = a^x.$$

**Activity 18.1**
We notice that, in the definition of a logarithm, both $y$ and $a$ have to have positive values and $a \neq 1$, why is this?

**Example 18.1**
Find the following logarithms, without using a calculator.
   (a) $\log_{10}(10)$.
   (b) $\log_{10}(10000)$.
   (c) $\log_{100}(100)$.
   (d) $\log_{100}(10000)$.
   (e) $\log_{10}(1)$.
   (f) $\log_{100}(1)$.

(g) $\log_4(2)$.

(h) $\log_4\left(\frac{1}{2}\right)$.

**Solution:**

(a) 10 has to be raised to the power 1 to give 10. Therefore, $\log_{10}(10) = 1$.

(b) $10000 = 10^4$, so $\log_{10}(10000) = 4$.

(c) 100 has to be raised to the power 1 to give 100. Therefore, $\log_{100}(100) = 1$.

(d) $10000 = 10^4 = 100^2$, so $\log_{100}(10000) = 2$.

(e) Any value raised to the power 0 gives 1, hence $\log_{10}(1) = 0$.

(f) Similarly, $\log_{100}(1) = 0$.

(g) 2 is the (positive) square root of 4. Hence, $\log_4(2) = \frac{1}{2}$.

(h) $\frac{1}{2} = 2^{-1} = (4^{1/2})^{-1} = 4^{-1/2}$. Hence, $\log_4\left(\frac{1}{2}\right) = -\frac{1}{2}$.

In the above example, two very important properties of logarithms were revealed:

**Formulae 18.1**

For any base $a > 0$, $a \neq 1$, we have

$$\log_a(a) = 1,$$
$$\log_a(1) = 0.$$

In addition to these, we know that, as $x \to \infty$, $\log_a(x) \to \infty$ and as $x \to 0$, $\log_a(x) \to -\infty$.

**Example 18.2**

Find the following logarithms without using a calculator.

(a) $\log_3(27)$;

(b) $\log_5\left(\frac{1}{5}\right)$;

(c) $\log_2\left(\frac{2^4}{4^2}\right)$;

(d) $\log_3((\sqrt{3})^3)$;

(e) $\log_4(4^6 \times 4^3)$;

(f) $\log_3(\log_4(64))$.

**Solution:**

(a) $27 = 3^3$, so $\log_3(27) = 3$.

(b) $\frac{1}{5} = 5^{-1}$, hence, $\log_5\left(\frac{1}{5}\right) = -1$.

(c) $\frac{2^4}{4^2} = \frac{16}{16} = 1$. Thus, $\log_2\left(\frac{2^4}{4^2}\right) = 0$.

(d) $4^6 \times 4^3 = 4^9$ and $\log_4(4^6 \times 4^3) = 9$.

(e) $64 = 4^3 = 4^{3^1}$, hence $\log_3(\log_4(64)) = 1$.

Two bases are much more common than others. The first is the base e. We have already seen that $e^x$ is a very important exponential function and similarly its inverse is also very useful. Logarithms to the base e are called *natural logarithms* and have their own special notation $\log_e(x) = \ln(x)$.

The other frequently occurring logarithmic base is base 10. As we shall see later in this chapter, many physical quantities are measured on a *logarithmic scale*, with base 10.

The graph of the inverse of a function can be found by reflecting the graph of the original function in the line $y = x$. As $\ln(x)$ is the inverse of $e^x$, we can plot the graph of $y = \ln(x)$

as in Figure 18.3.

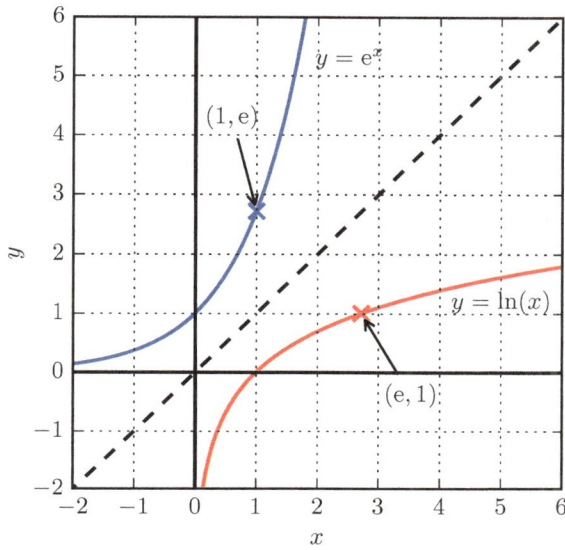

Figure 18.3: The graph of $y = \ln(x)$ is the reflection of $y = e^x$ in the line $y = x$.

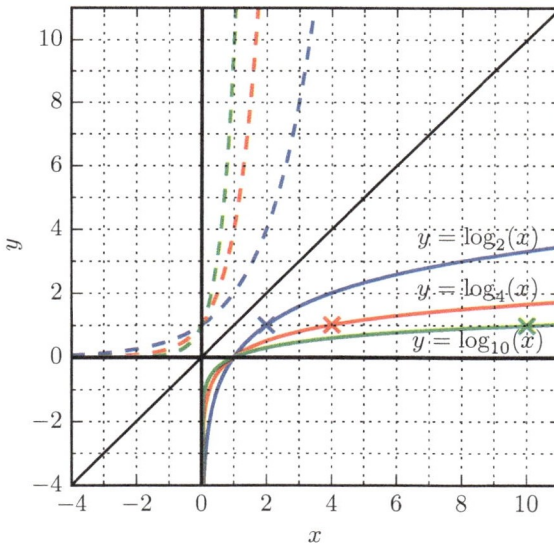

Figure 18.4: Graphs of $y = \log_2(x)$, $y = \log_4(x)$ and $y = \log_{10}(x)$ and the corresponding exponential curves (dashed lines).

Plots of $\log_a(x)$ can similarly be found by reflecting the curves of $a^x$ in the line $y = x$. Some examples of this are shown in Figure 18.4.

---

**Activity 18.2**

We saw in Chapter 17 that the gradient of $e^x$ is given by $e^x$. Using, Figure 18.3, suggest a formula for the gradient of $\ln(x)$.

---

**Exercise 18.1**

Q1. Find the values of the following logarithms without using a calculator.
  (a) $\log_{10}(1\,000\,000)$;
  (b) $\log_{10}(0.1)$;
  (c) $\log_{10}(\sqrt{10})$;
  (d) $\log_7(1)$;
  (e) $\log_9(9)$;
  (f) $\log_2\left(\frac{1}{4}\right)$;
  (g) $\log_3(81)$;
  (h) $\log_9(81)$;
  (i) $\log_3\left(\frac{1}{27}\right)$;
  (j) $\log_4\left(\frac{1}{\sqrt[3]{4}}\right)$.

Q2. Sketch the graphs of the following functions.
  (a) $f(x) = \log_3(x)$;
  (b) $f(x) = \log_{\frac{1}{2}}(x)$.

Q3. Find the following logarithms without using a calculator.
  (a) $\log_2\left(2^3 \times 2^7\right)$;
  (b) $\ln\left(e^3\right)$;
  (c) $\ln\left(e^{16x}\right)$;
  (d) $\ln\left(\left(e^{4x}\right)^3\right)$;
  (e) $\log_4\left(\left(\sqrt{16}\right)^3\right)$;
  (f) $\log_{100}\left(\log_2(1024)\right)$.

## 18.2  Laws of Logarithms

As taking logarithms is the inverse of exponentiation, we can expect logarithms to have the *reverse* of some of the laws of indices. These properties are called the *laws of logarithms*. Let us begin with the product of two number $x$ and $y$. We have, by the definition of logarithms, that

$$xy = a^{\log_a(xy)}, \quad x = a^{\log_a(x)} \quad \text{and} \quad y = a^{\log_a(y)}.$$

By rearranging the above and using the laws of indices for exponential functions, we have

$$a^{\log_a(xy)} = xy$$
$$= a^{\log_a(x)} \cdot a^{\log_a(y)}$$
$$= a^{\log_a(x) + \log_a(y)}.$$

It follows immediately that

$$\log_a(xy) = \log_a(x) + \log_a(y).$$

This result is known as the *multiplication law* of logarithms.

---

**Formula 18.2 — Multiplication Law**
If $x > 0$, $y > 0$ and base $a > 0$, then

$$\log_a(x) + \log_a(y) = \log_a(xy). \tag{18.1}$$

---

**Remark**
That $a^{\log_a(xy)} = a^{\log_a(x)+\log_a(y)} \Rightarrow \log_a(xy) = \log_a(x) + \log_a(y)$, depends on the fact that exponential functions are *one-to-one* functions. That is, for every value of $y > 0$, there can only be one value of $x$ such that $y = a^x$. If we had, for example, $a^2 = b^2$, then we cannot be sure $a = b$, because $y = x^2$ is not a one-to-one function.

---

In an almost identical manner, we can consider $\frac{x}{y}$ and note that

$$\frac{x}{y} = a^{\log_a\left(\frac{x}{y}\right)}.$$

We then have

$$a^{\log_a\left(\frac{x}{y}\right)} = \frac{x}{y}$$
$$= a^{\log_a(x)} / a^{\log_a(y)}$$
$$= a^{\log_a(x) - \log_a(y)}.$$

It follows immediately that

$$\log_a(x) - \log_a(y) = \log_a\left(\frac{x}{y}\right).$$

This is called the *division law* for logarithms.

---

**Formula 18.3 — Division Law**
If $x > 0$, $y > 0$ and a base $a > 0$, then

$$\log_a(x) - \log_a(y) = \log_a\left(\frac{x}{y}\right). \tag{18.2}$$

---

**Example 18.3**
Write the following as single logarithms.
  (a) $\log_4(7) + \log_4(3)$;
  (b) $\log_3(12) - \log_3(4)$;
  (c) $-\log_7(9) - \log_7(12)$;
  (d) $-\ln(5) + \ln(3)$.

**Solution:**

(a) We use the multiplication law to find

$$\log_4(7) + \log_4(3) = \log_4(7 \times 3)$$
$$= \log_4(21).$$

(b) We use the division law to find

$$\log_3(12) - \log_3(4) = \log_3(12 \div 4)$$
$$= \log_3(3).$$

(c) We rearrange and use the multiplication law as follows.

$$-\log_7(9) - \log_7(12) = -(\log_7(9) + \log_7(12))$$
$$= -\log_7(9 \times 12)$$
$$= -\log_7(108).$$

(d) We rearrange and use the division law:

$$-\ln(5) + \ln(3) = \ln(3) - \ln(5)$$
$$= \ln\left(\frac{3}{5}\right).$$

Our third law is called the *power law*. For an $x$ and exponent $k$, using the laws of indices gives:

$$a^{\log_a(x^k)} = x^k$$
$$= \left(a^{\log_a(x)}\right)^k$$
$$= a^{k\log_a(x)}.$$

Thus,

$$\log_a(x^k) = k\log_a(x).$$

---

**Formula 18.4 — Power Law**

For $x > 0$, a base $a > 0$ and an index $k$,

$$\log_a(x^k) = k\log_a(x). \tag{18.3}$$

---

As a special case of the division law, or the power law with $k = -1$, we obtain the following result.

**Formula 18.5**

For $x > 0$ and a base $a > 0$,

$$\log_a\left(\frac{1}{x}\right) = \log_a(x^{-1}) = -\log_a(x). \tag{18.4}$$

$$a = b^{\log_b(a)}.$$

Hence, for any $x > 0$, using the definition of logarithms and the laws of indices

$$x = a^{\log_a(x)}$$
$$= \left(b^{\log_b(a)}\right)^{\log_a(x)}$$
$$= b^{\log_b(a)\log_a(x)},$$
$$\Rightarrow \quad \log_b(x) = \log_b(a)\log_a(x),$$
$$\Rightarrow \quad \log_a(x) = \frac{\log_b(x)}{\log_b(a)}.$$

**Formula 18.6 — Change of Base**

Suppose that $a > 0$, $a \neq 1$, $b > 0$ and $b \neq 1$, then, for $x > 0$, we have

$$\log_a(x) = \frac{\log_b(x)}{\log_b(a)}. \tag{18.5}$$

**Tip**

Formulae 18.1, 18.2, 18.3, 18.4, 18.5 and 18.6 should be memorised and their use should come as second nature.

**Example 18.4**

Write the following as single logarithms.

(a) $4\log_2(3) + 2\log_2(4)$;
(b) $\frac{1}{2}\log_{10}(9) + 2\log_{10}(2)$;
(c) $\frac{1}{4}\log_2(5) - \frac{1}{3}\log_2(2)$;
(d) $2\ln(4) - 3\ln\left(\frac{1}{3}\right)$;
(e) $\log_2(5) + \log_4(25)$;

**Solution:**

(a) We use the power law, followed by the multiplication law:

$$4\log_2(3) + 2\log_2(4) = \log_2(3^4) + \log_2(4^2)$$
$$= \log_2(3^4 \times 4^2)$$
$$= \log_2(1296).$$

(b) Again, we use the power law first, then the multiplication law:

$$\frac{1}{2}\log_{10}(9) + 2\log_{10}(2) = \log_{10}(9^{1/2}) + \log_{10}(2^2)$$
$$= \log_{10}(9^{1/2} \times 2^2)$$
$$= \log_{10}(3 \times 4) = \log_{10}(12).$$

(c) Here we use the power law followed by the division law:

$$\frac{1}{4}\log_2(5) - \frac{1}{3}\log_2(2) = \log_2(5^{1/4}) - \log_2(2^{1/3})$$
$$= \log_2\left(\frac{5^{1/4}}{2^{1/3}}\right).$$

(d) There are a variety of way to solve this problem, we first apply Formula 18.5 then the power law and finally the multiplication law.

$$2\ln(4) - 3\ln\left(\frac{1}{3}\right) = 2\ln(4) + 3\ln(3)$$
$$= \ln(4^2) + \ln(3^3)$$
$$= \ln(4^2 \times 3^3)$$
$$= \ln(16 \times 9) = \ln(144).$$

(e) Here, the bases of the logarithms are different, but we know that $\log_2(x) = 2\log_4(x)$. Using this fact and then the power and multiplication laws gives

$$\log_2(5) + \log_4(25) = \log_2(5) + \frac{1}{2}\log_2(25)$$
$$= \log_2(5) + \log_2\left(25^{\frac{1}{2}}\right)$$
$$= \log_2(5 \times 5) = \log_2(25).$$

**Example 18.5**
Write the following in terms of $\log_a(x)$ and $\log_a(y)$.
(a) $\log_a(x^2 y^4)$;
(b) $\log_a(x^3\sqrt{y})$;
(c) $\log_a\left(\frac{y^4}{x^3}\right)$;
(d) $\log_a\left(\frac{(ax)^3}{y}\right)$.

**Solution:**

(a) We use the multiplication and power laws to split the logarithms:

$$\log_a(x^2 y^4) = \log_a(x^2) + \log_a(y^4)$$
$$= 2\log_a(x) + 4\log_a(y) = 2(\log_a(x) + 2\log_a(y)).$$

(b) Again, using the multiplication and power laws, we obtain

$$\log_a(x^3\sqrt{y}) = \log_a(x^3) + \log_a(\sqrt{y})$$
$$= 3\log_a(x) + \frac{1}{2}\log_a(y).$$

(c) In this case, we use the division law and the power law to find:

$$\log_a\left(\frac{y^4}{x^3}\right) = \log_a(y^4) - \log_a(x^3)$$
$$= 4\log_a(y) - 3\log_a(x).$$

(d) This requires use of the multiplication, division and power laws and the fact that $\log_a(a) = 1$.

$$\log_a\left(\frac{(ax)^3}{y}\right) = 3\log_a(ax) - \log(y)$$
$$= 3(\log_a(a) + \log_a(x)) - \log(y)$$
$$= 3(1 + \log_a(x)) - \log(y).$$

**Exercise 18.2**

Q1. Write the following as single logarithms.
   (a) $\log_2(6) + \log_2(8)$;
   (b) $\log_3(2) + \log_3(7)$;
   (c) $\log_4(10) - \log_4(5)$;
   (d) $1 + \ln(5)$;
   (e) $2 - \log_{10}(5)$;
   (f) $\log_3(27) - \log_9(729)$;
   (g) $\log_4(12) + \log_4(6) + \log_4(3)$;
   (h) $-\log_3(6) - \log_3(10) + \log_3(20)$.

Q2. In the following, simplify the answers as much as possible.
   (a) Rewrite $\log_{10}(4)$ as a logarithm to the base 4;
   (b) Rewrite $\log_4(13)$ as a logarithm to the base 2;
   (c) Rewrite $\log_{10}(x)$ as a logarithm to the base 100.

Q3. Write the following as single logarithms.
   (a) $3\log_4(3) + 2\log_4(3)$;
   (b) $\frac{3}{2}\log_3(7) - \frac{1}{2}\log_3\left(7^2\right)$;
   (c) $\ln(8) - 2\ln\left(\frac{1}{4}\right)$;
   (d) $-3\log_{10}(5) - 4\log_{10}(2)$.

Q4. Write the following as single logarithms.
   (a) $\log_{10}(x+3) + \log_{10}(x+2)$;
   (b) $\log_2(x^2 + 2x - 8) - \log_2(x - 2)$;
   (c) $\ln(x^3 - 1) - \ln(x - 1)$;
   (d) $2\log_{10}\left(\frac{y}{x}\right) + 4\log_{10}(x\sqrt{y})$.

### 18.3  Solving Equations with Logarithms

If we have equations with logarithmic terms, we can use the laws presented in the previous section to help solve the equations, as the following examples show.

---

**Example 18.6**

Find all possible values of $x$ which satisfy the equation

$$\log_4(8) + \log_4(x^3) = 3.$$

**Solution:**

$$\log_4(8) + 3\log_4(x) = 3,$$
$$\Rightarrow \quad \log_4(8) + \log_4(x^3) = 3, \qquad \text{(power law)}$$
$$\Rightarrow \quad \log_4(8x^3) = 3, \qquad \text{(multiplication law)}$$
$$\Rightarrow \quad 8x^3 = 4^3, \quad \text{(definition of logarithms)}$$
$$\Rightarrow \quad 8x^3 = 64,$$
$$\Rightarrow \quad x^3 = 8,$$
$$\Rightarrow \quad x = 2.$$

Hence, there is only one solution $x = 2$ to this problem.

---

**Example 18.7**

Find all possible values of $x$ which satisfy the equation

$$\log_2(x - 1) + \log_2(x + 2) = 2.$$

**Solution:**

$$\log_2(x - 1) + \log_2(x + 2) = 2,$$
$$\Rightarrow \quad \log_2((x - 1)(x + 2)) = 2, \qquad \text{(multiplication law)}$$
$$\Rightarrow \quad (x - 1)(x + 2) = 4, \quad \text{(definition of logarithms)}$$
$$\Rightarrow \quad x^2 + x - 2 = 4,$$
$$\Rightarrow \quad x^2 + x - 6 = 0,$$
$$\Rightarrow \quad (x - 2)(x + 3) = 0,$$

The final quadratic has solution $x = 2$ and $x = -3$. However, these are not both solutions of the original equation, because $\log_2(-3 + 1)$ is undefined. Therefore, $x = 2$ is the only solution of the equation.

---

The examples above are not the only types of problems which can be solved with logarithms. Indeed, one of the most important applications of logarithms is in solving equations of the form

$$a^x = b,$$

where $a$ and $b$ are known, positive values. Perhaps the most obvious thing to try in an attempt to solve this problem is to take logarithms with base $a$ of both sides. In which case, we have

$$\log_a(b) = \log_a(a^x) = x.$$

Sometimes, logarithms of bases, $a$, other than e and 10 are not readily available. Indeed, until recently $\log_{10}$ and ln were the only logarithms that could be found on calculators. Fortunately, we can take logarithms of any base, $c$ say, and can still solve the equation using Formula 18.6:

$$x = \log_a(b)$$
$$= \frac{\log_c(b)}{\log_c(a)}.$$

---

**Remark**

If $c = a$ in the above, then, because $\log_a(a) = 1$, we return to the solution

$$x = \log_a(b).$$

---

**Use of Technology 18.1 — Finding Logarithms**

In the digital book, the links below show how to find logarithms on a variety of calculators.

---

**Example 18.8**

Find the solution $x$ to the equation

$$2^x = 1024.$$

**Solution:**

1024 is an exact power of 2, with $x = 10$ (consider the number of bytes in a MB). If this is not obvious, it is possible to take logarithms with base 2 on a calculator to find the result. If our calculator cannot take logarithms with base 2, we use logarithms with different bases to give the same result.

We first use the natural logarithms to find the solution. We know that

$$x = \frac{\ln(1024)}{\ln(2)} = \frac{6.931471\ldots}{0.6931471\ldots} = 10.$$

Now, let us use logarithms to the base 10 to find the solution. We have

$$x = \frac{\log_{10}(1024)}{\log_{10}(2)} = \frac{3.010299\ldots}{0.3010299\ldots} = 10.$$

In both cases, the correct value of 10 has been found.

**Example 18.9**

Solve the equation

$$3^{7x-3} = 32$$

**Solution:**

Taking logarithms of base 3 of both sides we find

$$7x - 1 = \log_3(32),$$

$$\Rightarrow \qquad x = \frac{\log_3(32) + 1}{7}$$

$$= 0.5935 \text{ to 4 d.p.}$$

**Example 18.10**

Find the solutions of the following equation

$$7^{2x} - 5(7^x) + 6 = 0.$$

**Solution:**

Let $y = 7^x$, then the equation can be rewritten as

$$y^2 - 5y + 6 = 0,$$

$$\Rightarrow \quad (y - 2)(y - 5) = 0,$$

Hence, $y = 2$ or $y = 5$ and then

$$y = 2,$$

$$\Rightarrow \quad 7^x = 2,$$

$$\Rightarrow \quad x = \log_7(2) = 0.3562 \text{ to 4 d.p.}$$

and

$$y = 5,$$

$$\Rightarrow \quad 7^x = 5,$$

$$\Rightarrow \quad x = \log_7(5) = 0.8271 \text{ to 4 d.p.}$$

The examples above are relatively straightforward because only one term was raised to a power. In the next example we consider a problem where three different terms are raised to a power.

**Example 18.11**

Find the solution to the equation

$$4^x = 5^{x-1}3^{x+1}.$$

**Solution:**

Recalling Example 18.8, we remember that it did not matter what the base of the logarithm we used was, we still got the same result. In this case, we take logarithms of both sides, let us use natural logarithms. We find

$$\ln(4^x) = \ln(5^{x-1}3^{x+1}),$$
$$\Rightarrow \qquad \ln(4)x = \ln(5^{x-1}) + \ln(3^{x+1})$$
$$\Rightarrow \qquad = (x-1)\ln(5) + (x+1)\ln(3),$$
$$\Rightarrow \quad (\ln(4) - \ln(5) - \ln(3))x = \ln(3) - \ln(5),$$
$$\Rightarrow \qquad x = \frac{\ln(3) - \ln(5)}{\ln(4) - \ln(5) - \ln(3)}$$
$$= 0.3865 \text{ to 4 d.p.}$$

Using a logarithm of any base will produce the same result.

**Exercise 18.3**

Q1. Find the solutions to the following equations (calculators may be used where necessary).
   (a) $\log_{10}(20) + \log_{10}(x^4) = 3$;
   (b) $\ln(3x^2) - \ln(7x^3) = 2$;
   (c) $\log_2(x + 5) + \log_2(x + 3) = 4$;
   (d) $3\log_2(2x) + 2\log_2(3x^2) = 3$.
Q2. Find the solutions to the following equations (calculators may be used where necessary).
   (a) $4^x = 912$;
   (b) $3^x = \frac{1}{12}$;
   (c) $e^{2x} = 20$;
   (d) $10^{5x} = 100$;
   (e) $3^{4x-1} = 81$;
   (f) $2^{(x-1)^2} = 10$;
   (g) $5^{x^2-x+1} = 125$;
   (h) $4^{2^x} = 80$;
   (i) $3^x \cdot 4^{x-1} = 12$;
   (j) $2^x \cdot 3^{\frac{1}{x}} = 10$.

## 18.4 Graphing with Logarithmic Scales

Suppose that we have an equation of the form $y = Ax^b$ and we want to plot $y$ against $x$. We are used to using graph paper with a *linear* scale, *i.e.* a scale where the distance between grid-lines represents a fixed change in a variable. In the previous section, we saw the important result that $\log_a(Ax^b) = b\log_a(x) + \log(A)$, for any base $a > 0$. Taking logarithms ("taking logs") of both sides of $y = Ax^b$ thus yields

$$\log_a(y) = b\log_a(x) + \log(A).$$

This shows that $\log_a(y)$ is a linear function of $\log_a(x)$. Hence, if we plot $\log_a(y)$ against $\log_a(x)$ for any $a > 0$, we will see a straight line graph.

---

**Example 18.12**

Let $y = 2x^3$. If we plot $y$ against $x$ for this function, we obtain the graph shown in Figure 18.5(a). A quick glance at this graph might gives us an idea that $y$ is cubic, but without careful inspection of the values, we are not able to tell for sure. If however, we plot $\log_{10}(y)$ against $\log_{10}(x)$, we obtain the graph shown in Figure 18.5(b).

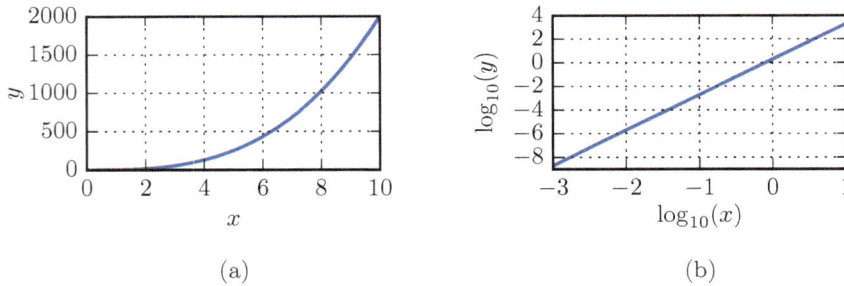

Figure 18.5: Plot of $y = 2x^3$ (a) and $\log_{10}(y) = 3\log_{10}(x) + 2$ (b).

By obtaining the straight line plot in Figure 18.5(b), we know immediately that we have an equation of the form $y = Ax^n$. We can calculate the gradient of the straight line to find $n = 3$.

---

**Remark**

The scale on the plot in Figure 18.5(b) can only take positive values, because $\log(x)$ for $x \leq 0$ is undefined.

---

Motivated by the above, it can be very useful to use graph paper with a *logarithmic* scale. With this type of scale, the distance between grid-lines represents a multiplication by a base, rather than an increment by a fixed value. It is most common for this to be a multiplication by a base of 10, but any positive base is allowable. To understand this further, consider the number line show in Figure 18.5. Here, the $x$-values range from $10^{-3}$ to $10^3$, with the distance between grid-lines representing a step up by a factor of 10 each time.

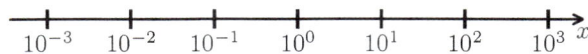

Figure 18.6: An example of a number line with a logarithmic (base 10) scale.

In order to be able to accurately place values on such a scale, we should further divide up the number line between the grid-lines. As we are working with a logarithmic scale, an equally spaced division is no longer meaningful. Instead, we choose a division such that the first *minor* grid-line has twice the value of the previous *major* grid line, the second minor grid-line has three times the value, and so on. Figure 18.7 shows an example of the resulting divisions between the major grid-lines $x = 1$ and $x = 10$.

Figure 18.7: Further division between gridlines with a logarithmic (base 10) scale.

Let us now plot the function $y = 2x^3$ for $10^{-3} \leq x \leq 10^1$ on a graph with logarithmic scales on both the $x$- and $y$-axes . The result is shown in Figure 18.8. We see that, with this logarithmic scale, the cubic function becomes a line. Again, by changing the scale of the axes, the nature of the function can be revealed much more easily.

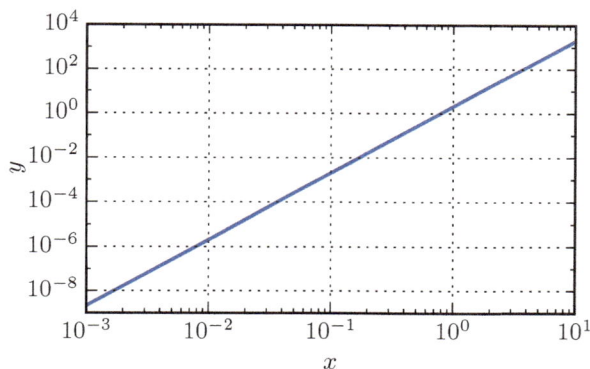

Figure 18.8: Plot of $y = 2x^2$ on a graph with logarithmic axes.

Graphs with logarithmic scales can also be used to plot functions of the form $y = a^{kx}$, for some $a > 0$. In this case, we already know from Equation 18.3 that

$$\log_{10}(y) = \log_{10}(a^{kx}) = k \log_{10}(a)x.$$

Hence, $\log_{10}(y)$ is a linear function of $x$. Thus, using a graph with a linear scale and plotting $\log_{10}(y)$ against $x$ will yield a straight line curve. Alternatively, plotting $y$ against $x$, with a linear scale for $x$ and a logarithmic scale for $y$ ,will also yield a straight line.

**Interactive Activity 18.1 — Using Logarithmic Scales to Find Model Parameters**
In this activity we investigate using logarithmic scales to approximate parameters $a$ and
$k$ in models of the form $y = ax^k$ and $y = ae^{kt}$. Each Geogebra activity below contains
three sets of data, use the sliders to find the appropriate values of $k$ and $a$ in each case.

**Example 18.13**
The two figures below show the graphs of $y = e^{2x}$. The first uses a linear scale for
both $x$ and $y$, while the second uses a logarithmic scale for $y$, but a linear one for $x$.
We see that the exponential curve is represented as a straight line in the second plot.
Another noticeable feature of the second graph is that for small $x$-values, it is must
easier to estimate the corresponding $y$-value, whereas this is almost impossible for the
linear scaled graph.

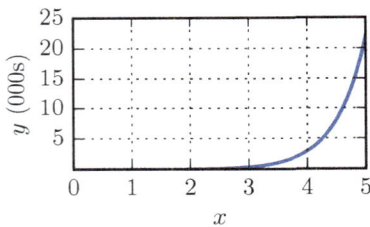

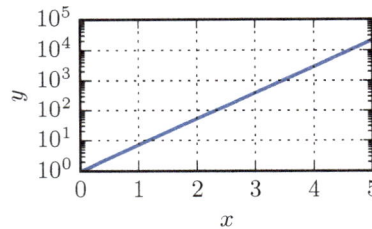

**Remark**
Even though a logarithmic scale of base 10 was used in Example 18.13 and the function
was an exponential of base e, the function still appeared as a straight line in the graph
with logarithmic scaling. This is a particular feature of logarithms.

**Example 18.14**
A small room is filled with radioactive radon gas. Over the course of 10 days, the number of radioactive incidents per second is measured using a Geiger counter at midday.
These results are shown in the table below.
It is suggested that radon has an exponential rate of decay. By plotting the Counts
per Second against Day on a graph where the $y$-axis has a logarithmic scale, suggest
whether or not radon does decay exponentially and, if so, what its half-life is.

| Day ($d$) | Counts per Second ($C$) |
|:---:|:---:|
| 0 | 1013 |
| 1 | 860 |
| 2 | 690 |
| 3 | 586 |
| 4 | 501 |
| 5 | 396 |
| 6 | 343 |
| 7 | 290 |
| 8 | 210 |
| 9 | 201 |
| 10 | 161 |

**Solution:**

A plot of Counts per Second against Day is shown below. With these linear-logarithmic axes, the points appear to lie on a straight line and we can draw the red line of best-fit through the points. This indicates that the radon does exhibit exponential decay, i.e., $C = Ae^{-kd}$, where $A$ and $k$ can be determined from the graph.

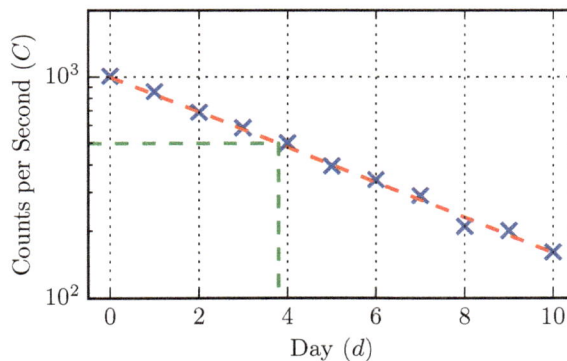

The half-life is the time taken for half of the radon to have decayed, or, equivalently, for the radiation intensity to have dropped by a half. At $d = 0$, the red line passes though $C = 1000$. We therefore look for the time when $C = 500$. The green line shows that this occurs when $d \approx 3.8$, hence, the half life of radon is approximately 3.8 days.

We now approximate $A$ and $k$. $A$ is immediately given by the value of $C$ when $d = 0$, hence $A \approx 1000$. Then, at $d \approx 3.8$, we have $C = 500$, therefore,

$$500 = 1000e^{-3.8k},$$

$$\Rightarrow \qquad \frac{1}{2} = e^{-3.8k},$$

$$\Rightarrow \quad \ln\left(\frac{1}{2}\right) = -3.8k,$$

$$\Rightarrow \qquad k = -\ln\left(\frac{1}{2}\right)/3.8,$$

$$\Rightarrow \qquad k = 0.1824 \text{ to 4 d.p.}$$

Thus,

$$C \approx 1000e^{-0.1824d}.$$

**Exercise 18.4**

Q1. Plot the following:
    (a) $4\log_{10}(x) + 2$ against $\log_{10}(x)$ for $10^{-4} \leq x \leq 10^3$, when both axes have linear scales;
    (b) $y = 100x^4$ when both axes have logarithmic (base 10) scales and $10^{-4} \leq x \leq 10^3$.
    What is noticeable about these plots and how can this be explained?

Q2. Make a copy of the logarithmic (base 10) scale below. Indicate the locations of the following $x$ values
    (a) 100;
    (b) 2;
    (c) 20;
    (d) $\frac{3}{10}$;
    (e) $\log_2(16)$;
    (f) $7 \times 10^{-2}$.
    (g) $10^{5/2}$.

Q3. Make a copy of the logarithmic (base 4) scale below. Indicate the locations of the following $x$ values
    (a) 8;
    (b) 48;
    (c) $\frac{1}{2}$;
    (d) $\frac{3}{16}$;
    (e) $4^{5/4}$;

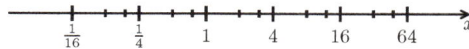

**Q4.** A particle is moving in the $(x, y)$-plane. At various times, the position of the particle is recorded and tabulated as below.

| $x$ | $y$ |
|---|---|
| 1 | 24 |
| 2 | 82 |
| 4 | 245 |
| 8 | 811 |
| 16 | 2540 |
| 32 | 8051 |

It is suggested the path of the particle can be modelled using the equation $y \approx Ax^b$. By plotting $\log_{10}(x)$ against $\log_{10}(y)$, determine whether this model is reasonable and, if so, find approximate values for $A$ and $b$.

**Q5.** Acceleration $g$ due to gravity has been measured at various distances from the centre of the Earth $r$, with the first measurement being taken on the Earth's surface and subsequent measurements being taken further away.

| $r$ (km) | $g$ (ms$^{-2}$) |
|---|---|
| 6400 | 9.80 |
| 10 000 | 4.01 |
| 20 000 | 1.00 |
| 30 000 | 0.45 |
| 40 000 | 0.25 |

Newton's universal law of gravitation applied to Earth says that $g = G\frac{m}{r^2}$, where $g$ is the acceleration due it gravity in ms$^{-2}$, $m$ is the mass of the Earth in kg, $r$ is the distance from the centre of the Earth in m and $G = 6.7 \times 10^{-11}$ N(mkg)$^2$ is the gravitational constant.

By plotting $\log_{10}(g)$ against $\log_{10}(r)$ on axes with linear scales, show that the measurements fit the law. Then, use the law to find an approximation to the mass of the Earth in kg.

Note, Newton's law of gravitation is an example of an *inverse square law*.

**Q6.** The number of bacteria in a petri dish is estimated at two-hourly intervals over the course of half a day, the results are tabulated below.

| Hour $(t)$ | Number of Bacteria $(N)$ |
|:----------:|:------------------------:|
| 0          | 308                      |
| 2          | 720                      |
| 4          | 1760                     |
| 6          | 4100                     |
| 8          | 9890                     |
| 10         | 23 450                   |
| 12         | 55 780                   |

Suggest a model for the number of bacteria, $N$, in terms of the hour $t$. By plotting the results on a graph with a suitable scales, show this model is appropriate and find the parameters of the model. Find and approximation to the length of time required for the number of bacteria to double.

Q7. A thermistor is a electrical resistor, whose resistance varies with temperature. The resistance, $R$, of one thermistor is measured at various temperatures, $T$, and the results tabulated below.

| $T$ ($\kappa$) | $R$ ($\Omega$) |
|:--------------:|:--------------:|
| 240            | 120            |
| 273            | 112            |
| 300            | 107            |
| 350            | 100            |
| 400            | 94             |
| 450            | 91             |

A model for how resistance depends on temperature is

$$R = Ae^{B/T},$$

for some unknown values $A$ and $B$. By plotting the tabulated values on a graph with suitable scales, find approximate values for $A$ and $k$.

## Activity 18.3 — Logarithmic Scales

We are familiar with measuring quantities on linear scales, for example, it is usual to measure the speed of a car in terms of miles per hour (mph) or kilometres per hour (kph). This means that the difference in speed between a car travelling at 20mph and one travelling at 10mph is the same as the difference in speed between two cars travelling at 70mph and 60mph. Temperature is also usually measured on linear scales, for example, the centigrade, Fahrenheit and Kelvin scales are all linear. Sometimes though, the range over which a quantity is measured is so large that using a linear scale means it is difficult for us to appreciate the size of the quantities involved. For example, an earthquake can be measured in terms of the amplitude of the seismic waves it produces (as measured on a seismograph), these can vary from micrometres (0.000001m) or smaller up to tens or even hundreds of metres. Instead, the Richter scale uses a logarithmic scale to the base 10 to measure the magnitude of earthquakes. This means that an earthquake of magnitude 2.0 will actually have ten times the shaking amplitude of one of magnitude

1.0. Similarly, an earthquake of magnitude 3.0 will have one hundred times the strength of a magnitude 1.0 earthquake. The equation for the magnitude $M_r$ of an earthquake is

$$M_r = \log_{10}(A) - \log_{10}(A_0), \tag{18.6}$$

where $A$ is the measured shaking amplitude and $A_0$ is some baseline amplitude. Richter took $A_0 = 1.0 \times 10^{-6}$m because this was the smallest amplitude that could be recorded on the seismographs at the time. This means that a magnitude 0 earthquake is possible. In fact, with improvements in technology, negative magnitude earthquakes are also possible, although, of course, these are far too small to cause any damage.

Q1. An earthquake is measured to have a shaking amplitude of 0.001m on a seismograph. What is the magnitude of the earthquake on the Richter scale?

Q2. How many times larger is the shaking amplitude of a 7.5 magnitude earthquake compared with a 4.2 magnitude earthquake?

Q3. The energy released by an earthquake scales with the (3/2) power of the shaking amplitude. How many more times as energetic is a magnitude 5 earthquake, compared with a magnitude 3 earthquake?

Q4. Investigate what other commonly used scales are logarithmic.

**Remark**

Although the term 'Richter Scale' is commonly mentioned when we hear reports of earthquakes, it is actually no longer in use. Instead, the Moment Magnitude Scale (MMS) is now used, which, in certain circumstances, gives the same values as the Richter scale. However, the name Richter Scale persists.

**Key Facts — Logarithms**

- $\log_a(x) = y \Leftrightarrow x = a^y$;
- $\log_a(a) = 1$;
- $\log_a(1) = 0$;
- $\ln(x) = \log_e(x)$;
- $\log_a(x) + \log_a(y) = \log_a(xy)$;
- $\log_a(x) - \log_a(y) = \log_a\left(\frac{x}{y}\right)$;
- $\log_a(x^k) = k\log_a(x)$;
- $\log_a\left(\frac{1}{x}\right) = \log_a(x^{-1}) = -\log_a(x)$;
- $\log_a(x) = \frac{\log_a(x)}{\log_b(a)}$.

**Chapter Assessment — Logarithms**

Download and sit the 30 minute assessment for this chapter from the digital book.

$$\sin^2(\theta) + \cos^2(\theta) = 1.$$

$$\tan(\theta) = \frac{\sin(\theta)}{\cos(\theta)},$$

# 19. Trigonometric Equations

Trigonometric functions are fundamental to many physical systems, and also to many branches of mathematics.

We have seen that trigonometric functions can be defined in terms of right angled triangles, and also by considering a point moving around a circle. Almost any object we choose to measure will involve triangles, circles or, very often, both. Trigonometric functions are closely related to triangles *and* circles and so they occur frequently in practical problems. However, trigonometric functions can also occur in situations where there are no triangles or circles. Suppose we take a spring, and stretch it. The further we stretch it, the greater the force pulling against us. Similarly, the more we compress it, the greater the force pushing back out again.

This situation – where an object is pulled towards some point with a force proportional to its distance from that point – describes the behaviour of springs, pendulums, waves and other common physical phenomena. Objects which are subject to these kinds of forces move according to trigonometric functions, such as $x = \sin(t)$ or $x = \cos(t)$, as shown in Figure 19.1.

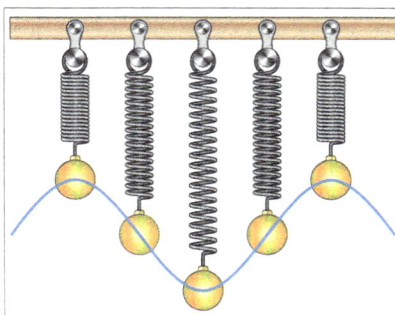

Figure 19.1: A spring undergoing simple harmonic motion.

Suppose we wished to find the times at which the mass is a certain distance away from its equilibrium point, then we would have to solve an equation of the form $x = A\sin(t)$, where $x$ is a known number and we are solving for $t$.

Any kind of scientific or mathematical investigation is likely to encounter trigonometric equations which must be solved. In this chapter, we will look at some ways of solving equations containing trigonometric functions.

## 19.1 Pythagoras' Theorem

Figure 19.2 shows how we can map any angle to a point on the unit circle. We start at the point $(1, 0)$ and trace an angle $\theta$ around the unit circle, moving anticlockwise for positive angles and clockwise for negative angles. This takes us to some point $C$ with coordinates $(x, y)$.

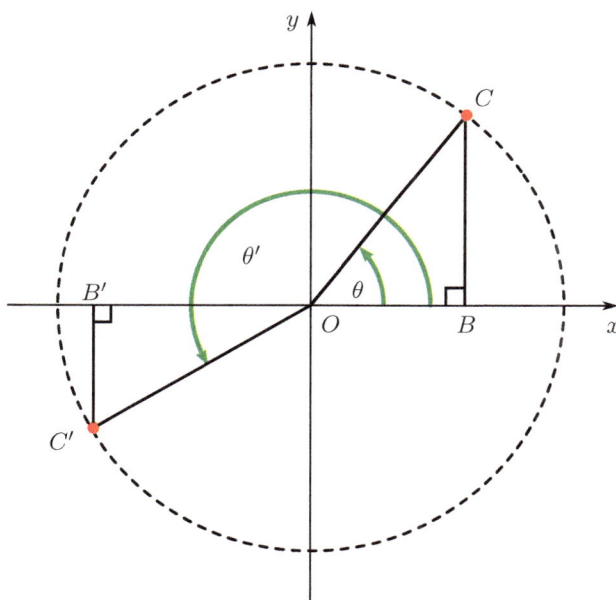

Figure 19.2: Mapping angles to point on the unit circle.

Figure 19.2 shows two examples. The angle $\theta$ in the first quadrant corresponds to point $C$ on the circle. The angle $\theta'$ is greater than $180°$ and corresponds to the point $C'$ in the third quadrant.

The trigonometric functions are defined to have the following values,

$$\sin(\theta) = y, \quad \cos(\theta) = x, \quad \text{and} \quad \tan(\theta) = \frac{\sin(\theta)}{\cos(\theta)} = \frac{y}{x}.$$

These definitions can be applied to any angle, including negative angles and angles greater than $360°$. For angles greater than a full turn, positive or negative, we simply go around the unit circle as many times as necessary.

Given any angle $\theta$ we can construct a right angled triangle as follows.

From the point $C$, we draw a perpendicular line to meet the $x$-axis at a point $B$. This forms a right-angled triangle $OBC$. For any such triangle $OBC$, the hypotenuse $OC$ is the radius of the circle and, therefore, has length 1. In Figure 19.3, the length of $OB$ is equal to $x$ and the length of $BC$ is equal to $y$. More generally, $x$ or $y$ may be negative, and so the lengths are equal to $|x|$ and $|y|$. That is,

$$|OB| = |x| = |\cos(\theta)| \quad \text{and} \quad |BC| = |y| = |\sin(\theta)|.$$

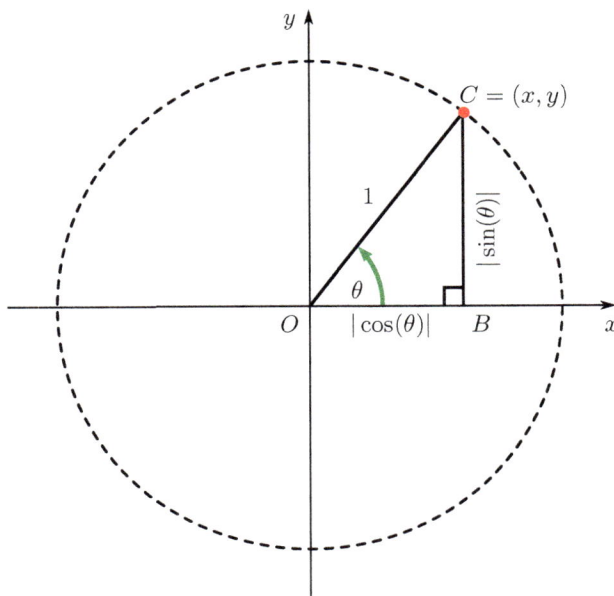

Figure 19.3: A right angled triangle in the unit circle.

We can apply Pythagoras' Theorem to this triangle to give us the following result.

**Formula 19.1 — Pythagoras' Theorem or the Pythagorean Identity.**
For any angle $\theta$ the following identity is true.

$$\sin^2(\theta) + \cos^2(\theta) = 1.$$

**Tip**
It is common to omit the brackets when writing trigonometric functions. Instead of writing $\sin(\theta)$, we will often see simply $\sin\theta$. However, this means that writing an expression such as $\sin\theta^2$ is ambiguous. We might mean either the square of $\sin(\theta)$ or the sine of $\theta^2$. Even when brackets are used, it would be very easy to confuse the expressions $(\sin(\theta^2))$ and $((\sin\theta)^2)$.
To avoid this confusion, if we are writing the power of a trigonometric function we always write the power immediately after the function. For example, the expression $\sin(\theta) \times \sin(\theta)$ is written as $\sin^2(\theta)$ or $\sin^2\theta$.

The identity $\sin^2(\theta) + \cos^2(\theta) = 1$ is simply another way of stating Pythagoras' theorem. This trigonometric identity and the theorem are equivalent.

> **Activity 19.1 — Connections with Other Results.**  (a) Using the cosine rule, prove the identity $\sin^2(\theta) + \cos^2(\theta) = 1$ for angles in the range $0° < \theta < 90°$.
>   (b) Use the identity $\sin^2(\theta) + \cos^2(\theta) = 1$ to prove Pythagoras' theorem for any right angled triangle.

## 19.2  Solving Trigonometric Equations

A linear equation, such as $3x - 12 = 0$, has just one solution. For a polynomial, such as $x^2 - x - 2 = 0$, there may be multiple solutions. Trigonometric functions are periodic and so equations containing them usually have infinitely many solutions.

For example, a simple equation such as $\cos(\theta) = 0.4$ has solutions

$$\theta = \ldots, -426.42°, -293.58°, -66.42°, 66.42°, 293.58°, 426.42°, 653.58°, \ldots.$$

When we solve an equation, we must choose which value, or values, to give as the answer. The question will usually specify the range of answers, or it will allow us to deduce the appropriate range.

> **Tip**
> The inverse of a trigonometric function has infinitely many solutions, but a calculator or computer will only show us one at a time. This will usually be a solution in the first quadrant, $0° \le \theta < 90°$, if there is one. If there is no solution in the first quadrant, we will usually be given an answer in the range $-90° < \theta \le 180°$.
> In either case, we must use the given value to work out all of the solutions.

---

**Example 19.1**

Give all solutions to $\cos(\theta) = 0.4$ in the range $270° \le \theta < 450°$.

**Solution:**

We have already seen a list of some of the solutions to this equation.

$$\theta = \ldots, -426.42°, -293.58°, -66.42°, 66.42°, 293.58°, 426.42°, 653.58°, \ldots.$$

We will look at how this list is generated.

First, a calculator will give us the solution $\theta = 66.42°$, to two decimal places. Figure 19.4 shows this angle and the corresponding point $C$ on the unit circle. The line $OC$ makes an angle of $66.42°$ with the $x$-axis. The point $C$ lies in the first quadrant, and we can reach it by tracing an angle of $66.42°$ anticlockwise around the circle from the point $(1, 0)$.

There are other ways of reaching the same point $C$. For example, starting from the point $(0, 1)$, we could trace $360°$ round the circle, back to the point $(0, 1)$. Then we could keep tracing round for a further $66.42°$ to reach the point $C$. This angle is also shown in Figure 19.4.

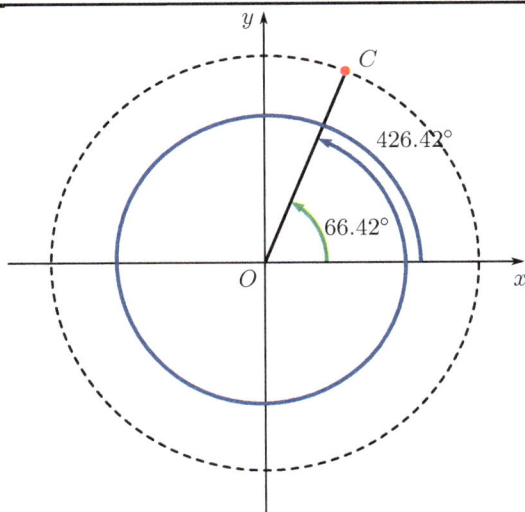

Figure 19.4: Comparison of angles which differ by $360°$.

Any angle which can be written in the form $\theta = (360n + 66.42)°$ for some integer $n$ will bring us to the point $C$. The angle $360n°$ represents $n$ complete circuits around the origin back to the point $(0, 1)$, and the angle $66.42°$ represents the path from the point $(0, 1)$ to the point $C$.

Every value $\theta = (360n + 66.42)°$ corresponds to the same point $C$, and since $\cos(\theta)$ is defined by the coordinates of $C$, $\cos(\theta)$ has the same value for each of these angles. Therefore, all of the values $\theta = (360n + 66.42)°$ are solutions to the equation.

Alternatively, we can look at these solutions on a graph of $\cos(\theta)$. The cosine function is periodic, and so by adding or subtracting multiples of $360°$, we can locate every solution at the equivalent point on the curve.

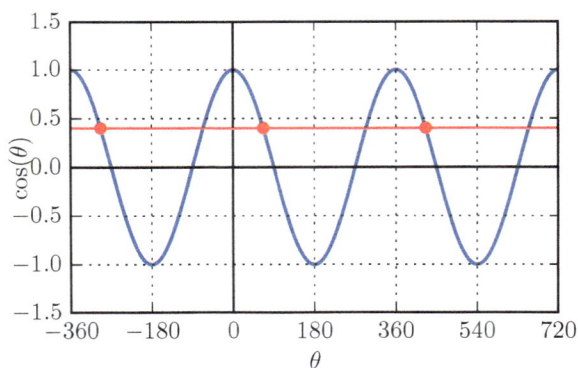

We can see from this graph that we have only found half of the solutions. There is another set of solutions. Using the symmetry of the cosine function, we can calculate that the value to the left of the $y$-axis is equal to $-66.42°$. The other values are a multiple of $360°$ from this value.

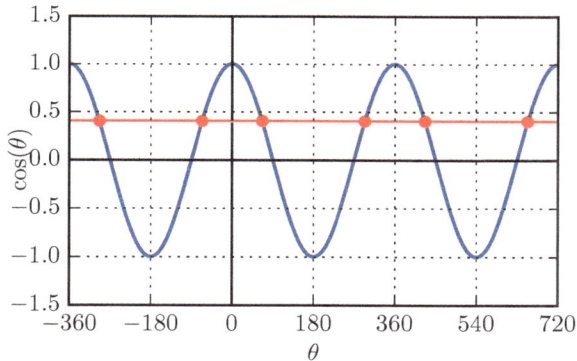

To find this second set of solutions from the unit circle diagram, we must determine which quadrant they lie in. Earlier, in Chapter 14, we saw methods for remembering which trigonometric functions are positive in which quadrants. All of the functions are positive in the first quadrant. The other solutions to $\cos(\theta) = 0.4$ lie in the other quadrant where the cosine function is positive, namely, the fourth quadrant.

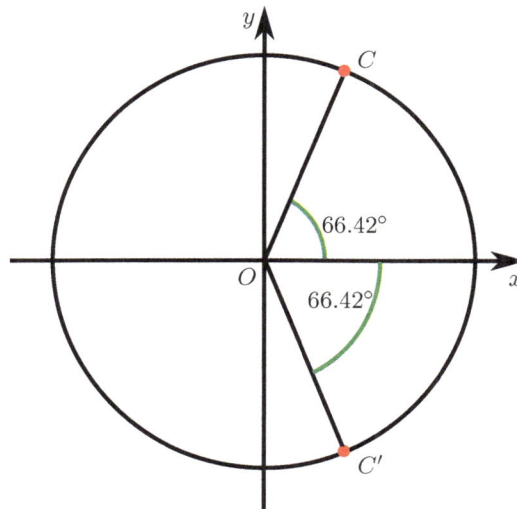

To find the angle, we can sketch out the unit circle, and draw an angle in the fourth quadrant which makes an angle of $66.42°$ with the $x$-axis. We do not need to do this with great accuracy. We will not be measuring the angle, but simply using the diagram to ensure we perform the correct calculation to find the solution.

We can see that the second set of solutions can be generated from the point $C'$, which is reached by tracing out an angle of 66.42° clockwise. Moving clockwise around the circle means that we have traced out a negative angle, and so the point $C'$ is described by the angle −66.42°.

We now have two solutions, $\theta = 66.42°$ and $\theta = -66.42°$. By adding or subtracting multiples of 360° from these two values, we can generate the full list of solutions. There are two solutions in the requested range, $\theta = 293.58°$ and $\theta = 426.42°$.

**Tip**

We can visualise the behaviour of trigonometric functions via a graph or with the unit circle. Given one solution to an equation, either of these two methods can help us calculate all of the other solutions.

If we use the graph method, we need to be able to draw a rough sketch of each function. We can then use the symmetry of the functions to find the other solutions.

If we use the unit circle method, we need to know which functions are positive in each quadrant. The solution is given by the point in the correct quadrant that makes the same angle with the $x$-axis as the known solution.

**Example 19.2**

Find all solutions to the equation $2 - 3\sin(\theta) = 1$ in the range $-180° \leq \theta < 180°$.

**Solution:**

We can rearrange this equation as we would any linear equation. Our aim is always to get the equation into the form $f(\varphi) = x$ where $f$ is some trigonometric function, $\varphi$ represents some angle, and $x$ is a real number. For this equation, we obtain $\sin(\theta) = \frac{1}{3}$.

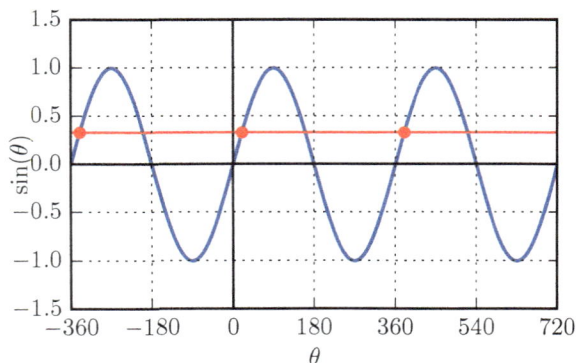

A calculator will give us a solution to this equation. The solution in the first quadrant is given by $\theta = 19.47°$. From this, we can add and subtract multiples of 360° to obtain an infinite set of solutions. As we would expect, this has found only half of the solutions. By the symmetry of the sine function, we can see that there is another solution at $180° - 19.47° = 160.53°$.

Alternatively, we could have found this value from the unit circle. We know that the sine function is positive in the first and second quadrants, and so we can sketch the angle 19.47° in the first quadrant, and then sketch another angle in the second quadrant which makes the same angle with the $x$-axis. This angle is $180° - 19.47° = 160.53°$.

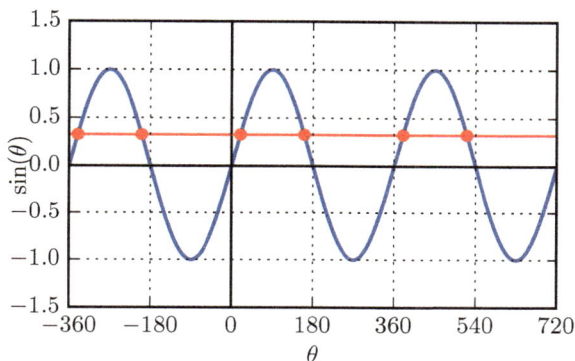

We now have our two base solutions, 19.47° and 160.53°. From these, we can generate all solutions by adding and subtracting multiples of 360°. This periodicity can be seen from the graph.

We find that there are no more solutions lying in the requested range and so $\theta = 19.47°$ or $\theta = 160.53°$.

**Tip**

In the previous example, we used the periodicity of the sine function to find all of the solutions we required. We saw that to find the full set of solutions, we required two starting values, 19.46° and 160.53°. From these, we calculated the other solutions by adding or subtracting multiples of 360°.

This same method is also used for the cosine function.

For the tangent function, although this same method will work, a simpler one is available. Since the period of $\tan(\theta)$ is 180°, we can calculate just one solution and then add and subtract multiples of 180°.

**Example 19.3**

Find all solutions to $\tan(\theta) = -6$ in the range $[-360°, 360°]$.

**Solution:**

Taking the inverse tangent of both sides of the equation, we calculate one solution to be $\theta = -80.54°$. Since the tangent function has a period of $180°$, we can calculate that $\theta = -260.54°$, $\theta = 99.46°$ and $\theta = 279.46°$ are also solutions.

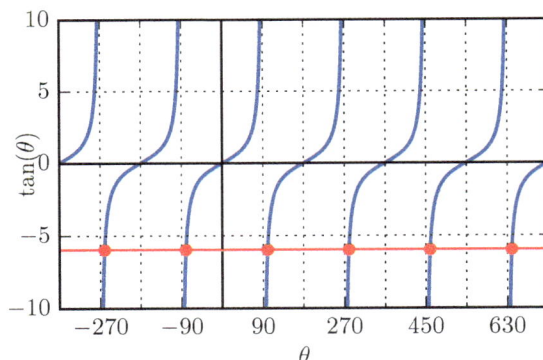

A sketch of the tangent function shows that there are no other solutions, and so the only values in the requested range are $\theta \in \{-260.54°, -80.54°, 99.46°, 279.46°\}$.

---

**Example 19.4**

Find all solutions to the equation $\sin^2(\theta) = \frac{1}{4}$ in the range $0° \leq \theta < 360°$.

**Solution:**

This equation is a quadratic equation in $\sin(\theta)$. Solving it will be a two stage process. First, we solve the quadratic to get linear expressions in $\sin(\theta)$ and then we find solutions to each of those expressions.

We can solve the quadratic by taking the square root of both sides to obtain $\sin(\theta) = \frac{1}{2}$ or $\sin(\theta) = -\frac{1}{2}$. Values of $\theta$ which solve either of these expressions will be solutions of the original equation, and so we must find the solutions to both.

One solution to $\sin(\theta) = \frac{1}{2}$ is given by $\theta = 30°$. We can find other solutions by adding and subtracting multiples of $360°$. The values found by this method represent only half of the solutions to $\sin(\theta) = \frac{1}{2}$. As in the previous example, we can sketch a graph or a unit circle to find the remaining solutions. We find that there is a solution in the second quadrant and that it is given by $180° - 30° = 150°$. Therefore, the solutions in the required range found by solving $\sin(\theta) = \frac{1}{2}$ are given by $\theta = 30°$ and $\theta = 150°$.

Now we calculate the solutions found by solving $\sin(\theta) = -\frac{1}{2}$.

A calculator may give the solution $\theta = -30°$, representing a solution in the fourth quadrant. Either by sketching the graph or by drawing a unit circle, we can find the corresponding solution in the third quadrant, $\theta = -150°$. We now need to add $360°$ to obtain results in the specified range, $\theta = 210°$ and $\theta = 330°$.

Therefore, the full solution is given by $\theta \in \{30°, 150°, 210°, 330°\}$.

## Example 19.5

Find all solutions to the equation $2\sin^2(\theta) = 3\cos(\theta)$ in the range $0° \le \theta < 360°$.

## Solution:

First, let us rewrite the equation as $3\cos(\theta) - 2\sin^2(\theta) = 0$. This equation contains the expressions $\cos(\theta)$ *and* $\sin(\theta)$ and is likely to be easier to solve if we have only one or the other. Since the sine term is squared, we can use the identity $\sin^2(\theta) + \cos^2(\theta) = 1$ to replace it with a cosine term.

$$
\begin{aligned}
0 = 3\cos(\theta) - 2\sin^2(\theta) &= 3\cos(\theta) - 2\left[\sin^2(\theta)\right] \\
&= 3\cos(\theta) - 2\left[1 - \cos^2(\theta)\right] \\
&= 2\cos^2(\theta) + 3\cos(\theta) - 2.
\end{aligned}
$$

This is now a quadratic expression in $\cos(\theta)$. If we write $x = \cos(\theta)$, then we have $2x^2 + 3x - 2 = 0$. We can factorise this expression as $(2x - 1)(x + 2) = 0$ to give the solutions $x = \frac{1}{2}$ and $x = -2$.

Now, $\cos(\theta)$ can only take values in the range $[-1, +1]$ and so we can discard the solution $x = -2$, and find solutions to $\cos(\theta) = \frac{1}{2}$ in the given range.

The solution in the first quadrant is $\theta = 60°$. From this, we can find the solution in the fourth quadrant, $\theta = (360 - 60)° = 300°$. The cosine function is negative in the second and third quadrants, and so these are the only two solutions, $\theta \in \{60°, 300°\}$.

## Exercise 19.1

Q1. Find the solutions, in the given range, to the following equations. Give your answers to one decimal place.
   (a) $\sin(\theta) = 0.338$ for $0° \le \theta \le 360°$.
   (b) $\cos(2\theta) = 0.848$ for $-180° < \theta < 180°$.
   (c) $\cos(\theta - 20°) = 0.5$ for $-180° < \theta < 180°$.
   (d) $\tan(\theta - 60°) = -5.5$ for $\theta \in [180°, 540°]$.
   (e) $\sin(4\theta + 25°) = -0.13$ for $0° \le \theta \le 45°$.
   (f) $2\sin(2\theta) + 3\cos(2\theta - 90°) = 2.4$ for $|\theta| < 90°$.

Q2. For some value $a \in \mathbb{N}$, $\cos(a\theta + 25°) = 0.6$.
   There is exactly one solution for $\theta$ in the range $[0°, 10°]$ and there are exactly two solutions with $\theta \in [0°, 100°]$.
   Find $a$ and the two solutions $\theta \in [0°, 100°]$.

Q3. Find all solutions in the range $[-180°, +180°]$ for the following equations.
   (a) $9\cos^2(\theta) = 4$.
   (b) $3\sin^2(\theta) + 5\sin(\theta) - 2 = 0$.
   (c) $3\tan^2(\theta) + 5\tan(\theta) - 2 = 0$.
   (d) $\cos^2(\theta) - 1 = 0$.
   (e) $6\cos^2(\theta) - \sin(\theta) - 4 = 0$.

     (f) $6\sin^2(\theta) - 5\cos(\theta) - 7 = 0$.

     (g) $\tan^2(2\theta) + \tan(2\theta) - 6 = 0$.

Q4. Find the solution $\theta$ with the smallest absolute value, $|\theta|$.

$$3\tan(\theta) + \frac{2}{\cos(\theta)} = 2\left[2 + 3\sin(\theta)\right].$$

Q5. Calculate all of the solutions to the equation $\cos(2\theta) = 3\sin(2\theta)$ in the range $0° < \theta < 360°$.

Q6. (a) Using the identity $\sin^2(\theta) + \cos^2(\theta) = 1$, or otherwise, prove that

$$\sin(45°) = \frac{1}{\sqrt{2}}.$$

     (b) Give all solutions to the equation $2\sin^2(\theta) = 1$ for $\theta \in (0°, 360°)$.

## 19.3 Trigonometric Identities

Consider the trigonometric equation,

$$\frac{\tan(\theta) + \sin(\theta)}{\tan(\theta)} = \frac{2}{3}. \tag{19.1}$$

It is not immediately obvious how this can be solved using the techniques discussed in this chapter and we shall return to this question later. With equations like these, however, it is often worth attempting to simplify the left hand side.

To do this simplification, we shall apply the following two trigonometric identities.

$$\tan(\theta) = \frac{\sin(\theta)}{\cos(\theta)}, \tag{19.2}$$

$$\sin^2(\theta) + \cos^2(\theta) = 1. \tag{19.3}$$

---

**Example 19.6**

Simplify the expression $\sin^4(\theta) + \sin^2(\theta)\cos^2(\theta)$.

**Solution:**

We use (19.3)

$$\begin{aligned}\sin^4(\theta) + \sin^2(\theta)\cos^2(\theta) &= \sin^4(\theta) + \sin^2(\theta)\left(1 - \sin^2(\theta)\right), \\ &= \sin^4(\theta) + \sin^2(\theta) - \sin^4(\theta), \\ &= \sin^2(\theta).\end{aligned}$$

---

**Remark**

As mentioned previously the word "simplify" is ambiguous. In cases like this the goal is to find an expression that uses only one trigonometric function.

We now return to the solution of (19.1).

---

**Example 19.7**

Solve, finding all solutions in the range $0° \leq \theta \leq 360°$,

$$\frac{\tan(\theta) + \sin(\theta)}{\tan(\theta)} = \frac{2}{3}. \qquad (19.4)$$

**Solution:**

We first consider the left hand side,

$$\frac{\tan(\theta) + \sin(\theta)}{\tan(\theta)} = \frac{\frac{\sin(\theta)}{\cos(\theta)} + \sin(\theta)}{\frac{\sin(\theta)}{\cos(\theta)}},$$

$$= \frac{\frac{\sin(\theta) + \cos(\theta)\sin(\theta)}{\cos(\theta)}}{\frac{\sin(\theta)}{\cos(\theta)}},$$

$$= \frac{\cos(\theta)\left(\sin(\theta) + \cos(\theta)\sin(\theta)\right)}{\cos(\theta)\sin(\theta)},$$

$$= \frac{\sin(\theta) + \cos(\theta)\sin(\theta)}{\sin(\theta)},$$

$$= \frac{\sin(\theta)\left(1 + \cos(\theta)\right)}{\sin(\theta)},$$

$$= 1 + \cos(\theta).$$

Using this result,

$$\frac{\tan(\theta) + \sin(\theta)}{\tan(\theta)} = \frac{2}{3},$$

$$\Rightarrow \qquad 1 + \cos(\theta) = \frac{2}{3}.$$

Solving (19.1) is therefore equivalent to solving,

$$\cos(\theta) = \frac{1}{3}$$

Using a calculator we find that $\theta \approx 70.53°$. By the periodicity of the $\cos(\theta)$ function there is a further solution at $270 + (90 - 70.53) = 289.47°$.
Therefore, the full solution is given by $\theta \in \{70.53, 289.47\}$.

---

**Example 19.8**

Given that $x = 4\sin(\theta)$ and $y = 5\cos(\theta)$ show that $25x^2 + 16y^2 = 400$.

**Solution:**

We first rearrange,

$$x = 4\sin(\theta),$$
$$\Rightarrow \quad \frac{x}{4} = \sin(\theta),$$
$$y = 5\cos(\theta),$$
$$\Rightarrow \quad \frac{y}{5} = \cos(\theta).$$

Using Equation (19.3), we have

$$\sin^2(\theta) + \cos^2(\theta) = 1,$$
$$\Rightarrow \quad \left(\frac{x}{4}\right)^2 + \left(\frac{y}{5}\right)^2 = 1,$$
$$\Rightarrow \quad \frac{x^2}{16} + \frac{y^2}{25} = 1,$$
$$\Rightarrow \quad 25x^2 + 16y^2 = 400.$$

---

**Exercise 19.2**

Q1. Express $\frac{\sin^2(\theta)}{\tan^2(\theta)}$ in terms of powers of $\sin(\theta)$ only.

Q2. Simplify $f(x) = \tan^2(x)\cos^4(x)$.

Q3. Solve $\cos(\theta)\tan(\theta) = \frac{\sqrt{3}}{2}$, giving all solutions in the range $0° \le \theta \le 360°$.

Q4. For $0° \le \theta \le 360°$ solve the following equation,

$$\frac{1 - 2\cos^2(\theta) + \cos^4(\theta)}{\sin^2(\theta)} = \frac{1}{2}.$$

Q5. (a) Prove

$$\frac{\cos^4(\theta) - \sin^4(\theta)}{\cos^2(\theta)} \equiv 1 - \tan^2(\theta)$$

(b) Hence, or otherwise, solve,

$$\frac{\cos^4(\theta) - \sin^4(\theta)}{\cos^2(\theta)} = \frac{1}{2},$$

giving all solutions in the range $0° \le \theta \le 540°$.

Q6. Express, solely in terms of $\cos(x)$,

$$f(x) = (\cos(x) + \sin(x))^3 - \sin(x)\left(2\cos^2(x) + 1\right)$$

**Key Facts — Trigonometric Equations**

- The tangent function is defined to be

$$\tan(\theta) = \frac{\sin(\theta)}{\cos(\theta)}$$

- The Pythagorean identity states that:

$$\sin^2(\theta) + \cos^2(\theta) = 1.$$

- It is important to ensure that all solutions in the given range are found when solving trigonometric equations.
- To ensure all solutions have been found, either sketch the curve and use the periodicity of the trigonometric function, or use the CAST diagram.
- When proving trigonometric identities start from either the left hand side or right hand side of the identity we are trying to show and work towards the result.

**Chapter Assessment — Trigonometric Equations**

Download and sit the 30 minute assessment for this chapter from the digital book.

$-\sin(x)$

$$\tan(\theta) = \frac{\sin(\theta)}{\cos(\theta)},$$

$$\sin^2(\theta) + \cos^2(\theta) = 1.$$

# 20. Tangents, Normals and Stationary Points

The techniques discussed in this chapter have wide reaching applications in a range of fields from Business to Physics. We shall start by learning how to find tangents and normals to curves. When studying motion, an object such as the roller coaster car shown in Figure 20.1, moving along a curved path will always have a velocity tangent to that path and when in contact with a surface that surface will exert a reaction force normal to the point of contact.

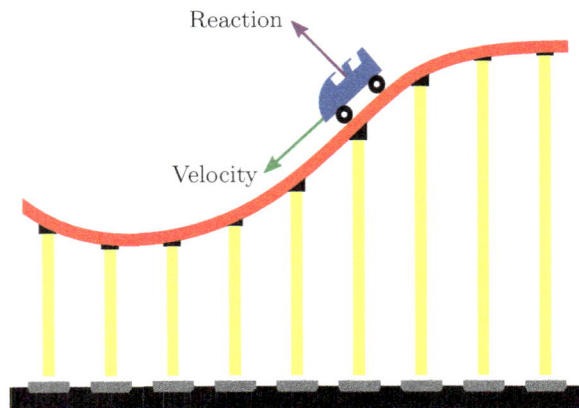

Figure 20.1: The velocity and reaction force are tangent and normal to the path of motion of a roller coaster car, respectively.

Later in the chapter we look at how to find stationary points of a curve which can be useful for maximising profit or minimising cost in business.

## 20.1  Tangents and Normals

We have learned how to differentiate a function to find its gradient function. This tells us the gradient of a curve for any give value of $x$. We can now use this information to find the equations of tangents and normals to a curve.

---

**Definition 20.1 — Tangents and Normals**

A *tangent* to a curve $f(x)$ at point $p$ is a straight line touching the curve at $x = p$ with gradient equal to $f'(p)$. A *normal* is perpendicular to the tangent.

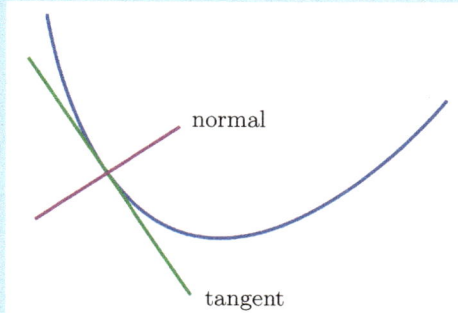

---

**Remark**

Recall that, from previous study, two straight lines are perpendicular if their gradients multiply together to give negative one.

---

**Example 20.1**

Find the equation of the tangent to the curve $y = 3x^2 - 2x - 1$ at the point where $x = \frac{2}{3}$.

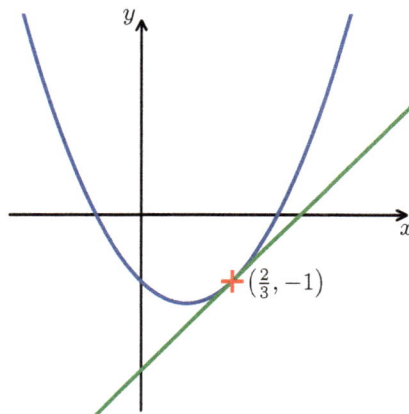

**Solution:**
By differentiating $y(x)$ and substituting $x = \frac{2}{3}$ we can find the gradient of the tangent. We will also need the $y$ value at this point.

$$\frac{\mathrm{d}y}{\mathrm{d}x} = 6x - 2.$$

Substituting $x = \frac{2}{3}$ we get $y = -1$ and $\frac{\mathrm{d}y}{\mathrm{d}x} = 2$.
Using $y - y_1 = m(x - x_1)$ to find the straight line equation we obtain

$$y - (-1) = 2\left(x - \frac{2}{3}\right),$$

$$\Rightarrow \qquad y + 1 = 2x - \frac{4}{3},$$

$$\Rightarrow \qquad y = 2x - \frac{7}{3}.$$

**Example 20.2**
Find the equation of the normal to the curve $y = \frac{1}{4}x^{-2}$ at the point where $x = \frac{1}{2}$. This normal line intersects the curve twice more at points $P$ and $Q$. Find the $x$ coordinates of $P$ and $Q$.

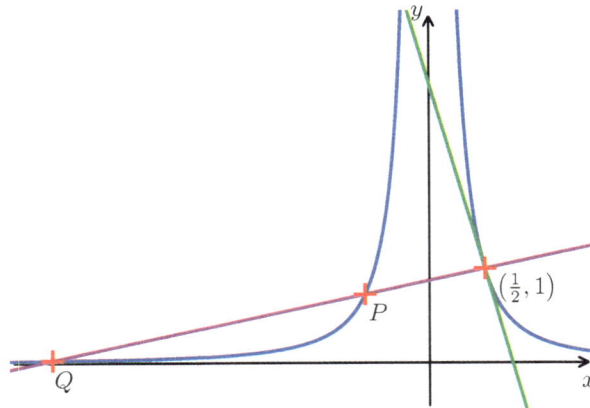

**Solution:**
By differentiating $y(x)$ and substituting $x = \frac{1}{2}$ we can find the gradient of the tangent.

$$\frac{\mathrm{d}y}{\mathrm{d}x} = -\frac{1}{2}x^{-3}.$$

Substituting $x = \frac{1}{2}$ we get $\frac{\mathrm{d}y}{\mathrm{d}x} = -4$. Since the normal line is perpendicular to the tangent, the gradient of the normal will be $\frac{1}{4}$.

Using $y - y_1 = m(x - x_1)$ to find the straight line equation we obtain

$$y - 1 = \frac{1}{4}\left(x - \frac{2}{3}\right),$$

$$\Rightarrow \quad y - 1 = \frac{1}{4}x - \frac{1}{8},$$

$$\Rightarrow \quad y = \frac{1}{4}x + \frac{7}{8}.$$

To find the other intersection points, we can solve $y = \frac{1}{4}x^{-2}$ and $y = \frac{1}{4}x + \frac{7}{8}$ as simultaneous equations.

$$\frac{1}{4}x^{-2} = \frac{1}{4}x + \frac{7}{8},$$

$$\Rightarrow \quad 2x^{-2} = 2x + 7,$$

$$\Rightarrow \quad 2 = 2x^3 + 7x^2,$$

$$\Rightarrow \quad 0 = 2x^3 + 7x^2 - 2.$$

We already know that $x = \frac{1}{2}$ is a solution to this equation, hence by the factor theorem, $(2x - 1)$ is a factor. It can then be shown that the quadratic factor is $(x^2 + 4x + 2)$ and using the quadratic formula or completing the square, the remaining two solutions are $x = -2 + \sqrt{2}$ and $x = -2 + \sqrt{2}$ at $P$ and $Q$ respectively.

---

## Activity 20.1 — Tangents to Cubics

(a) For the cubic $y = x^3 - x^2 - 9x + 9$:
   i. Find the location of all three roots, $x_1$, $x_2$, and $x_3$.
   ii. Define $p_1 = \frac{x_1 + x_2}{2}$, $p_2 = \frac{x_2 + x_3}{2}$ and $p_3 = \frac{x_3 + x_1}{2}$. Find the equations of the tangents to the cubic at $p_1$, $p_2$ and $p_3$, and calculate where these tangents intersect the $x$-axis.
   iii. What can be observed?
(b) Show that the property observed in part (a)iii. holds for a general cubic function.

---

## Exercise 20.1

Q1. Find the equation of the tangent to the following curves at the specified point giving your answer in the form $ax + by = c$ where $a$ and $b$ are integers.
   (a) $y = x^2 + 5$ when $x = 2$;
   (b) $y = 3x + x^2 - x^3$ when $x = -1$;
   (c) $y = x^2 - \frac{2}{x^2}$ when $x = -2$;
   (d) $y = \sin(x)$ when $x = \frac{\pi}{3}$ ;
   (e) $y = 4\ln(x) - x$ when $x = 1$.

Q2. Find the equation of the normal to the following curves at the specified point giving your answer in the form $ax + by = c$.

(a) $y = (x + 2)(x - 3)$ when $x = -1$;

(b) $y = x^2 + 2x - 3$ when $x = \frac{1}{2}$;

(c) $y = \frac{x^2 - 3x}{\sqrt{x}}$ when $x = 4$;

(d) $y = e^{2x}$ when $x = 0$;

(e) $y = \sin(2x) - \cos(x)$ when $x = \frac{\pi}{6}$.

Q3. Find the equations of tangents to $y = (x^2 + x + 2)(x + 3)$ where the curve intersects the coordinate axes. Hence, find the point at which these two tangents intersect each other.

Q4. Find the area bounded by tangent to the curve $y = \frac{x^3}{9} - x + 1$ at $x = 3$ and the coordinate axes.

Q5. Find the value of $k$ such that $y = 15x + k$ is a tangent to the curve $y = \frac{1}{x^2} - x$.

Q6. Find the equation of the normal to $y = x^3 + 2x^2 - 1$ at $x = -1$, hence, find the $x$-coordinates of the points where this normal intersects the curve again.

Q7. The curve $y = x^3 - 9x^2 + 25x - 17$ intersects the $x$-axis once at $x = 1$. There are exactly two tangents to this curve which pass through $x = 1$, one is clearly at $x = 1$, the other is at $x = p$. Find the value $p$.

## 20.2  Stationary Points

On some curves there are points at which a tangent will be parallel to the $x$-axis. These points often characterise important features of physical systems such as minimum or maximum values.

### Definition 20.2 — Stationary Points

At *stationary points*, tangents to a curve are horizontal and their gradients are zero, so $\frac{dy}{dx} = 0$. The points $P$, $Q$ and $R$ are all stationary points.

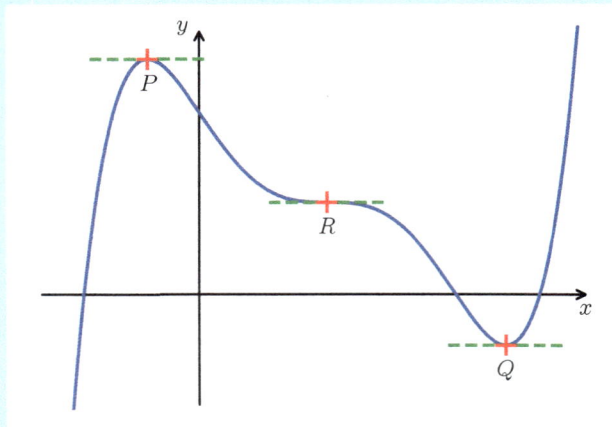

**Example 20.3**

Find the coordinates of all stationary points on the curve $y = x^3 + \frac{3}{2}x^2 - 6x + 1$.

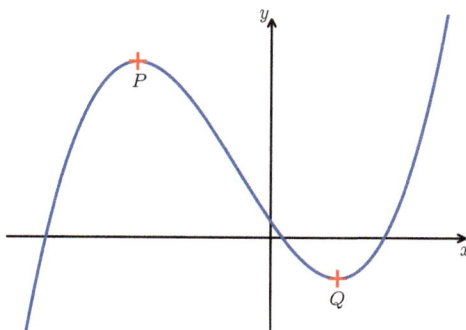

**Solution:**

By differentiation we obtain $\frac{dy}{dx} = 3x^2 + 3x - 6$. Setting $\frac{dy}{dx} = 0$ and solving for $x$ gives,

$$0 = 3x^2 + 3x - 6$$
$$0 = 3(x^2 + x - 2),$$
$$0 = 3(x - 1)(x + 2) \quad \Rightarrow \quad x = -2, \ x = 1.$$

Substituting these into $y = x^3 + \frac{3}{2}x^2 - 6x + 1$ we can find the $y$ values at these points to obtain coordinates $(-2, 11)$ and $\left(1, -\frac{5}{2}\right)$ of $P$ and $Q$ respectively.

**Exercise 20.2**

Q1. For each of the following curves find the $x$-coordinates of any stationary points that exist.

    (a) $y = 2x^2 + 7x - 3$;
    (b) $y = (x^2 + 3)(x + 5)$;
    (c) $y = x^3 - 6x^2 + 3x - 5$;
    (d) $y = 4x^3 - 11x^2 - 20x - 7$;
    (e) $y = 3x^4 + 4x^3 - 12x^2$;
    (f) $y = x^5 - 5x$;
    (g) $y = \frac{4x^2 - 3x + 2}{x}$;
    (h) $y = \ln(7x) - x^2 + x$.

Q2. Consider the polynomial expression $g(x) = x^2 - 6x + 11$.

    (a) Calculate the derivative of $g(x)$ and hence find the turning point of $y = g(x)$.
    (b) Calculate the value of $g(x)$ at its turning point.
    (c) Explain why $g(x)$ does not cross the $x$-axis.

## 20.3 Classifying Stationary Points

Now that we are able to find stationary points we shall look at how to determine their nature. It is often useful to know if a stationary point is a local maxima or minima, particularly when dealing with a real world problem such as maximising volume. Figure 20.2 shows the three types of stationary points which may occur.

At $P$ there is a *local maximum* and at $R$ there is a *local minimum*. We call these *local maxima* and minima because in a small neighbourhood they are the greatest or smallest values that the function can attain, however, there may be other points which are larger or smaller which may or may not be stationary points. $P$ and $Q$ may also be called *turning points* as the gradient changes direction, or 'turns', from positive to negative or vice versa. At $R$ there is a horizontal *inflection* point. $R$ is not a turning point since the gradient does not change sign. There are two more inflection points, one between $P$ and $Q$ and another between $Q$ and $R$ although these are not stationary points. We shall discuss two methods to identify the nature of a stationary point, the first will be to consider the sign of the gradient near the stationary point.

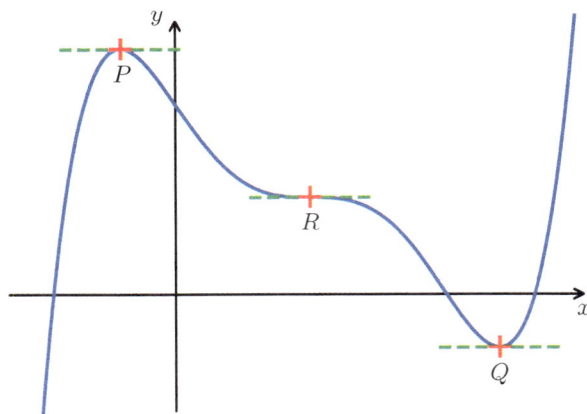

Figure 20.2: Types of stationary point.

Figure 20.3(a) shows that, at a local maximum point, the gradient is positive to the left of the stationary point and negative to the right of the stationary point. This behaviour is summarised in associated table. In contrast, at a local minimum, the gradient is negative to the left of the stationary point and positive to the right of the stationary point, as highlighted in Figure 20.3(b).

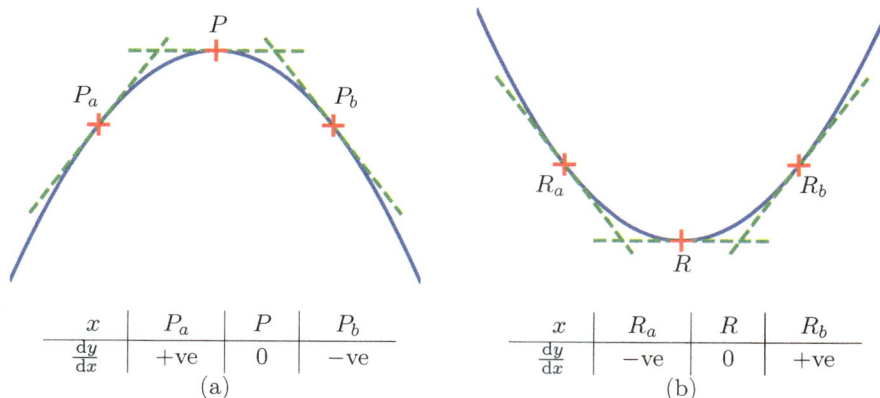

| $x$ | $P_a$ | $P$ | $P_b$ |
|---|---|---|---|
| $\frac{dy}{dx}$ | +ve | 0 | $-$ve |

(a)

| $x$ | $R_a$ | $R$ | $R_b$ |
|---|---|---|---|
| $\frac{dy}{dx}$ | $-$ve | 0 | +ve |

(b)

Figure 20.3: Gradient behaviour at a local maximum (a) and at a local minimum (b).

Near an inflection point the gradient has the *same sign* on the left and right of the stationary point. In this case, negative on both sides. This behaviour is shown in Figure 20.4.

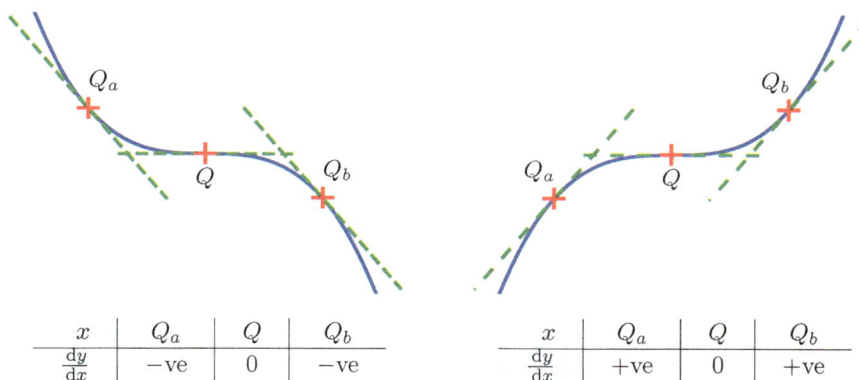

| $x$ | $Q_a$ | $Q$ | $Q_b$ |
|---|---|---|---|
| $\frac{dy}{dx}$ | $-$ve | 0 | $-$ve |

| $x$ | $Q_a$ | $Q$ | $Q_b$ |
|---|---|---|---|
| $\frac{dy}{dx}$ | +ve | 0 | +ve |

Figure 20.4: Gradient behaviour at a point of inflection.

---

**Example 20.4**

Find the coordinates of all stationary points on the curve $y = x^4 + x^3 - 5x^2$ and determine their nature, hence, also find the regions in which the curve is increasing or decreasing.

**Solution:**

Differentiating we obtain $\frac{dy}{dx} = 4x^3 + 3x^2 - 10x$. Setting $\frac{dy}{dx} = 0$ and solving for $x$,

$$0 = 4x^3 + 3x^2 - 10x$$
$$0 = x(4x^2 + 3x - 10)$$
$$0 = x(4x - 5)(x + 2) \quad \Rightarrow \quad x = -2, \ x = 0, \ x = \frac{5}{4}.$$

By considering the gradient near each of these points we can determine which type of stationary points they are.

| $x$ | $x < -2$ | $x = -2$ | $-2 < x < 0$ | $x = 0$ | $0 < x < \frac{5}{4}$ | $x = \frac{5}{4}$ | $x > \frac{5}{4}$ |
|---|---|---|---|---|---|---|---|
| $\frac{dy}{dx}$ | $-$ | $0$ | $+$ | $0$ | $-$ | $0$ | $+$ |

We can see that there are minimum points at $x = -2$ and $x = \frac{5}{4}$ and a maximum point at $x = 0$. If we also calculate the $y$ values at these points it will help us to draw the graph. Stationary points are at $(-2, -12)$, $(0, 0)$ and $\left(\frac{5}{4}, -\frac{875}{256}\right)$. See the plot below,
Now that we know the shape of the curve, it is clear which regions are *increasing* or *decreasing*. For $x < -2$ the value of the function is *decreasing*, for $-2 < x < 0$ the function is *increasing*, for $0 < x < \frac{5}{4}$ the function is *decreasing* and finally, for $x > \frac{5}{4}$ the function is *increasing*.

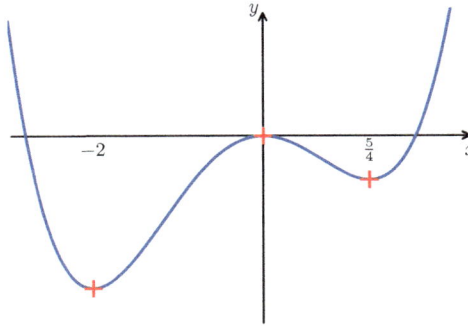

Another way to identify local minima and maxima is to look at the *second derivative* of the function. We have already seen that the value of $\frac{dy}{dx}$ in the neighbourhood of a minimum point is first negative, then zero at the minimum point and then positive, *i.e.* $\frac{dy}{dx}$ is increasing, hence the value of $\frac{d^2y}{dx^2}$ is positive. Similarly, for a maximum point, the value of $\frac{dy}{dx}$ is decreasing from positive through zero to negative so $\frac{d^2y}{dx^2}$ is negative. At an inflection point, while it is true that $\frac{d^2y}{dx^2} = 0$, it is not a sufficient condition to show this. It is possible for $\frac{d^2y}{dx^2}$ to be equal to zero at a minimum or a maximum, consider $y = x^4$ which has a minimum point at $x = 0$. If we find $\frac{d^2y}{dx^2} = 0$ this is inconclusive and we should check the value of $\frac{dy}{dx}$ near the stationary point as in the example above.

**Example 20.5**

Using the same curve as Example 20.4 we can check our solution using the second derivative. Differentiating a second time gives $\frac{d^2y}{dx^2} = 12x^2 + 6x - 10$.

| | $x$ | $-2$ | $0$ | $\frac{5}{4}$ |
|---|---|---|---|---|
| | | $26$ | $-10$ | $\frac{65}{4}$ |
| $\frac{d^2y}{dx^2}$ | | positive | negative | positive |
| Conclusion | | minimum | maximum | minimum |

**Remark**

Below is a quick summary of the concepts discussed regarding types of stationary points.

|  | Local Minimum | Inflection Point | Local Maximum |
|---|---|---|---|
| $\frac{dy}{dx}$ | -0+ | −0− or +0+ | +0− |
| $\frac{d^2y}{dx^2}$ | positive | zero (*inconclusive*) | negative |

Remember that if $\frac{d^2y}{dx^2} = 0$ we *cannot* conclude there is an inflection point and we must check the value of $\frac{dy}{dx}$ near the stationary point.

We have briefly looked at identifying where a function is increasing or decreasing. More formally, a function is *increasing* when $\frac{dy}{dx} > 0$ and *decreasing* when $\frac{dy}{dx} < 0$

**Interactive Activity 20.1 — Categorise the Stationary Points**

Categorise the stationary points of the functions given in the Tarquin interactive activity shown below (in the digital book) by dragging them into the correct category.

**Interactive Activity 20.2 — Categorising Turning Points**

Categorise the turning points in the interactive activity below as local maxima or local minima by dragging them into the correct container.

**Exercise 20.3**

Q1. For each of the curves in Exercise 20.2 investigate the nature of their stationary points.

Q2. For each of the following functions, find the $x$-coordinates of any stationary points and investigate their nature.

(a) $y = x^5 - x^4 - 4x^3$;

(b) $y = 16x^2 - 3x^{\frac{1}{3}}$;

(c) $y = \frac{1}{x} - \frac{1}{x^3}$;

(d) $y = e^{-x} + 2x$.

Q3. Determine the interval on which the function $y = 3x - x^3$ is increasing.

Q4. Determine the intervals on which the function $y = x^4 - 2x^2$ is decreasing.

Q5. A factory is mass producing cakes and would like to minimise the fondant they use to cover the cakes. Each cake has a specified recipe that results in a cake with volume $3000\,\text{cm}^3$. If the cakes are cylindrical in shape and require fondant covering the curved edge and top, find the minimum surface area needed to cover this volume.

Q6. Find the equation of a tangent to the curve $y = (x - 2)^2$ at $x = p$ and the area bounded by this tangent and the coordinate axes in terms of $p$. Hence, or otherwise, find the maximum area of the region bounded by a tangent and the coordinate axes for $-2 \leq p \leq 2$.

Q7. From a $30 \, \text{cm}$ square of card a smaller square of length $x \, \text{cm}$ is removed from each corner and the sides folded up to form an open box. Find the value of $x$ which maximises the volume of the resulting box.

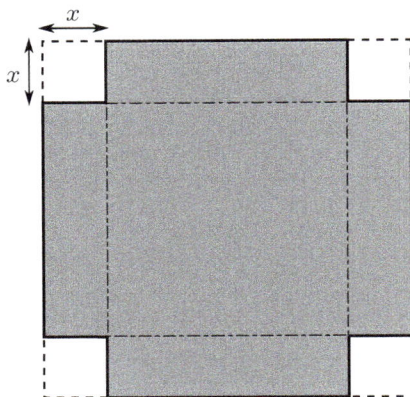

**Key Facts — Tangents, Normals and Stationary Points**

- A *tangent* to a curve $f(x)$ at point $p$ is a straight line touching the curve at $x = p$ with gradient equal to $f'(p)$. A *normal* is perpendicular to the tangent.
- At *stationary points*, tangents to a curve are horizontal and their gradients are zero, so $\frac{dy}{dx} = 0$.
- Stationary points can be classified as *local maxima*, *local minima*, or *points of inflection*.
- The table below can be used to identify a type of stationary point.

|  | Local Minimum | Inflection Point | Local Maximum |
|---|---|---|---|
| $\frac{dy}{dx}$ | -0+ | $-0-$ or $+0+$ | +0− |
| $\frac{d^2y}{dx^2}$ | positive | zero (*inconclusive*) | negative |

- Remember, if $\frac{d^2y}{dx^2} = 0$, we *cannot* conclude there is an inflection point and we must check the value of $\frac{dy}{dx}$ near the stationary point.
- A function is *increasing* when $\frac{dy}{dx} > 0$ and *decreasing* when $\frac{dy}{dx} < 0$.

**Chapter Assessment — Tangents, Normals and Stationary Points**

Download and sit the 30 minute assessment for this chapter from the digital book.

As discussed in Chapter 16, we can use integration to find the area between a curve and an axis. In this chapter we formalise this by introducing the concept of a *definite* integral, where the limits of integration relate to the area we wish to find.

As an example, suppose that a remote control helicopter manufacturer is investigating the cross sectional area of a range of airfoils that are being considered for use as a main rotor blade. They have decided to use a symmetrical airfoil, as defined by a 4 digit NACA airfoil description. The NACA airfoils were developed by the American National Advisory Committee for Aeronautics (NACA) as a way of standardising the description of airfoil shapes. Any 4 digit symmetric NACA airfoil has a designation of the form "NACA00$xy$" where $xy$ indicates the ratio of maximum thickness to chord length. The figure below shows a NACA0012 airfoil with chord length 1.

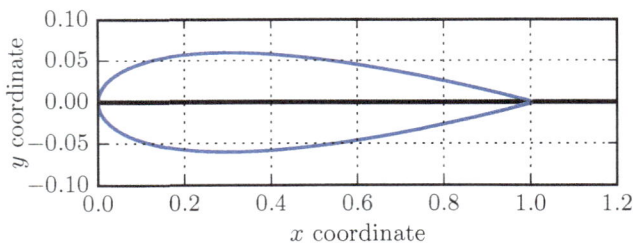

In this case, the 12 appearing in the designation indicates that the airfoil is at most 12% as thick as it is long. By default, the position of maximum thickness is 30% of the chord length from the leading edge of the wing, *i.e.*, in the airfoil shown above, the position of maximum thickness occurs where the $x$-coordinate is 0.3.

The $y$-coordinate of the upper edge of the airfoil can be calculated using the formula below.

$$y = 5t \left[ 0.2969\sqrt{\frac{x}{c}} - 0.1260 \left(\frac{x}{c}\right) - 0.3516 \left(\frac{x}{c}\right)^2 + 0.2843 \left(\frac{x}{c}\right)^3 - 0.1015 \left(\frac{x}{c}\right)^4 \right] \quad (21.1)$$

In this formula, $c$ is the chord length, $t$ is the maximum thickness as a percentage of the chord length and $x$, which is used to compute the $y$-coordinate, can take values between 0 and $c$.

> **Remark**
> As our airfoil is symmetric, the $y$-coordinate of the lower surface is the negative of the $y$-coordinate for the upper surface.

For the NACA0012 airfoil shown above, Equation (21.1) becomes,

$$y = 0.6 \left[ 0.2969\sqrt{x} - 0.1260x - 0.3156x^2 + 0.2843x^3 - 0.1015x^4 \right].$$

Letting $I$ denote the definite integral of this function between $x = 0$ and $x = c$ will enable the manufacturer to find the cross-sectional area of the rotor since, due to symmetry, it is $2I$. We shall return to this problem in an example later in this chapter.

> **Remark**
> Many old helicopters used symmetrical airfoils for their rotors, as they were easy to produce and had acceptable performance over the wide range of airspeeds that rotor blades are subject to. More modern helicopters use asymmetric rotors, some of which include a twist, to increase performance.

## 21.1  Notation

We begin by introducing terminology that is necessary for us to discuss definite integration. The "definite integral" of $f(x)$ between the limits $x = a$ and $x = b$ is denoted by

$$\int_a^b f(x)\, \mathrm{d}x.$$

The combined $\int_a^b [\ldots]\, \mathrm{d}\cdot$ symbol is the " definite integral sign".
We also introduce the following notation

$$\left[ F(x) \right]_a^b = F(x) \Big|_a^b = F(b) - F(a),$$

which will be useful later.

## 21.2  The Fundamental Theorem of Calculus

The Fundamental Theorem of Calculus is an important result establishing the relationship between integration and differentiation. Although we have already discussed this theorem in 16.1.2, the formal statement requires the notation used in definite integration, therefore we revisit it briefly here.

> **Theorem 21.1 — The Fundamental Theorem of Calculus**
> If $f$ is continuous on $[a, b]$, and $F$ is defined by
>
> $$F(x) = \int_a^x f(t)\, \mathrm{d}t$$

for all $x$ in $[a, b]$, then $F$ is differentiable on $[a, b]$ and $F'(x) = f(x)$, or

$$\frac{\mathrm{d}}{\mathrm{d}x} \int_a^x f(t) \, \mathrm{d}t = f(x)$$

for all $x$ in $[a, b]$.

**Remark**

The above is often referred to as Part I of the Fundamental Theorem of Calculus. It is a formal statement of the result we already proved in Section 16.1.2 when we derived that the derivative of the area under $f(x)$ between $x = a$ and $x$ is equal to $f(x)$.

This theorem and its proof give us two important characteristics of integrals:

1. The definite integral of $f(x)$ between the limits $x = a$ and $x = b$ gives us the area under $f(x)$ between $x = a$ and $x = b$;
2. Integration is the reverse of differentiation; that is, if $F'(x) = f(x)$ then integrating $f(x)$ yields $F(x)$.

**Tip**

When we are solving a problem requiring us either to compute an integral by reversing differentiation, or use a definite integral to calculate an area, we should state that we are using the Fundamental Theorem of Calculus.

**Example 21.1**

Given that

$$\frac{\mathrm{d}}{\mathrm{d}x} (\sin(x)) = \cos(x),$$

where $x$ is measured in radians, find

$$\int \cos(x) \, \mathrm{d}x.$$

**Solution:**

From the question we know that

$$\frac{\mathrm{d}}{\mathrm{d}x} (\sin(x)) = \cos(x).$$

Therefore, *by the Fundamental Theorem of Calculus,*

$$\int \cos(x) \, \mathrm{d}x = \sin(x).$$

We will examine integrals of trigonometric functions further in Year 2.

## 21.3 Evaluating Definite Integrals

The following theorem gives us a methodology for evaluating a definite integral.

---

**Theorem 21.2**

If $F'$ is continuous on $[a, b]$, then

$$\int_a^b F'(x)\,\mathrm{d}x = F(b) - F(a).$$

In other words, if $F$ is the integral of $f$ on $[a, b]$, then

$$\int_a^b f(x)\,\mathrm{d}x = F(b) - F(a).$$

---

**Proof:**

First we define a function $G(x)$ that is continuous on the interval $[a, b]$ such that

$$G(x) = \int_a^x F'(t)\,\mathrm{d}t.$$

Then, by the Fundamental Theorem of Calculus,

$$G'(x) = F'(x)$$

If the derivatives are equal, then integrating each side would give us

$$G(x) + c_1 = F(x) + c_2.$$

Therefore, $G(x)$ and $F(x)$ differ by a constant, $c$. In other words,

$$G(x) + c = F(x)$$

for all $x \in [a, b]$. Therefore, taking the right hand side

$$F(b) - F(a) = (G(b) + c) - (G(a) + c) = G(b) - G(a) = \int_a^b F'(t)\,\mathrm{d}t - \int_a^a F'(t)\,\mathrm{d}t.$$

We know that

$$\int_a^a F'(t)\,\mathrm{d}t = 0,$$

and so,

$$F(b) - F(a) = \int_a^b F'(t)\,\mathrm{d}t$$

as required.

---

**Remark**

This theorem is often referred to as the Fundamental Theorem of Calculus Part II. Its proof is here for completeness, but it is not examinable.

---

We now explore how we use this methodology to find a definite integral in the worked example below.

**Example 21.2**

Find

$$\int_0^2 x^3 \, \mathrm{d}x.$$

**Solution:**

First, we find

$$F(x) = \int x^3 \, \mathrm{d}x.$$

We know from Section 16.2 that

$$F(x) = \int x^3 \, \mathrm{d}x = \frac{1}{4}x^4 + c.$$

Theorem 21.2 tells us that

$$\int_0^2 x^3 \, \mathrm{d}x = F(x)\Big|_0^2 = F(2) - F(0).$$

Therefore,

$$\int_0^2 x^3 \, \mathrm{d}x = \left[\frac{1}{4}2^4 + c\right] - \left[\frac{1}{4}0^4 + c\right]$$

$$= \frac{16}{4} - \frac{0}{4}$$

$$= 4.$$

**Tip**

When evaluating a definite integral, the constant of integration will always disappear. This is because both $F(b)$ and $F(a)$ include the same constant of integration and we subtract one from the other. Therefore, evaluating a definite integral will always include the calculation $c - c = 0$, where $c$ is the constant of integration.

We also need to be able to solve more complex problems. The following theorem enables us to reduce complex problems to simpler problems that we can solve easily.

**Theorem 21.3**

The definite integral between $x = a$ and $x = b$ of the sum of two functions is equal to the sum of the definite integrals of each function between the same limits:

$$\int_a^b f(x) + g(x) \, \mathrm{d}x = \int_a^b f(x) \, \mathrm{d}x + \int_a^b g(x) \, \mathrm{d}x \qquad (21.2)$$

In other words, if $F$ is the integral of $f$ and $G$ is the integral of $g$:

$$\int_a^b f(x) + g(x)\,\mathrm{d}x = F(b) - F(a) + G(b) - G(a).$$

The definite integral between $x = a$ and $x = b$ of the difference between two functions is equal to the sum of the definite integrals of each function where the limits on the second function are reversed:

$$\int_a^b f(x) - g(x)\,\mathrm{d}x = \int_a^b f(x)\,\mathrm{d}x + \int_b^a g(x)\,\mathrm{d}x \tag{21.3}$$

In other words, if $F$ is the integral of $f$ and $G$ is the integral of $g$:

$$\int_a^b f(x) - g(x)\,\mathrm{d}x = F(b) - F(a) + G(a) - G(b).$$

---

**Activity 21.1**

Prove Theorem 21.3.

---

**Tip**

In summary, the method for evaluating a definite integral is:
1. Simplify the problem.
2. Integrate as required.
3. Substitute $a$ and $b$ into the integral.
4. Calculate the numeric solution.

---

**Example 21.3**

Find

$$\int_0^2 3x^2 + 8x^3\,\mathrm{d}x. \tag{21.4}$$

**Solution:**

By equation (21.2) we have

$$\int_0^2 3x^2 + 8x^3\,\mathrm{d}x = \int_0^2 3x^2\,\mathrm{d}x + \int_0^2 8x^3\,\mathrm{d}x. \tag{21.5}$$

Integrating first, we get

$$\int 3x^2\,\mathrm{d}x = 3\int x^2\,\mathrm{d}x = 3 \cdot \frac{1}{3}x^3 + c_1 = x^3 + c_1 \text{ and}$$

$$\int 8x^3\,\mathrm{d}x = 8\int x^3\,\mathrm{d}x = 8 \cdot \frac{1}{4}x^4 + c_2 = 2x^4 + c_2.$$

Next we use Theorem 21.2 to find the two definite integrals:

$$\int_0^2 3x^2 \, dx = \left(2^3 + c_1\right) - \left(0^3 + c_1\right) = 8,$$

$$\int_0^2 8x^3 \, dx = \left(2(2)^4 + c_2\right) - \left(2(0)^4 + c_2\right) = 32.$$

Substituting back into (21.5) we

$$\int_0^2 3x^2 + 8x^3 \, dx = 8 + 32 = 40.$$

---

**Example 21.4**

Find

$$\int_1^5 5x^4 - 4x^7 \, dx.$$

**Solution:**

From equation (21.3)

$$\int_1^5 5x^4 - 4x^7 \, dx = \int_1^5 5x^4 \, dx - \int_5^1 4x^7 \, dx. \qquad (21.6)$$

Integrating first, we obtain

$$\int 5x^4 \, dx = 5 \int x^4 \, dx = 5 \cdot \frac{1}{5}x^5 + c_1 = x^5 + c_1 \text{ and}$$

$$\int 4x^7 \, dx = 4 \int x^7 \, dx = 4 \cdot \frac{1}{8}x^8 + c_2 = \frac{1}{2}x^8 + c_2.$$

Now we use Theorem 21.2 to find the definite integrals:

$$\int_1^5 5x^4 \, dx = \left(5^5 + c_1\right) - \left(1^5 + c_1\right) = 3124 \text{ and}$$

$$\int_5^1 4x^7 \, dx = \left(\frac{1}{2}1^8 + c_2\right) - \left(\frac{1}{2}5^8 + c_2\right) = -\frac{390624}{2} = -195312.$$

Substituting back into (21.6) we get

$$\int_1^5 5x^4 - 4x^7 \, dx = 3124 + (-195312) = -192188.$$

---

**Remark**

Notice that the solution to this example is negative. This is confusing because we said earlier that the definite integral gives the area under a curve, however, the area must, by definition, be positive. We will investigate this further in the next section.

## 21.4 Using Definite Integrals to Calculate Areas

We saw in Section 16.1.2 that an antiderivative can be used to calculate the area under a curve. The method used then depended on our ability to deduce an antiderivative from our knowledge of differentiation. If, instead, we are able to use the formal terminology and techniques that we have learnt for integration we will be able to solve a wider range of problems quickly and easily.

### 21.4.1 Curves Above the Horizontal Axis

Theorem 21.1, the Fundamental Theorem of Calculus, confirms that the area under a curve can be found from the definite integral of the curve. We state this formally in the following formula.

**Formula 21.1**

The area under a curve $f(x)$ between $x = a$ and $x = b$ is given by

$$A = \int_a^b f(x)\,dx$$

when $f(x) > 0$ on $[a, b]$.

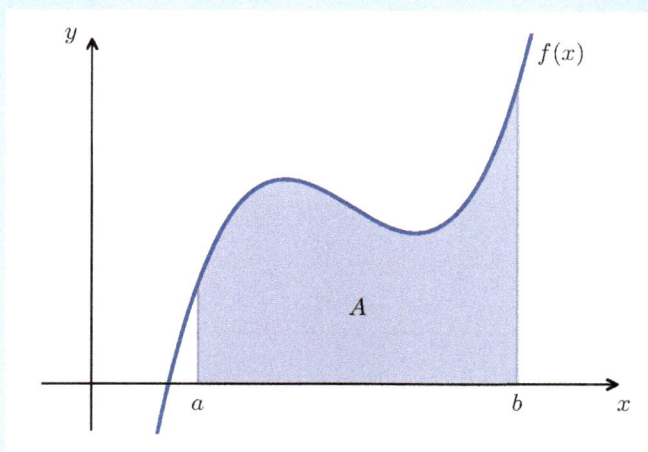

**Example 21.5**

Use definite integrals to find the area under the curve $y = x^2$ between the limits $x = 1$ and $x = 2$.

**Solution:**

The Fundamental Theorem of Calculus tells us that the area under $y = x^2$ between $x = 1$ and $x = 2$ is given by

$$\int_1^2 x^2\,dx.$$

First we find the integral of $y = x^2$:

$$\int x^2 \, \mathrm{d}x = \frac{1}{3}x^3 + c.$$

Using Theorem 21.2 we have

$$\int_1^2 x^2 \, \mathrm{d}x = \left[\frac{1}{3}2^3 + c\right] - \left[\frac{1}{3}1^3 + c\right]$$
$$= \frac{8}{3} + c - \frac{1}{3} - c$$
$$= \frac{7}{3}.$$

**Remark**

Looking back at Example 16.3 we notice that both the solution and the calculations used to find it are the same here. Therefore, the definite integral formalises the antiderivative method.

**Example 21.6**

Returning to finding the cross-sectional area of a NACA 0012 airfoil, we wish to compute

$$I = \int_0^1 0.6 \left(0.2969\sqrt{x} - 0.1260x - 0.3156x^2 + 0.2843x^3 - 0.1015x^4\right) \, \mathrm{d}x,$$
$$= \left[0.11876x^{\frac{3}{2}} - 0.0378x^2 - 0.06312x^3 + 0.042645x^4 - 0.01218x^5\right]_0^1,$$
$$= 0.048305$$

Hence, the cross sectional area of the rotor blade is 0.09661 square units.

### 21.4.2 Curves Below the Horizontal Axis

far we have only discussed the area under a curve or, in other words, the area between the curve and the $x$-axis when the curve is above the $x$-axis. In the following example we examine what happens when the curve is below the $x$-axis.

**Example 21.7**

Use integration to find the area between the $x$-axis, the curve $f(x) = x^3 - 3x^2$, and the limits $x = 0$ and $x = 3$.

**Solution:**

It is helpful to start with a sketch of the curve.

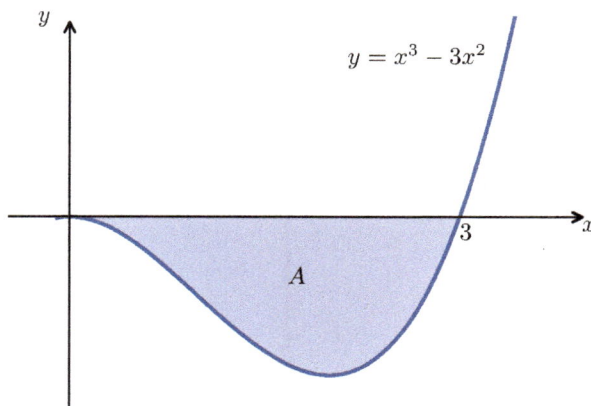

We can see that $f(x) \leq 0$ on the whole interval $x = 0$ to $x = 3$. Formula 21.1 required that $f(x) > 0$. The problem set out in the formula is a reflection of the problem here in the $x$-axis. Therefore, if we reflect the curve in the $x$-axis we will be able to use the formula. The reflection of the curve is given by $g(x) = -f(x) = -x^3 + 3x^2$. Therefore, we need to find

$$\int_0^3 3x^2 - x^3 \, \mathrm{d}x.$$

By equation (21.3) we can simplify this as follows:

$$\int_0^3 3x^2 - x^3 \, \mathrm{d}x = \int_0^3 3x^2 \, \mathrm{d}x + \int_3^0 x^3 \, \mathrm{d}x.$$

First, we find the indefinite integrals:

$$\int 3x^2 \, \mathrm{d}x = 3 \int x^2 \, \mathrm{d}x = 3 \cdot \frac{1}{3} x^3 + c_1 = x^3 + c_1 \text{ and}$$

$$\int x^3 \, \mathrm{d}x = \frac{1}{4} x^4 + c_2.$$

Using Theorem 21.2 we can find the definite integrals:

$$\int_0^3 3x^2 \, \mathrm{d}x = (3^3 + c_1) - (0^3 + c_1) = 27 \text{ and}$$

$$\int_3^0 x^3 \, \mathrm{d}x = \left( \frac{1}{4} 0^4 + c_2 \right) - \left( \frac{1}{4} 3^4 + c_2 \right) = -\frac{81}{4}.$$

Therefore,

$$\int_0^3 3x^2 - x^3 \, \mathrm{d}x = 27 + \left( -\frac{81}{4} \right) = \frac{27}{4},$$

and the area we are seeking is $\frac{27}{4}$.

Let us now check what the definite integral of $f(x)$ is and see how that relates to the area we have found. We need to find

$$\int_0^3 x^3 - 3x^2 \, \mathrm{d}x.$$

By equation (21.3) we have:

$$\int_0^3 x^3 - 3x^2 \, \mathrm{d}x = \int_0^3 x^3 \, \mathrm{d}x + \int_3^0 3x^2 \, \mathrm{d}x.$$

Integrating first, we obtain

$$\int x^3 \, \mathrm{d}x = \frac{1}{4}x^4 + c_3$$

$$3 \int x^2 \, \mathrm{d}x = 3 \cdot \frac{1}{3}x^3 + c_4 = x^3 + c_4.$$

Now we use Theorem 21.2 to find the definite integrals

$$\int_0^3 x^3 \, \mathrm{d}x = \left( \frac{1}{4}3^4 + c_3 \right) - \left( \frac{1}{4}0^4 + c_3 \right) = \frac{81}{4}$$

$$\int_3^0 3x^2 \, \mathrm{d}x = \left( 0^3 + c_4 \right) - \left( 3^3 + c_4 \right) = -27.$$

Therefore,

$$\int_0^3 x^3 - 3x^2 \, \mathrm{d}x = \frac{81}{4} - 27 = -\frac{27}{4}.$$

The definite integral of $f(x)$ is the negative of the area we found earlier! Rather than reflecting our initial curve $f(x)$, we could have simply found the definite integral of $f(x)$ and taken the modulus to find the area.

---

This leads us to the following formula for finding the area from definite integrals for curves below the $x$-axis.

**Formula 21.2**

The area above a curve $f(x)$ between $x = a$ and $x = b$ is given by

$$A = -\int_a^b f(x) \, \mathrm{d}x$$

when $f(x) < 0$ on $[a, b]$.

### 21.4.3  Curves that are both Above and Below the Horizontal Axis

now explore how to calculate the area between a curve and the $x$-axis when part of the curve is above the $x$-axis and part of the curve is below the $x$-axis.

---

**Example 21.8**

Find the area between the $x$-axis and the curve given by $f(x) = x^4 - 4x^3 + 3x^2$.

**Solution:**

We begin by sketching the curve.

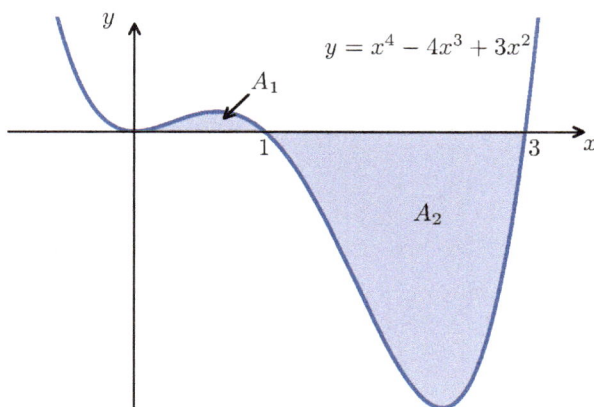

We can see from the figure that the area between the curve and the $x$-axis between $x = 0$ and $x = 3$ is split with part of the area above the $x$-axis and part of it below the $x$-axis. The formulae we have previously used to calculate the area, 21.1 and 21.2 only apply when the whole area is either above or below the $x$-axis. We can work around this by splitting the problem into two, and using Formula 21.1 to find the area between $x = 0$ and $x = 1$ and using Formula 21.2 to find the area between $x = 1$ and $x = 3$. First we find the indefinite integral of the full equation as we can then use this in both parts of the problem.

$$\int x^4 \, \mathrm{d}x = \frac{1}{5}x^5 + c_1$$

$$\int -4x^3 \, \mathrm{d}x = -4 \int x^3 \, \mathrm{d}x = -4\frac{1}{4}x^4 + c_2 = -x^4 + c_2$$

$$\int 3x^2 \, \mathrm{d}x = 3 \int x^2 \, \mathrm{d}x = 3\frac{1}{3}x^3 + c_3 = x^3 + c_3.$$

Therefore the indefinite integral is

$$\int x^4 - 4x^3 + 3x^2 \, \mathrm{d}x = \frac{1}{5}x^5 - x^4 + x^3 + k,$$

where $k = c_1 + c_2 + c_3$ is a constant. To find the area between $x = 0$ and $x = 1$ we use

Formula 21.1 as follows:

$$A_1 = \int_0^1 x^4 - 4x^3 + 3x^2 \, \mathrm{d}x$$

$$= \left[ \frac{1}{5}x^5 - x^4 + x^3 + k \right]_0^1$$

$$= \left( \frac{1}{5}1^5 - 1^4 + 1^3 + k \right) - \left( \frac{1}{5}0^5 - 0^4 + 0^3 + k \right)$$

$$= \frac{1}{5}.$$

Now we use Formula 21.2 to find the area between $x = 1$ and $x = 3$:

$$A_2 = - \int_1^3 x^4 - 4x^3 + 3x^2 \, \mathrm{d}x$$

$$= - \left[ \frac{1}{5}x^5 - x^4 + x^3 + k \right]_1^3$$

$$= - \left( \left( \frac{1}{5}3^5 - 3^4 + 3^3 + k \right) - \left( \frac{1}{5}1^5 - 1^4 + 1^3 + k \right) \right)$$

$$= - \left( \frac{243}{5} - 81 + 27 - \frac{1}{5} \right)$$

$$= - \left( -\frac{28}{5} \right)$$

$$= \frac{28}{5}.$$

The total area is given by

$$A = A_1 + A_2 = \frac{1}{5} + \frac{28}{5} = \frac{29}{5}.$$

**Activity 21.2**
Compare the area found in the example above with the definite integral for $f(x) = x^4 - 4x^3 + 3x^2$ between the limits $x = 0$ and $x = 3$.

The examples above show that a definite integral can give a negative result where all or part of a curve is below the $x$-axis. Another way in which we might obtain a negative answer from a definite integral is if the upper limit is less than the lower limit.

**Activity 21.3**
Find

$$\int_3^1 6x^3 \, \mathrm{d}x,$$

and use the answer to find the area under the curve $y = 6x^3$ between the limits $x = 1$ and $x = 3$.

### 21.4.4 Calculating the Area under a Non-continuous or Non-smooth Function

All the above examples are for smooth continuous functions, *i.e.* functions whose derivatives are also continuous. We now consider how to find the area for non-smooth functions.

---

**Example 21.9**

The function $y = |x|$ is sketched below.

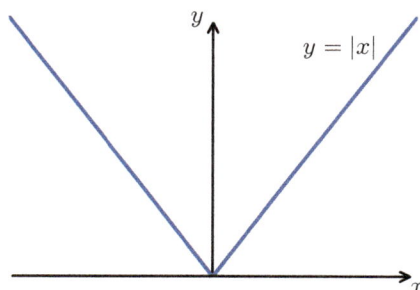

This function is defined as follows,

$$|x| = \begin{cases} x, & x \geq 0, \\ -x, & x < 0, \end{cases}$$

and is called the *modulus* function; it is very important and will be studied further in Year 2. It is an example of a function that is continuous, but is not smooth, due to the derivative being undefined at $x = 0$.

---

**Example 21.10**

The function $y = H(x)$ is defined as follows,

$$H(x) = \begin{cases} 0, & x < 0, \\ 1, & x \geq 0. \end{cases}$$

This function is called the *Heaviside step function*; it is an example of a function that is discontinuous due to the *jump* at $x = 0$.

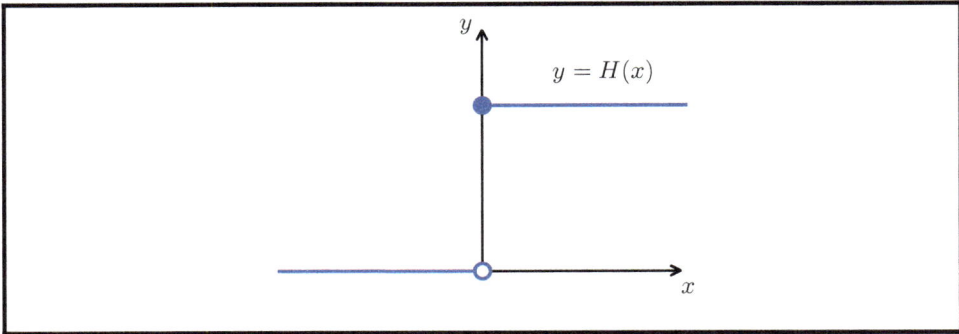

**Activity 21.4**
Find the area under $f(x) = |x|$ between the limits $x = -2$ and $x = 2$.

**Example 21.11**
Find the area under $f(x) = \lfloor x \rfloor$ between $x = 1.5$ and $x = 3.5$. Generalise this in order to find

$$\int_0^a \lfloor x \rfloor \, dx.$$

**Solution:**
The floor of $x$, or $\lfloor x \rfloor$, is the greatest integer that is less than or equal to $x$. In mathematical notation:

$$\lfloor x \rfloor = n,$$

where $n \in \mathbb{N}$ and $x - 1 < n \leq x$. The graph of $f(x) = \lfloor x \rfloor$ for $x$ in $[0, 4]$ is

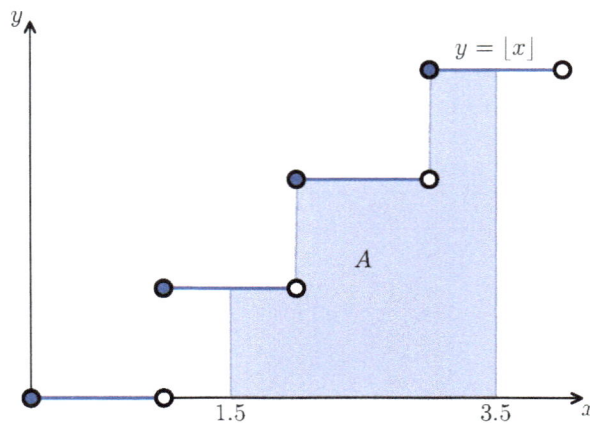

The filled circle denotes that the point is included in the connected line. The empty circle means that the point is not included in the connected line, but everything up to that point is.

We can see from the plot of $\lfloor x \rfloor$ that the floor function is not continuous across the region $[1.5, 3.5]$. The formula we have previously used to calculate the area under a curve required that $f(x)$ be continuous. Therefore, before we can use integration to find the area we need to divide the problem into a series of continuous problems that can be added together to give the solution. That is,

$$A_1 = \int_{1.5}^{2} \lfloor x \rfloor \, dx,$$

$$A_2 = \int_{2}^{3} \lfloor x \rfloor \, dx,$$

$$A_3 = \int_{3}^{3.5} \lfloor x \rfloor \, dx.$$

The overall area between $x = 1.5$ and $x = 3.5$, $A$, will be given by

$$A = A_1 + A_2 + A_3.$$

First, let us consider $A_1$. On the region $[1.5, 2)$, $f(x) = \lfloor x \rfloor = 1$. Therefore,

$$
\begin{aligned}
A_1 &= \int_{1.5}^{2} \lfloor x \rfloor \, dx \\
&= \int_{1.5}^{2} 1 \, dx \\
&= \left[ x + c \right]_{1.5}^{2} \\
&= (2 + c) - (1.5 + c) \\
&= 2 - 1.5 \\
&= 0.5.
\end{aligned}
$$

Now we look at $A_2$. On the region $[2, 3)$, $f(x) = \lfloor x \rfloor = 2$. Therefore,

$$
\begin{aligned}
A_2 &= \int_{2}^{3} \lfloor x \rfloor \, dx \\
&= \int_{2}^{3} 2 \, dx \\
&= \left[ 2x + c \right]_{2}^{3} \\
&= (6 + c) - (4 + c) \\
&= 6 - 4 \\
&= 2.
\end{aligned}
$$

Finally we consider $A_3$. On the region $[3, 3.5]$, $f(x) = \lfloor x \rfloor = 3$. Therefore

$$
\begin{aligned}
A_3 &= \int_3^{3.5} \lfloor x \rfloor \, dx \\
&= \int_3^{3.5} 3 \, dx \\
&= \Big[ 3x + c \Big]_3^{3.5} \\
&= (10.5 + c) - (9 + c) \\
&= 10.5 - 9 \\
&= 1.5.
\end{aligned}
$$

Now we can find the total area $A$:

$$
A = A_1 + A_2 + A_3 = 0.5 + 2 + 1.5 = 4.
$$

Looking at our original graph we can see that the area $A$ is in fact a series of rectangles, and that $A_1$, $A_2$, and $A_3$ are those rectangles. Therefore we can use geometric calculations to check the answer we found with integrals:
$A_1$ is a rectangle of width 0.5 and height 1, therefore $A_1 = 0.5 \times 1 = 0.5$.
$A_2$ is a rectangle of width 1 and height 2, therefore $A_2 = 1 \times 2 = 2$.
$A_3$ is a rectangle of width 0.5 and height 3, therefore $A_3 = 0.5 \times 3 = 1.5$.
Taking the sum, we obtain

$$
A = A_1 + A_2 + A_3 = 0.5 + 2 + 1.5 = 4
$$

and we can confirm the solution we found with integration.
In order to generalise, notice that the total area under $f(x) = \lfloor x \rfloor$ between $x = 0$ and $x = a$ is the sum of a set of rectangles. Each rectangle is of width 1 and of height $f(i) = \lfloor i \rfloor = i$, where $i = 0, ..., a - 1$. We know that the area under $f(x)$ between $x = 0$ and $x = a$ is equal to the definite integral of $f(x)$ between $x = 0$ and $x = a$. Therefore,

$$
\int_0^a \lfloor x \rfloor = \sum_{i=0}^{a-1} 1i = \sum_{i=0}^{a-1} i.
$$

**Exercise 21.1**

Q1. Which is the correct solution to

$$
\int_0^4 x^2 \, dx \; ?
$$

(a) $\dfrac{1}{3} x^3 + c$,

(b) $\dfrac{64}{3} + c$,

(c) $\dfrac{64}{3}$, or

(d) 32.

Q2. Which is the correct solution to

$$\int_1^2 x + 1 \, dx \ ?$$

(a) $\dfrac{3}{2} + c$,

(b) $\dfrac{1}{2}x^2 + c$,

(c) $\dfrac{5}{2}$, or

(d) $-\dfrac{5}{2}$.

Q3. Find the following integrals:

(a)

$$\int_1^4 x \, dx,$$

(c)

$$\int_2^1 x^3 \, dx,$$

(e)

$$\int_1^2 x + x^4 \, dx.$$

(b)

$$\int_0^3 -x^2 \, dx,$$

(d)

$$\int_2^3 x^3 - x^2 \, dx,$$

(f)

$$\int_4^6 x^- 4 \, dx$$

Q4. Find the area under $y = x$ between the limits $x = 2$ and $x = 50$. Confirm your answer using geometric formulae.

Q5. Find the area between $y = \sqrt{x+1}$ and the $y$-axis between the limits $y = 1$ and $y = 2$. Use your answer to find the area between $y = \sqrt{x+1}$ and the $x$-axis.

Q6. Find the area between $f(x) = -3x^{1/3}$ and the $x$-axis and the limits $x = 1$ and $x = 8$.

Q7. A distribution company charges £4.30 per cubic metre for transporting a package. The company has been asked to transport a package with an irregular cross-section (shown below, and known from manufacturing). Given that the depth of the package is 0.75m and the cross-section is the same throughout the depth, how much will it cost to transport the package?

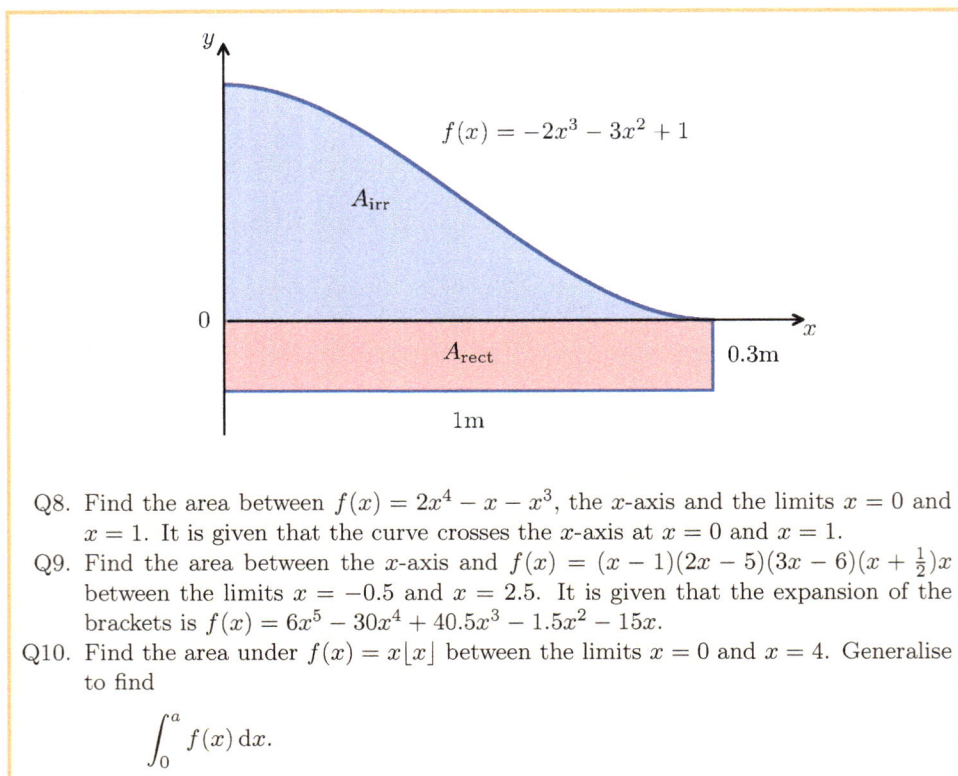

Q8. Find the area between $f(x) = 2x^4 - x - x^3$, the $x$-axis and the limits $x = 0$ and $x = 1$. It is given that the curve crosses the $x$-axis at $x = 0$ and $x = 1$.

Q9. Find the area between the $x$-axis and $f(x) = (x - 1)(2x - 5)(3x - 6)(x + \frac{1}{2})x$ between the limits $x = -0.5$ and $x = 2.5$. It is given that the expansion of the brackets is $f(x) = 6x^5 - 30x^4 + 40.5x^3 - 1.5x^2 - 15x$.

Q10. Find the area under $f(x) = x\lfloor x \rfloor$ between the limits $x = 0$ and $x = 4$. Generalise to find

$$\int_0^a f(x)\,dx.$$

## Key Facts — Integration 2 - Definite Integrals

- The Fundamental Theorem of Calculus states that, if $f(x)$ is continuous, and if $F(x)$ satisifies

$$F(x) = \int_a^x f(t)\,dt,$$

then $F$ is differentiable on $[a, b]$ and $F'(x) = f(x)$, or

$$\frac{d}{dx}\int_a^x f(t)\,dt = f(x).$$

- If $F'$ is continuous on $[a, b]$, then

$$\int_a^b F'(x)\,dx = F(b) - F(a).$$

Alternatively, if $F$ is the integral of $f$, then

$$\int_a^b f(x)\,dx = F(b) - F(a).$$

- For two continuous functions $f(x)$ and $g(x)$ we have

$$\int_a^b f(x) + g(x)\, \mathrm{d}x = \int_a^b f(x)\, \mathrm{d}x + \int_a^b g(x)\, \mathrm{d}x,$$

$$\int_a^b f(x) - g(x)\, \mathrm{d}x = \int_a^b f(x)\, \mathrm{d}x + \int_b^a g(x)\, \mathrm{d}x.$$

## Chapter Assessment — Integration 2 - Definite Integrals

Download and sit the 30 minute assessment for this chapter from the digital book.

$$\bar{X} = \frac{1}{N} \sum_{i=1}^{N} X_i.$$

# 22. Populations and Samples

In May 2015, the United Kingdom held a General Election. Almost 30 million people voted to elect their Member of Parliament and decide who would form the next government. In the weeks before the election, opinion polls predicting the outcome were being published every day. The polls predicted a very close result, probably leading to a coalition government.

In fact, the Conservative party won comfortably and was able to form a government on its own.

Opinion polls gather data from just a small sample of people, typically one or two thousand, but the election gathers data from the entire population. The difference between the polls and the 2015 election result shows the difficulty in predicting results about the entire population from sample statistics.

It has been suggested that the close opinion polls led politicians to promise to hold a referendum on Britain's membership of the European Union, and that the election result meant they had to keep that promise. If this is true, then mistakes made in statistical sampling played a direct role in one of the most significant political events for a generation.

## 22.1 Populations

The purpose of statistics is to summarise large amounts of data so that they can be understood, analysed and used to make predictions and decisions. The set of things that we collect the data from, which could be a group of people or animals, or a set of events, is called the *population*. The definition of the population depends on what we are studying. The values, such as the mean or the variance, which describe this entire data set are called *population parameters*.

**Example 22.1**
Consider the following population parameters.
   (a) The mean exam score of pupils in a certain school.
   (b) The median height of teenagers living in Newcastle.
   (c) The probability of a chess grandmaster beating a new chess-playing computer.
Identify the population related to each parameter.

The populations for these statistics are as follows:
   (a) All pupils in the school who are taking the exam.
   (b) All teenagers living in Newcastle.
   (c) All possible chess games between the computer and the chess grandmaster.

**Remark**
If the entire data set is known and is small enough to analyse, such as in the case of exam results in a school, then we can calculate and report the population parameters. For larger populations, it may not be possible to gather data from all members. It would be difficult to collect data from every teenager in a whole city. The median height is a well-defined quantity, but we have no way of finding out what it is. We must use sampling techniques to produce some kind of estimate. Even though the median height is a fixed value, using sampling introduces some randomness into our calculations. Some populations are more abstract. When we talk about the probability $p$ of some event, we mean that, if an experiment is repeated $N$ times then as $N$ gets larger, the event will occur roughly $Np$ times. Theoretically, to know the true value, we must run the experiment an infinite number of times. To find the probability of the grandmaster beating the computer, they would have to play an infinite number of games. The population in this case is every game they could play.

For large populations we may never know the true values of its parameters. The best we can do is collect data from a subset of the population, and make an estimate of the parameter. Any such subset is called a *sample*.

## 22.2 Samples

Suppose that we want to estimate how many times a year, people living in a certain town attend events at the theatre. We do not have the resources to ask every inhabitant: we can only select a sample of people to ask.

**Activity 22.1 — Sampling Strategies**
How might we choose the sample of people to interview? Which of the following strategies might be useful?
   • Select random landline telephone numbers in the town and contact them one Monday afternoon.
   • Place a notice in the local paper, inviting people to respond to a questionnaire.
   • Select people who come into the theatre's lobby on a Saturday morning.
   • Interview one person from each house on a randomly chosen street.

By choosing different sample selection strategies, we obtain different kinds of samples. Different samples will give us different estimates for the population parameter.

## 22.3  Constructing a Sample

Although almost all of the opinion polls before the 2015 General Election were inaccurate, an unpublished poll completed on the day before the election predicted the actual result to within 1%. The sample size was just 0.0035% of the population. This demonstrates how much care modern political polling organisations take in designing their samples.

Suppose that a business would like to survey its ten thousand employees in face-to-face interviews, to find out about their work-life balance. They decide that they will interview a sample of one hundred members of staff.

They could construct the sample using *simple random sampling*. In this kind of sampling, every choice is made independently, and every employee has the same probability of being chosen.

One way to achieve this is to assign the numbers 1 to 10 000 randomly to the employees and then interview those people assigned numbers 1 to 100. This can be adjusted to select any proportion of the population that we wish.

> **Activity 22.2 — Simple Random Sampling**
> Why do we have to allocate the numbers randomly? Could we just use the employee number?
> Alternatively, could we ask each Head of Department to select 1% of their staff randomly?

Simple random sampling is a good technique to use if we have very little information about the population. However, it requires us to have a complete list of the population, and we must be able to obtain data from whichever individuals we select for the sample.

> **Use of Technology 22.1 — How to Sample Randomly**
> To ensure random sampling, once every member of the population has been allocated a number a computer or calculator should be used to select the items in the sample. This is because humans can never be truly random.

Suppose that we want to perform quality checks on 1% of the items coming off a production line. We could start by checking a randomly chosen item and then go on to check every hundredth item. Each item has an equal chance of being selected, but the selections are not independent. The entire sample is determined by our first selection. This is called *systematic sampling*.

> **Tip**
> Note the difference between simple random sampling and systematic sampling. In simple random sampling, the choices are independent, because we have assigned the numbers

randomly. In systematic sampling, if the order of the individuals is not random, then the choices will not be independent.

Both of these methods rely on our having access to any member of the population that we select. If this is not possible, or if it is too costly, we can use *opportunity sampling*. Opportunity sampling collects data from those members of the population for which data is readily available. This is usually a poor method for selecting samples, since we are making no effort to make our sample representative of the population. However, there may be no alternative. If we wanted to describe the distribution of heights of a certain kind of dinosaur, then our population has been dead for millions of years. Our sample can only contain those dinosaurs whose remains have, by chance, been fossilised and discovered. We will not have very high confidence that statistics calculated using opportunity sampling will be close to the population parameters.

**Remark**
When we are choosing a sampling technique, we must be practical. We must take into account the time and expense of gathering the data from the chosen sample. In some situations, it is better to gather data quickly and cheaply, even if this means that the data is of lower quality.

We can develop more sophisticated sampling techniques if we consider the characteristics of the population. For example, in selecting a sample of one hundred employees from a company of ten thousand, if there are a total of 200 engineers in the company, we would choose two to be included in our sample. Generally, we could make the sample reflect the proportions of different categories of employees in the whole company. This is called *stratified sampling*. We divide the population into different classes, called *strata*, and randomly select the same proportion of elements from each one.

The advantage of this kind of sampling is that, if the individuals within each of the strata are similar, we can obtain more accurate results with a smaller sample. However, we need to start with a good understanding of the population in order for our smaller sample to be properly representative of the population.

**Remark**
We know that there is always a chance that our selected sample is not representative of the population. Our goal is to make sure that we have given ourselves the best chance of obtaining a representative sample. This means that we try not to introduce any new source of bias.

Stratified sampling divides the whole population into a number of strata, and then selects the same proportion from each. A process called *quota sampling* does something similar, but does not rely on a knowledge of the entire population. For example, if we are trying to survey shoppers, then we could divide the population into strata according to their age. Although we do not know the actual proportion of people in each age band, we set quotas according to the needs of our survey, such as twenty people in their twenties, or ten people aged eighty and above.

The advantage here is that we can design the survey without knowing as much about the whole population. However, there is no guarantee that our sample is representative of that population.

Stratified and quota sampling both involve dividing the population into groups which we think will have less variation than the overall population. In *cluster sampling*, we do the opposite and create groups which we think have the same variation as the whole population. For example, suppose we want to survey surgeons in UK hospitals. We might like to use simple random sampling, but this would require travelling to very many different hospitals. Depending on our survey, we might decide that the range of results are likely to be roughly the same in each hospital. To make the study cheaper, we can select some hospitals at random, and then survey all of the surgeons in those hospitals.

If each cluster has a large number of elements, then we can choose the clusters and then use simple random sampling to select just some of the individuals from the cluster.

This type of sampling can reduce the costs of taking measurements. However, we assumed that the range of results from different clusters were likely to be the same, and this assumption may not be correct. In our example, perhaps we were measuring surgery mortality rates. We are likely to obtain different figures from a large, specialist hospital, handling the most complex cases from around the country, compared to a smaller hospital which handles only routine cases.

## 22.4 Estimating Parameters

Recall that our aim is to estimate some parameters for a population which is too large to measure completely. If we can construct a sample which has the same characteristics as the total population, we can take measurements from it and hope that from these we can estimate the population parameters.

Suppose we construct a sample so that we can estimate the mean of a population. The population mean has a fixed, well-defined value, but we cannot measure it. Our sample also has a fixed, well-defined value, which we *can* measure. However, we used a random process to generate the sample. Sometimes, this random process will give us a value which is very close to the population parameter, other times, it will be further away.

---

**Example 22.2**

Six friends each throw three darts. There scores are shown in this table.

| 6 | 26 | 44 | 56 | 168 | 180 |
|---|----|----|----|----|-----|

What is their mean score?

Calculate the highest and lowest possible sample means that could be obtained from a sample size of two, constructed using simple random sampling.

**Solution:**

The population mean score is

$$\frac{6 + 26 + 44 + 56 + 168 + 180}{6} = 80.$$

Using simple random sampling, we might select any two scores to form the sample. Therefore, the lowest possible sample mean is given by

$$\frac{6 + 26}{2} = 16.$$

---

The highest possible sample mean is equal to

$$\frac{168 + 180}{6} = 174.$$

**Remark**

For the above example, we can calculate all of the possible sample means we get by selecting the different possible pairs of scores.

|     | 6  | 26  | 44  | 56  | 168 |
| --- | -- | --- | --- | --- | --- |
| 26  | 16 |     |     |     |     |
| 44  | 25 | 35  |     |     |     |
| 56  | 31 | 41  | 50  |     |     |
| 168 | 87 | 97  | 106 | 112 |     |
| 180 | 93 | 103 | 112 | 118 | 174 |

With such widely varying scores, and for such a small sample, the sample mean is not very likely to be close to the population mean.

Notice that the average value of the sample mean is equal to the population mean.

$$\frac{\text{sum of possible sample means}}{\text{number of different sample means}} = \frac{1200}{15} = 80.$$

This means that if many people each took a random sample of size two, then the mean value of those sample means would be close to the population mean. However, most of the individual sample means would not be close.

**Interactive Activity 22.1 — Taking Samples**

The Geogebra interactive below (in the digital book) can be used to examine the effect of changing the sample size and/or the number of samples taken when estimating the mean of a population by the mean of a sample.

The sample mean is an example of a *statistic*; we can define the concept of a *statistic* more formally as below.

**Definition 22.1 — Statistic**

A statistic is a random variable which is a function of a sample.

For example, for a sample $\{X_1, X_2, \ldots, X_N\}$, the sample mean is a statistic:

$$\bar{X} = \frac{1}{N} \sum_{i=1}^{N} X_i.$$

If a statistic is used to estimate a parameter, it is called an *estimator*.

**Activity 22.3 — Identifying Statistics**

A population has mean $\mu$. For a sample $X_1, X_2, \ldots, X_N$, which of the following are statistics?

- $\frac{1}{N} \sum_{i=1}^{N} X_i$;
- $\sum_{i=1}^{N} (X_i - \bar{X})^2$;
- $\sum_{i=1}^{N} (X_i - \mu)^2$;
- $X_5 - N$.

The roll of a die is a random process. Once we have rolled it, it takes some definite value, such as four. Before we roll it, we can investigate its probabilistic behaviour, such as the probability of rolling a three. A statistic is exactly the same. When we have a sample, we can calculate the value of the statistic. Before we have chosen the sample, we can investigate the behaviour of the statistic.

**Tip**

The population can be described with population parameters: these are fixed, but unknown, values. Our goal is to construct a sample and select statistics that will give us good estimations of the parameters. We can end up with poor estimations if we choose our sample badly, or simply by chance.

**Example 22.3**

A researcher plans to travel to the Forest of Dean and select fifty oak trees. She will make each choice independently of the other choices. She will measure the heights of the trees she has chosen $X_1, X_2, \ldots, X_{50}$ and calculate the sample mean $\bar{X} = \sum_{i=1}^{50} X_i / 50$. Express the expected value of $\bar{X}$ in terms of $\mu$, the mean height of all the oak trees in the forest.

**Solution:**

Suppose there are $N$ oak trees in the forest. If we choose one at random, then the probability of choosing any given tree is $1/N$. The expected height $H$ of our chosen tree is

$$\mathbb{E}(H) = \sum_{i=1}^{N} \mathbb{P}(\text{choose tree } i) \times (\text{height of tree } i) = \sum_{i=1}^{N} \frac{1}{N} h_i = \mu.$$

Since our researcher has chosen fifty trees randomly and independently, the expected value of each one is also $\mu$. Therefore,

$$\mathbb{E}(\bar{X}) = \mathbb{E}\left(\frac{1}{50} \sum_{i=1}^{50} X_i\right) = \frac{1}{50} \sum_{i=1}^{50} \mathbb{E}(X_i) = \frac{1}{50} \sum_{i=1}^{50} \mu = \mu.$$

The expected value of the sample mean is equal to the population mean.

> **Remark**
>
> There is one possibility that we have not mentioned. If we are choosing fifty trees from a forest, and each choice is independent, then there should be some (very small) probability of choosing the same tree twice. If we choose the second tree from the set of all possible trees *except for the first tree*, then these choices are not independent.
>
> To be independent, we should make our selection *with replacement*. If the same tree is randomly selected twice, we should include it in our sample twice. This is only really a concern when the sample is a significant proportion of the entire population.
>
> The alternative is to make our selection *without replacement*. In this case, we would randomly select fifty different trees. Our selections are no longer independent, since if the first element in the sample is $x_1$ then the second element can be any member of the population apart from $x_1$.

> **Definition 22.2 — Unbiased Estimator**
>
> If we have a population parameter $\rho$, and a sample statistic $R$, such that $\mathbb{E}(R) = \rho$ then we say that $R$ is an *unbiased estimator* for $\rho$.
>
> For example, the sample mean $\bar{X}$ is an unbiased estimator for the population mean $\mu$.

We have seen that for a randomly chosen sample, the expected value of the sample mean is equal to the population mean. This is true even if the sample has only one element, even though it seems intuitively clear that we should get a better estimate with a larger sample size. This is indeed the case: although both statistics have the correct expected value, the one from the smaller sample size has a larger variance. That is, the smaller the sample size, the more often the sample mean will be far from the population mean. Even if we start with a carefully constructed sample, not all sample statistics will give good estimates of the corresponding population parameter. If the expected value of the sample statistic is not equal to the population parameter, it is called a *biased estimator*.

We might consider using the variance of the sample as an estimate of the population variance but, in fact, this is a biased estimator. Extreme values tend to be under-represented in samples, simply because we are less likely to select rare values. As a result, the sample variance tends to be lower than the population variance. If we take a sample $\{X_1, X_2, \ldots, X_N\}$, and calculate its mean $\bar{X}$ and its variance,

$$s^2 = \frac{1}{N} \sum_{I}^{N} \left( X_i - \bar{X} \right)^2,$$

then the expected value of $s^2$ is equal to $\frac{(N-1)}{N}\sigma^2$.

We can correct this bias by using

$$s^2 = \frac{1}{(N-1)} \sum_{i}^{n} \left( X_i - \bar{X} \right)^2,$$

instead. Sometimes this is called the *unbiased sample variance* to make it clear when the factor $(N-1)$ is being used. If $N$ is used, then we can call it the *biased sample variance*.

**Exercise 22.1**

Q1. A large cheese manufacturer ages one of their products by storing them in long rows of shelves with sixty cheeses on each shelf. When the cheese has reached the correct age, some of the cheese is quality tested.

   (a) The Head of Quality at the cheese company says that he tests a sample of the population. Explain the terms *population* and *sample*.

   (b) He goes on to say that he tests the first cheese, and then tests every twentieth cheese along the shelf. Suggest a possible problem with his method. How could he improve his sample?

Q2. As part of the development of a new drug, it must be tested on healthy people to check for any side effects.

   (a) Suggest a suitable method for constructing a sample of the population to test the drug. What are the advantages and disadvantages of the method <chosen?

   (b) The first trials for a new drug are usually conducted on a very small sample group. Suggest a reason why this might be the case and comment on the strength of any conclusions you might draw from such a study.

Q3. An exam board produces an English syllabus which includes course work to be marked by teachers at each school. To ensure consistency between schools, the board re-marks a sample of each school's work and compares those marks with the original marks given by the teacher.

One school has five classes of thirty pupils entered in the exam. The school is asked to provide a sample of twenty pieces of work.

   (a) Describe a method the school could use to construct a sample using simple random sampling.

   (b) The school constructs the sample by choosing four pupils from each class at random. Explain why this method is *not* a form of simple random sampling.

Q4. A landowner owns a large lake. She believes there are several hundred carp fish in the lake. She sets a trap in one part of the lake, which allows her to capture, weigh and release fish. On one day, she records the following weights, given in pounds.

| | | | | | |
|---|---|---|---|---|---|
| 16.3 | 16.6 | 17.0 | 17.3 | 17.1 | 17.4 |

   (a) Give an estimate for the mean weight of all carp in the lake.

   (b) Give one reason why this estimate may be accurate.

   (c) Give one reason why this estimate may be inaccurate, and suggest what could be done to overcome the problem you have identified.

Q5. A researcher interviews people in a certain town to find out how many times they visit their General Practitioner each year. He conducts two sessions, one on a Tuesday afternoon, when he interviews thirty people and one on a Wednesday evening, when he interviews twenty people.

On Tuesday, he records a mean value of 4.5 visits per year. On Wednesday, he records a mean value of 3.8 visits per year.

Give *two* possible reasons for the different results and suggest one way in which he could improve his survey.

**Key Facts — Populations and Samples**

- The population is the set of items that we can collect data from. Values which describe the population, such as the mean and variance are known as population parameters.
- A sample is a subset of the population, obtained in some way.
- The following methods can all be used to extract a sample from the population.
  - (a) Simple random sampling.
  - (b) Systematic sampling.
  - (c) Opportunity sampling.
  - (d) Stratified sampling.
  - (e) Quota sampling.
  - (f) Cluster sampling.
- Parameters for the population can be estimated using a sample. However, there is no guarantee that this estimate is close to the true value.
- A statistic is a random variable which is a function of a sample. If a statistics is used to estimate a parameter, it is called an estimator.
- If we have a population parameter $\rho$, and a sample statistic $R$, such that $\mathbb{E}(R) = \rho$ then we say that $R$ is an *unbiased estimator* for $\rho$.

**Chapter Assessment — Populations and Samples**

Download and sit the 30 minute assessment for this chapter from the digital book.

# 23. Statistics - Presenting and Interpreting Data

Statistical tools and techniques allow us to summarise and simplify large collections of numbers, make predictions and estimate values which are hard to measure directly. The word "statistics" has the same root as the word "state": these tools were developed to help in the running and governing of countries and nations. As probability theory developed and computing power increased, data was collected in greater amounts. Today, statistical techniques are used to make and justify decisions that affect many aspects of our lives.

We see the results of statistical analysis every day. We see them in adverts and news reports. They are used to persuade people to buy everything from a new detergent to a new power station. Sometimes they give a fair and reasonable view of the underlying data: sometimes the analysis is used, deliberately or not, to draw the wrong conclusions. If we have an understanding of the tools and how they should be used, we can make better decisions for ourselves and avoid being misled.

## 23.1 Averages and Variation

A group of friends all live on a very long, straight street. They are deciding where along the street they will meet. They want to keep the amount of walking to a minimum so they agree to meet "in the middle", but they have different opinions about where the middle actually is. One idea is to meet at the point halfway between the furthest two houses. This keeps the furthest any one person has to walk down to a minimum. Another idea is to meet at the house where half the group live on one side and half live on the other. This choice minimises the total distance walked by everyone. Finally, it is suggested that everyone meets at the apartment block where several of the group of friends all live. This minimises the total number of people who have to walk anywhere at all.

**Question**

What are the advantages and disadvantages of these ideas? What would happen to the chosen location if someone new, who lives at the very far end of the street, joins the group?

Notice how a meeting place is chosen:

- First we choose a way to measure the distance to a suggested meeting place, such as"the furthest distance anyone has to travel" or "the sum of the distances travelled".
- Then we select the point for which this distance is smallest. This is our meeting place.

The problem of choosing where to meet is exactly the same problem as choosing the average, or the representative value, or the centre, of a set of data points.

- To minimise the *maximum* distance from the average, we use the *midpoint*.
- To minimise the *total* distance from the average, we use the *median*.
- To minimise the number of data points *not equal* to the average, we use the *mode*.

The mode is only suitable for discrete data where many data points have the same value. Where all, or most, of the data points are different, it does not make sense to ask which is the most common value.

The median is usually found by listing the data in increasing order and then selecting the middle value, rather than by finding the point which minimises the total distance from the average. We need to show that these two definitions are the same. Suppose that there are five friends in the group, living at numbers 12, 16, 20, 24 and 28. For simplicity, suppose that the houses are all equally spaced, so that the distance from house $n$ to house $m$ is $|m - n|$. For each house number $m$, we can plot the total distance walked by all of the friends if they meet up at house number $m$. The graph consists of a series of line segments, as shown in Figure 23.1.

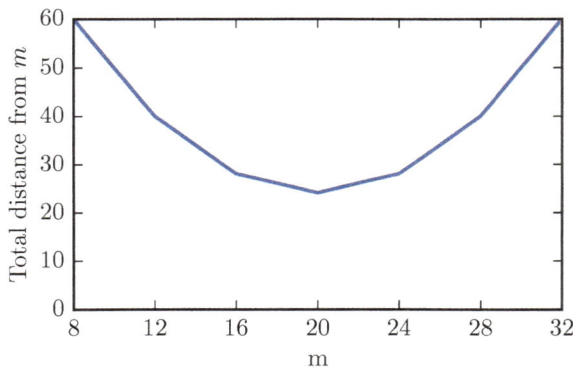

Figure 23.1: Plot of total distance walked.

---

**Question**

The group would like to minimise the total distance walked.

The current suggested meeting place is $m_0$.

For what values of $m_0$ is $m_0 + 1$ a better meeting place than $m_0$? When is $m_0 - 1$ better than $m_0$?

How does this show that the middle data value minimises the total distance walked?

---

The median is not uniquely defined by this method. If a sixth friend, living at number 32, joins the group then the meeting place will change. The total distance travelled by group members is minimised if they meet anywhere between houses 20 and 24. By convention, we define the median to be the midpoint of the two central data points. In our example, the median is 22.

We have not yet mentioned one commonly used average: the mean. It is slightly more complex to derive than the median or the midpoint. Suppose that one of the six friends living on the street is a mathematics student. She favours the idea of meeting at the midpoint, because she does not want anyone to have to walk a very long way. On the other hand, she is concerned that the midpoint ignores almost all of the data. She comes up with an idea.

Another new member joins the group: he lives at number 162. As a result of the new member, all of the averages are pulled towards house number 162. The midpoint is calculated using the extreme points only, and so it jumps out to house number 87. The median is increased just a little, to house number 24.

Our student realises that she does not have to use $|m - n|$ as the measure of distance between two houses. She could choose an expression that makes bigger distances seem very big. One way of doing this is to use $(m - n)^2$, which grows much faster than $|m - n|$. Using this measure of distance in our example, the average moves from 22 to 42. This average has moved further than the median, but not as far as the midpoint. All of the data points are still considered, but the extreme data point has had more effect on this type of average than it had on the median.

Measuring the variation in this way, using the square of the distance, is very widely used. The average we obtain is the *mean*, which is found by summing the values $\{X_1, X_2, \ldots, X_N\}$ and dividing by $N$. We will study this further when we look at the variance and standard deviation of a data set.

---

**Remark**

These questions reintroduce, with a slightly different interpretation, the familiar concepts of midpoint, mode, median and mean and may help students to understand the behaviour of these averages, and their corresponding measures of variation.

The mode is generally not a good representative value for ordered data. It is unchanged by the addition of even the most extreme outlier. Conversely, if the data contains a few outliers with the same value, then this unrepresentative value could be chosen as the average. The mode makes most sense when the data is unordered, such as {strawberry, vanilla, chocolate}, where the "distance" between any two points can indeed be considered to be equal. In our example, it would perhaps make sense if the friends were all determined to drive to their destination, so the cost (in terms of time and effort) of leaving the house was large compared to the distance travelled. The mode does have the advantage that the chosen place is the home of someone in the group (that is, the mode is a value in the data set).

The midpoint (and the range, its associated measure of spread) is more useful when dealing with ordered data but is not very robust: a single outlier can have a large effect on it. On their own, these measures tell us very little about the data.

The median would seem to be the best definition of "middle" in our example. Minimising the total distance walked is a reasonable objective, and the group's new member would not have a disproportionate effect on the chosen meeting place.

The mean is slightly more complex, since the quantity being minimised is not simply a measured distance, but the sum of the *squares* of the measured distance. However, if students can understand that the average value is defined by the choice of the measure of spread, then they may get a more intuitive understanding of why the mean is particularly affected by outliers.

When the data is roughly symmetrical around a single mode, the averages will be almost equal and all give a good measure of the centre of the data. When the data is asymmetrical the averages diverge, and this is why it is important to consider the behaviour of each average when there are outliers in the data.

---

**Formulae 23.1 — Averages**

For a set of data $\{X_1, X_2, X_3, \ldots, X_N\}$ written in increasing order, the following averages are defined.

$$\text{midpoint} = \frac{1}{2}(X_1 + X_N),$$

$$\text{mean} = \frac{1}{N} \sum_{i=1}^{N} X_i,$$

$$\text{median} = \begin{cases} X_n & \text{for odd } N = 2n - 1, \\ \frac{1}{2}(X_n + X_{n+1}) & \text{for even } N = 2n. \end{cases}$$

mode – The most frequent value in the data set.

Data sets with a single mode are called *unimodal*.

---

The *variation* of the data is the amount that it is spread out from the average. The more the data is spread out from the average, the greater the variation.

---

**Question**

Which of the different averages might be appropriate for the following types of data?

- Time per day spent watching television.
- Favourite social networking site.
- Number of cars per household.
- Time spent travelling to place of work.
- Ranking of different schools from "outstanding" to "inadequate".

**Use of Technology 23.1 — Calculating Summary Data Statistics**

Many calculators now allow the user to input lists of data and calculate summary statistics.

## 23.2  Bar Charts

A researcher asks a class of students how many people under the age of eighteen live in their household. He asks us to analyse the data for him, and he gives us the following data set:

$$\{1, 2, 2, 2, 3, 1, 1, 2, 2, 4, 1, 2, 2, 3, 1, 2, 1, 2, 1, 2, 1, 3, 1, 2, 2, 1, 3, 12, 2, 1, 3\}.$$

**Activity 23.1**

The data set has been presented as a raw list of values.

Q1. What can be said about the data?

Q2. What properties are easier to see? Which are harder?

Q3. What comments might we make to the researcher?

Our researcher checks and corrects his results. He presents them again, this time in the form of a bar chart, see Figure 23.2.

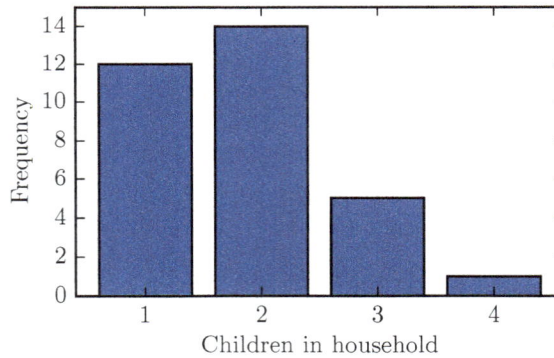

Figure 23.2: Bar chart showing number of children in a household.

**Question**

What are the advantages of this presentation of the data?

> **Tip**
> Bar charts are used for discrete data. They are not suitable for showing continuous data.
> Always put a gap between each bar, and make all the bars the same width.

Bar charts are a way of communicating discrete data simply and clearly. For example, they show us where the mode of the data is. Sometimes data has multiple distinct peaks. If these peaks are clearly separate from each other, we can call each peak a local mode. However, we can assume that all of the data sets we will analyse will have just one mode. Bar charts also help us to identify the variation of the data and any outliers. An outlier is a value which is a long way from the average value, compared to the variation.

**Example 23.1**

Consider the data shown in the bar chart, representing the number of skiing holidays taken within a ten year period by members of a sample group. Identify any modes and outliers in the data.

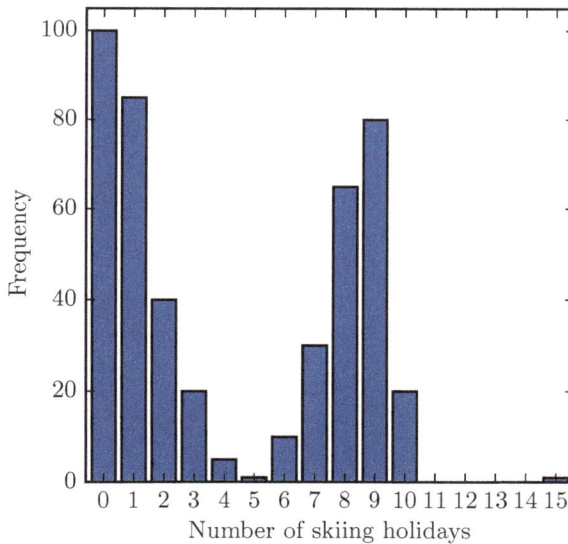

**Solution:**

We can see that the overall mode of the data is zero. However, there is another, smaller peak in the data representing those people who took nine holidays in the previous ten years. We would describe this data as *bimodal*: it has two modes. Even though the nine-year peak is lower than the other, it is a clear local maximum and is still referred to as a mode.

This sample contains just one outlier, representing someone who took more than one holiday a year.

In this example, although it may not have been apparent to the person conducting the survey, the sample consists of two populations: people who ski regularly and those who do not. The infrequent skiers either never skied, or did so just a few times. The committed skiers went almost every year. These features would be difficult to see from the raw data, but the bar chart brings them out very clearly.

**Exercise 23.1**

Q1. For the bar chart in Figure 23.2, calculate the mode, the median and the mean number of children per household.

Q2. One hundred householders were asked to state the main fuel source for their hot water. Fifty-seven respondents had gas fired boilers, 32 used an electric heater, three had a liquefied petroleum gas (LPG) system and the rest used domestic heating oil.

Draw a bar chart showing these results.

Q3. Draw a bar chart of the following exam results. What is the median grade?

| Grade | A* | A | B | C | D |
|---|---|---|---|---|---|
| Frequency | 61 | 131 | 184 | 223 | 103 |

Q4. The number of people in each car that passes a given point are recorded. After recording data for the first seventeen cars we notice that:
- no car has more than three occupants;
- the mode is less than the median;
- the number of cars whose occupants equalled the mode was twice the number of cars whose occupants equalled the median.

Draw a bar chart to illustrate these findings.

Q5. Identify any modes and outliers in the bar charts shown below. Compare the variation of the two unimodal data sets.

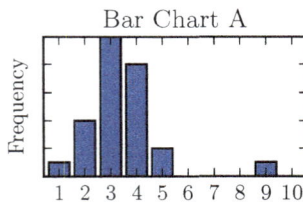

Bar Chart A

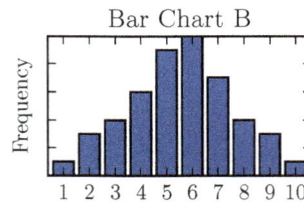

Bar Chart B

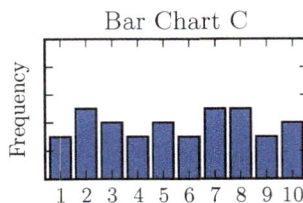

Bar Chart C

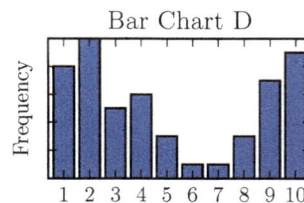

Bar Chart D

## 23.3  Dot plots

Dot plots are an alternative way of presenting small data sets. Instead of representing the frequency with a bar, we use a column of individual dots. Each dot represents a single data point.

Suppose we have a bar chart showing the frequencies $\{1, 3, 5, 9, 7, 6, 3, 2\}$.

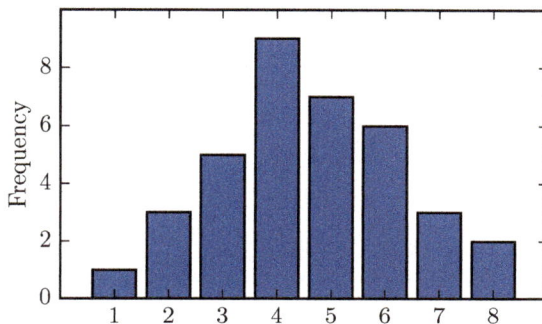

The exact frequency for a given value may not be obvious when we look only at the bar itself. For example, if we try to see the frequency of the value 2 in this bar chart, we would probably glance over at the scale on the $y$-axis to confirm that the frequency is three.

If we draw a dot plot of the same data, we can see immediately that there are three dots corresponding to the value 2.

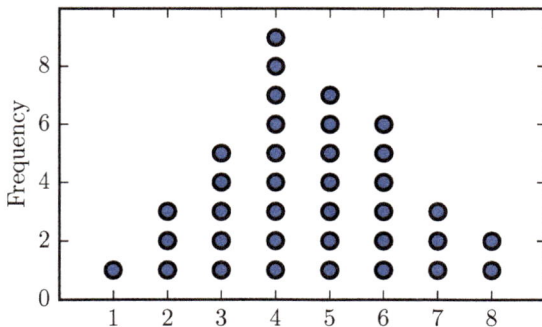

For larger frequencies, the exact value is not quite as clear. We would probably refer back to the scale to check that the frequency of the value 4 is equal to nine. However, the difference between adjacent values can still be clearly seen. We can easily see that there is a difference of two between the frequencies of 4 and 5.

**Tip**

If the data is discrete, and the data set is small, consider whether a dot plot or a bar chart would be easier to read.

**Remark**

Dot plots have other advantages over bar charts.

Since we have a separate dot for each data point, it is possible to attach additional information to the dots. For example, dots representing countries could be labelled with country codes such as UK, FR and DE. Alternatively, additional data could be indicated with dots of different colours. For example, in a dot plot showing exam results of a small sample, different colours could be used to represent different genders.

In newspapers, and other settings where a less formal chart is appropriate, the dot can be replaced with other symbols. For example, where the data represents the number of people in different categories, small stick figures could be used.

Presenting data is a form of communication. We should choose the type of chart which will communicate the correct level of detail as clearly and easily as possible.

**Example 23.2**

A local company monitors intruder alarms at business and residential properties. Over the course of a week, they record the number of times they are called out to an incident.

| Day | Mon | Tues | Wed | Thur | Fri | Sat | Sun |
|---|---|---|---|---|---|---|---|
| Business | 4 | 2 | 2 | 1 | 3 | 1 | 0 |
| Residential | 2 | 2 | 3 | 2 | 5 | 4 | 3 |

(a) Plot this data on a single dot plot, using different colours or styles to distinguish between business and residential callouts.

(b) Identify the company's busiest day from the data given.

**Solution:**

(a) In this dot plot, business callouts are shown in blue, and residential ones in red.

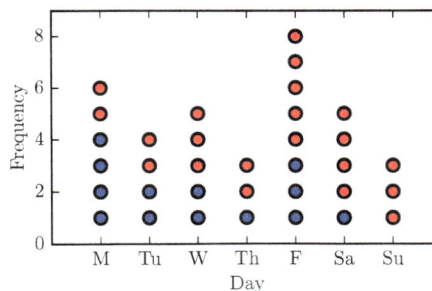

(b) We can see from the dot plot, that Friday had the largest number of callouts.

**Exercise 23.2**

Q1. Show the data given in this bar chart in the form of a dot plot.

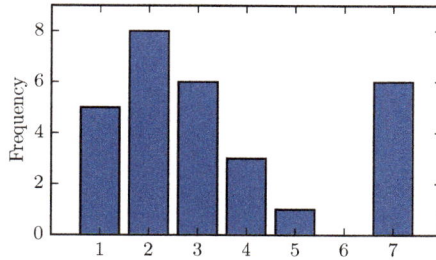

Q2. The number of people in an office who cycled to work over the course of a week was recorded. These are the results.

| Day | Mon | Tues | Wed | Thur | Fri |
|-----|-----|------|-----|------|-----|
| Count | 6 | 4 | 3 | 5 | 2 |

(a) Show these results in a dot plot.
(b) Calculate the mean number of people who cycle to work each day.
(c) Give the day of the week for which the frequency is equal to the median.

Q3. A class of students is given a short test. These are their scores.

| Boys | 4 | 5 | 3 | 9 | 3 | 1 | 4 |
|------|---|---|---|---|---|---|---|
|      | 5 | 3 | 7 | 4 | 5 | 4 | 6 |
| Girls | 5 | 3 | 4 | 2 | 7 | 8 | 10 |
|       | 6 | 2 | 7 | 6 | 3 | 9 | 4 |

(a) Plot this data on a single dot plot, using different colours or styles to distinguish between the results for the boys and the girls.
(b) State the modal test score.
(c) Give the median score for the girls in the class.

Q4. A delivery company offers a three day delivery service. A quality inspector chose a random sample of parcels and tabulated the actual delivery time.

| Time (days) | 1 | 2 | 3 | 4 | 5 | 6 |
|-------------|---|---|---|---|---|---|
| Frequency | 4 | 7 | 9 | 4 | 0 | 1 |

(a) Show this data on a dot plot
(b) Calculate the mean delivery time.
(c) What percentage of parcels were delivered within the three day target?
(d) Explain why the mean is not a useful measure of the company's performance and suggest a better measurement.

## 23.4　Stem-and-leaf Plots

Statistical techniques are designed to summarise data. If the range is continuous, or if there are many more values in the range than there are data points, then a bar chart will simply show a series of very short bars indicating where each data point happens to fall. This is not a helpful way of presenting the data: there is too much detail.

For example, suppose that we interviewed shoppers walking down a particular street and asked them to count the loose change in their pockets. In theory, we will obtain a discrete data set which could be shown in a bar chart. In practice, showing the number of people who had exactly £2.43 is not relevant and the bar chart will be unhelpful in illustrating the overall shape of the data. An example of this is shown in Figure 23.3.

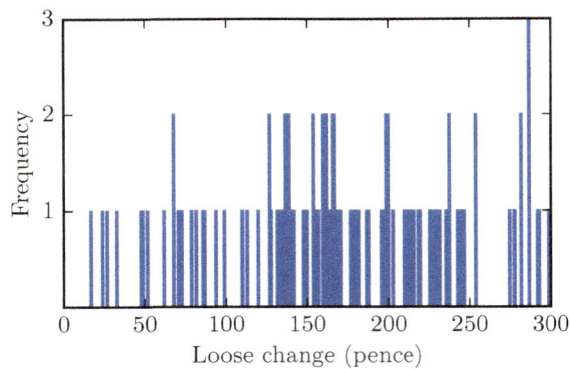

Figure 23.3: An unhelpful bar chart.

We need to group the data points together in some way so that we can see the general pattern. If the data set is not too large, then a simple way of representing the data, while still keeping some of the detail, is by using a *stem-and-leaf plot*.

To create the plot, we write the data in order to a selected level of accuracy (for example, to two decimal places or to the nearest hundred). Then we list the values in a column, leaving out the last significant digit. Finally, for each data point, we add in its last significant digit along the row corresponding to the value of that data point.

---

**Example 23.3**

A shop manager records the number of customers between 12pm and 1pm over twenty days. She obtains the following data.

$$\{135, 152, 121, 141, 168, 122, 148, 145, 118, 133,$$
$$148, 110, 140, 186, 167, 134, 156, 110, 145, 131\}$$

Represent this as a stem-and-leaf plot. What is the median value?
The manager collects another set of data six months later.

$$\{115, 161, 149, 138, 142, 173, 166, 162, 127, 153,$$
$$157, 172, 146, 146, 135, 158, 168, 151, 133, 158\}$$

---

Make a back-to-back stem-and-leaf plot of the two data sets. What is the median of the new data?

**Solution:**
We obtain the following stem-and-leaf plot, where 11|0 represents the value 110.

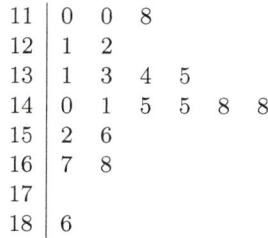

```
11 │ 0  0  8
12 │ 1  2
13 │ 1  3  4  5
14 │ 0  1  5  5  8  8
15 │ 2  6
16 │ 7  8
17 │
18 │ 6
```

The middle two values are easily read from a stem and leaf plot. They are 140 and 141. The median is given by the mean of these two values and so is equal to 140.5.

To draw a back-to-back stem-and-leaf plot, we put the new data set on the left hand side of the same set of rows as the first set.

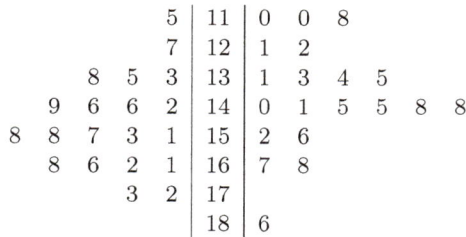

```
            5 │ 11 │ 0  0  8
            7 │ 12 │ 1  2
        8 5 3 │ 13 │ 1  3  4  5
        9 6 6 2 │ 14 │ 0  1  5  5  8  8
    8 8 7 3 1 │ 15 │ 2  6
      8 6 2 1 │ 16 │ 7  8
          3 2 │ 17 │
              │ 18 │ 6
```

From this plot, we can see that the number of customers has increased since six months ago. This is confirmed by calculating the median of the new data. The middle two values are now 151 and 153, and so the new median is 152.

---

### Activity 23.2 — Presentation of Data 1

Suppose that our researcher from earlier went back to the class of students and asked them to tell him their household income. These are his results:

{13 715, 14 850, 19 952, 23 100, 25 000, 26 527, 26 527, 26 800,

28 331, 29 577, 30 000, 30 000, 31 072, 31 692, 32 833, 34 950,

35 300, 36 452, 36 560, 37 405, 37 752, 40 900, 41 011, 43 582,

44 000, 45 500, 47 600, 47 750, 52 078, 58 000, 85 000, 95 000}.

Comment on the quality and presentation of this set of data.

**Tip**

There must be no gaps between the rows in a stem-and-leaf plot. In the first plot in the example above, there are no values in the range 170 to 179 and so the line 17| is included with no leaves listed against it.

**Exercise 23.3**

Q1. (a) Use the household income data given above to draw a stem-and-leaf plot. Round the values to the nearest £1 000 to draw the plot.

   (b) Use the plot to provide an estimate of the median income.

Q2. A hospital records the weight of newborn babies in kg. Draw a stem-and-leaf plot of the data.

$$\{3.06, 3.08, 3.14, 3.16, 3.17, 3.19, 3.20, 3.20, 3.21,$$
$$3.23, 3.26, 3.28, 3.32, 3.38, 3.38, 3.44, 3.49, 3.60\}.$$

Q3. A monitoring station measures the amount of nitrogen dioxide in the air. Draw a stem-and-leaf plot for the following readings taken at noon on consecutive days, given in $\mu g\, m^{-3}$.

$$\{103, 124, 100, 109, 82, 84, 112, 63, 91, 113, 104, 91, 108, 98\}$$

Calculate the median and the mean of the data.

Q4. Present the results of the 2016 Olympics men's 100 m semi-final results as a stem-and-leaf plot.

| Heat | Results (s) | | | | | | | |
|------|------|------|------|------|------|------|------|------|
| 1 | 9.94 | 10.01 | 10.07 | 10.08 | 10.13 | 10.16 | 10.17 | |
| 2 | 9.86 | 9.92 | 10.01 | 10.01 | 10.05 | 10.12 | 10.13 | |
| 3 | 9.95 | 9.97 | 9.98 | 10.03 | 10.05 | 10.08 | 10.11 | 10.23 |

Give the mean and the median of the data, for simplicity, assume $\sum_i X_i = 221.06$.

Q5. A group of volunteers is given a tonic which claims to increase mental agility. Another group is given a placebo. After a month, the two groups take a logic puzzle test.

Show the scores in a back-to-back stem-and-leaf plot and calculate the median for each group.

| Placebo group | | | | | | Tonic group | | | | | |
|------|------|------|------|------|------|------|------|------|------|------|------|
| 84 | 86 | 90 | 91 | 96 | 103 | 91 | 94 | 102 | 102 | 108 | 110 |
| 104 | 108 | 112 | 113 | 114 | 118 | 114 | 115 | 119 | 122 | 122 | 123 |
| 118 | 123 | 124 | 127 | 130 | 132 | 124 | 127 | 128 | 130 | 133 | 138 |
| 143 | 147 | 154 | | | | 139 | 140 | 141 | | | |

## 23.5 Histograms

Stem-and-leaf plots are only suitable when the data set is fairly small, since every data point is represented by an individual digit in the plot. A stem-and-leaf plot with hundreds of data points would be unwieldy. In cases like this, it can be more helpful to present the data as a *histogram*.

To create a histogram, we first divide the data into smaller ranges. These smaller ranges are often called *classes* or *bins*. When we constructed our stem-and-leaf plots in the previous section, we formed classes by taking the stem of each data value. For a histogram, we can choose the class boundaries in any way that gives a suitable presentation of the data. When we have decided on the classes, we write down the frequencies: the number of values that occur in each class. For our household income data we might split the data into evenly spaced classes, as shown in Table 23.1.

| Income (£000s) | Frequency |
|---|---|
| $10 \leq x < 20$ | 3 |
| $20 \leq x < 30$ | 7 |
| $30 \leq x < 40$ | 11 |
| $40 \leq x < 50$ | 7 |
| $50 \leq x < 60$ | 2 |
| $60 \leq x < 70$ | 0 |
| $70 \leq x < 80$ | 0 |
| $80 \leq x < 90$ | 1 |
| $90 \leq x < 100$ | 1 |

Table 23.1: Frequency table with equal class widths.

The histogram given by these classes is shown in Figure 23.4.

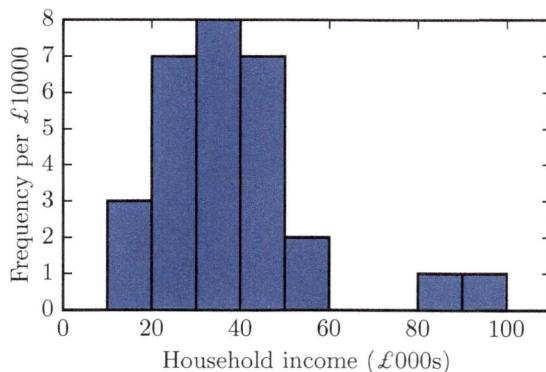

Figure 23.4: Histogram of household income with equal class sizes.

**Remark**

Histograms whose bars are all the same width look similar to bar charts, but there are important differences.

The $x$-axis of a bar chart consists of a list of distinct values. Each bar represents a single value and its height indicates how many data points have that value.

We can see in the histogram above that the $x$-axis represents a continuous range of values. Each bar represents a range of values (for example, £40 000 to £50 000) and the height represents the number of data points which take any of the values in that range. It is easy to tell what type of chart we are dealing with – a bar chart has gaps between the bars and each one is labelled with a single value; there are no gaps in the values in a histogram and each bar covers a range of values.

**Remark**

Our goal in choosing the classes is to simplify the data, but not to over-simplify it. If the classes are too big, our histogram will have very little detail in it. If they are too small, we will not be able to see the overall shape of the data. We can see both kinds of problem in the histograms shown here. These histograms all show the same data. For comparison, the first histogram shows the results when the classes are evenly spaced.

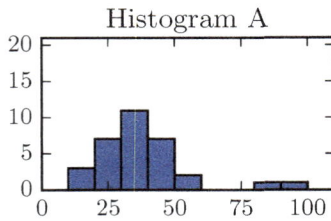

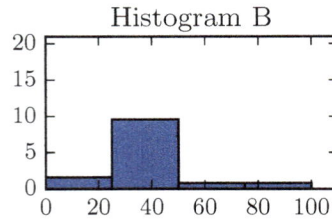

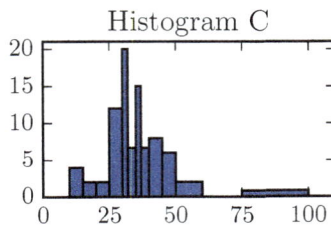

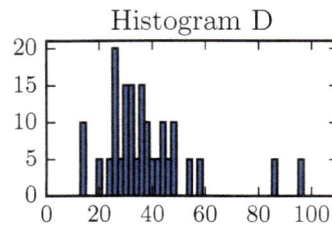

One of the ways in which histograms are more flexible than stem-and-leaf plots is that the class widths do not have to be equal. This can help us to provide more detail in parts of the range where there are many data points, without also having the same level of detail in places where the data is more spread out, such as at the edges of the range.

Whenever we have a histogram with unequal class widths, we must remember that the $y$-axis represents the frequency density, not the total frequency. As an example, suppose that we decided to merge the highest two classes in Table 23.1. We replace the classes from 80 to 90 and from 90 to 100 with a single class covering the range 80 to 100. The last two lines of the table would be replaced with a single line:

| $80 \leq x < 100$ | 2 |
|---|---|

This change to the class boundaries will give us a different histogram, with a single bar from 80 to 100 along the $x$-axis to represent the two data points in this class. If we gave this bar a height of two (on the grounds that the frequency of that class is two), it would appear that the number of outliers had doubled. The purpose of the histogram is to give a visual representation of the data, but doubling the area of the bar makes these data points much more prominent. To avoid this problem, the height of the bar is always proportional to the *frequency density*. The frequency density is found by dividing the total frequency by the width of the class.

> **Tip**
> When drawing a histogram, remember that the $y$-axis represents the density of data points.
> The *area* of each bar is proportional to the total number of data points in that class.

Some popular spreadsheet software programs are very good at drawing bar charts, but do not support histograms. If the software is not designed to produce them, the results can be misleading. We must always ensure that the $y$-axis represents the frequency density, that the area of the bar represents the frequency and that adjacent classes have no gaps between them.

**Example 23.4**
Draw a histogram with class boundaries at 10, 20, 30, 35, 40, 45, 50, 60 and 100 for the data shown in the following stem-and-leaf plot.

```
1 | 3  5
2 | 3  4  7  9
3 | 1  1  4  5  6  7  7  8
4 | 0  2  2  3  4  8  9
5 | 2  4  6
6 |
7 | 0
8 |
9 | 2
```

**Solution:**

We start by calculating the frequencies for each class and then the frequency densities. The density is found by dividing the total frequency by the width of the class.

| Range | Frequency | Width | Frequency density |
|---|---|---|---|
| $10 \leq x < 20$ | 2 | 10 | 0.2 |
| $20 \leq x < 30$ | 4 | 10 | 0.4 |
| $30 \leq x < 35$ | 3 | 5 | 0.6 |
| $35 \leq x < 40$ | 5 | 5 | 1.0 |
| $40 \leq x < 45$ | 5 | 5 | 1.0 |
| $45 \leq x < 50$ | 2 | 5 | 0.4 |
| $50 \leq x < 60$ | 3 | 10 | 0.3 |
| $60 \leq x < 100$ | 2 | 40 | 0.05 |

The height of each bar is proportional to its frequency density.

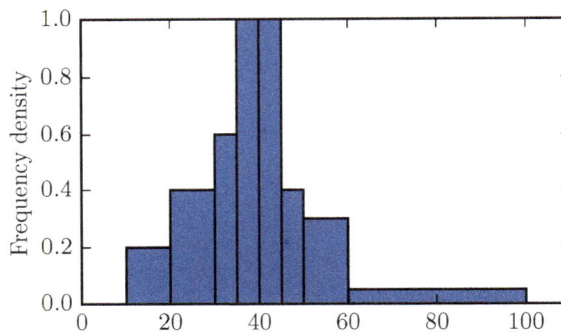

**Exercise 23.4**

Q1. Plot a histogram of the following data.

| Range | Frequency |
|---|---|
| $0 \leq x < 10$ | 3 |
| $10 \leq x < 20$ | 7 |
| $20 \leq x < 25$ | 11 |
| $25 \leq x < 35$ | 15 |
| $35 \leq x < 40$ | 6 |
| $40 \leq x < 50$ | 6 |
| $50 \leq x < 70$ | 6 |
| $70 \leq x < 100$ | 3 |

Q2. A histogram of the following data has been drawn.

| Range | Frequency |
|---|---|
| $0 \leq x < 20$ | 24 |
| $20 \leq x < 30$ | 28 |
| $30 \leq x < 40$ | $x$ |
| $40 \leq x < 50$ | 16 |
| $50 \leq x < 70$ | 16 |
| $70 \leq x < 100$ | $y$ |

The height of the first bar is 6 cm. The width of the final bar is 1.5 cm. There are 120 data points in total.

(a) What is the height of the bar representing the range 20 to 30?
(b) The bar representing the range 30 to 40 is 12 cm high. Find $x$.
(c) What is the width of bar representing the range 50 to 70?
(d) What is the height of the final bar?

Q3. A company tracks how long visitors spend on their website and records the following data.

| Range (s) | Frequency |
|---|---|
| $0 \leq x < 5$ | 18 673 |
| $5 \leq x < 10$ | 16 131 |
| $10 \leq x < 15$ | 17 097 |
| $15 \leq x < 30$ | 20 848 |
| $30 \leq x < 60$ | 17 204 |
| $60 \leq x < 120$ | 10 047 |

Plot a histogram of this data.

Q4. In 1848, at the request of a Parliamentary commission, Mr William Miller, a clerk at the Bank of England, "weighed singly, with the weighing machines in use at the Bank, several thousands of new sovereigns". Each gold coin was supposed to weigh 123.274 grains (about 8 g). Here is a summary of the deviations from the required weight that Mr Miller recorded.

| Deviation (grains) | Cumulative Frequency |
|---|---|
| $d < -0.265\,25$ | 209 |
| $-0.265\,25 \leq d < -0.2$ | 571 |
| $-0.2 \leq d < -0.1$ | 1832 |
| $-0.1 \leq d < 0$ | 4855 |
| $0 \leq d < 0.1$ | 7486 |
| $0.1 \leq d < 0.2$ | 9219 |
| $0.2 \leq d < 0.265\,25$ | 9755 |
| $0.265\,25 \leq d$ | 10 000 |

Table 23.2: Deviation of sovereigns from standard weight

The minimum legal weight of a sovereign was 122.5 grains. In his evidence to the Commissioners, Mr Miller said that the lightest coin weighed "0.207, nearly 0.208, about 0.207" grains above this minimum. Choose suitable limits for the outside classes and plot a histogram of his findings. State any assumptions made.

**Remark**

The deviation of $\pm 0.265\,25$ grains in Table 23.2 represents the allowable "remedy". Coins produced by the Mint were selected at random and put into a box, called the pyx. These coins were then tested at the Trial of the Pyx, a ceremony which began in the twelfth century and is still held today. The Master of the Mint could be put "at the Prince's mercy or will in life and members" if the coins being tested deviated by more than the remedy.

We can see from Mr Miller's data that roughly 5% of the coins in 1848 had deviation greater than the remedy. However, since many coins are weighed at once, the probability of the mean deviation being greater than the remedy was very small. A reasonably competent and honest Master of the Mint would be in little danger.

The data is taken from the Report of the Commissioners Appointed to Inquire into the Constitution, Management and Expense of the Royal Mint (published by William Clowes and Sons, London for Her Majesty's Stationery Office, 1849).

## 23.6  Linear Interpolation

Sometimes we will be working with data sets which include the complete set of exact values: other times we must deal with data which has already been grouped into classes. This might be because the data set is very large, so that the complete set is difficult to analyse. Or it may be that the data was only collected to a certain level of accuracy in the first place. In this case, we need a technique for analysing the data, even though we do not know all of the actual values.

Suppose that we have the following data set.

| Range | Frequency | Cumulative frequency |
|---|---|---|
| $0 \leq x < 10$ | 2 | 2 |
| $10 \leq x < 20$ | 4 | 6 |
| $20 \leq x < 30$ | 7 | 13 |
| $30 \leq x < 40$ | 10 | 23 |
| $40 \leq x < 50$ | 15 | 38 |
| $50 \leq x < 60$ | 23 | 61 |
| $60 \leq x < 70$ | 18 | 79 |
| $70 \leq x < 80$ | 9 | 88 |
| $80 \leq x < 90$ | 7 | 95 |
| $90 \leq x < 100$ | 5 | 100 |

The third column shows the cumulative frequency: the number of data points in all of the classes up to and including the current row. We can plot these cumulative frequency points on a graph and then draw a smooth curve between them, as shown in Figure 23.5.

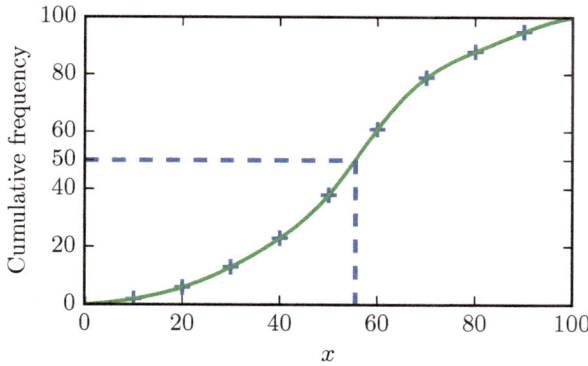

Figure 23.5: Cumulative frequency with freehand interpolation.

If we plot the line $y = 50$, we can see that it intersects the curve at approximately $x = 55$. That is, we estimate that there are 50 data points less than $x = 55$. Therefore there are also $100 - 50 = 50$ data points greater than $x = 55$. We know that the median is the value which divides the data points into two equal halves, and so this cumulative frequency curve gives us $x = 55$ as an estimate of the median.

It is only an estimate because we do not know the cumulative frequency at all of the points. We have plotted the points we do know, and estimated the others. The process of filling in unknown points based on known values is called *interpolation*. Here, we completed the graph by drawing a smooth curve between the known points. Another method is to join them with straight lines. This is called *linear interpolation*. We can use linear interpolation to calculate an estimate for the median, just as we did with the freehand curve.

Suppose that we have the following data set.

| Range | Frequency | Cumulative frequency |
|---|---|---|
| $0 \leq x < 10$ | 55 | 55 |
| $10 \leq x < 20$ | 182 | 237 |
| $20 \leq x < 30$ | 410 | 647 |
| $30 \leq x < 40$ | 684 | 1331 |
| $40 \leq x < 50$ | 522 | 1853 |
| $50 \leq x < 60$ | 283 | 2136 |
| $60 \leq x < 70$ | 185 | 2321 |

To draw the cumulative frequency curve using linear interpolation, we first plot the known points: $(0, 0)$, $(10, 55)$, $(20, 237)$, $(30, 647)$, $(40, 1331)$, $(50, 1853)$, $(60, 2136)$ and $(70, 2321)$. Then we complete the graph by joining those points with straight lines. This is our estimate of the cumulative frequency of the data over all values of $x$ and is shown in Figure 23.6.

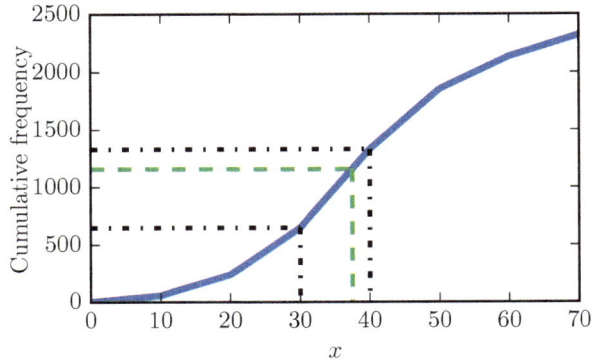

Figure 23.6: Cumulative frequency with linear interpolation.

Since there are 2321 data points, the median is given by the value at position 1160. The cumulative frequency at $x = 30$ is 647. At $x = 40$ it is 1331. Therefore, the median lies somewhere in the 30 to 40 class.

To estimate the median value, we find the value of $x$ when the cumulative frequency $y = 1160$. This is on the line segment which passes through $(30, 647)$ and $(40, 1331)$. Since 1160 is three quarters of the way between 647 and 1331, when $y = 1160$ then $x$ must be three quarters of the way between 30 and 40. This gives us $x = 37.5$ as our estimated value for the median.

---

**Tip**

Always check which kind of interpolation you have been asked to use: linear interpolation or a smooth curve.

---

We can calculate the estimate of the median more formally as follows. If a line passes through the points $(x_1, y_1)$ and $(x_2, y_2)$ then $(y_2 - y_1) = m(x_2 - x_1)$ where $m$ is the gradient of the line. This means that if the line also goes through the point $(x_3, y_3)$ then

$$m = \frac{(y_2 - y_1)}{(x_2 - x_1)} = \frac{(y_3 - y_1)}{(x_3 - x_1)}.$$

If we take $(x_1, y_1) = (30, 647)$ and $(x_2, y_2) = (40, 1331)$, then we can use this to find $(x_3, 1160)$. We have

$$\frac{40 - 30}{1331 - 647} = \frac{x_3 - 30}{1160 - 647} \Rightarrow \frac{10}{684} = \frac{x_3 - 30}{513}$$

$$\Rightarrow \frac{10 \times 513}{684} = \frac{15}{2} = x_3 - 30,$$

which gives us our estimated median of 37.5.

**Example 23.5**

Use linear interpolation to estimate the median of the following data.

| Range | Frequency |
|---|---|
| $0 \le x < 400$ | 1 |
| $400 \le x < 800$ | 4 |
| $800 \le x < 1000$ | 4 |
| $1000 \le x < 1100$ | 5 |
| $1100 \le x < 1200$ | 6 |
| $1200 \le x < 1300$ | 4 |
| $1300 \le x < 1500$ | 6 |
| $1500 \le x < 1800$ | 3 |

First, we will extend the table with a column for the cumulative frequency.
From this, we can plot the cumulative frequency graph for this data as show below. The graph also indicates the method we will use for estimating the median.

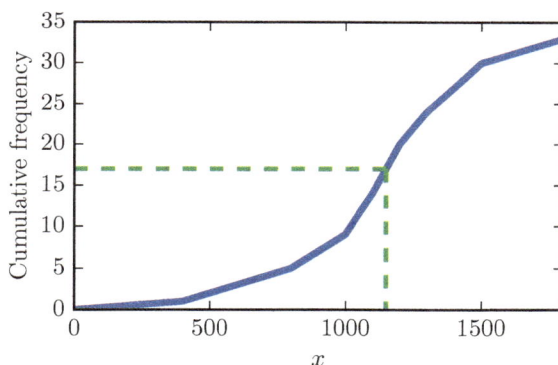

If we take the ordered data values to be $\{X_1, X_2, \ldots X_{33}\}$ then our estimate for the median will be our estimate of the data value $X_{17}$. This lies in class $1100 \le x < 1200$. Using linear interpolation, our estimate of the cumulative frequency in this range is a straight line between $(1100, 14)$ and $(1200, 20)$. The value $y = 17$ lies in the middle of this line segment, and therefore our estimate for the median is 1150.

Alternatively, using the equation $(y_2 - y_1) = m(x_2 - x_1)$ with the points $(x_1, y_1) = (1000, 14)$ and $(x_2, y_2) = (1100, 20)$, we can calculate the point $(x_3, 17)$.

$$\frac{20 - 14}{1200 - 1100} = \frac{17 - 14}{x_3 - 1100} \Rightarrow \frac{3}{50} = \frac{3}{x_3 - 1100}$$

$$\Rightarrow 50 = x_3 - 1100 \Rightarrow x_3 = 1150.$$

**Exercise 23.5**

Q1. For each data set, draw a cumulative frequency graph, interpolating the data with a smooth curve. Use the graphs to estimate the median of each data set. Which estimate is likely to be more accurate?

| Range | Data set A Frequency | Data set B Frequency |
|---|---|---|
| $0 \leq x < 25$ | 80 | 2435 |
| $25 \leq x < 50$ | 134 | 1354 |
| $50 \leq x < 75$ | 186 | 843 |
| $75 \leq x < 100$ | 497 | 244 |
| $100 \leq x < 125$ | 1534 | 358 |
| $125 \leq x < 150$ | 3975 | 1358 |
| $150 \leq x < 175$ | 2865 | 2079 |
| $175 \leq x < 200$ | 729 | 1329 |

Q2. Use linear interpolation to estimate the median of the following data.

| Range | Frequency |
|---|---|
| $0 \leq x < 100$ | 23 |
| $100 \leq x < 200$ | 107 |
| $200 \leq x < 250$ | 284 |
| $250 \leq x < 300$ | 435 |
| $300 \leq x < 400$ | 348 |
| $400 \leq x < 500$ | 231 |
| $500 \leq x < 600$ | 95 |

Q3. Use linear interpolation to estimate the median of the following data.

| Range | Cumulative Frequency |
|---|---|
| $0 \leq x < 4$ | 28 |
| $4 \leq x < 8$ | 162 |
| $8 \leq x < 12$ | 418 |
| $12 \leq x < 16$ | 598 |
| $16 \leq x < 20$ | 784 |
| $20 \leq x < 24$ | 836 |

Q4. Estimate the median weight (in grains) of a sovereign coin, using the data in Table 23.2.

## 23.7  Box-and-whisker Plots

Histograms are ideal for showing the shape of a data set, but sometimes we want to show less detail. For example, we might want to compare different data sets side by side. One helpful way of presenting the data in a more compact form is to use the *box-and-whisker plot*, sometimes simply called a box plot.

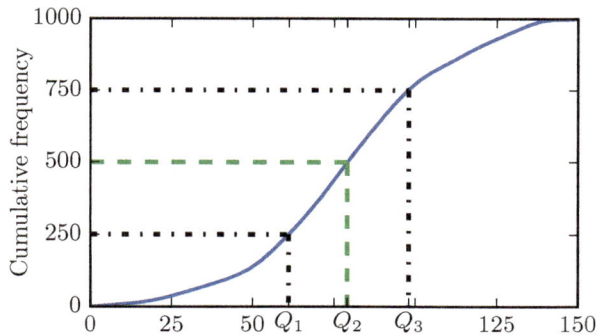

Figure 23.7: Cumulative frequency showing quartiles.

A box plot shows the median value and gives an idea of the spread of the data around the average by showing the position of the *quartiles*. The quartiles divide the ordered data into four equal parts, in the same way that the median divides it into two equal parts. The larger the gap between the quartiles, the bigger the spread of the data. Figure 23.7 shows a cumulative frequency graph with the values of the three quartiles $Q_1$, $Q_2$ and $Q_3$ marked.

The first quartile, $Q_1$, lies one quarter of the way along the data and is also called the lower quartile. The second quartile, $Q_2$, is the median. The third (or upper) quartile, $Q_3$, lies three-quarters of the way along the data.

In Figure 23.8 we have presented the summary statistics from Figure 23.7 in a box plot. The range of the data is 10 to 140 and the quartiles are $Q_1 = 61$, $Q_2 = 79$ and $Q_3 = 98$.

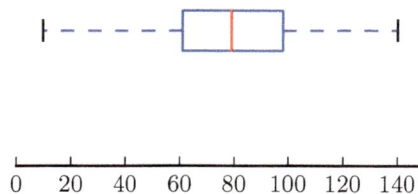

Figure 23.8: Range and quartiles shown on box-and-whisker plot.

**Tip**

The simplest type of box plot indicates five values:
- the lowest point in the range;
- $Q_1$, the lower quartile;
- $Q_2$, the median;
- $Q_3$, the upper quartile;
- and the highest point in the range.

The box stretches from the lower to the upper quartile and shows the middle half of the data. The size of the box indicates the spread of the data in a way which does not change very much when extreme values are added to the data set. The longer the box, the greater the spread of the data. The length of the box is called the *interquartile range*. Given two similar distributions, we can use this value to compare the spread of each data set around the median.

**Definition 23.1 — Interquartile Range**

The interquartile range is equal to $Q_3 - Q_1$, the distance between the upper and lower quartiles.

The line in the middle of the box shows the position of the median, $Q_2$, and the whiskers extend either side of the box to the show the full range of the data.

**Remark**

Each whisker, and each section of the box, contains one quarter of the data. The smaller the section or whisker, the greater the density of data points in that part of the range.

We already know how to calculate the median value of a data set. We can calculate the other two quartiles by finding the median of the lower and upper halves of the data.

**Example 23.6**

Draw a box plot for the following data, and calculate the interquartile range.

$$\begin{array}{cccccccccc} 12 & 21 & 23 & 30 & 34 & 38 & 40 & 42 & 43 & 48 \\ 50 & 52 & 58 & 62 & 63 & 67 & 82 & 84 & 88 & 94 \end{array}$$

**Solution:**

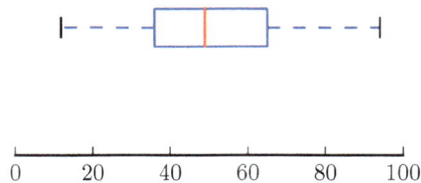

The median of the data is given by the mean of the two central values, 48 and 50, and so $Q_2 = 49$. The lower quartile is the median of the lower half of the data, so $Q_1 = 36$. Similarly, $Q_3$ is equal to 65, the median of the upper half of the data. The interquartile range is equal to $Q_3 - Q_1 = 65 - 36 = 29$.

Now that we have a definition of an average (the median) and a corresponding variation (the interquartile range), we can use this to decide when data points are sufficiently far from the average to be classed as *outliers*.

One common definition for outliers is to say that when we draw a box plot, we will never draw whiskers which are more than one and a half times the length of the box. Any data points which lie further from the box than this are plotted as individual data points.

**Tip**
The whiskers of a box-and-whisker plot must finish at data points. If there are outliers, draw the whiskers out to the furthest point which is not an outlier, then plot each outlier separately.

**Definition 23.2 — Outliers**
A data point $X$ will often be considered to be an outlier if

$$X < Q_1 - \frac{3}{2}(Q_3 - Q_1) \qquad \text{or} \qquad X > Q_3 + \frac{3}{2}(Q_3 - Q_1),$$

where $Q_1$ and $Q_3$ are the lower and upper quartiles, and $Q_3 - Q_1$ is the interquartile range.

**Remark**
Remember that different definitions of outliers are possible. For example, instead, we could consider points to be outliers if they are more than twice the interquartile range below $Q_1$ or above $Q_3$. In this case, outliers would be any points where

$$X < Q_1 - 2(Q_3 - Q_1) \qquad \text{or} \qquad X > Q_3 + 2(Q_3 - Q_1).$$

Make sure it is clear which definition you are using.
If a particular definition is specified, make sure that you use it.

We have seen that a box plot shows us how far the data spreads out from the median. It also shows us whether this spread is symmetrical. A unimodal data distribution has a peak with a *tail* on each side. If the data set is perfectly symmetrical, then the two tails will be the same length and have the same density of data points. In this case, when we draw a box plot, the median will divide the box into two equal sections.

Now, suppose that we take the perfectly symmetrical distribution and stretch out the data points in one of the tails, pulling them all further away from the median. In the box plot, this will stretch the box on that side, without changing the position of the median or the length of the box on the other side. As a result, the median will no longer be in the centre of the box.

In Figure 23.9, we can see a data set where the data is asymmetric. The data in the right hand tail stretches out much further than the data in the left hand tail. We say that the data is *positively skewed*, or *right skewed*. This can be seen in the box plot underneath the histogram. The right hand side of the box is longer than the left hand side.

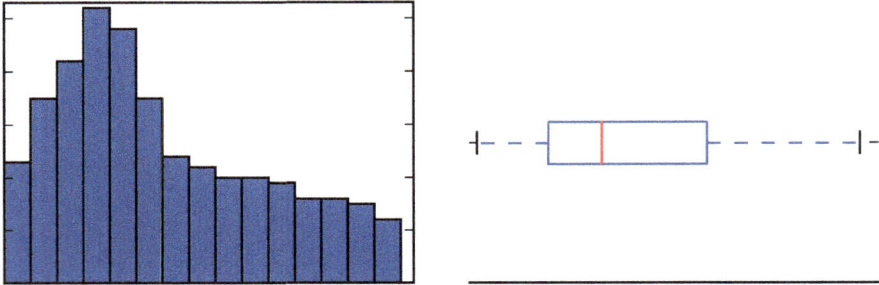

Figure 23.9: Histogram and box plot of positively skewed data.

Similarly, if the left hand tail is longer, or fatter, we say that the data is negatively skewed.

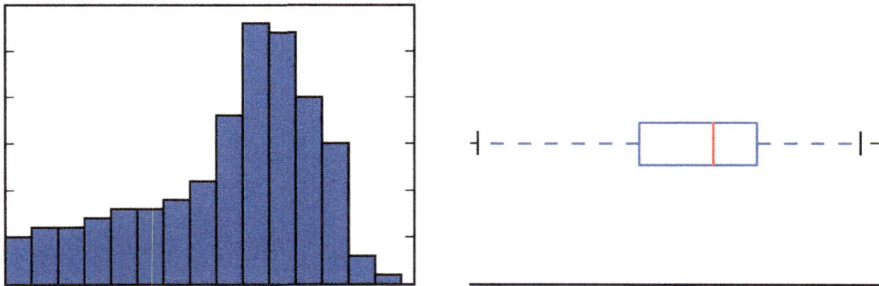

Figure 23.10: Histogram and box plot of negatively skewed data.

This allows us to identify the skew of the data by comparing the distance of the upper and lower quartiles from the median. If $(Q_3 - Q_2) > (Q_2 - Q_1)$ then the right hand side of the box is longer and the data is positively skewed. If $(Q_3 - Q_2) < (Q_2 - Q_1)$ then the left hand side is longer and the data has negative skew.

If $(Q_3 - Q_2) = (Q_2 - Q_1)$ then the data has no skew. This might be because the data is very symmetrical, or because unevenness of the two tails balances out. If the data is not unimodal, there is not always a clear interpretation of the skew.

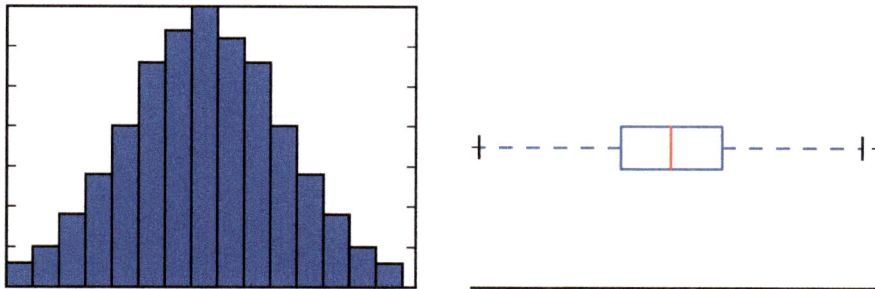

Figure 23.11: Histogram and box plot of symmetric data with zero skew.

We can attach a number to this idea of skew by calculating the difference in length of the two sections of the box: $(Q_3 - Q_2) - (Q_2 - Q_1) = Q_3 - 2Q_2 + Q_1$. This will be positive if the skew is positive, and negative if the skew is negative.

Skew is a description of the shape of the data: whether one tail is fatter or longer than the other. The shape of the data does not change if we change the scale, so we would prefer the skew to be unchanged if the data is rescaled. For example, the skew should not change if the data is recorded in inches rather than centimetres. We can make our value for skew independent of the scale of the data by dividing by the length of the whole box.

**Definition 23.3 — Skew coefficient 1**
One way of measuring the skew of data is by using the following skew coefficient:

$$\text{skew} = \frac{Q_3 - 2Q_2 + Q_1}{Q_3 - Q_1}.$$

The median lies somewhere between the lower and upper quartiles, and this puts limits on the range of the skew coefficient.

$$Q_1 \leq Q_2 \leq Q_3$$
$$-2Q_3 \leq -2Q_2 \leq -2Q_1$$
$$Q_1 - Q_3 \leq Q_3 - 2Q_2 + Q_1 \leq Q_3 - Q_1$$

If we now divide by $(Q_3 - Q_1)$ throughout, we see that $-1 \leq \text{skew} \leq +1$.

**Remark**
The interquartile range has the same units as the data, but the skew coefficient has no units. It is a dimensionless number between $-1$ and $+1$. Perfectly symmetrical data has zero skew.

**Example 23.7**

This stem and leaf plot shows house prices in a certain area, where 20|2 represents £202 000. Draw a box plot of the data and comment on its skew.

```
20 | 2  7
21 | 5  7
22 | 2  6
23 | 0  8  9
24 | 0  2  3  8
25 | 2  4  4
26 | 1  7
27 | 2  4  8
28 | 2
29 | 1
```

There are 23 prices given, so the median is equal to the central value, £243 000.

To calculate the lower and upper quartiles, we calculate the median of the lower 12 and the upper 12 data points. For the lower quartile, this is the mean of £226 000 and £230 000. Therefore $Q_1 = £228\,000$. The upper quartile is the mean of £261 000 and £267 000 and so $Q_3 = £264\,000$.

The range of the data is from £202 000 to £291 000.

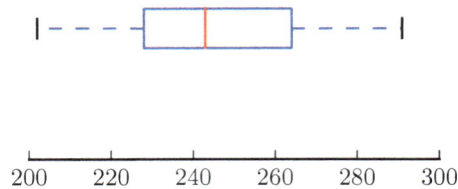

```
200   220   240   260   280   300
```

The distance from the median to the lower quartile is £243 000 − £228 000 = £15 000. The distance from the upper quartile to the median is £264 00 − £243 000 = £21 000. Since £21 000 > £15 000, the data has positive skew.

**Remark**

If a data set is perfectly symmetrical, then it has zero skew. If we take a symmetrical, unimodal distribution, and we "stretch" the right hand tail (so that each point is moved some distance further to the right) then it will have positive skew. Similarly, if we stretch the left hand tail, the distribution will have negative skew.

In these cases, we can use the skew coefficient to represent the extent of the stretching away from a symmetrical distribution.

Beyond this, the interpretation of skew becomes less clear. For example, a skew of zero does *not* imply that the data is symmetrical. The distribution shown below is not symmetrical: the left hand tail is longer than the right, but the right hand tail is fatter. The result is that the upper and lower quartiles, $Q_1$ and $Q_3$, are the same distance away from the median. This gives us an asymmetric distribution with zero skew.

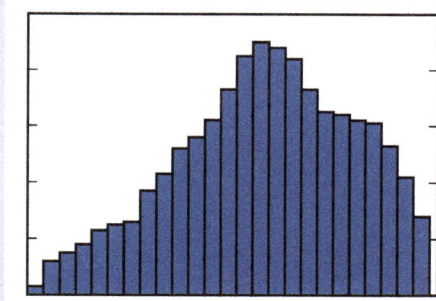

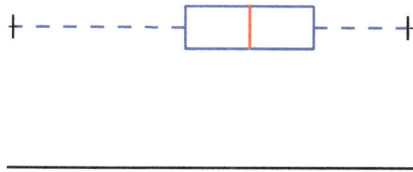

Always consider the shape of the data before interpreting the skew coefficient.

**Exercise 23.6**

Q1. A school sets the same test for each of three classes.

| Class A | 18 | 20 | 43 | 44 | 46 | 48 | 50 | 51 | 53 | 59 |
|---|---|---|---|---|---|---|---|---|---|---|
|  | 60 | 63 | 63 | 64 | 64 | 65 | 66 | 66 | 68 | 69 |
|  | 71 | 73 | 75 | 77 | 78 | 80 | 82 | 83 | 84 | 87 |
| Class B | 27 | 31 | 33 | 42 | 46 | 50 |  |  |  |  |
|  | 51 | 53 | 53 | 55 | 63 | 72 |  |  |  |  |
|  | 74 | 78 | 82 | 83 | 84 | 86 | 92 |  |  |  |
| Class C | 18 | 20 | 37 | 43 | 48 | 53 | 58 | 60 | 62 | 63 |
|  | 65 | 68 | 68 | 69 | 75 | 77 | 78 | 78 | 80 | 82 |
|  | 83 | 84 | 85 | 87 | 87 | 89 | 90 | 91 | 93 | 95 |

(a) Draw box-and-whisker plots comparing the results of the classes. Indicate any outliers clearly.

(b) State which class had the best results and justify this answer.

Q2. A new fertiliser claims to increase the yield of potato plants. A gardener planted two rows of plants and used the new fertiliser on one of the rows. At harvest time, he weighed the potatoes produced by each plant. Here are his results.

| Without fertiliser (kg) | 0.2 | 0.5 | 0.5 | 0.6 | 0.9 | 0.9 | 1.1 | 1.2 | 1.2 | 1.2 |
|---|---|---|---|---|---|---|---|---|---|---|
|  | 1.3 | 1.3 | 1.4 | 1.4 | 1.4 | 1.5 | 1.5 | 1.5 | 1.6 | 1.6 |
|  | 1.7 | 1.7 | 1.7 | 1.8 | 1.8 | 1.9 | 1.9 | 2.0 | 2.0 |  |
| With fertiliser (kg) | 0.6 | 0.7 | 0.7 | 0.8 | 0.9 | 0.9 | 1.0 | 1.0 | 1.1 | 1.1 |
|  | 1.1 | 1.2 | 1.2 | 1.2 | 1.2 | 1.3 | 1.3 | 1.5 | 1.6 | 1.6 |
|  | 1.8 | 1.8 | 2.0 | 2.1 | 2.4 | 2.6 |  |  |  |  |

Draw box plots comparing the yields from the two rows of potatoes. Describe the effect of the fertiliser on the potato yield.

Q3. A company that makes metal washers is considering buying a new machine. It is claimed that the new machine is more accurate. To test the claim, the factory manager produces a batch of 15 mm washers from each machine and measures them accurately. The results are shown in the following stem-and-leaf plot, where 7|14.5|6 indicates a measurement of 14.57 mm from the old machine and 14.56 mm on the new machine.

| Old machine | | New machine |
|---:|:---:|:---|
| 7 | 14.5 | 0 6 |
| 6 3 | 14.6 | 1 2 4 7 9 |
| 9 9 5 3 | 14.7 | 1 3 3 6 |
| 9 9 7 6 4 4 2 | 14.8 | 0 2 7 |
| 9 5 3 2 2 0 0 | 14.9 | 2 4 6 |
| 7 4 3 0 0 | 15.0 | 3 4 6 6 8 9 |
| 9 5 3 1 1 | 15.1 | 2 2 4 5 6 8 9 |
| 3 0 0 | 15.2 | 1 1 3 4 6 |
| 4 | 15.3 | |

Show these data sets in box plots. Based on these results, should we advise the manager to buy the new machine?

Q4. A commuter records the length of her journey to work each morning for ten weeks. The fastest journey was 46 min and the slowest took 103 min. She made the following table to summarise her results.

| Journey time (minutes) | Frequency |
|:---:|:---:|
| $40 \leq x < 50$ | 6 |
| $50 \leq x < 60$ | 5 |
| $60 \leq x < 70$ | 15 |
| $70 \leq x < 80$ | 9 |
| $80 \leq x < 90$ | 7 |
| $90 \leq x < 105$ | 5 |
| $100 \leq x < 110$ | 3 |

(a) Use linear interpolation to calculate the median $Q_2$ and the lower and upper quartiles, $Q_1$ and $Q_3$.

(b) One way of measuring the skew of a set of data is to use the following quantity

$$\frac{Q_3 - 2Q_2 + Q_1}{Q_3 - Q_1}.$$

Calculate this coefficient, and comment on the skew of the data.

## 23.8   Mean and Variance

For a set of data $\{X_1, X_2, \ldots, X_N\}$, we defined the median to be the point $m$ where $\sum_{i=1}^{N} |X_i - m|$ takes its minimum value. The median, along with the interquartile range and the skew coefficient, give us a way of describing the centre, the spread and the shape of the data.

However, the function $|x|$ is difficult to analyse: in particular, it has no well-defined gradient at zero. For this reason, we often prefer to use $x^2$ instead of $|x|$. The value $\mu$ which minimises $\sum_{i=1}^{N}(X_i - \mu)^2$ is the *mean* of the data:

$$\text{mean} = \mu = \frac{1}{N}\sum_{i=1}^{N} X_i.$$

In this section we will look at the mean, and some related statistics, which give us another way of describing the centre, the spread and the shape of a data set.

We noted earlier that the median is not significantly affected if the data contains outliers: we say that it is *robust*. The same is true of the quartiles $Q_1$ and $Q_3$. Since the quartiles are robust, the interquartile range and the skew coefficient are also robust. Robust statistics usually give a better representation of the data than non-robust statistics: rare but extreme events do not have a significant effect on them. As a result of this, the median and its associated statistics are good choices for describing data in many situations.

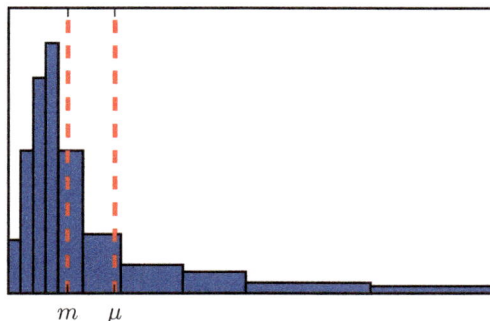

Figure 23.12: Example of skewed data on median $m$ and mean $\mu$.

As $x$ grows, $x^2$ grows much faster than $|x|$. This means that if we add outliers to a data set, the mean and its associated statistics change much more than the median and the interquartile range. For this reason, although they are widely used, the statistics we will see in this section are *not robust*. Figure 23.12 shows how a skewed data set can pull the mean away from the part of the range where the data is densest.

---

**Remark**

Non-robust statistics must be used carefully. If the data is highly asymmetrical, or is not unimodal, they can give results which are not a good description of the data. However, there are many situations where we know that the data is unimodal and sufficiently symmetric to use non-robust statistics.

> **Remark**
> It is sometimes said that one advantage of the mean is that all of the data points are used in its calculation. In some sense, this is true: all of the values appear in its formula, and its value changes if a single data point is altered. However, although the median is not sensitive to changes to an arbitrary data point, it is not true that it ignores any of the data. The median minimises the mean absolute distance from the average, and the mean minimises the mean square distance from the average. Both of these calculations consider the entire data set.

The measure of spread that we associate with the mean is called the *variance*. The variance is equal to the distance we used to define the mean, $\sum(X_i-\mu)^2$, divided by the total number of data points.

$$\text{Variance} = \sigma_X^2 = \frac{1}{N}\sum_{i=1}^{N}(X_i - \mu)^2.$$

> **Activity 23.3 — Mean as the Point of Minimum Variance**
> We have defined the mean to be the point which minimises the sum of the square distances from each data point.
> For a given data set $\{X_1, X_2, \ldots, X_N\}$, let $V(x) = \sum_{i=1}^{N}(X_i - x)^2/N$ be the variance of the data around the point $x$.
> Show that $V(x)$ takes its minimum value when $x = \mu$.

There is another formula for finding the variance, which is often more convenient to use.

$$\begin{aligned}
\sigma_X^2 &= \frac{1}{N}\sum_{i=1}^{N}(X_i - \mu)^2 \\
&= \frac{1}{N}\sum_{i=1}^{N}(X_i^2 - 2X_i\mu + \mu^2) \\
&= \frac{1}{N}\sum_{i=1}^{N}X_i^2 - 2\mu\left(\frac{1}{N}\sum X_i\right) + \frac{1}{N}\sum\mu^2 \\
&= \frac{1}{N}\sum_{i=1}^{N}X_i^2 - 2\mu^2 + \mu^2 \\
&= \frac{1}{N}\sum_{i=1}^{N}X_i^2 - \mu^2.
\end{aligned}$$

In this version of the formula, we do not need to subtract the mean from every data point. We only need to calculate the mean and the sum of the squares of the data points.

> **Tip**
> This formula for the variance can be stated as "the mean of the squares minus the square of the mean".

An alternative, closely related measure of spread is the *standard deviation*, $\sigma_X$. This is equal to the square root of the variance. The units of the standard deviation are the same as the units of the underlying data and this measure has a more intuitive meaning. For example, if a set of distances are normally distributed with a standard deviation of 5 cm, there is a result which says that approximately 68% of the data points fall within 5 cm of the mean. We will discuss the normal distribution further in the Year 2 book. The variance of this data set is 25 cm², and this does not have a simple direct interpretation from the data. When reporting the spread of the data, we usually give the standard deviation.

---

**Definition 23.4 — Variance and Standard Deviation**

The variance of a set of data $\{X_1, X_2, \ldots, X_N\}$ is given by

$$\text{variance} = \sigma_X^2 = \frac{1}{N}\sum_{i=1}^{N}(X_i - \mu)^2 = \frac{1}{N}\sum_{i=1}^{N}X_i^2 - \mu^2.$$

The standard deviation is the square root of the variance:

$$\text{standard deviation} = \sigma_X = \sqrt{\frac{1}{N}\sum_{i=1}^{N}(X_i - \mu)^2}.$$

---

**Example 23.8**

A garage records the carbon dioxide emissions for twenty vehicles. Calculate the mean and standard deviation of the results, given here in $\text{g km}^{-1}$.

| | | | | | | | | | |
|---|---|---|---|---|---|---|---|---|---|
| 37 | 75 | 101 | 103 | 105 | 108 | 112 | 118 | 123 | 124 |
| 128 | 135 | 137 | 137 | 140 | 145 | 152 | 161 | 164 | 182 |

**Solution:**

To find the mean and variance, we must find the sum of the data points, and the sum of the squares. If we write $X_i$ for the data points, then $\sum_{i=1}^{N} X_i = 2487$ and $\sum_{i=1}^{N} X_i^2 = 329283$.

Therefore, for the $N = 20$ data points, the mean is equal to

$$\mu = \frac{1}{N}\sum_{i=1}^{N}X_i = \frac{2487}{20} = 124.35 \text{ g km}^{-1}$$

and the standard deviation is equal to

$$\sigma = \sqrt{\frac{1}{N}\sum_{i=1}^{N}X_i^2 - \mu^2} = \sqrt{\frac{329283}{20} - 15462.92} = \sqrt{1001.23} = 31.64 \text{ g km}^{-1}$$

to two decimal places.

**Use of Technology 23.2 — Calculating Variance and Standard Deviation**

As with other summary statistics, many calculators will calculate the variance and standard deviation from a list of user inputted numbers.

The example above shows that finding the standard deviation can lead to calculations involving very large numbers. We can make the calculations simpler by *coding* the data. For each point $X_i$ in the data set, we transform it to $(X_i - a)/b$ where $b > 0$, and use these values instead. If we choose the values $a$ and $b$ correctly, then we can simplify the calculation. We can use the mean and standard deviation of the transformed data to find the corresponding values for the original data set.

**Activity 23.4 — Mean and Variance Identities**

Given the data set $\{X_1, X_2, \ldots, X_N\}$, let $Y_i = (X_i - a)/b$ where $b > 0$. Let $\mu_X$ and $\sigma_X$ be the mean and standard deviation of $X_i$, and $\mu_Y$ and $\sigma_Y$ be the mean and standard deviation of $Y_i$.

Prove the following identities.

$$\mu_X = b\mu_Y + a \qquad \text{and} \qquad \sigma_X = b\sigma_Y.$$

**Remark**

The first identity, $m_X = b\mu_Y + a$, tells us that if we add the same amount to every data point, the mean changes by that same amount. Also, if we multiply every data point by a fixed value, the mean is also multiplied by that value, just as we would expect.

The second identity, $\sigma_X = b\sigma_Y$, is about the shape of the data. If we add the same amount to every data point then the shape of the data is unchanged. As we would expect, this does not affect the standard deviation. However, if we multiply every data point by a constant value, the distance between the mean and each point is multiplied by this same value. This causes the standard deviation to be scaled in the same way.

**Example 23.9**

Using a coding of $(X_i - 137)/10$, recalculate the mean and standard deviation of the carbon dioxide emissions, given in $\text{g km}^{-1}$.

$$
\begin{array}{cccccccccc}
37 & 75 & 101 & 103 & 105 & 108 & 112 & 118 & 123 & 124 \\
128 & 135 & 137 & 137 & 140 & 145 & 152 & 161 & 164 & 182
\end{array}
$$

**Solution:**

We encode the data and calculate the summary statistics

This gives the results $\sum_{i=1}^{N} Y_i = -25.3$ and $\sum Y_i^2 = 232.25$.

Therefore, for the $N = 20$ data points, the mean of the coded data is equal to

$$\mu_Y = \frac{1}{N} \sum_{i=1}^{N} Y_i = \frac{-25.3}{20} = -1.265 \text{ g km}^{-1}$$

and the standard deviation is equal to

$$\sigma_Y = \sqrt{\frac{1}{N}\sum_{i=1}^{N} Y_i^2 - \mu_Y^2} = \sqrt{\frac{232.25}{20} - 1.6} = \sqrt{10.012} = 3.164\,\mathrm{g\,km^{-1}}.$$

Now we can undo the coding to calculate the mean and standard variation of the original data.

$$\mu_X = 10\mu_Y + 137 = -12.65 + 137 = 124.35\,\mathrm{g\,km^{-1}},$$
$$\sigma_X = 10\sigma_Y = 31.64\,\mathrm{g\,km^{-1}},$$

both to two decimal places. These results agree with the results we calculated without using coding.

---

**Example 23.10**

Using the coding $Y_i = \frac{1}{12}(X_i - 1843)$, calculate the mean and standard deviation of the following values.

1867    1879    1891    1903    1939    1987    1999    2011    2071    2083

**Solution:**

First, we can calculate the values of the transformed data.

| $X_i$ | 1867 | 1879 | 1891 | 1903 | 1939 | 1987 | 1999 | 2011 | 2071 | 2083 |
|---|---|---|---|---|---|---|---|---|---|---|
| $Y_i$ | 2 | 3 | 4 | 5 | 8 | 12 | 13 | 14 | 19 | 20 |
| $Y_i^2$ | 4 | 9 | 16 | 25 | 64 | 144 | 169 | 196 | 361 | 400 |

This gives us $\sum_{i=1}^{N} Y_i = 100$ and $\sum_{i=1}^{N} Y_i^2 = 1388$. From this we can calculate $\mu_Y = 10$ and $\sigma_Y^2 = (1388/10) - 10^2 = 38.8$. Therefore, $\sigma_Y = 6.229$ to three decimal places. Finally, we can undo the coding to find the mean and standard deviation of the original values.

$$\mu_X = 12\mu_Y + 1843 = 1963,$$
$$\sigma_X = 12\sigma_Y = 74.75 \text{ to two decimal places.}$$

---

When we wanted to calculate the median for data which had already been grouped into classes, we used linear interpolation. This method makes the simplifying assumption that the points are evenly spaced within each class. This is a reasonable thing to do since we have no other information about the distribution of the data points within each class. To calculate the mean or the variance, we will also make an assumption which simplifies our calculations. We will assume that the data points are equal to the midpoint of the class. For example, if the data has the following line:

| Range | Frequency |
|---|---|
| $40 \le x < 50$ | 12 |

then we will treat this as representing 12 data points, all with the same value 45.

**Example 23.11**

Estimate the mean and variance of the following data.

| Range | Frequency $(f)$ |
|---|---|
| $-40 \leq x < -30$ | 7 |
| $-30 \leq x < -20$ | 12 |
| $-20 \leq x < -15$ | 8 |
| $-15 \leq x < -10$ | 20 |
| $-10 \leq x < 0$ | 43 |
| $0 \leq x < 10$ | 45 |
| $10 \leq x < 15$ | 18 |
| $15 \leq x < 20$ | 14 |

**Solution:**

First, we extend the table to show the midpoint, $m$, of each class. From that, we can estimate the sum of the data, and the square of the data, for each class.

| Range | Freq. $(f)$ | $m$ | $fm$ | $fm^2$ |
|---|---|---|---|---|
| $-40 \leq x < -30$ | 7 | $-35$ | $-245$ | 8575 |
| $-30 \leq x < -20$ | 12 | $-25$ | $-300$ | 7500 |
| $-20 \leq x < -15$ | 8 | $-17.5$ | $-140$ | 2450 |
| $-15 \leq x < -10$ | 20 | $-12.5$ | $-250$ | 3125 |
| $-10 \leq x < 0$ | 43 | $-5$ | $-215$ | 1075 |
| $0 \leq x < 10$ | 45 | 5 | 225 | 1125 |
| $10 \leq x < 15$ | 18 | 12.5 | 225 | 2812.5 |
| $15 \leq x < 20$ | 14 | 17.5 | 245 | 4287.5 |

Our estimate of the mean is given by

$$\mu = \frac{\sum fm}{\sum f} = \frac{-455}{167} = -2.72 \text{ to two decimal places.}$$

The estimate for the variance is

$$\sigma^2 = \frac{\sum fm^2}{\sum f} - \mu^2 = \frac{30950}{167} - (-2.72)^2 = 177.91 \text{ to two decimal places.}$$

**Exercise 23.7**

Q1. Use the coding $(x - 972)/133$ to calculate the mean and standard deviation of the following data set.

| 440 | 706 | 972 | 1371 | 1770 | 2036 | 2302 |
|---|---|---|---|---|---|---|

Q2. A company reports the hourly rates of its employees, including senior management.

| Hourly rate (£) | Frequency |
|---|---|
| $8 \leq x < 10$ | 30 |
| $10 \leq x < 30$ | 26 |
| $30 \leq x < 40$ | 16 |
| $40 \leq x < 50$ | 12 |
| $50 \leq x < 70$ | 8 |
| $70 \leq x < 100$ | 6 |
| $100 \leq x < 200$ | 0 |
| $200 \leq x < 300$ | 2 |

(a) Calculate the mean hourly rate, and its standard deviation.

(b) State, with reasons, whether the mean is an appropriate way to report the average hourly rate.

Q3. This table gives the distribution of word lengths used by two Presidents of the United States in an inauguration speech: Richard Nixon in 1969 and Donald Trump in 2017.

| Word length | Frequency | | Word length | Frequency | |
|---|---|---|---|---|---|
| | Nixon | Trump | | Nixon | Trump |
| 1 | 54 | 18 | 8 | 91 | 88 |
| 2 | 482 | 236 | 9 | 58 | 52 |
| 3 | 444 | 352 | 10 | 32 | 35 |
| 4 | 352 | 244 | 11 | 25 | 9 |
| 5 | 244 | 183 | 12 | 4 | 3 |
| 6 | 171 | 116 | 13 | 4 | 4 |
| 7 | 142 | 114 | 14 | 1 | 3 |

(a) Use the figures to calculate the mean and standard deviation of the word lengths used in the speeches.

(b) In 1863, Abraham Lincoln's speech was 3623 words long. The sum of the word lengths was 16 932 and the sum of the squares of the lengths was 106 966. Use these summary statistics to calculate the mean and standard deviation of the word lengths in Lincoln's speech.

(c) Do your results suggest that the words used in inauguration speeches are getting either longer or shorter? Give a reason for your answer.

Q4. A traffic camera records vehicle speeds. From the following results, estimate the mean and standard deviation of the speed of the vehicles.

| Speed (mph) | Frequency ($f$) |
|---|---|
| $10 \leq x < 20$ | 23 |
| $20 \leq x < 30$ | 242 |
| $30 \leq x < 35$ | 156 |
| $35 \leq x < 40$ | 184 |
| $40 \leq x < 45$ | 160 |
| $45 \leq x < 50$ | 76 |
| $50 \leq x < 60$ | 62 |
| $60 \leq x < 65$ | 8 |

Q5. This table presents the demand for electricity for Great Britain during January 2012, with one reading taken every five minutes throughout the month. A researcher finds the midpoint $X_i$ of each class and transforms it using the coding $Y_i = a(X_i - b)$.

| Demand (GW)       | Frequency $(f)$ | Coded midpoint $(Y_i)$ |
|-------------------|-----------------|------------------------|
| $25 \leq x <\ 30$ | 487             | $-15$                  |
| $30 \leq x <\ 35$ | 2245            | $-5$                   |
| $35 \leq x <\ 40$ | 1251            | 5                      |
| $40 \leq x <\ 45$ | 1476            | 15                     |
| $45 \leq x <\ 47$ | 1303            | 22                     |
| $47 \leq x <\ 50$ | 783             | 27                     |
| $50 \leq x <\ 55$ | 649             | 35                     |

The researcher uses the values $\sum_i Y_i = 82\,387$ and $\sum_i Y_i^2 = 2\,525\,560$.
Find the values of $a$ and $b$ and hence obtain estimates for the mean and standard deviation of the electricity demand.

Although there are disadvantages to using non-robust statistics such as the mean, they do give us another way to measure skew.

In a perfectly symmetrical data set, the median, $\mu$, and the mean, $m$, are equal. In Figure 23.12 we can see that skewed data has a greater effect on the mean than it has on the median. If we skew the data to the right, the mean will be greater than the median. If we stretch the left hand tail instead, then the mean will be less than the median. This suggests that we can measure the skew by looking at the difference between these two different averages, $(\mu - m)$. For data which is skewed to the right, this will be positive: for left skewed data, it will be negative. As before, we want to define the skew coefficient in such a way that it is unchanged if we scale the data. For this version of skew, we define the coefficient as follows:

**Definition 23.5 — Skew Coefficient 2**
Another way of measuring the skew of data is by using the following skew coefficient:

$$\text{skew} = \frac{3(\mu - m)}{\sigma}$$

where $\mu$ is the mean, $m$ is the median and $\sigma$ is the standard deviation of the data.

**Remark**
The value in Definition 23.5 is known as Pearson's *second* skewness coefficient. Pearson's *first* skewness coefficient is defined as

$$\frac{\text{mean} - \text{mode}}{\text{standard deviation}}.$$

Note the surprising inclusion of a factor of three in the definition of the second skew coefficient. Pearson had observed that $(\text{mean} - \text{mode}) \approx 3(\text{mean} - \text{median})$. This holds if the data is slightly skewed: it is not a general result. When the relationship does hold, we can therefore say that

$$\frac{3(\mu - m)}{\sigma} \approx \frac{\text{mean} - \text{mode}}{\sigma}.$$

In this way, Pearson could make his coefficients approximately equal by adding a factor of three.

Some years after Pearson had defined his coefficients, it was proved that $|\mu - m| \leq \sigma$. This means that the second skewness coefficient takes values in the range $[-3, +3]$ rather than the more natural range $[-1, +1]$.

The fact that $|\mu - m| \leq \sigma$ has a simple proof which is of interest to us, since it uses our definition that the median is the value which minimises $\sum_{i=1}^{N} |X_i - m|$. It relies on Jensen's inequality, one form of which says that if $f$ is a convex function, then $f(\frac{1}{N} \sum_{i=1}^{N} X_i) \leq \frac{1}{N} \sum_{i=1}^{N} f(X_i)$. (This result is a generalisation of the observation that a line drawn between two points on a convex curve will lie above that curve.) Since the functions $|x|$ and $x^2$ are both convex, we can substitute these directly into Jensen's inequality as the function $f$ to obtain the following two results:

$$\left| \frac{1}{N} \sum_{i=1}^{N} (X_i - m) \right| \leq \frac{1}{N} \sum_{i=1}^{N} |X_i - m| \text{ and}$$

$$\left[ \frac{1}{N} \sum_{i=1}^{N} |X_i - \mu| \right]^2 \leq \frac{1}{N} \sum_{i=1}^{N} |X_i - \mu|^2.$$

By definition, the median is the value $m$ which minimises $\sum |X_i - m|$ and so

$$\frac{1}{N} \sum_{i=1}^{N} |X_i - m| \leq \frac{1}{N} \sum_{i=1}^{N} |X_i - \mu|.$$

Putting these together gives us the result that

$$|\mu - m| = \left| \frac{1}{N} \sum_{i=1}^{N} (X_i - m) \right| \leq \frac{1}{N} \sum_{i=1}^{N} |X_i - m|$$

$$\leq \left( \frac{1}{N} \sum_{i=1}^{N} |X_i - \mu| \right) \leq \sqrt{\frac{1}{N} \sum_{i=1}^{N} |X_i - \mu|^2} = \sigma$$

and so

$$-1 \leq \frac{\mu - m}{\sigma} \leq +1.$$

**Remark**
We have already seen that we must take care to consider the shape of the data before we interpret the skew coefficient of a distribution. Now that we have two different definitions of the skew coefficient, we find another reason to use these values carefully: the two definitions do not necessarily agree.

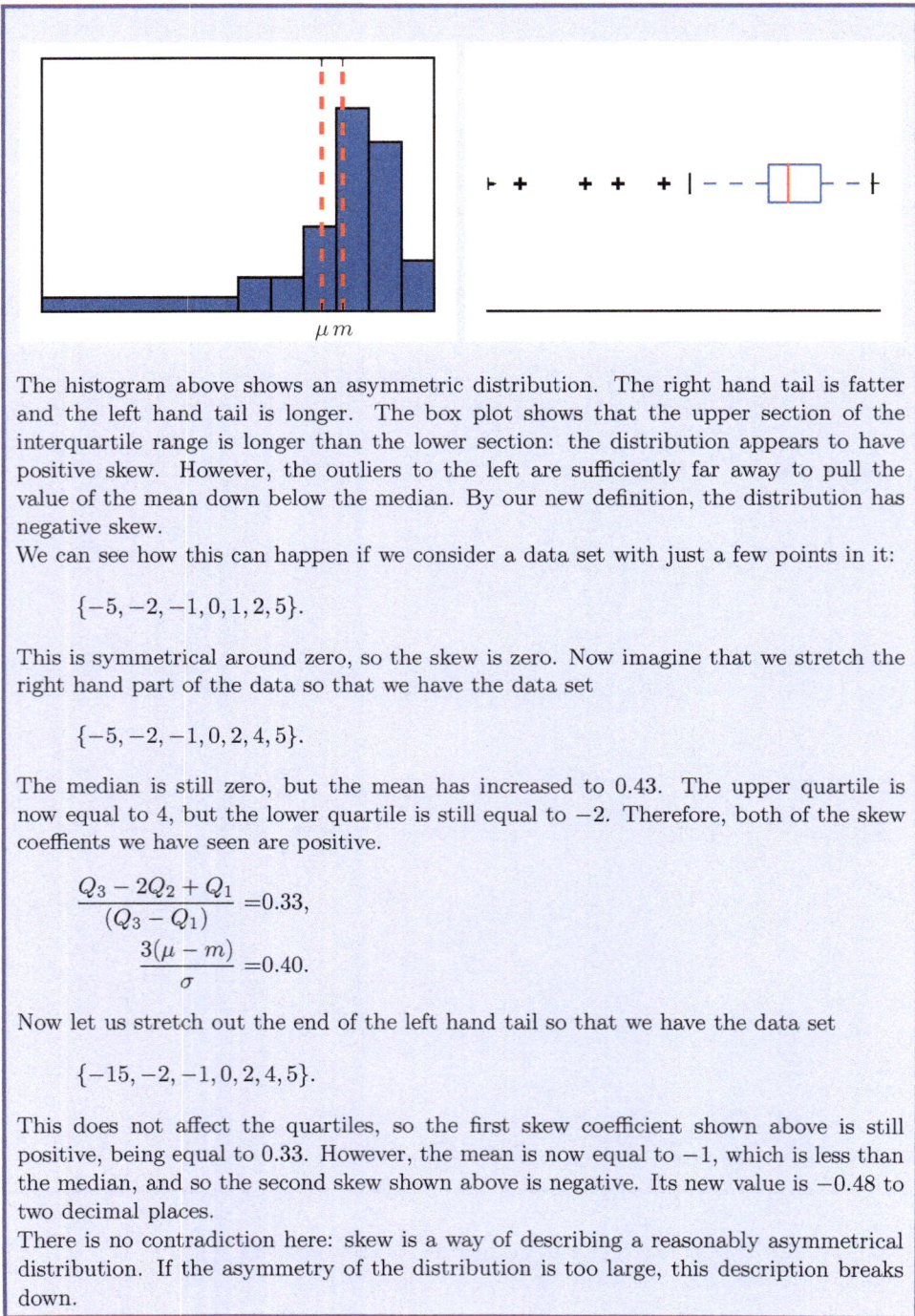

The histogram above shows an asymmetric distribution. The right hand tail is fatter and the left hand tail is longer. The box plot shows that the upper section of the interquartile range is longer than the lower section: the distribution appears to have positive skew. However, the outliers to the left are sufficiently far away to pull the value of the mean down below the median. By our new definition, the distribution has negative skew.

We can see how this can happen if we consider a data set with just a few points in it:

$$\{-5, -2, -1, 0, 1, 2, 5\}.$$

This is symmetrical around zero, so the skew is zero. Now imagine that we stretch the right hand part of the data so that we have the data set

$$\{-5, -2, -1, 0, 2, 4, 5\}.$$

The median is still zero, but the mean has increased to 0.43. The upper quartile is now equal to 4, but the lower quartile is still equal to $-2$. Therefore, both of the skew coeffients we have seen are positive.

$$\frac{Q_3 - 2Q_2 + Q_1}{(Q_3 - Q_1)} = 0.33,$$

$$\frac{3(\mu - m)}{\sigma} = 0.40.$$

Now let us stretch out the end of the left hand tail so that we have the data set

$$\{-15, -2, -1, 0, 2, 4, 5\}.$$

This does not affect the quartiles, so the first skew coefficient shown above is still positive, being equal to 0.33. However, the mean is now equal to $-1$, which is less than the median, and so the second skew shown above is negative. Its new value is $-0.48$ to two decimal places.

There is no contradiction here: skew is a way of describing a reasonably asymmetrical distribution. If the asymmetry of the distribution is too large, this description breaks down.

The skew coefficient $3(\mu - m)/\sigma$ is based on a comparison between a robust and a non-robust statistic. If the range of the data is discrete, then we can use the mode instead of the median as the robust statistic. This gives us our third type of skew coefficient.

**Definition 23.6 — Skew Coefficient 3**

A third way of measuring the skew of data is by using the following skew coefficient:

$$\text{skew} = \frac{(\mu - M)}{\sigma},$$

where $\mu$ is the mean, $M$ is the mode and $\sigma$ is the standard deviation of the data.

**Exercise 23.8**

Q1. A class of twenty students obtains the following results in a test, expressed as percentages.

| 12 | 17 | 25 | 35 | 48 | 65 | 68 | 73 | 85 | 86 |
|----|----|----|----|----|----|----|----|-----|-----|
| 86 | 86 | 87 | 90 | 95 | 95 | 96 | 98 | 100 | 100 |

(a) Calculate the mean and median of the results.

(b) State what is meant by *negative skew* and give one way of identifying the skew of this data set.

(c) What is causing the negative skew in this data set?

Q2. The concept of statistical skew was first developed by Karl Pearson. In 1895 he delivered a paper to the Royal Society where he gave the theoretical basis for skew and examined many different data sets, from cases of typhoid fever to people's attempts to place nine varying shades of paint in order. The following data, taken from his paper, gives the number of teeth found in 915 specimens of palaemon varians, also known as the common ditch shrimp.

| Teeth | 1 | 2 | 3 | 4 | 5 | 6 | 7 |
|-------|---|----|-----|-----|-----|----|---|
| Frequency | 2 | 18 | 123 | 372 | 349 | 50 | 1 |

(a) Calculate the mean and standard deviation of the number of teeth.

(b) The skew of discrete data can be measured using the following formula:

$$\text{skew} = \frac{(\text{mean} - \text{mode})}{\text{standard deviation}}.$$

Find the value of the skew coefficient for this data set.

Q3. The Hat4Life company offers a lifetime guarantee on its products. When a customer returns a faulty or worn-out hat made by the company, they are entitled to a replacement. Hat4Life records the age of the returned hats.

| Age (years) | Frequency ($f$) |
|---|---|
| $x < 1$ | 18 |
| $1 \leq x < 2$ | 23 |
| $2 \leq x < 3$ | 328 |
| $3 \leq x < 4$ | 85 |
| $4 \leq x < 5$ | 121 |
| $5 \leq x < 6$ | 276 |
| $6 \leq x < 7$ | 321 |
| $7 \leq x < 8$ | 355 |
| $8 \leq x < 9$ | 340 |
| $9 \leq x < 10$ | 133 |

(a) Use linear interpolation to estimate the median lifetime of the company's hats.

(b) For these results, the company estimates that $\sum X_i = 11\,376$ and $\sum X_i^2 = 82\,171$. From these summary statistics, describe the skew of the data using the following formula:

$$\text{skew} = \frac{(\text{mean} - \text{median})}{\text{standard deviation}}.$$

---

**Key Facts — Presenting and Interpreting Data**

- For a set of data $\{X_1, X_2, X_3, \ldots, X_N\}$ written in increasing order, the following averages are defined.

$$\text{midpoint} = \frac{1}{2}(X_1 + X_N),$$

$$\text{mean} = \frac{1}{N} \sum_{i=1}^{N} X_i,$$

$$\text{median} = \begin{cases} X_n & \text{for odd } N = 2n - 1, \\ \frac{1}{2}(X_n + X_{n+1}) & \text{for even } N = 2n. \end{cases}$$

     mode – The most frequent value in the data set.

         Data sets with a single mode are called *unimodal*.

- Bar charts are used for discrete data. They are not suitable for showing continuous data. Always put a gap between each bar, and make all the bars the same width.
- Dot plots are an alternative way of presenting small data sets. Instead of representing the frequency with a bar, we use a column of individual dots. Each dot represents a single data point.
- For data sets that are not too large, a stem and leaf plot can be used. These provide a visual representation of the shape of the data whilst retaining the detail.
- A histogram is used for large data sets with grouped data.
- Unlike bar charts there are no gaps between the bars of a histogram.
- If we have a histogram with unequal class widths then the $y$-axis represents fre-

quency density, not the total frequency.
- The area of each bar is proportional to the total number of data points in that class.
- Cumulative frequency graphs can be either a smooth curve or piecewise linear.
- Box and whisker plots are useful when comparing different sets of data. They show the following pieces of information.
    - the lowest point in the range;
    - $Q_1$, the lower quartile;
    - $Q_2$, the median;
    - $Q_3$, the upper quartile;
    - and the highest point in the range.
- The interquartile range is equal to $Q_3 - Q_1$, the distance between the upper and lower quartiles.
- The whiskers of a box-and-whisker plot must finish at data points. If there are outliers, draw the whiskers out to the furthest point which is not an outlier, then plot each outlier separately.
- There are multiple definitions of an outlier. It is important to be clear about which definition you are using.
- Box plots can be used to visualise the skew of a set of data. Data can show negative, zero or positive skew.
- There are different ways of measuring the skew of a set of data.
-

$$\text{variance} = \sigma_X^2 = \frac{1}{N} \sum_{i=1}^{N} (X_i - \mu)^2 = \frac{1}{N} \sum_{i=1}^{N} X_i^2 - \mu^2.$$

-

$$\text{standard deviation} = \sigma_X = \sqrt{\frac{1}{N} \sum_{i=1}^{N} (X_i - \mu)^2}.$$

- Given the data set $\{X_1, X_2, \ldots, X_N\}$, let $Y_i = (X_i - a)/b$ where $b > 0$. Let $\mu_X$ and $\sigma_X$ be the mean and standard deviation of $X_i$, and $\mu_Y$ and $\sigma_Y$ be the mean and standard deviation of $Y_i$, then

$$\mu_X = b\mu_Y + a \qquad \text{and} \qquad \sigma_X = b\sigma_Y.$$

**Chapter Assessment — Presenting and Interpreting Data**
Download and sit the 30 minute assessment for this chapter from the digital book.

# 24. Presenting Multivariate Data

In this chapter, we will study ways of presenting and interpreting data with more than one variable. We have already looked at ways of describing and displaying data sets with just a single variable: for example, the price of some set of items, or the heights of a group of plants.

There are many occasions when it is useful to compare two associated variables, such as a person's weight and their height, or their diet and the likelihood they suffer a certain disease. Data with two variables is called *bivariate data*.

Analysing bivariate data is especially important if we want to gather evidence before we make a decision. We cannot make decisions from levels of air quality or cancer survival rates on their own: we need to see how air quality varies with nearby traffic speeds, or how survival rates are affected by different treatments. The first question to ask about bivariate data is whether the two variables are connected at all. For example, we could make a better guess at a person's weight if we were first told their height. We say that their height and weight are *correlated*. On the other hand, if we wanted to guess how many books a person has read this year, it would not be helpful to know their height: height and book reading are *uncorrelated*.

## 24.1  Scatter Graphs

So far we have looked at ways of describing and displaying data sets with just a single variable: for example, the price of some item, or the height of some plants. There are many occasions when it is useful to compare two associated variables, such as a person's weight and their height, or their diet and the likelihood they suffer a certain disease. Data with two variables are called *bivariate data*.

Analysing bivariate data is especially important if we want to gather evidence before we make a decision. We cannot make decisions from levels of air quality or cancer survival rates on their own: we need to see how air quality varies with nearby traffic speeds, or how survival rates are affected by different treatments.

The first question to ask about bivariate data is whether the two variables are connected at all. For example, you could make a better guess at a person's weight if you were first told their height. We say that their height and weight are *correlated*. On the other hand, if you wanted to guess how many books a person has read this year, it would not be helpful to know their height: height and book reading are *uncorrelated*.

---

**Definition 24.1 — Correlation**

Two variables are *positively correlated* if when one is larger, the other one also tends to be larger.

They are *negatively correlated* if one tends to be smaller when the other is larger.

If there is no relation between the variables, they are *uncorrelated*.

---

We can display bivariate data using a *scatter graph* by plotting each pair of data points $(X_i, Y_i)$. Scatter graphs help us see the extent to which the data is correlated. Figure 24.1 shows examples of scatter graphs. In the first plot, the data is uncorrelated. In the second plot, there is a positive correlation between the two variables. However, the points are fairly widely scattered, so we would say that this is a *weak correlation*. The correlation in the final plot is much stronger: the variables have a strong negative correlation.

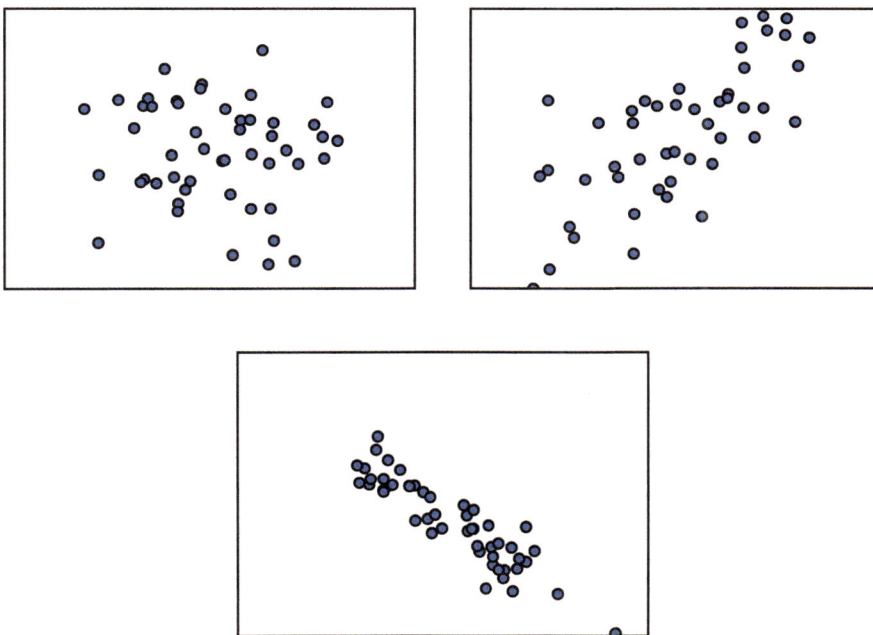

Figure 24.1: Examples of scatter graphs.

If there is a linear correlation between the variables, we can use the scatter graph to estimate the relationship by drawing a straight line through the points. From this we can predict values which are not represented in the data set.

**Example 24.1**

A hotel owner records the average amount of time that guests use air conditioning against the maximum temperature for that day and records the following data.

| Temperature (°C) | Minutes |
|:---:|:---:|
| 22.4 | 166.4 |
| 20.7 | 165.2 |
| 23.8 | 161.8 |
| 24.2 | 167.7 |
| 23.9 | 171.1 |
| 25.3 | 170.8 |
| 28.4 | 173.8 |
| 27.2 | 184.4 |
| 29.8 | 180.6 |
| 31.3 | 183.8 |
| 29.0 | 168.8 |
| 32.2 | 185.8 |
| 31.8 | 180.4 |

Table 24.1: Average use of air conditioning against maximum daily temperature.

Show this data in a scatter plot. Use the plot to estimate how long a typical guest would use the air conditioning if the maximum temperature was equal to 27 °C.

**Solution:**

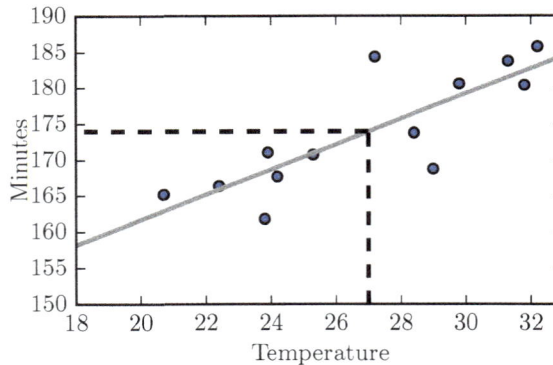

From the scatter plot, we can see that there is a fairly strong positive correlation between the two variables. We can draw a line of best fit through these data points to estimate the relationship between them.

From this, we can estimate that for a maximum temperature of 27 °C, we would expect a typical guest to use the air conditioning for around 174 minutes.

Sometimes, two variables are correlated because one causes the other. For example, we would expect outdoor temperature and ice cream sales to be correlated because it is reasonable to think that the higher temperature *causes* people to want an ice cream. We would also expect sales of sunglasses and ice cream to be correlated: people are more likely to buy both in hot weather. However, the sunglasses sales are not causing the ice cream sales: both are being caused by a third variable, the weather.

**Remark**

Some variables show a correlation because one of them causes the other. Some are correlated because they are both affected by a third variable. Variables which have no connection at all can, by chance, still show some correlation. If we search through a large number of data sets, we will always be able to find two which are very closely correlated.

For example, below is a scatter plot of the number of divorces per thousand people in the state of Maine, plotted against the per capita margarine consumption in the United States in the years 2000 to 2009.

The correlation here is close to perfect, yet there is presumably no connection between the two variables. It would be absurd to suggest that a campaign to reduce margarine consumption in the United States would have a favourable affect on the divorce rate in Maine. It is important, when designing surveys or scientific experiments, to design the experiment first and then collect the data. If we collect the data first, we will always be able to find some patterns in it, but this is not a reliable method for finding genuine connections between variables.

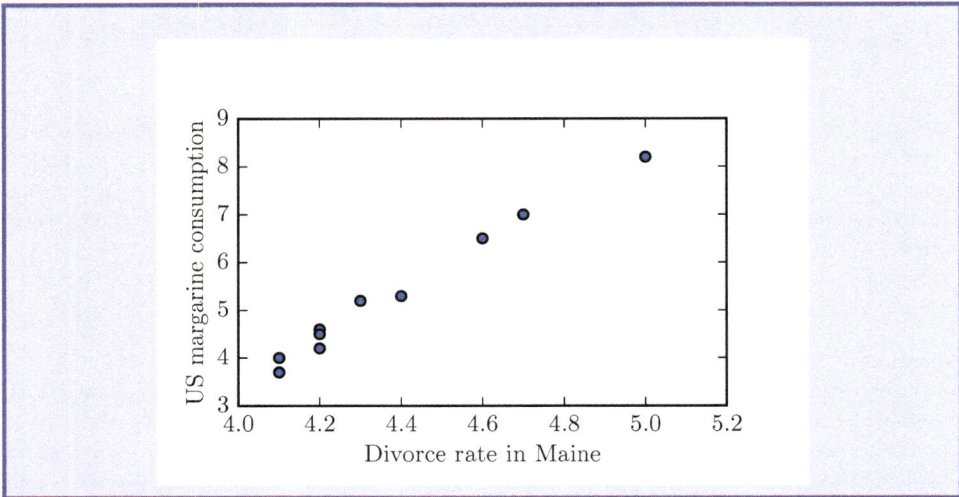

When we gather bivariate data, we often choose the value of one variable and then measure the value of the other. For example, we might select areas near certain types of road and then measure the air quality there, or we might set up a medical trial with certain drug doses and then measure the results.

The variable that we measure is sometimes called the *response variable* or the *dependent variable*. The variable that we choose can be called the *explanatory* or *independent* variable. We must take care with these names, since they imply that one variable responds to or depends on the other. However, statistical analysis can never show that one thing is causing another.

**Remark**

Many things that we accept as simple statements of fact are derived from correlations. Most people would agree that "smoking causes cancer", but not everyone who smokes develops cancer and not everyone with cancer is a smoker.

A connection between smoking and lung cancer was first made in 1912, and four centuries ago James I of England called smoking "loathsome to the eye . . . dangerous to the lungs". However, through most of the twentieth century smoking was widespread in the United Kingdom and was sometimes advertised as being recommended by doctors or dentists, or as a healthy alternative to sugary snacks.

In 1951, Dr Richard Doll and Professor Austin Bradford Hill published a paper showing a strong correlation between smoking and lung cancer: of 1357 men with lung cancer, 99.5% were smokers.

As a result of their findings, many doctors stopped smoking. This enabled the two scientists to publish a paper showing that there was a significant drop in lung cancer cases among the doctors who had stopped, compared to those who had not. Later still, in 1994 Doll, now Sir Richard, released figures showing that around half of all smokers eventually die from a smoking related disease.

It took a long time to gather this statistical evidence to convince people that smoking was a significant health risk. When enough people had been convinced, the sale and marketing of cigarettes became heavily regulated and the proportion of people smoking went down.

Major political decisions, affecting many people's lives, are made on the basis of whether one thing does (or does not) cause another.

### Exercise 24.1

Q1. For each of the following data sets, state whether you expect the variables to be positively correlated, negatively correlated or uncorrelated. Where there is a correlation, identify the response variable.

(a) For a class of students with similar predicted grades, the number of hours they spend revising, and their score in a test.

(b) For a given cinema, the number of letters in a film's title, and that film's takings over its opening week.

(c) For each employee at a company, the time they spend exercising per week and the number of days sick leave taken in the last year.

Q2. A class of students had a chemistry test and a physics test in the same week. The following table shows the amount of time each student spent studying for the chemistry test and the score that they got on the *physics* test.

| Study time (hours) | Score |
|--------------------|-------|
| 1                  | 46    |
| 1                  | 32    |
| 2                  | 47    |
| 2.5                | 42    |
| 2.5                | 53    |
| 3                  | 61    |
| 3                  | 91    |
| 3.25               | 45    |
| 3.5                | 72    |
| 4                  | 50    |
| 5                  | 34    |
| 5                  | 84    |

Plot a scatter graph of this data and describe the correlation. How do you explain the correlation?

Q3. High-density lipoproteins (HDL) are molecules which transport fat around the body. Studies have suggested that HDL levels can be raised by reducing alcohol intake, stopping smoking or taking up exercise.

(a) Explain what is meant by the following statement:
*There is a strong negative correlation between HDL levels and incidence of heart disease.*

(b) Drugs which are designed to raise HDL levels have not been shown to reduce the chances of heart disease. Explain why this is consistent with the correlation.

Q4. In 1929, the astronomer Edwin Hubble published a paper describing a relationship between the distance of a galaxy and the speed that it is moving away from the earth. Here are some of his results. The distances are given in millions of light years, and the velocities in kilometres per second.

| Distance | Velocity |
|----------|----------|
| 1.04     | 170      |
| 1.11     | 290      |
| 14.67    | 200      |
| 16.30    | 290      |
| 16.30    | 270      |
| 20.54    | 200      |
| 26.08    | 300      |
| 29.34    | 650      |
| 29.67    | 500      |
| 35.86    | 450      |
| 45.64    | 500      |
| 65.20    | 500      |
| 65.20    | 850      |
| 65.20    | 800      |

(a) Plot a scatter graph of Hubble's data. Describe the correlation between the distance of a galaxy from the earth and its velocity.

(b) It is now accepted that within the range of distances we are considering, a galaxy's velocity is almost exactly proportional to its distance from us. What does your previous answer tell you about the quality of the data that Hubble was using?

(c) Use the graph to estimate the speed of a galaxy 50 million light years from earth.

## 24.2  Covariance

We have seen how scatter graphs can be used to find correlation between two variables and how we can estimate the relationship between them by drawing a straight line through the data. These lines are called *regression lines*, and we will now see how we can calculate the equation of these lines from the data.

> **Remark**
> The term "regression line" comes from Francis Galton's 1886 paper, *Regressions towards mediocrity in Hereditary Stature*. In this paper he observed that, although tall parents tend to produce tall children, the heights of those children tend to be closer to the mean. That is, if there are extreme heights in one generation, the heights of the next generation tend to be less extreme. Galton referred to this as "regression to the mean", and the lines which he used to show the relationship between the heights became known as "regression lines". The term became associated with these lines, even when it was not relevant to the data.

Suppose that we have a data set of two variables $(X_i, Y_i)$ and we code the data by subtracting the mean of $X_i$ from the each value $X_i$ and the mean of $Y_i$ from each value $Y_i$. We now have a data set $(X_i - \mu_X, Y_i - \mu_Y)$ which has the same distribution as the original, but has been translated so that it is centred on the origin, as shown in the example in Figure 24.2.

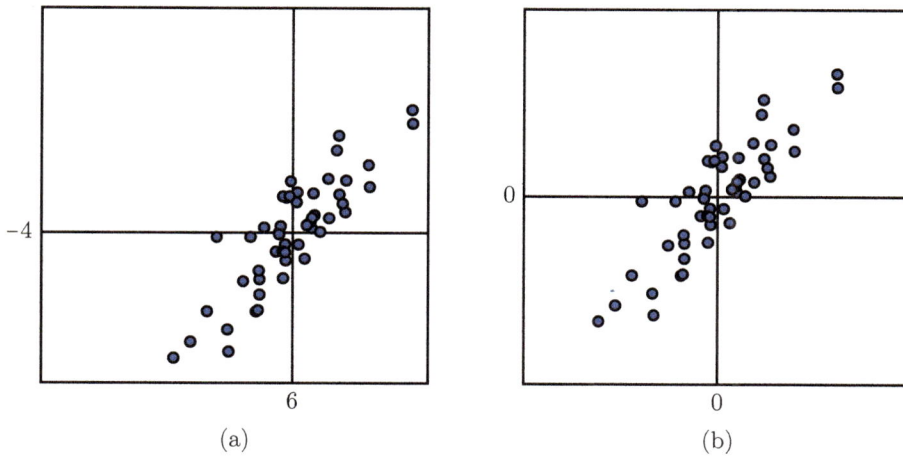

Figure 24.2: Example of a dataset (a) and its translated counterpart (b).

If the data is positively correlated, as in the example shown in Figure 24.2, then the points will tend to lie in the upper right and the lower left quadrants. In both of these quadrants, the product of the coded data, $(X_i - \mu_X)(Y_i - \mu_Y)$, is positive. If the data is negatively correlated, then most of the data will be found in the other two quadrants, where the product is negative, $(X_i - \mu_X)(Y_i - \mu_Y) < 0$.

If we sum all of these values, then this should give us some idea about the correlation of the data. We will get a positive value if the data is positively correlated, a negative value if there is a negative correlation, and close to zero if there is no correlation.

---

**Definition 24.2 — Covariance**

The covariance of $X_i$ and $Y_i$ is given by

$$\sigma_{XY} = \operatorname{cov}(X, Y)$$

$$= \frac{1}{N} \sum_{i=1}^{N} (X_i - \mu_X)(Y_i - \mu_Y),$$

where $\mu_X$ is the mean of $X_i$ and $\mu_Y$ is the mean of $Y_i$.

Equivalently,

$$\sigma_{XY} = \frac{1}{N} \sum_{i=1}^{N} X_i Y_i - \mu_X \mu_Y$$

Note that, although variances and standard deviations are always positive, covariance can be negative.

---

**Activity 24.1 — Definitions of Covariance**

Show that the two definitions of covariance in Definition 24.2 are equivalent.

**Remark**
The formula for covariance is similar to the formula for variance.
Notice that if $X_i = Y_i$, then the covariance $\sigma_{XY}$ is equal to the variance $\sigma_X^2$.

**Example 24.2**
Calculate the covariance of the maximum daily temperature and the average number of minutes of air conditioner use, as given in Table 24.1.

**Solution:**
To find the covariance, we need to find the product of each pair of variables.

| $X_i =$ Temperature (°C) | $Y_i =$ minutes | $x_i y_i$ |
|---|---|---|
| 22.4 | 166.4 | 3727.36 |
| 20.7 | 165.2 | 3419.64 |
| 23.8 | 161.8 | 3850.84 |
| 24.2 | 167.7 | 4058.34 |
| 23.9 | 171.1 | 4089.29 |
| 25.3 | 170.8 | 4321.24 |
| 28.4 | 173.8 | 4935.92 |
| 27.2 | 184.4 | 5015.68 |
| 29.8 | 180.6 | 5381.88 |
| 31.3 | 183.8 | 5752.94 |
| 29.0 | 168.8 | 4895.20 |
| 32.2 | 185.8 | 5982.76 |
| 31.8 | 180.4 | 5736.72 |

From this table, we can calculate $\sum X_i = 350$, $\sum Y_i = 2260.6$ and $\sum X_i Y_i = 61\,167.81$. There are $N = 13$ data points, so the mean values of the two variables are $\mu_X = 26.92$ and $\mu_Y = 173.89$. The covariance is given by

$$\sigma_{XY} = \frac{\sum X_i Y_i}{N} - \mu_X \mu_Y = \frac{61167.81}{13} - 4681.72 = 24.10 \min {}^{\circ}\text{C}$$

to two decimal places.

**Remark**
It is important to be careful with rounding when performing these calculations. For example, incorrectly rounding $\mu_Y$ to 173.90 results in a covariance, $\sigma_{XY} = 23.83$.

Just like variance, covariances will change if we measure the data in different units. For example, if we double the value of every $X_i$, then the value of the covariance will double. However, the property of correlation does not change when we scale the data. To measure the correlation, we divide the correlation by the standard deviation of the variables. This value is constant when the data is scaled.

**Definition 24.3 — Pearson's Correlation Coefficient**

The correlation coefficient of $X_i$ and $Y_i$ is a dimensionless number in the range $[-1, +1]$. It is given by

$$\rho_{XY} = \frac{\sigma_{XY}}{\sigma_X \sigma_Y},$$

where $\sigma_X$ and $\sigma_Y$ are the standard deviations of $X_i$ and $Y_i$, and $\sigma_{XY}$ is their covariance. This value is known as Pearson's correlation coefficient or Pearson's product-moment correlation coefficient.

**Example 24.3**

Some data has been collected on the length and weight of 10 pigs.

| length (cm) | 0.9 | 1.2 | 1.7 | 1.0 | 1.2 | 1.5 | 1.7 | 1.4 | 1.8 | 1.3 |
|---|---|---|---|---|---|---|---|---|---|---|
| weight (g) | 53 | 121 | 308 | 70 | 154 | 186 | 322 | 137 | 287 | 119 |

Compute the correlation coefficient and interpret it.

**Solution:**

Plotting this data in a scatter diagram,

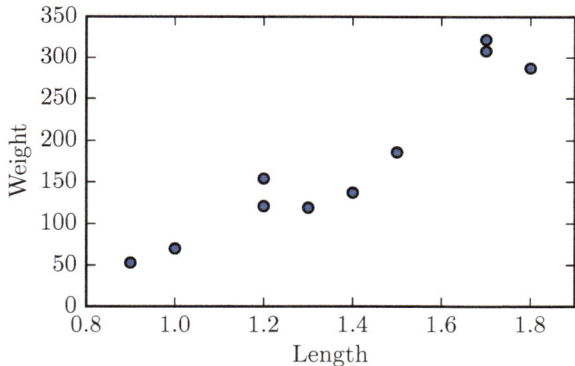

Visually we can tell that the two sets of data are positively correlated. Computing Pearson's product-moment correlation coefficient we obtain a value of $r = 0.9444643449 \approx 0.944$, indicating strong positive correlation.

**Remark**

We use $\rho$ to denote the product-moment correlation coefficient of a sample and $r$ to denote the product moment correlation coefficient of a population.

**Use of Technology 24.1 — Calculating Correlation Coefficients**

Many scientific calculators will now calculate correlation coefficients. The step-by-step guides below, in the digital book, show how some calculators can be used to find regression lines.

**Learning Resource 24.1 — Interpreting Correlations**

Download the card sort from the digital book. Each complete set should include the raw data, an explanation of the data, a scatter plot and the value of the correlation coefficient. Any missing cards should be completed.

**Activity 24.2 — Coding and Covariance**

We know that if we use a coding $W_i = aX_i + b$, then $\mu_W = |a|\mu_X + b$ and $\sigma_W = |a|\sigma_X$. What is the effect of coding on covariance? Express the covariance of $(aX_i + b, Y_i)$ in terms of the covariance of $(X_i, Y_i)$. Use this to show that the correlation coefficient is not changed by coding.

**Interactive Activity 24.1 — Influencing the Correlation**

Using the Geogebra applet in the digital book, explore the effect varying individual data points has on the correlation coefficient.

**Remark**

The correlation coefficient gives us a way of quantifying the linear relationship between two variables. If the variables are independent, then the coefficient will be close to zero. If the variables lie very close to a straight line, the coefficient will be very close to $+1$ or $-1$.

However, if the variables have some other relationship, then the correlation coefficient is not useful. In the example here, the two variables are clearly not independent, but the correlation is almost zero.

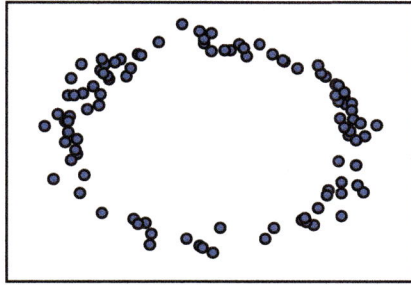

The relationship is not linear and so it is not captured by the correlation coefficient. First check that the data has a linear relationship, then calculate the correlation coefficient to determine the strength of that relationship.

Suppose that the variables $X_i$ and $Y_i$ are perfectly positively correlated, so $\rho_{XY} = +1$. Perfect linear correlation means that the points all lie exactly on some straight line, $y = mx + c$ for some $m > 0$. That is, $Y_i = mX_i + c$. If we take the standard deviation of both sides of this equation, we see that $\sigma_Y = m\sigma_X$. This means that the data lies on a line with a gradient equal to $\sigma_Y/\sigma_X$. Similarly, if the variables are perfectly negatively correlated, the gradient is equal to $-\sigma_Y/\sigma_X$. If the correlation is not perfect, then the points lie near, but not always on, some straight line $y = mx + c$. Let us say that $Y_i = mX_i + c + \epsilon_i$, where $\epsilon_i$ represents the vertical distance of the point $Y_i$ away from the line. We want to choose our regression line so that these distances $\epsilon_i$ are as small as possible.

We will do this by choosing values of $m$ and $c$ which minimise $E = \sum \epsilon_i^2$. We consider the data points $X_i$ and $Y_i$ to be fixed values so that $E$ only depends on $m$ and $c$.

We can use differentiation to find the place at which $E$ takes its minimum value.

$$\frac{dE}{dc} = \frac{d}{dc} \sum_{i=1}^{N} \epsilon_i^2 = \frac{d}{dc} \sum_{i=1}^{N} (Y_i - mX_i - c)^2 = -2\sum_{i=1}^{N} (Y_i - mX_i - c).$$

We can set this to zero to find a relationship between $m$ and $c$.

$$\sum_{i=1}^{N} (Y_i - mX_i - c) = 0 \to \frac{1}{N} \sum_{i=1}^{N} (Y_i - mX_i - c) = 0 \to \mu_Y = m\mu_X + c.$$

Now we can repeat this for $m$.

$$\frac{dE}{dm} = \frac{d}{dm} \sum_{i=1}^{N} \epsilon_i^2 = \frac{d}{dm} \sum_{i=1}^{N} (Y_i - mX_i - c)^2 = -2\sum X_i (Y_i - mX_i - c).$$

Again, we set this to zero.

$$\sum X_i (Y_i - mX_i - c) = 0 \Rightarrow \frac{1}{N} \sum_{i=1}^{N} X_i (Y_i - mX_i - c) = 0,$$

$$\Rightarrow \frac{1}{N} \sum_{i=1}^{N} X_i Y_i - \frac{m}{N} \sum_{i=1}^{N} X_i^2 - c\mu_X = 0,$$

$$\Rightarrow \frac{1}{N} \sum_{i=1}^{N} X_i Y_i - \frac{m}{N} \sum_{i=1}^{N} X_i^2 - (\mu_Y - m\mu_X)\mu_X = 0,$$

$$\Rightarrow \frac{1}{N} \sum_{i=1}^{N} X_i Y_i - \mu_X \mu_Y = m \left( \frac{1}{N} \sum_{i=1}^{N} X_i^2 - \mu_X^2 \right),$$

$$\Rightarrow \sigma_{XY} = m\sigma_X^2,$$

$$\Rightarrow m = \frac{\sigma_{XY}}{\sigma_X^2}.$$

Finally, we can substitute this back into the equation $\mu_Y = m\mu_X + c$ to derive

$$c = \mu_Y - \frac{\sigma_{XY}}{\sigma_X^2} \mu_X.$$

---

**Remark**
It will not be necessary to reproduce this derivation: the values will either be supplied, or can be found using a scientific calculator.

---

**Use of Technology 24.2 — Calculating Regression Lines**
The step-by-step guides below, in the digital book, show how some calculators can be used to find regression lines.

---

**Example 24.4**
The number of days absent from Sixth Form of 10 students was compared with the total UCAS tarriff points they obtained at the end of Year 13.

| length (cm) | 10 | 26 | 82 | 43 | 2 | 17 | 38 | 19 | 22 | 60 |
|---|---|---|---|---|---|---|---|---|---|---|
| weight (g) | 168 | 128 | 56 | 104 | 32 | 120 | 88 | 136 | 100 | 104 |

By finding the regression line estimate the number of UCAS points a student who has had 30 days off would achieve. What are limitations of this model?

**Solution:**
With this data, letting $d$ be the number of days absent at $p$ being the UCAS points obtained, we find the following regression line.

$$p = 118.62 - 0.469d$$

We can plot these data points on a scatter plot together with the regression line.

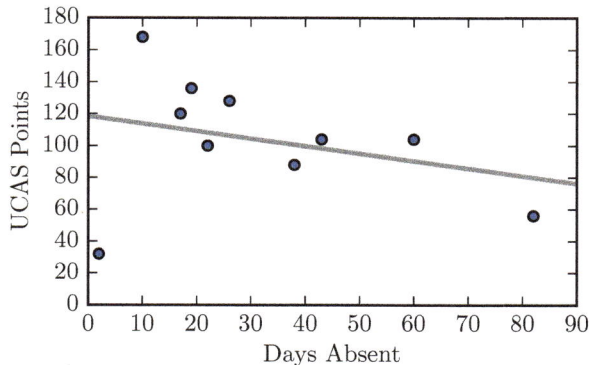

With this model if a student had 30 days of absence we would predict their number of UCAS points to be 105.

One limitation of this model is that we have assumed a continuous variation in the number of points achieved. In actuality, UCAS points increase at distinct points based on the grades achieved. This model can predict a number of UCAS points that is not possible. Whilst the data overall shows negative correlation it would appear that there is at least one outlier point, namely, $(2, 32)$. Removal of this outlier could have resulted in a regression line that better matched the remaining data.

**Remark**

The gradient of the regression line is equal to the covariance of $X$ and $Y$ divided by the variance of $X$. Since

$$\sigma_{XY} = \frac{1}{N} \sum_{i=1}^{N} (X_i - \mu_X)(Y_i - \mu_Y) \text{ and } \sigma_X^2 = \frac{1}{N} \sum_{i=1}^{N} (X_i - \mu_X)^2,$$

we can put these together to obtain an alternative expression for the gradient.

$$m = \frac{\sum_{i=1}^{N} (X_i - \mu_X)(Y_i - \mu_Y)}{\sum (X_i - \mu_X)^2}.$$

We complete the equation $y = mx + c$ by setting $c = \mu_Y + m\mu_X$, as before.

**Remark**

In 1973 the statistician Francis Anscombe published a paper concerning the importance of visualising statistical data. In this paper he considered the four data sets shown below.

| Set 1 | | Set 2 | | Set 3 | | Set 4 | |
|---|---|---|---|---|---|---|---|
| $X_1$ | $Y_1$ | $X_2$ | $Y_2$ | $X_3$ | $Y_3$ | $X_4$ | $Y_4$ |
| 10.0 | 8.04 | 10.0 | 9.14 | 10.0 | 7.46 | 8.0 | 6.58 |
| 8.0 | 6.95 | 18.0 | 8.14 | 8.0 | 6.77 | 8.0 | 5.76 |
| 13.0 | 7.58 | 13.0 | 8.74 | 13.0 | 12.74 | 8.0 | 7.71 |
| 9.0 | 8.81 | 9.0 | 8.77 | 9.0 | 77.11 | 8.0 | 8.84 |
| 11.0 | 8.33 | 11.0 | 9.26 | 11.0 | 7.81 | 8.0 | 8.47 |
| 14.0 | 9.96 | 14.0 | 8.10 | 14.0 | 8.84 | 8.0 | 7.04 |
| 6.0 | 7.2 | 6.0 | 6.13 | 6.0 | 6.08 | 19.0 | 5.25 |
| 4.0 | 4.26 | 4.0 | 3.10 | 4.0 | 5.39 | 19.0 | 12.50 |
| 12.0 | 10.84 | 12.0 | 9.13 | 8.15 | 8.0 | 8.0 | 5.56 |
| 7.0 | 4.82 | 7.0 | 7.26 | 7.0 | 6.42 | 8.0 | 7.91 |
| 5.0 | 5.68 | 5.0 | 94.74 | 5.0 | 5.73 | 8.0 | 6.589 |

When visualised, these datasets appear very different.

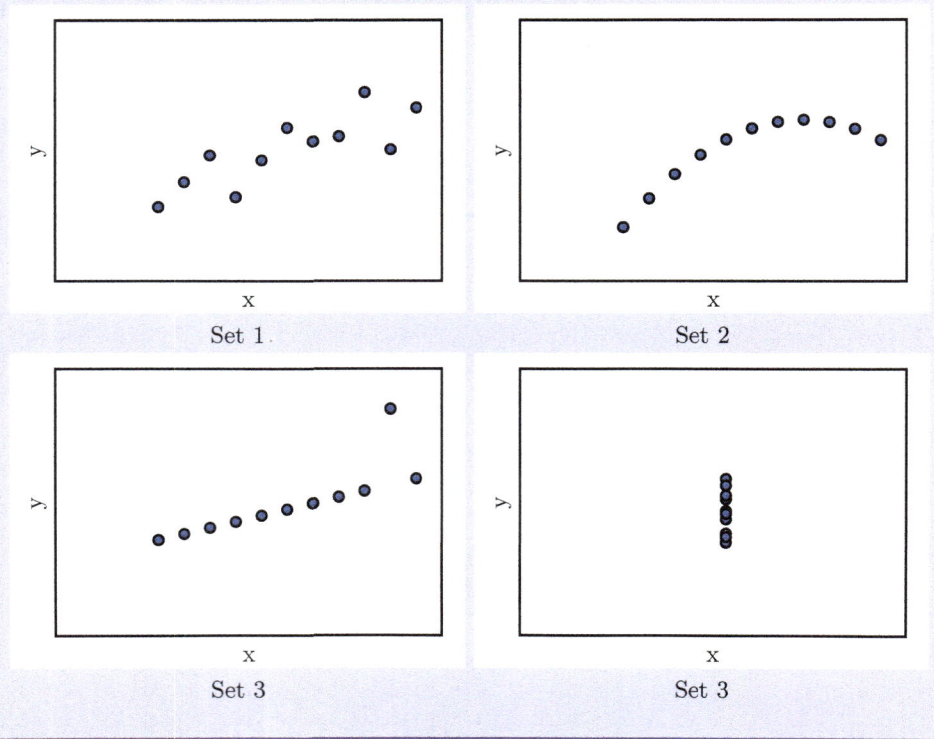

Set 1

Set 2

Set 3

Set 3

Despite this very different quantitative behaviour the following quantities for each dataset are the same:

- The mean of $x$.
- The mean of $y$.
- The sample variance of $x$.
- The sample variance of $y$.
- The correlation between $x$ and $y$.
- The equation of the linear regression line.

These datasets highlight the importance of visualising data in addition to computing summary statistics.

**Learning Resource 24.2 — Anscombe's Quartet**

Download the activity below from the digital book and explore Anscombe's quartet with the use of a statistical software package.

**Interactive Activity 24.2 — Exploring the Datasaurus Set**

In 2017 Justin Matejka and George Fitzmaurice of Autodesk Research published a paper entitled "Same Stats, Different Graphs: Generating Datasets with Varied Appearance and Identical Statistics through Simulated Annealing". In this paper they have produced a modern successor to Anscombe's quartet. Use the Geogebra applet in the digital book to explore this data.

**Exercise 24.2**

Q1. A number of grass frogs are captured, weighed, measured and released.

| length (cm) | 6.1 | 6.3 | 6.7 | 7.2 | 7.4 | 7.6 | 7.7 | 8.2 | 8.5 | 8.9 |
|---|---|---|---|---|---|---|---|---|---|---|
| weight (g) | 14 | 20 | 22 | 20 | 22 | 29 | 24 | 33 | 32 | 43 |

(a) Plot a scatter graph of the data and describe the correlation of the data.

(b) A scientist describes the relationship between the length, $l$, and weight, $w$, of typical grass frog using the equation

$$w = 8.39l - 36.7.$$

Use this equation to estimate the length of a frog which weighs $30\,\text{g}$.

(c) The scientist estimates that a frog measuring $10.5\,\text{cm}$ will weigh about $51\,\text{g}$. Comment on this estimate.

Q2. A garage sold a number of a certain model of car and recorded the age of the car in months, and the price it was sold for in thousands of pounds.

| Age ($A_i$)   | 40   | 42 | 51 | 53  | 60  | 62  | 64  | 65  | 65  | 70  |
|---------------|------|----|----|-----|-----|-----|-----|-----|-----|-----|
| Price ($P_p$) | 11.2 | 10 | 7  | 6.5 | 6.5 | 5.9 | 6.5 | 5.9 | 5.3 | 5.5 |

(a) Which variable is the response variable? Plot a scatter graph and describe the correlation of the data.

(b) Calculate the covariance of the data.

(c) The gradient of the regression line is given by

$$m = \frac{\sigma_{AP}}{\sigma_A^2}.$$

Use the value $\sigma_A = 9.72$ months to calculate the gradient of the regression line. What does the gradient represent?

Q3. In a previous exercise, we suggested that the length of a film's title was not correlated with its financial success. We will now try to confirm this from data. The following table shows the length of film titles of all the films nominated in 2015 for an Academy Award for Best Director, Best Film, or Best Actor or Actress in a Leading or Supporting Role. The table also shows the reported box office takings for those films, in millions of dollars.

| Title length ($l_i$) | 8     | 8     | 7    | 39    | 10    | 4     | 16    | 12    |
|----------------------|-------|-------|------|-------|-------|-------|-------|-------|
| Box Office ($b_i$)   | 49    | 84.4  | 44.5 | 103.2 | 13.6  | 52.5  | 233.6 | 213.1 |
| Title length ($l_i$) | 21    | 14    | 10   | 15    | 21    | 8     | 5     |       |
| Box Office ($b_i$)   | 174.8 | 547.4 | 43.9 | 9.0   | 123.7 | 369.3 | 66.8  |       |

(a) Plot this data on a scatter graph, and describe the correlation of the variables.

(b) Using $\sum l_i = 166$, $\sum b_i = 2128.8$ and $\sum l_i b_i = 26\,536.4$, calculate the covariance of the data.

(c) The correlation coefficient is given by

$$\rho = \frac{\sigma_{lb}}{\sigma_l \sigma_b}.$$

Using $\sum l_i^2 = 2234$ and $\sum b_i^2 = 613\,430$, find the correlation coefficient. Does its value confirm the description of the correlation?

---

## Key Facts — Presenting Multivariate Data

- Scatter graphs are used to visually display the relationship between two sets of data.
- The covariance of $X_i$ and $Y_i$ is given by

$$\sigma_{XY} = \text{cov}(X, Y)$$

$$= \frac{1}{N} \sum_{i=1}^{N} (X_i - \mu_X)(Y_i - \mu_Y),$$

where $\mu_X$ is the mean of $X_i$ and $\mu_Y$ is the mean of $Y_i$.

Equivalently,

$$\sigma_{XY} = \frac{1}{N} \sum_{i=1}^{N} X_i Y_i - \mu_X \mu_Y.$$

- Two variables are positively correlated if when one is larger, the other one also tends to be larger.
- They are negatively correlated if one tends to be smaller when the other is larger.
- If there is no relation between the variables, they are uncorrelated.
- The correlation coefficient of $X_i$ and $Y_i$ is a dimensionless number in the range $[-1, +1]$. It is given by

$$\rho_{XY} = \frac{\sigma_{XY}}{\sigma_X \sigma_Y},$$

where $\sigma_X$ and $\sigma_Y$ are the standard deviations of $X_i$ and $Y_i$, and $\sigma_{XY}$ is their covariance.
- Correlation does not imply causation.
- If the points on a scatter plot appear to be correlated and lie fairly close to a straight line, we can plot a line of best fit through the data and use this to make predictions.
- Take care when making predictions from data. If the correlation is weak, we cannot expect our predictions to be very accurate.
- We have more confidence in making predictions within the range of the observed data than for predictions outside the range.

**Chapter Assessment — Presenting Multivariate Data**

Download and sit the 30 minute assessment for this chapter from the digital book.

# 25. Analysing Large Data Sets

The variety and volume of statistical data has grown dramatically over the last few decades, and this growth has transformed our society.

In 2013, the European Space Agency launched the Gaia space observatory. It will record the position and movement of a billion stars, both within our own galaxy and in other galaxies in our Local Group.

Governments and businesses can monitor our behaviour, with or without our knowledge, and create databases of our movements and activities. Every time we send a text message, visit a web site, make an electronic payment, or perform one of many other daily activities, a piece of data is created. Each one is stored, somewhere, possibly for months or years.

Many activities in modern society, from advertising to astronomy, require us to work with data sets that are much larger than anything we could analyse by hand. We need computers; first to collect and store these data sets, and then to analyse them.

In this chapter, we will examine ways of analysing and understanding large data sets, using data taken from the BP Statistical Review of World Energy 2016. We will see that real world data does not always, or easily, give clear answers, but requires honest and skilful interpretation.

## 25.1 Energy

Everything we do, make, or build requires energy. For much of human history, almost all of this energy has come either from human or animal muscle power directly, or from biomass, such as wood and peat. Since muscle power comes from food, and food comes from other animals or from plants, all of these energy sources can be traced back to the energy contained in sunlight falling on a plant.

The energy embodied in fresh vegetables fell on the earth within the last year. If we burn some wood, that heat energy comes from sunlight that arrived on earth probably within the last decade or two. For hundreds of thousands of years, all of human life – its civilisations and empires – was powered almost entirely by this yearly ration of sunlight.

As countries began to industrialise, people realised that the new factories, which had been powered by water wheel, could also be powered by coal. Coal is more controllable than a river, and it allowed factories to be built anywhere.

Instead of being restricted to energy from sunlight that fell within the last few years, human society could access energy from sunlight that fell hundreds of millions of years ago. Figure 25.1 shows how fossilised carbon compounds became the dominant fuel source[1]. Not only did the share of fossil fuels grow from around 5% in 1820 to over 80% in 2003, but in little more than the span of two human lifetimes, the total amount of energy used by society grew fifty-fold.

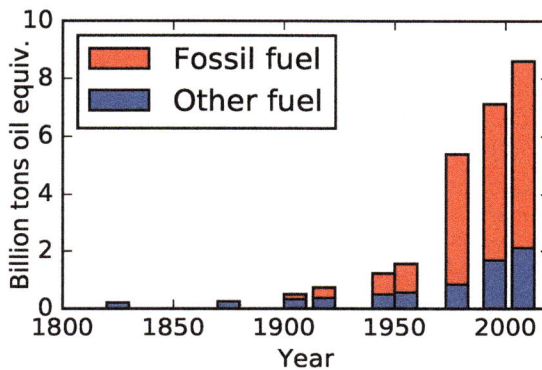

Figure 25.1: World Consumption of Primary Energy for Selected Years 1820 to 2003.

Two hundred years ago, society's energy came from agriculture. The energy to transport goods and build buildings came from muscle power. Muscles were fed with food, and the food was provided by farms. If the effort of farming was greater than the energy provided by its food, a community would starve. Everyone in the society knew the possible consequences of a failed harvest. For the poor, hunger and hardship: for the rich, the risk of civil disorder and revolution.

Today, agriculture does not supply us with any energy. The amount of energy we use to plant, tend, harvest, process and transport our food, far outweighs the energy it embodies. Most of our energy comes from fossil fuels: oil, coal and natural gas.

**Remark**
One way of measuring the cost of energy is by calculating how much fuel can be produced for each unit of energy expended. In the 1970s, with the energy from one barrel of oil, sixty or more barrels could be extracted from North American oil fields. We say that the energy return on energy invested (EROEI) was 60.

This number can be pushed higher by improving efficiency, but as the most easily available resources become depleted the value will tend to fall. In 2007, the EROEI for North American oil had dropped to somewhere between 10 and 15. If the EROEI for a

---

[1]Non-fossil fuels include hydro, wind, nuclear and biomass, such as wood and peat. The data is taken from *Contours of the World Economy 1-2030 AD: Essays in Macro-Economic History* by Angus Maddison (Oxford University Press 2007).

resource falls below one, then it can no longer supply any net energy.

There are different ways of calculating the EROEI, and the results can be controversial. Energy companies would like to report a high value. If the fuel is easy to extract, they will expect to make a bigger profit, and the company's value will increase. However, for anyone who opposes the energy companies, such as environmentalists or those living near a proposed development, their position would be supported by a *lower* EROEI.

The UK government, in an attempt to reduce carbon emissions, requires about 5% of vehicle fuel to come from renewable sources. One way suppliers meet this requirement is by mixing petrol with ethanol made from plants, such as wheat or sugar cane. The EROEI for ethanol depends on the plant being used, but some studies have shown that the EROEI for ethanol can be less than one. If this calculation is correct, then the process of making ethanol can require more energy from other fuels than we obtain from the finished product[a].

---

[a]Values are taken from *EROI of different fuels and the implications for society* by Charles Hall, Jessica Lambert, Stephen Balogh (Elsevier 2014).

The supply of energy to developed countries has become so reliable, that we no longer fear the results of a poor harvest. However, we know that there are threats to our energy supply. These include resource depletion, political instability and the need to change our habits to try to avoid the worst effects of pollution and climate change.

In this chapter, we will look at the single fuel which provides most of the world's energy: oil. We will look at who has it, how much they are producing, who uses it, and what effect it has on the world around us.

## 25.2 Oil

Extracting oil from the ground, in theory, follows a simple basic pattern. Oil companies find a region where there is some good quality oil which can be extracted easily enough to make a profit. As the company drills more wells, and builds more infrastructure in the area, more oil is produced. Eventually, the oil which is still in the ground becomes more difficult, and more expensive, to extract. The activity slows down, fewer new wells are drilled, and the volume of oil produced falls. Finally, no more oil can be extracted from the area and the oil company moves on to a new, more profitable region.

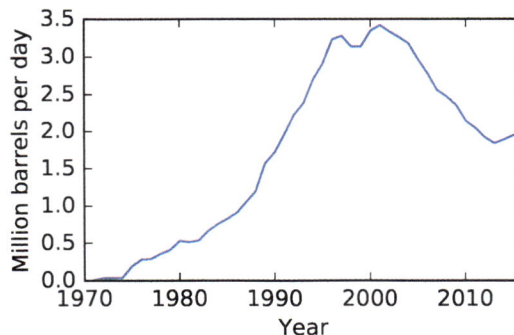

Figure 25.2: Norwegian Oil Production, 1971 to 2015.

Figure 25.2 shows us the basic pattern very clearly for Norway's oil production, from the time that the North Sea oil reserves were first exploited, through the peak production in 2001 and into the clear declining phase.

The graph also shows that smooth production curves are always affected by other factors. For example, in 1998, a drop in the oil price prompted many countries, including Norway, to cut back on production in an attempt to maintain the price. This can be seen as a small dip just before the peak.

In recent years, oil prices have been very volatile. The price of a barrel of oil hit an all time high of $147 in July 2008. By December, the price had dropped to $30. Prices climbed again, and a barrel cost around $100 for the next few years, but then dropped back to around $40[2].

When the oil price is high, oil companies increase their spending and start to develop more costly reserves. This shows up as an increase in production a few years later. The high prices around 2008 explains the uptick in production at the end of the graph of Norwegian production. Since prices did not stay high, this is not expected to last, and the downward trend is forecast to resume within the next few years.

## 25.3  Interpreting Data

Data can only be interpreted correctly when the real thing being described is sufficiently well understood.

For example, our data set includes historical values for oil reserves, broken down by year and country. We might think that these figures tell us how much oil each country has underneath its land and waters, and that this data can be easily checked, but the situation quickly becomes much more complicated than this.

The oil in the ground is referred to as the *resources*, but most of it will never be recovered. If the financial or physical cost of extracting and processing the oil is too high, it cannot be used as a source of energy. Trying to run a society on this fuel would be the equivalent of a rural society whose farming required more energy than the energy in the food it produced.

> **Tip**
> When interpreting the data set, remember that oil reserves and production do not depend only on the size of the oil resources in the ground.
> The reported reserves also depend on the following.
> - The cost, in terms of money and energy, in extracting the oil.
> - The current price of oil.
> - The cost of the current technology. Advances in technology may cause the reserves to increase.
> - Political or commercial pressures on the oil company to under- or over-report the reserves.

The data set contains the *reserves* or, more strictly, the *proven reserves*. This is the amount of oil that the engineers are almost certain can be extracted at today's prices with today's technology. We can see that this is not a directly measurable number, but will require some judgement on the part of those making this estimate.

---

[2]For consumers, these price changes show up first in the price of petrol and diesel. In the United Kingdom, tax makes up a large part of the fuel price, and so large changes in the crude oil price lead to smaller percentage changes in the fuel price.

Furthermore, the reserves data for an oil company will affect the confidence that others place in that company. For publicly traded companies, such as BP, they are most likely to gain investors' confidence if they are honest and follow stock market guidelines. For this reason, figures from these companies are considered to be fairly reliable[3].

Governments that control their oil industries directly do not have the same incentive to publish accurate reserve figures. Countries which depend very heavily on their oil revenue can find that larger reserves brings them more political influence. In some cases, information about oil reserves are state secrets. This does not mean that the figures are incorrect, but it does mean that there is no way of verifying them independently.

---

**Tip**

The amount of oil being extracted partly depends on the following.

- The expense involved in extracting the oil and transforming it into a usable fuel.
- The price of crude oil, both now, and while the area was being developed.
- If there are any wars or conflicts in the region.
- Whether the country is trying to alter the oil price by altering its output.

---

**Example 25.1**

Annual oil production in the United Kingdom peaked in 1999.

(a) Calculate the median annual percentage production drop for UK oil production from its peak through to 2014.

(b) Suppose that a politician, speaking at the beginning of 2015, had claimed that "without additional investment, by the middle of the next decade our oil production will be half what it is today."

Use your previous answer to explain whether this is a reasonable claim.

**Solution:**

(a) We can calculate the percentage drop for each year from the data set.

---

[3]Mistakes can be still made, however. In 2004, the oil company Shell overestimated their confidence in how much oil they could extract. As a result, the company faced large fines, its value dropped by billions of pounds and the chairman resigned.

| Year | Production | Decline |
|------|-----------|---------|
| 1999 | 2930.2 |  |
| 2000 | 2714.1 | 7.37 |
| 2001 | 2518.0 | 7.22 |
| 2002 | 2504.3 | 0.54 |
| 2003 | 2295.8 | 8.32 |
| 2004 | 2063.8 | 10.10 |
| 2005 | 1842.8 | 10.70 |
| 2006 | 1666.3 | 9.57 |
| 2007 | 1658.7 | 0.45 |
| 2008 | 1555.0 | 6.25 |
| 2009 | 1476.9 | 5.02 |
| 2010 | 1361.2 | 7.83 |
| 2011 | 1115.9 | 18.02 |
| 2012 | 948.9 | 14.96 |
| 2013 | 866.7 | 8.66 |
| 2014 | 855.0 | 1.34 |

The production is given in thousand barrels per day, and the decline is given as a percentage.

If we put these values into order, we find that the median value is 7.83%.

(b) If we assume that the yearly percentage decline will be equal to the median in coming years, then we can calculate the drop from 2014 to 2025.

$$\text{(Output in 2025)} = \text{(Output in 2014)} \times (1 - 0.0783)^{11}$$
$$= 855.0 \times (0.9217)^{11} = 148.7.$$

That is, the output would drop to around 41% of its value. This would tend to support the politician's claim. In fact, output may have fallen to half its current level well before the middle of the decade.

It could be argued that the range of data values is very wide, with values between 0.45% and 18.02% and that, therefore, the median value may not be a good guide to the next ten years.

Instead of calculating the median, we could calculate a constant annual percentage decline which would produce the same drop in output from 1999 to 2014. Suppose that the decline had been a constant $r\%$ every year. In that case,

$$\text{(Output in 1999)} \times \left(1 - \frac{r}{100}\right)^{15} = \text{(Output in 2014)}$$
$$2930.2 \times \left(1 - \frac{r}{100}\right)^{15} = 855.0$$
$$\left(1 - \frac{r}{100}\right)^{15} = \frac{855.0}{2930.2} = 0.2918$$
$$\left(1 - \frac{r}{100}\right) = (0.2918)^{1/15} = 0.921$$
$$r = 7.9.$$

This gives us an alternative value of 7.9% which is close enough to the median to justify using either value as a guide to the future annual depletion rate.

**Remark**

When analysing data which is growing or shrinking at a steady rate, it can be helpful to be aware of the Rule of 70, or one of its variations. This rule gives a quick estimate of how long it will take for the quantity to halve, or to double.

This doubling time is often easier to understand than the percentage. It is not easy to understand the implication of a country using 2% more resources each year. It is simpler to understand that in 35 years, twice as many power plants will be needed.

The Rule of 70 says that if some quantity grows by $r\%$ in one time unit, it will double after approximately $(70/r)$ time units. Similarly, if the quantity shrinks by $r\%$ in one time unit, it will halve after $(70/r)$ time intervals. For example, a population which grows at 2% per year will double in approximately $70/2 = 35$ years.

These estimates are more accurate for small percentages. For values up to 12%, the error is less than 5% of the actual value.

The rule is derived as follows. If some quantity starts with a value $x$ and grows at $r\%$ every unit of time, then after $k$ units of time its value will be $x(1 + r/100)^k$. If it has doubled in size after $T$ units, then

$$x \left(1 + \frac{r}{100}\right)^T = 2x.$$

We use two approximations to derive the rule. For small values of $r$, we will use the approximation $\log_e(1 + r/100) = r/100$. We will also use the approximation $\log_e 2 \simeq 0.7$.

$$\left(1 + \frac{r}{100}\right)^T = 2,$$
$$T \log_e \left(1 + \frac{r}{100}\right) = \log_e 2.$$

Applying the approximations above, we obtain $rT/100 \simeq 0.7$ and so $T \simeq 70/r$.

We obtain a similar result if the quantity shrinks to $(1 - r/100)$ in each time interval. For example, if oil production from an area reduces by 7% each year, it will halve approximately every ten years.

This confirms our calculation in the previous example, that a depletion rate of 7.83% would cause the output to drop by over a half in the course of eleven years.

**Exercise 25.1**

Q1. The following table shows the oil reserves and production in 2015 for a sample of countries.

| Country | Reserves ($10^9$ barrels) | Production ($10^3$ barrels per day) |
|---|---|---|
| Argentina | 2.4 | 636.6 |
| Colombia | 2.3 | 1007.6 |
| Peru | 1.4 | 112.6 |
| Azerbaijan | 7.0 | 840.6 |
| Italy | 0.6 | 114.7 |
| Norway | 8.0 | 1947.8 |
| United Kingdom | 2.8 | 965.1 |
| Uzbekistan | 0.6 | 64.2 |
| Qatar | 25.7 | 1898.0 |
| Syria | 2.5 | 27.0 |
| Yemen | 3.0 | 46.5 |
| Algeria | 12.2 | 1585.5 |
| Chad | 1.5 | 78.4 |
| Egypt | 3.5 | 722.7 |
| Gabon | 2.0 | 232.8 |
| Libya | 48.4 | 431.9 |
| Nigeria | 37.1 | 2352.1 |
| India | 5.7 | 876.1 |
| Malaysia | 3.6 | 693.0 |
| Vietnam | 4.4 | 361.9 |

(a) Draw a scatter diagram of the data.

(b) Describe the correlation between a country's oil reserves and its oil production.

(c) Identify any outliers in the data, and give one possible reason for their position.

(d) Oman had reserves of 5.3 thousand million barrels in 2015. Use your scatter diagram to estimate its production in that year.

(e) Venezuela had an estimated 300 thousand million barrels of oil reserves remaining in 2015. Its production for that year was around 2.6 million barrels per day. Use the scatter diagram to explain why this production rate could be considered to be very low.

Q2. Here is an extract from the data set.

(a) Calculate the mean oil production by the United States during the 1990s.

(b) Calculate the median level of oil reserves for the United States.

(c) Which type of average would be more suitable for stating Mexico's reported average oil reserves during the 1990s? Give one reason for your answer.

(d) Suggest one possible reason for the dramatic drop in the Mexican reserves.

| Year | Reserves ($10^9$ barrels) | | Production ($10^3$ barrels per day) |
|---|---|---|---|
| | Mexico | US | US |
| 1990 | 51.3 | 33.8 | 8914.3 |
| 1991 | 50.9 | 32.1 | 9075.5 |
| 1992 | 51.2 | 31.2 | 8868.1 |
| 1993 | 50.8 | 30.2 | 8582.7 |
| 1994 | 49.8 | 29.6 | 8388.6 |
| 1995 | 48.8 | 29.8 | 8321.6 |
| 1996 | 48.5 | 29.8 | 8294.5 |
| 1997 | 47.8 | 30.5 | 8268.6 |
| 1998 | 21.6 | 28.6 | 8010.8 |
| 1999 | 21.5 | 29.7 | 7731.5 |

Q3. The following graph shows the oil production of the United Kingdom and Kuwait from 1965 to 2015.

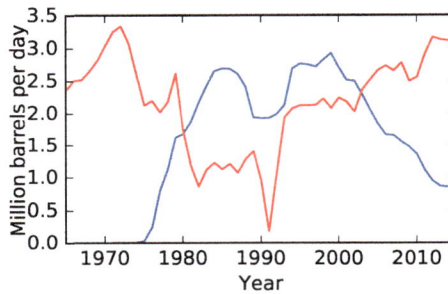

The United Kingdom exploited most of its North Sea oil reserves in the late twentieth century. Kuwait remains a significant producer of oil and its economy relies heavily on income from oil.

(a) Identify which line represents which country and describe the effects of the following events on oil production.
  • The attempt in the 1980s by OPEC members, including Kuwait, to raise the oil price.
  • The destruction of the North Sea oil platform Piper Alpha, with the loss of 167 lives, in 1988.
  • The Iraqi invasion of Kuwait in 1990.

(b) What might have caused the steep drop in the blue line from the year 2000 onwards?

(c) For every year from 2004 to 2015, Kuwait reported oil reserves of 101.5 thousand million barrels. Comment on the likely quality of this data.

Q4. The following table shows the oil consumption, in thousands of barrels per day, for three countries between 1990 and 2005.

| Year | France | Italy | India |
|------|--------|-------|-------|
| 1990 | 1894.6 | 1924.4 | 1211.1 |
| 1991 | 2001.1 | 1900.6 | 1232.7 |
| 1992 | 1995.7 | 1931.9 | 1296.1 |
| 1993 | 1925.5 | 1906.2 | 1312.5 |
| 1994 | 1864.2 | 1903.5 | 1411.7 |
| 1995 | 1879.3 | 1971.7 | 1579.5 |
| 1996 | 1916.5 | 1941.9 | 1699.0 |
| 1997 | 1936.3 | 1954.1 | 1829.5 |
| 1998 | 2002.5 | 1955.2 | 1965.8 |
| 1999 | 2030.2 | 1962.5 | 2138.3 |
| 2000 | 1994.1 | 1928.0 | 2258.8 |
| 2001 | 2009.9 | 1918.7 | 2285.5 |
| 2002 | 1953.4 | 1914.8 | 2413.4 |
| 2003 | 1951.5 | 1900.5 | 2485.3 |
| 2004 | 1963.3 | 1849.9 | 2555.5 |
| 2005 | 1946.2 | 1797.7 | 2605.6 |

(a) Explain how to use random sampling to generate a sample of five data points from the data for France. Use your method to calculate a sample mean.

(b) Explain why different students might obtain different sample means from this data.

(c) Use systematic sampling to select five data points from the data for Italy, and calculate the sample mean.

(d) India has been industrialising during the second half of the twentieth century. Calculate the median percentage yearly increase of India's oil consumption.

(e) Using the median percentage increase, estimate India's oil consumption in 2015.

(f) At this rate, how many years would it take for India's oil consumption to double? When would India be using 84 726 000 barrels per day, the equivalent of the entire global oil production for 2005?

Q5. (a) Use the large data set to draw a histogram showing Denmark's yearly $CO_2$ emissions from 1991 to 2015.

(b) The following box-and-whisker plot shows the $CO_2$ emissions from Denmark between 1965 to 1989 and 1991 to 2015.

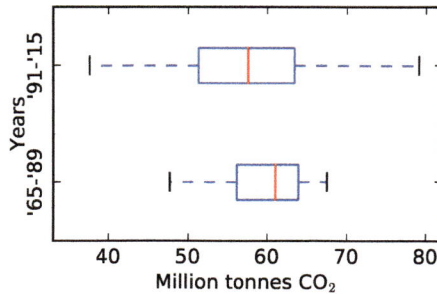

Plot the equivalent graph for Norway. Mark any outliers clearly.

(c) A newspaper article claims that "countries that produce more oil tend to use more oil."

Give one reason why these box plots for Norway and Denmark support this statement.

(d) Give one reason why the statement might not be true.

Q6. (a) Use the large data set to plot oil production of the United States between 1965 to 2015. Describe the features of the graph.

(b) Some US oil comes from tight oil formations (sometimes called shale oil). In 2000, the US produced less than one hundred thousand barrels of tight oil per day. In 2015, the figure was around five million barrels per day. Most of this oil came from the Eagle Ford field, discovered in 1962, and the Bakken field, discovered in 1953.

Give one reason these fields might not have been developed until decades after their discovery.

Q7. (a) Use the large data set to determine the country which emitted the largest amount of carbon dioxide in each year from 1965 to 2015.

(b) The Chinese government has tried to reduce their emissions by generating more renewable energy and by using less coal. The following table gives the relative size of the Chinese economy for the years 1965 to 2015 [a].

| 1965 to 1969 | 100 | 110 | 104 | 100 | 117 | |
|---|---|---|---|---|---|---|
| 1970 to 1974 | 139 | 149 | 155 | 167 | 171 | |
| 1975 to 1979 | 186 | 183 | 196 | 220 | 236 | |
| 1980 to 1984 | 255 | 268 | 292 | 323 | 373 | |
| 1985 to 1989 | 423 | 460 | 514 | 572 | 596 | |
| 1990 to 1994 | 619 | 677 | 773 | 881 | 995 | |
| 1995 to 1999 | 1105 | 1214 | 1326 | 1429 | 1539 | |
| 2000 to 2004 | 1670 | 1809 | 1974 | 2171 | 2390 | |
| 2005 to 2009 | 2663 | 3001 | 3428 | 3760 | 4114 | |
| 2010 to 2015 | 4550 | 4982 | 5375 | 5795 | 6218 | 6647 |

Use this data to draw a graph to illustrate whether or not the Chinese government's policies have succeeded in reducing carbon dioxide emissions without slowing economic growth.

---

[a]The figures are derived from the Gross Domestic Product data from the Chinese National Bureau of Statistics, as quoted on the Wikipedia website.

---

**Chapter Assessment — Analysing Large Data Sets**

Download and sit the 30 minute assessment for this chapter from the digital book.

# 26. Kinematics in One Dimension

In this chapter we use mathematics to predict the position, velocity and acceleration of an object which is free to move along a one dimensional path. Examples include Olympic sprinters or swimmers who must stay in their designated lane; here, the path is simply the straight line of the lane. Another example is a car travelling along a motorway. Although the motorway will not be straight, the car is constrained to stay on the motorway and, therefore, follows a one dimensional path.

Figure 26.1: Simple kinematics can be used to help model the motion of swimmers and dragsters.

Kinematics is the study of how we make these predictions and provides us with mathematical relationships between the object's position, its velocity and its acceleration. These relationships are called the *equations of motion*.

## 26.1  Definition of Terms

In order to derive the equations of motion, we first need some terminology.

Figure 26.2 shows an object, in this case, a toy car, which is free to move either to the left or right. The faded blue car shows its initial position, while the dark blue car is its position at some later time.

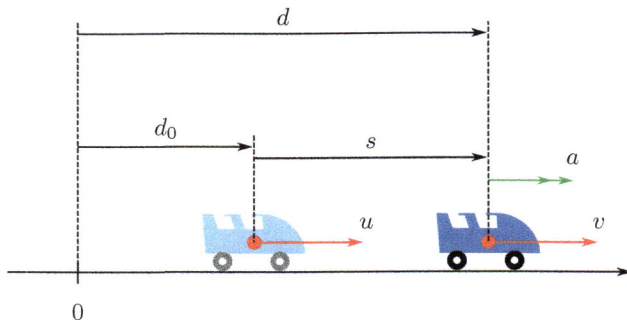

Figure 26.2: A toy car at an initial time $t = 0$ (faded blue) and at a time $t > 0$ (darker blue).

Figure 26.2 shows a number of different variables, which we describe below. These are the most important variables in kinematics and we should understand their meaning.

$t$:    The *time* relative to an initial time. $t = 0$ therefore represents the initial starting time of the car.

$r_0$:    The *initial position* of the object at time $t = 0$, relative to the origin.

$r$:    The *position* of the object at a later time $t$, relative to the origin.

$s$:    The *displacement* of the object at time $t$, relative to its initial position.

$u$:    The *initial velocity* of the object at time $t = 0$.

$v$:    The *velocity* of the object at time $t$.

$a$:    The *acceleration* at time $t$.

> **Remark**
> At time $t = 0$, we have $s = 0$ and $r = r_0$, hence, $r$ is just a translation of $s$ by $r_0$ units, with $r = s + r_0$.

> **Remark**
> An acceleration in the direction opposite to the velocity is termed a *deceleration*.

> **Definition 26.1 — Velocity, Speed and Acceleration**
> Consider an object with displacement $s(t)$ and position $r(t)$ at time $t$. Velocity, $v(t)$, is defined to be the (instantaneous) rate of change of displacement at time $t$. $v(t)$ is, therefore, the *time derivative* of $s$:
>
> $$v(t) = \frac{\mathrm{d}s}{\mathrm{d}t}(t). \tag{26.1}$$

As $s = r - r_0$, we also find that

$$v(t) = \frac{\mathrm{d}(r - r_0)}{\mathrm{d}t}$$
$$= \frac{\mathrm{d}r}{\mathrm{d}t} - \frac{\mathrm{d}r_0}{\mathrm{d}t}$$
$$= \frac{\mathrm{d}r}{\mathrm{d}t}. \tag{26.2}$$

Hence, velocity is equivalently the (instantaneous) rate of change of position.

Velocity is a vector quantity. In one-dimension, it can either be positively or negatively valued. Speed is defined to be the magnitude of the velocity $|v|$.

Acceleration, $a(t)$, is defined to be the (instantaneous) rate of change of velocity at time $t$:

$$a(t) = \frac{\mathrm{d}v}{\mathrm{d}t}(t). \tag{26.3}$$

Using the expressions for velocity (26.1) and (26.2), we find the alternative expressions for $a$:

$$a(t) = \frac{\mathrm{d}^2 s}{\mathrm{d}t^2}, \tag{26.4}$$
$$a(t) = \frac{\mathrm{d}^2 r}{\mathrm{d}t^2}. \tag{26.5}$$

In other words, acceleration is the second derivative of displacement, or position.

We remark that, as the car is free to move both to the left and to the right, the positions, displacements, velocities and acceleration can all take positive and negative values. For example, a negative velocity means the car is travelling towards the left.

## 26.1.1  S.I. Units

Whenever we discuss variables such as displacement, velocity and acceleration, it is important that we associate a unit of measurement to the variable. The International System of Units (more commonly known as *S.I. units*) provides a set of common *base* units from which the units for all other quantities can be derived. The S.I. unit of time is the second (s) and the universal unit of distance is the metre (m). Hence, displacement, $s$, will be measured in m. Now, as velocity is defined to be the rate of change of displacement, it has units $\mathrm{m\,s^{-1}}$, that is, *metres per second*, which is alternatively written as $\mathrm{ms^{-1}}$. This measure of velocity tells us how many metres an object will travel in one second. Similarly, since, acceleration is defined to be the rate of change of velocity, in S.I. units, acceleration will have units *metres per second per second*, or $\mathrm{m\,s^{-1}\,s^{-1}}$. Using the laws of indices, this becomes $\mathrm{ms^{-2}}$. When discussing the units of acceleration, it is common to say *metres per second squared*.

**Remark**
Originally, the second was defined as $\frac{1}{24 \times 60 \times 60} = \frac{1}{86\,400}$ of a day. However, this definition

of a second is not robust, due to the slowing down of the Earth's rotation: a second defined in this way would be longer today than it was in the past. There have been other definitions of a second over the years, but the current definition is $9\,192\,631\,770$ periods of the radiation of the caesium-133 atom.

Similarly, the metre had an original definition of $\frac{1}{10\,000\,000}$ of the meridian through Paris between the North Pole and the Equator. Again, this is not a robust measure of distance. The current definition of a metre is the distance travelled by light in a vacuum in $\frac{1}{299\,792\,458}$ of a second.

Other S.I. units of measure include the kilogram (kg) for mass and kelvin (K) for temperature. Note, it is the kilogram that is used for mass, not the gram, even though kilogram literally means 1000 grams.

Famously, in 1999, NASA's Mars Climate Orbiter accidentally entered Mars' atmosphere and broke up. The cause was that the control systems on board used Imperial units, while NASA's ground systems used S.I. units.

## 26.2  Displacement/Position-Time and Velocity/Speed-Time graphs

As is so often the case, plotting graphs of variables allows us to understand what is happening far more easily. We consider plots of displacement against time first, then velocity/speed-time graphs after that.

---

**Example 26.1**

Below are four displacement-time graphs. Match, with explanation, the graphs to the following scenarios.

   (a) A sprinter running the 100 m;

   (b) A swimmer completing four lengths of a swimming pool;

   (c) A train travelling along a track with five stations;

   (d) A lift moving in a three storey building with a basement, assuming the lift starts on the ground floor.

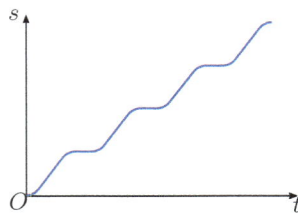
(i)

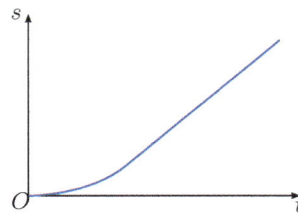
(ii)

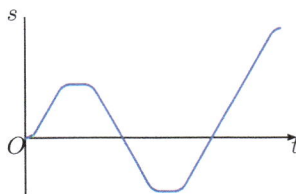
(iii)

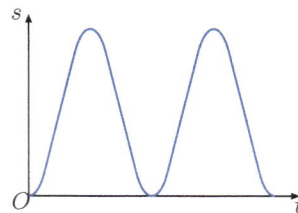
(iv)

**Solution:**

(a) The correct graph is (ii). A sprinter will start from rest, then their speed will increase until they reach a maximum speed at some point. They are likely to continue running at this speed until they cross the finish line. This period of constant speed appears as a straight line on the displacement-time graph.

(b) The correct graph is (iv). As with the sprinter, a swimmer will start from rest and their speed will increase to a maximum speed, for most of the length they will maintain this speed, before slowing down to turn around at the end of the lap, where their displacement is at a maximum. After turning around, the swimmer's speed will again increase to a maximum, but their velocity will be in the negative direction. Their displacement will reduce until they reach the starting point. As they are swimming four lengths, the same curve will then be reproduced again.

(c) The correct graph is (i). The train will start from rest and its speed will increase to a maximum before slowing down as it approaches the second station. The train will remain at rest in the station for a short period of time, which corresponds to the horizontal portion of the graph. Then the train will set off again, and the curve will repeat, albeit with the displacement from the starting point increasing.

(d) The correct graph is (iii). The motion of the lift is very similar to that of the train, however, the lift can move both up and down. Starting from the ground floor, the lift will move up to the first floor, then stop for a brief period, before moving down to the basement, where the displacement takes a negative value. Finally, it travels to the second floor.

We can plot position-time graphs as well. As we have seen, $s = r - r_0$, or $r = s + r_0$, hence the position-time graph will just be a translation in the $y$-direction by $r_0$ units. The next example demonstrates this.

**Example 26.2**

A pendulum is set in motion and the horizontal, $h$, and vertical position, $p$, of the bottom of the pendulum are given by

$$h(t) = \frac{1}{2}\left(1 - \frac{\sqrt{3}}{2}\right)(1 + \cos(180t)),$$

$$p(t) = \frac{1}{2}\cos(180t).$$

(a) On the same set of axes, sketch the position-time graphs for $h(t)$ and $p(t)$ for $0 \leq t \leq 4$.

(b) Let $s_p(t)$ and $s_h(t)$ be the horizontal and vertical displacements of the bottom of the pendulum, respectively. On the same set of axes, sketch the displacement-time graphs for $h(t)$ and $p(t)$ for $0 \leq t \leq 4$.

**Solution:**

(a) Sketches are shown below with $h(t)$ the blue curve and $p(t)$ the red curve.

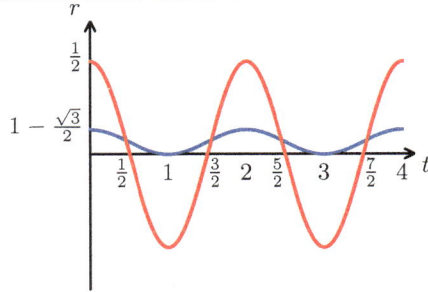

(b) We have that $s = r - r_0$. Hence,

$$s_h(t) = h(t) - h(0) = \frac{1}{2}\left(1 - \frac{\sqrt{3}}{2}\right)(1 + \cos(180t)) - \frac{1}{2}\left(1 - \frac{\sqrt{3}}{2}\right) \cdot 2$$

$$= \frac{1}{2}\left(1 - \frac{\sqrt{3}}{2}\right)(\cos(180t) - 1),$$

$$s_p(t) = p(t) - p(0) = \frac{1}{2}\cos(180t) - \frac{1}{2}.$$

Sketches of these are shown below, with $s_h(t)$ plotted in blue and $s_p(t)$ plotted in red.

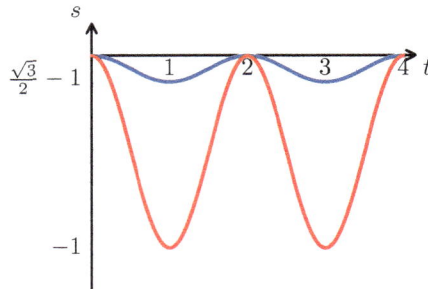

Instead of plotting displacement-time or position-time graphs, we could instead plot velocity against time or speed against time.

**Example 26.3**

The velocity of a particle is given by $v(t) = (2 - t)(t^2 + 1)$.
  (a) Plot the velocity of the particle against time.
  (b) Plot the speed of the particle against time.

**Solution:**
  (a) A plot of the velocity against time is shown below. As $v(t)$ has a factor of $(2 - t)$, the velocity changes sign at $t = 2$, passing from positive to negative.

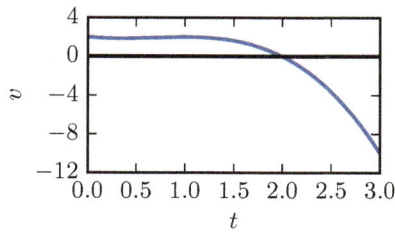

(b) Speed is given by $|v|$, hence, it can never take a negative value and the portion of the velocity graph below the $x$-axis is reflected about the $x$-axis. The plot is shown below.

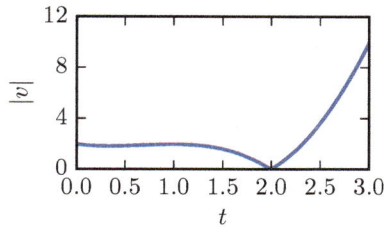

We have already seen that the velocity is the rate of change of displacement, or the rate of change of position. It follows that, if we have a displacement-time or position-time graph, then the velocity at some time $t$ is the gradient of the tangent to the graph. An example of this is shown in Figure 26.3. As such, the velocity-time graph can be recovered from the displacement-time graph, by plotting $\frac{\mathrm{d}s}{\mathrm{d}t}$ against $t$.

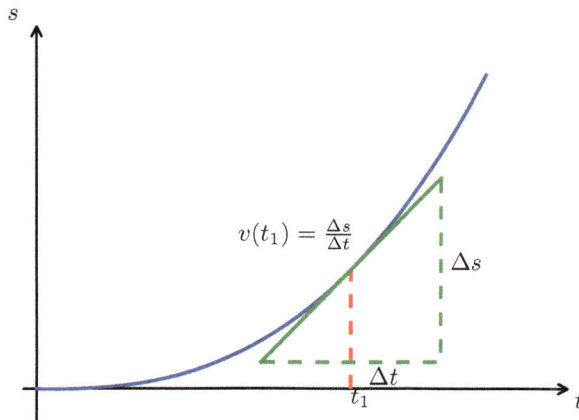

Figure 26.3: Velocity is the gradient of the tangent to a displacement-time graph.

Similarly, as acceleration is the rate of change of velocity, then, given a velocity-time graph, the acceleration at time $(t)$ is the gradient of the tangent to the curve.

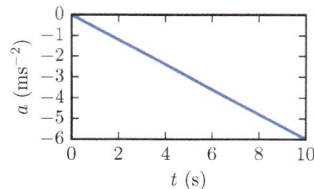

---

### Example 26.4

A commuter is travelling to work on a bus. She is currently $2\,\text{km}$ away from her work. Over the time interval $0 \leq t \leq 10$, where $t$ is measured in s, her position measured in m from her work is given by $r(t) = 30t - \frac{t^3}{10} - 2000$.

(a) How far from work is she when $t = 10\,\text{s}$?

(b) Plot her position against time over the interval $0 \leq t \leq 10$.

(c) Find expressions for the velocity and acceleration of the bus over the interval $0 \leq t \leq 10$ and plot these on separate axes.

(d) What is the velocity and acceleration of the bus at time $t = 5\,\text{s}$?

**Solution:**

(a) When $t = 10\,\text{s}$, we find $r(10) = -1800\,\text{m}$. Hence, she is $1800\,\text{m}$ from her work.

(b) The plot is shown below

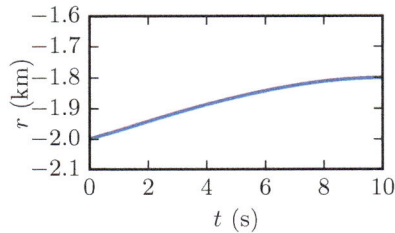

(c) We differentiate to find

$$v(t) = \frac{dr}{dt} = 30 - \frac{3t^2}{10},$$

$$a(t) = \frac{dv}{dt} = -\frac{3t}{5}.$$

The plots are shown below

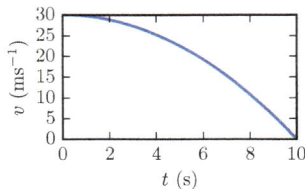

Velocity-time                    Acceleration-time

(d) We substitute $t = 5$ s into the above expressions to find $v(5) = 22\frac{1}{2}$ ms$^{-1}$ and $a(5) = -3$ ms$^{-2}$.

**Example 26.5**

A car has velocity $v$ ms$^{-1}$ at time $t$ s given by the following

$$v = \begin{cases} \frac{t^2}{30} & 0 \le t \le 30, \\ -\frac{(t-60)^2}{30} + 60 & 30 < t < 60, \\ 60 & \text{otherwise} \end{cases}$$

Plot a velocity-time graph for the car and find
(a) The velocity of the car at $t = 20$;
(b) The maximum velocity of the car and the time at which this first occurs.
(c) The time at which the car has maximum acceleration.

**Solution:**

The velocity-time graph is shown below.

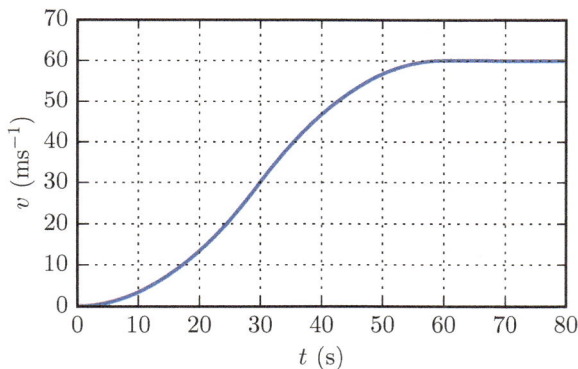

(a) The velocity of the car at $t = 20$ is

$$\frac{20^2}{30} = \frac{400}{30} = 13\frac{1}{3} \text{ ms}^{-1}.$$

(b) The maximum velocity is 60 ms$^{-1}$ and this first occurs at time $t = 60$ s.
(c) The maximum acceleration occurs where the velocity-time graph is steepest. Inspection of the graph reveals this to be at time $t = 30$ s.

**Example 26.6**

Suppose the velocity of a car has the same equation from Example 26.5.
(a) Find an expression for the acceleration of the car and plot this acceleration against

time;
(b) Evaluate the acceleration at the following times:
  i. $t = 10$ s;
  ii. $t = 50$ s;
  iii. $t = 80$ s;
  iv. $t = 60$ s;
  v. $t = 30$ s;

**Solution:**
(a) To find the acceleration, we differentiate the expression for $v$ in the three different regions:

$$a = \begin{cases} \frac{t}{15} & 0 \leq t \leq 30, \\ -\frac{t-60}{15} & 30 < t < 60, \\ 0 & \text{otherwise} \end{cases}$$

A plot of acceleration against time is shown below. We notice that the acceleration is a continuous function of time.

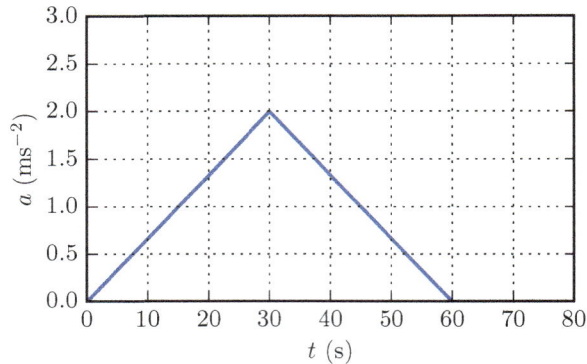

(b)  i. When $t = 10$ s, we use the first expression above to find $a = \frac{10}{15} = \frac{2}{3}$ ms$^{-2}$.
  ii. When $t = 50$ s, we use the second expression above to find $a = \frac{10}{15} = \frac{2}{3}$ ms$^{-2}$.
  iii. When $t = 80$ s, we use the third expression above to find $a = 0$ ms$^{-2}$.
  iv. As the acceleration is a continuous function at $t = 60$ s, we have two options, we could either use the second expression, or the third expression. Using the second expression gives $a = \frac{60-60}{15} = 0$ ms$^{-2}$. The third expression also gives $a = 0$ ms$^{-2}$.
  v. Acceleration is also a continuous function for $t = 30$ s, hence, using the first or second expression, we find $a = 2$ ms$^{-2}$.

**Interactive Activity 26.1 — Matching Kinematic Graphs to Real-Life Scenarios**
Follow the link below (in the digital book) and match the displacement-time and velocity-time graphs to their real life scenarios

**Exercise 26.1**
Q1. A gardener pushes a wheelbarrow 9 m from a flowerbed to the compost heap. He walks at 2mph and spends 7 s emptying the wheelbarrow. The return journey, with an empty wheelbarrow, takes 5 s.
   (a) Use the approximation 1mph=$0.45\,\mathrm{m\,s^{-1}}$ to draw a displacement-time graph.
   (b) Give the speed, in $\mathrm{m\,s^{-1}}$, of the gardener on his return to the flowerbed.
Q2. This graph shows the distance travelled by an object as it travels 30 m from point $A$ to point $B$. The time is measured in seconds.

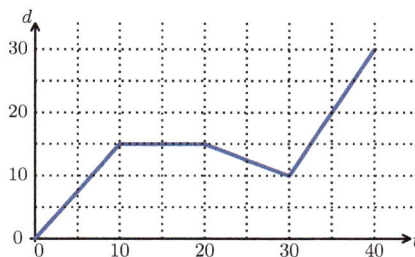

   (a) What is the object's velocity at time $t = 5$s?
   (b) What is the object's velocity when it is a distance of 5 m from point $B$?
   (c) Describe the behaviour of the object in the time interval $10 \leq t \leq 20$.
   (d) Give the velocity of the object at time $t = 25$s and describe the movement of the object at that time.
Q3. A toy clockwork car is wound up and released. It crosses the floor and strikes an obstacle. The following graph shows the car's velocity from the moment it is released until it comes to rest.
   (a) When is the forward acceleration greatest?
   (b) When does the car have the greatest speed?
   (c) State when the car hits the obstacle, and describe its behaviour after that time.

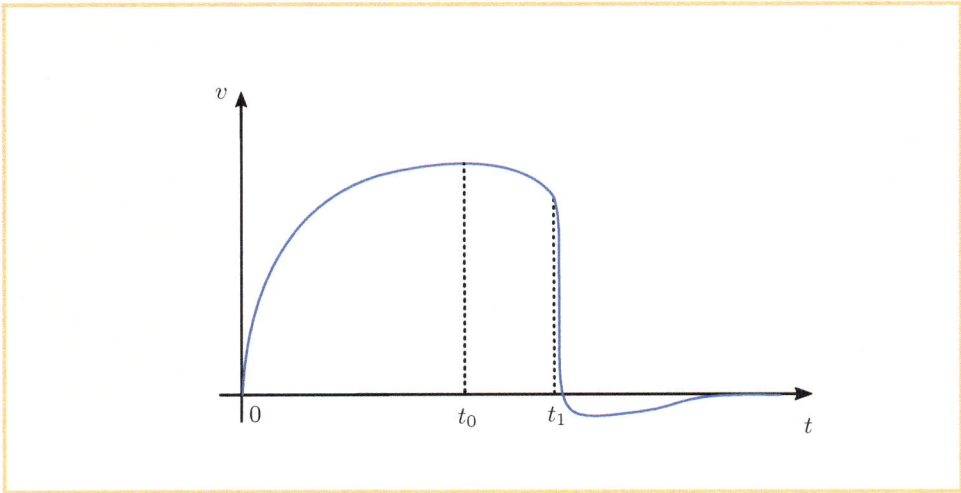

## 26.3 Equations of Motion - Constant Acceleration

We have seen that velocity is the gradient of the tangent to the displacement-time graph and that acceleration is the gradient of the tangent to the velocity-time graph. Hence, velocity can be found by differentiating displacement with respect to time and acceleration can be found by differentiating velocity with respect to time. In this section we work the other way and ask whether, given the acceleration of an object, it is possible to find the object's velocity, displacement and the distance the object has travelled at a time $t$. Specifically, in this section, we consider an object which has constant acceleration, before looking at the more general case in Section 26.4.

Let us consider the simplest case, when acceleration is zero. In this case, velocity would be a constant and, referring back to Figure 26.2, $u = v$ for all time. We would obtain a plot such as that in Figure 26.4.

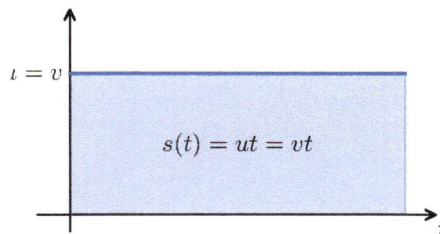

Figure 26.4: Velocity-time graph for constant acceleration $a$.

Now, as velocity is the rate of change of displacement, and velocity is a constant, we simply have

$$s(t) = ut = vt.$$

In this case, we notice that the displacement is the area under the velocity-time graph, between 0 and $t$, see Figure 26.4.

In the above, the displacement can have a positive or negative value, depending on the sign of the velocity. In contrast, the total distance travelled by the object must have a positive value. To find the total distance travelled we must instead work with the speed $|u| = |v|$. To find the total distance travelled at time $t$, $d(t)$, the speed is multiplied by $t$ and we obtain the equation below.

$$d(t) = |u|t = |v|t.$$

---

**Example 26.7**

An athlete has an initial velocity $u = -5\,\mathrm{ms}^{-1}$ and has zero acceleration.
  (a) What is the runner's speed?
  (b) What is the runner's displacement at time $t = 20\,\mathrm{s}$?
  (c) How far has the runner travelled at time $t = 25\,\mathrm{s}$?

**Solution:**
  (a) As there is no acceleration, the runner's velocity $v = u = -5\,\mathrm{ms}^{-1}$. The runner's speed is then given by $|v| = |-5| = 5\,\mathrm{ms}^{-1}$.
  (b) Displacement $s = vt$, hence, at $t = 20\,\mathrm{s}$, $s = -5 \cdot 20 = 100\,\mathrm{m}$.
  (c) The distance travelled, $d$, is given by $d = |v|t = 5 \cdot 25 = 125\,\mathrm{m}$.

---

We now consider the case of constant non-zero acceleration, see Figure 26.5(a) and again seek to find expressions for the velocity, displacement and distance travelled. At time $t = 0$, the velocity is $u$. Hence, as acceleration is rate of change of velocity, we must have the following equation for velocity, $v$:

$$v = u + at.$$

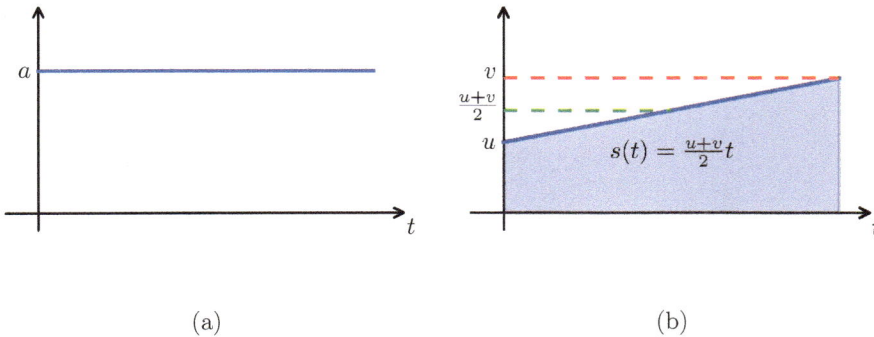

(a)                                    (b)

Figure 26.5: A constant acceleration-time graph (a) and corresponding velocity-time graph (b).

To find the displacement $s$, let us consider what $s/t$ represents. It is the average rate of change of displacement up to time $t$. In other words, it is the *average velocity*. If we consider the velocity-time graph Figure 26.5(b) , we can see that the average velocity will be $(u + v)/2$. Hence,

$$\frac{s}{t} = \frac{(u + v)}{2}.$$

By multiplying both sides by $t$, we obtain the following expression for $s$.

$$s = \frac{(u+v)}{2}t. \tag{26.6}$$

Now, by substituting the formula $v = u + at$ into (26.6), we obtain

$$\begin{aligned}
s &= \frac{(u+v)}{2}t \\
&= \frac{(u+u+at)}{2}t \\
&= ut + \frac{1}{2}at^2. \tag{26.7}
\end{aligned}$$

Glancing at Figure 26.5(b), we see that $s = \frac{(u+v)t}{2}$ is the area under the velocity-time graph in this case as well.

If we work slightly differently and rearrange $v = u + at$ to be $u = v - at$, and substitute this into (26.6), then we obtain the alternative relationship

$$\begin{aligned}
s &= \frac{(u+v)}{2}t \\
&= \frac{(v-at+v)}{2}t \\
&= vt - \frac{1}{2}at^2.
\end{aligned}$$

It is possible to eliminate time $t$ and obtain a relationship involving $u$, $v$, $s$ and $a$. We rewrite $v = u + at$ as

$$t = \frac{v-u}{a}.$$

Inserting this into (26.7), we find

$$\begin{aligned}
s &= u\left(\frac{v-u}{a}\right) + \frac{a}{2}\left(\frac{v-u}{a}\right)^2 \\
&= u\left(\frac{v-u}{a}\right) + \frac{1}{2a}\left(v^2 - 2uv + u^2\right) \\
&= \frac{1}{a}\left(uv - u^2 + \frac{v^2}{2} - uv + \frac{u^2}{2}\right) \\
&= \frac{1}{2a}\left(v^2 - u^2\right), \\
\Rightarrow \quad v^2 &= u^2 + 2as.
\end{aligned}$$

The results we have just found are called the SUVAT equations for constant acceleration and we summarise them in the formula box below.

**Formulae 26.1 — SUVAT Equations for Constant Acceleration**

The following equations apply for an object undergoing a constant acceleration $a$.

$$v = u + at, \tag{26.8}$$

$$s = \frac{(u+v)}{2}t, \tag{26.9}$$

$$s = ut + \frac{1}{2}at^2, \tag{26.10}$$

$$s = vt - \frac{1}{2}at^2, \tag{26.11}$$

$$v^2 = u^2 + 2as. \tag{26.12}$$

**Example 26.8**

A car is travelling at $20\,\mathrm{ms^{-1}}$. It then accelerates at a constant rate until, $20\,\mathrm{s}$ later, it has travelled a total of $800\,\mathrm{m}$ further down the road. What is the acceleration of the car and its final velocity?

**Solution:**

We know that $u = 20\,\mathrm{ms^{-1}}$. We also know that the car must be gaining speed, otherwise it would travel less than $20\,\mathrm{ms^{-1}} \times 20\,\mathrm{s} = 400\,\mathrm{m}$. In this case then, total distance travelled and displacement are the same and $s = 800\,\mathrm{m}$. We are required to find $a$, as we have information about $u$, $t$ and $s$, we can use equation (26.10).

$$800 = 20 \cdot 20 + \frac{a \cdot 20^2}{2}$$

$$= 400 + 200a,$$

$$\Rightarrow \quad a = \frac{800 - 400}{200} = 2\,\mathrm{ms^{-2}}.$$

Equation (26.8) can then be used to find

$$v = 20 + 2 \cdot 20 = 60\,\mathrm{ms^{-1}}.$$

**Example 26.9**

A stone is dropped from a height of $10\,\mathrm{m}$. Assuming the object undergoes constant acceleration of $a = -10\,\mathrm{ms^{-2}}$, find

(a) the velocity of the stone when it impacts the floor,

(b) the time taken for the stone to hit the floor.

**Solution:**

(a) As the acceleration has a negative value, a positive displacement will be in the direction above the starting point. Hence, impact will occur when $s = -10\,\mathrm{s}$. As we do not yet know the time at which impact occurs, we can use (26.12) to find the velocity at impact. The stone starts from rest, so $u = 0\,\text{ms}^{-1}$, hence,

$$v^2 = 2 \cdot (-10) \cdot (-10),$$
$$\Rightarrow \quad v = -\sqrt{200} = -10\sqrt{2}\,\text{ms}^{-1}.$$

Notice that, above we have chosen the negative value of the square root. This is because, as the stone is moving in a downwards direction, its velocity must be negative.

(b) To find the time to impact, we can then use $v = u + at = at$, in which case

$$-10\sqrt{2} = -10t,$$
$$\Rightarrow \quad t = \sqrt{2}\,\text{s}.$$

In the above example, finding the total distance travelled was relatively straightforward, as the velocity had the same sign throughout. The equations above for $s$ find the area between the velocity-time graph and the $x$-axis. If the velocity passes from positive to negative, or vice-versa, then the displacement will not be the same as the distance travelled. Instead, we must find the area under the speed-time graph.

**Example 26.10**
A particle starts with initial velocity $u = 5\,\text{ms}^{-1}$ and undergoes a constant acceleration of $a = -3\,\text{ms}^{-2}$. Find the velocity $v$, the displacement, $s$, and the total distance travelled, $d$, after a time $t = 4\,\text{s}$.

**Solution:**
To find the velocity $v$, we apply formula (26.8). In which case,

$$v = 5 - 3 \cdot 4 = 7\,\text{ms}^{-1}.$$

To find the final displacement, we can either apply (26.9), (26.10) or (26.11). As we have just calculated $v$, let us use (26.9) to find

$$s = \frac{4(5 + (-7))}{2} = 4\,\text{m}.$$

To find the total distance travelled, we first plot the velocity-time graph and the corresponding speed-time graph, see below. The shaded area represents the total distance travelled. We, therefore, need to find the area of two triangular regions, which requires us to compute the point $A$, where the velocity-time graph crosses the $x$-axis. In this case, $v = 0$ and hence

$$0 = 5 - 3t,$$
$$\Rightarrow \quad t = \frac{5}{3}\,\text{s}.$$

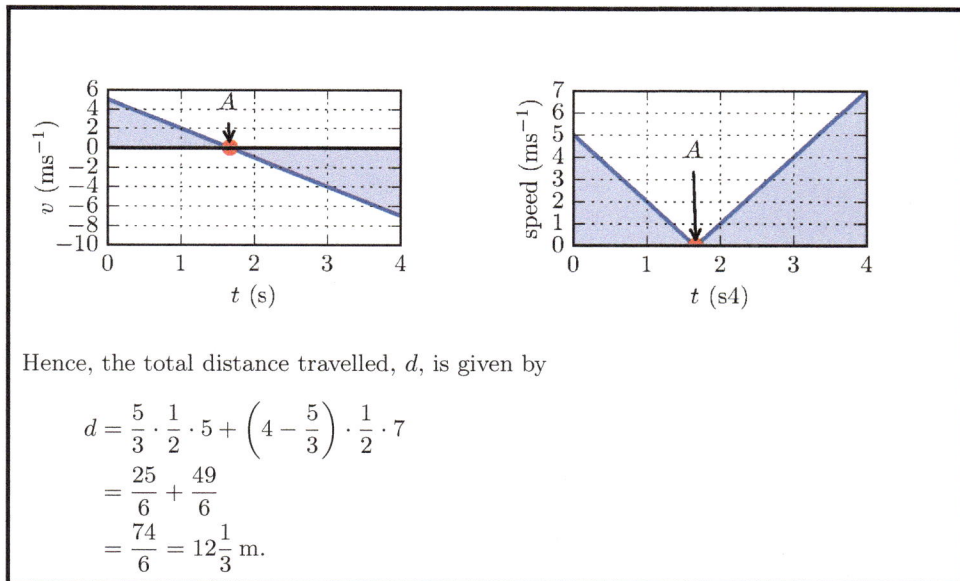

Hence, the total distance travelled, $d$, is given by

$$d = \frac{5}{3} \cdot \frac{1}{2} \cdot 5 + \left(4 - \frac{5}{3}\right) \cdot \frac{1}{2} \cdot 7$$
$$= \frac{25}{6} + \frac{49}{6}$$
$$= \frac{74}{6} = 12\frac{1}{3} \text{ m}.$$

**Exercise 26.2**

Q1. A car is travelling at $50 \text{ ms}^{-1}$. It decelerates at a constant rate and after $30\text{ s}$ it has travelled $800$ m. What is the car's rate of deceleration and what is its speed after $30\text{ s}$?

Q2. Particle $A$ and $B$ are moving in the same direction. Particle $A$ has initial speed of $20 \text{ ms}^{-1}$ and accelerates at a constant $2 \text{ ms}^{-2}$. Particle $B$ has constant velocity $30 \text{ ms}^{-1}$. If $A$ is initially $60$ m behind $B$, how long will it take for the particles to collide?

Q3. Particle $A$ and particle $B$ have initial velocities $10 \text{ ms}^{-1}$ and $-20 \text{ ms}^{-1}$, respectively and are moving towards each other. $A$ has constant acceleration $1 \text{ ms}^{-2}$ and $B$ has constant acceleration $-2 \text{ ms}^{-2}$. They collide after $20\text{ s}$. How far apart were the particles initially?

Q4. A bullet is fired directly upwards with a speed of $300 \text{ ms}^{-1}$. Assuming acceleration due to gravity is $g = 9.8 \text{ ms}^{-2}$. Ignoring air resistance, what is the maximum height of the bullet and how long is it in flight?

Q5. A sports car travels $200$ m in $4\text{ s}$ and the next $200$ m in $3\text{ s}$. Assuming the car is travelling with constant acceleration, find its acceleration and the time the car will take to travel the next $200$ m.

Q6. Particle $A$ is dropped from a height of $20$ m and Particle $B$ is fired directly upwards from the ground, simultaneously The two particles collide after $1\text{ s}$. What speed is particle $B$ fired at?

Q7. Two dragsters are competing in a race of $305$ m. Dragster $A$ accelerates at $205 \text{ kmh}^{-1}\text{s}^{-1}$ for $2.79\text{ s}$, while Dragster $B$ accelerates at $198 \text{ kmh}^{-1}/\text{s}^{-1}$ for $2.89\text{ s}$. Assuming both cars reach their maximum speed and do not decelerate again, which car will win the race and with what time gap?

Q8. A rubber ball is dropped from a height of $10$ m. Every time the ball hits the floor, the ball rebounds with $80\%$ of the speed it had on impact. At what time will the ball hit the ground for second time, assuming acceleration due to gravity is $9.8\,\text{ms}^{-1}$ and air resistance can be neglected?

Q9. The following graph shows the velocity of an object, in $\text{m\,s}^{-1}$, against time, measured in seconds.

The object travels $16.25\,\text{m}$ during the first 10 seconds.

The acceleration remains constant after time $17\,\text{s}$.

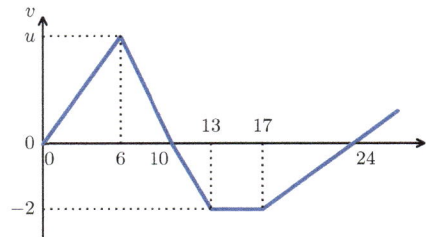

(a) What is the maximum speed of the object during the interval $0 \le t \le 24$?
(b) During the first $24\,\text{s}$, when is the object furthest from its starting point?
(c) What is the constant acceleration of the object in the time between $17\,\text{s}$ and $24\,\text{s}$?
(d) Prove that the object returns to its starting point at time $t = 20.5\,\text{s}$.
(e) Calculate the time $t_0 > 24\text{s}$ at which the object returns to its starting point for a second time.

Q10. A car is at point $P$. It accelerates in a straight line, at a uniform rate, for $8\,\text{s}$ and reaches a velocity of $15\,\text{m\,s}^{-1}$. It travels at a constant velocity for $t_0$. It comes to a stop in another $8\,\text{s}$ by decelerating at a uniform rate. The total distance covered is $420\,\text{m}$.

(a) Sketch a velocity-time graph of the car's journey and calculate $t_0$.
(b) At the time the car reaches its top speed, a motorbike sets off from point $P$ and accelerates at a uniform rate until time $t = 20\text{s}$. At this time, it has caught up with the car. Calculate the motorbike's rate of acceleration.
(c) At time $t = 20\text{s}$, the motorbike decelerates at a constant rate until its velocity reaches zero. The car and the motorbike both end at the same distance from the point $P$. Calculate the time when the motorbike stops and draw a velocity-time graph for the car and the motorbike.

## 26.4  Equations of Motion - General Case

In many real-life situations, it is unlikely that an object will move with constant acceleration. For example, a car cannot continue to accelerate: at some point air resistance and friction will be too much to overcome and the car will reach a maximum speed. Similarly, an object falling under gravity will eventually reach a *terminal velocity* due to air resistance. Resistance forces will be covered in more detail in Chapter 28, but for the time being we will assume that velocity $v$ has the general form $v = v(t)$, where $t$ is the time variable.

We can ask the same questions as we did in the case of constant acceleration in Section

26.3. What is the object's displacement at a given time and how far has it travelled?
We saw before that, in the case of constant acceleration, the displacement of the object at time $t$ was given by the area under the velocity-time graph, up to time $t$. The same is true for non-constant acceleration and this can be shown by application of the Fundamental Theorem of Calculus.

First, let us recall that velocity is defined to be the rate of change of displacement, hence

$$v = \frac{ds}{dt}.$$

The Fundamental Theorem of Calculus (Theorem 21.1) can immediately be used to reveal that, at some time $t = t_1$,

$$s(t_1) = \int_b^{t_1} \frac{ds}{dt} \, dt = \int_b^{t_1} v(t) \, dt,$$

where $b$ is some point in time. In particular, if we let $b = 0$, then

$$s(t_1) = \int_0^{t_1} \frac{ds}{dt} \, dt = \int_0^t v(t) \, dt, \tag{26.13}$$

The Fundamental Theorem of Calculus also tells us that finding a definite integral is equivalent to finding the area under the curve. Hence, the area under the velocity-time graph between 0 and $t_1$ yields the displacement at time $t_1$. An example of this is shown in Figure 26.6.

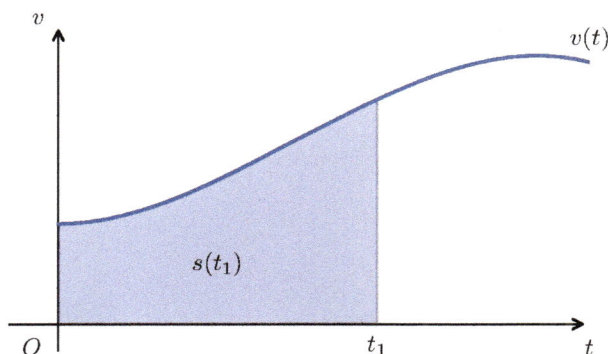

Figure 26.6: The area under a velocity-time graph is the displacement.

If the displacement is already known at some time $t = b$, then the displacement at time $t = t_1$ can be found by using the formula

$$
\begin{aligned}
s(t_1) &= \int_0^{t_1} v(t) \ dt \\
&= \int_0^a v(t) \ dt + \int_b^{t_1} v(t) \ dt \\
&= s(b) + \int_b^{t_1} v(t) \ dt. \tag{26.14}
\end{aligned}
$$

An example of this is shown in Figure 26.7.

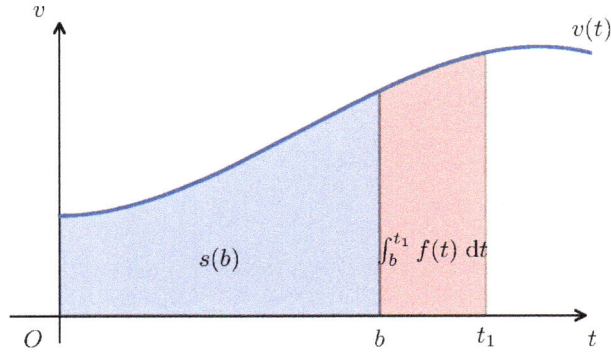

Figure 26.7: Displacement can be found by dividing the interval $[0, t_1]$ into multiple intervals.

---

**Example 26.11**

A particle has velocity $v = 2t^3 - 4t^2 + t + 3$.
  (a) Find an expression for the displacement $s$ of the particle in terms of time $t$.
  (b) How far from its starting point is the particle when $t = 10$?

**Solution:**

  (a) We use Equation (26.13) to find the displacement

  $$s(t) = \int_0^t v(t) \ dt$$

  $$= \int_0^t 2t^3 - 4t^2 + t + 3 \ dt$$

  $$= \left[ \frac{1}{2} t^4 - \frac{4}{3} t^3 + \frac{1}{2} t^2 + 3t \right]_0^t$$

  $$= \frac{1}{2} (t^4 + t^2) - \frac{4}{3} t^3 + 3t.$$

  (b) Substituting $t = 10$ into the result from part (a) gives

  $$s(10) = \frac{1}{2} (10^4 + 10^2) - \frac{4}{3} 10^3 + 30 = 3746 \frac{2}{3}.$$

---

**Example 26.12**

Suppose the velocity of a car has the same equation from Example 26.5.
  (a) Find an expression for the displacement of the car and plot this displacement against time;

**Solution:**

(a) We must consider the three different cases separately again. Using the formula for displacement, we have that for $0 \leq t \leq 30$

$$s(t) = \int_0^t \frac{t^2}{30} \, \mathrm{d}t = \left[ \frac{t^3}{90} \right]_0^t.$$

Now, we know that initial displacement is zero, hence $s(t) = \frac{t^3}{90}$ for $0 \leq t \leq 30$. For $30 \leq t \leq 60$, we can use Equation (26.14) with $b = 30$ so that

$$s(t) = s(30) + \int_{30}^t -\frac{(t-60)^2}{30} + 60 \, \mathrm{d}t$$

$$= \frac{30^3}{90} + \left[ -\frac{(t-60)^3}{90} + 60t \right]_{30}^t$$

$$= 300 - \frac{(t-60)^3}{90} + 60t - \left( -\frac{(30-60)^3}{90} + 60 \cdot 30 \right)$$

$$= 60t - \frac{(t-60)^3}{90} + 300 - 300 - 1800$$

$$= 60t - \frac{(t-60)^3}{90} - 1800$$

Similarly, for $t > 60$, we use Equation (26.14) with $b = 60$ so that

$$s(t) = s(60) + \int_6^t 0^t 60 \, \mathrm{d}t$$

$$= 60 \cdot 60 - \frac{(60-60)^3}{90} - 1800 + [60t]_{60}^t$$

$$= 1800 + 60t - 60 * 60$$

$$= 60t - 1800.$$

To summarise, the displacement $s$ satisfies the following

$$s(t) = \begin{cases} \frac{t^3}{90} & 0 \leq t \leq 30, \\ 60t - \frac{(t-60)^3}{90} - 1800 & 30 < t < 60, \\ 60t - 1800 & \text{otherwise} \end{cases}$$

The displacement-time graph is shown below

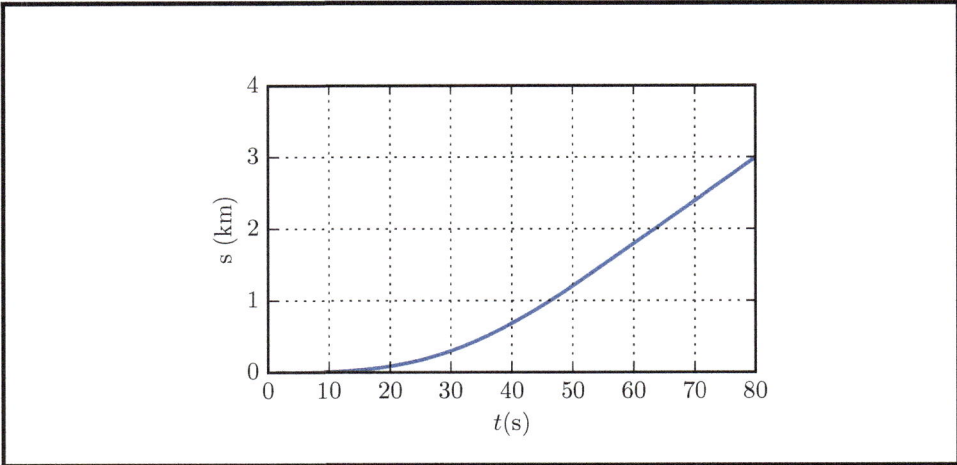

In exactly the same way, due to the fact that

$$a = \frac{\mathrm{d}v}{\mathrm{d}t},$$

the Fundamental Theorem of Calculus can be used to reveal that $v(t_1)$ is given by

$$v(t_1) = v(0) + \int_0^{t_1} \frac{\mathrm{d}v}{\mathrm{d}t}\,\mathrm{d}t = v(0) + \int_0^t a(t)\,\mathrm{d}t.$$

---

**Formula 26.2 — SUVAT Equations for Non-constant Acceleration**

For non-constant acceleration $a(t)$ and non-constant velocity $v(t)$, we have the following:

$$v(t_1) = v(0) + \int_0^{t_1} a(t)\,\mathrm{d}t, \tag{26.15}$$

$$s(t_1) = \int_0^{t_1} v(t)\,\mathrm{d}t. \tag{26.16}$$

---

**Exercise 26.3**

Q1. An object has initial velocity $u$ and constant acceleration $a$. Use (26.15) and (26.16) to show (26.8) and (26.10) hold.

Q2. A particle's velocity satisfies the equation $v(t) = -t^3 + 2t - 10$. Find an expression for the displacement of the particle. What is the particle's displacement at time $t = 2$?

Q3. A particle starts from rest and has acceleration $a(t) = 4t + 2$. Find expressions for both the particle's velocity $v(t)$ and its displacement $s(t)$.

Q4. Two particles $A$ and $B$ are initially both at rest and are 20 m apart. Particle $A$ accelerates with acceleration $a_A(t) = -t^{\frac{1}{2}}\,\mathrm{ms}^{-2}$ and particle $B$ accelerates with acceleration $a_B(t) = t^{\frac{3}{2}} + 3$. Assuming they begin to move away from each other, how far apart are the particles at time $t = 5\,\mathrm{s}$?

Q5. A sprinter is running the 100 m. For the first 2.5 s of the race, their acceleration is given by $a(t) = 2.1t^2$. For the remainder of the race they maintain a constant speed. How long do they take to complete the race?

## Key Facts — Kinematics in One Dimension

- We use the following notation to for properties of an object in motion.
  $t$:    The *time* relative to an initial time. $t = 0$ therefore represents the initial starting time of the car.
  $r_0$:    The *initial position* of the object at time $t = 0$, relative to the origin.
  $r$:    The *position* of the object at a later time $t$, relative to the origin.
  $s$:    The *displacement* of the object at time $t$, relative to its initial position.
  $u$:    The *initial velocity* of the object at time $t = 0$.
  $v$:    The *velocity* of the object at time $t$.
  $a$:    The *acceleration* at time $t$.
- Velocity is the rate of change of displacement:

$$v(t) = \frac{\mathrm{d}s}{\mathrm{d}t}(t).$$

- Acceleration is the rate of change of velocity:

$$a(t) = \frac{\mathrm{d}v}{\mathrm{d}t}(t).$$

- Displacement-time and velocity-time graphs can be drawn for easier understanding of an object's motion.
- Velocity is the gradient of the tangent to a displacement-time graph. Acceleration is the gradient of the tangent to a velocity-time graph.
- The following are equations of motion of an object which has constant acceleration:

$$v = u + at,$$
$$s = \frac{(u + v)}{2}t,$$
$$s = ut + \frac{1}{2}at^2,$$
$$s = vt - \frac{1}{2}at^2,$$
$$v^2 = u^2 + 2as.$$

- The following are equations of motion of an object with non-constant acceleration.

$$v(t_1) = v(0) + \int_0^{t_1} a(t)\,\mathrm{d}t,$$
$$s(t_1) = \int_0^{t_1} v(t)\,\mathrm{d}t.$$

**Chapter Assessment — Kinematics in One Dimension**

Download and sit the 30 minute assessment for this chapter from the digital book.

$$\sin(x)$$

$$R = \sqrt{1^2 + 2^2}$$
$$= \sqrt{1 + 4}$$
$$= \sqrt{5}$$
$$= 2.24\,N \text{ to 2 decimal places}$$

$$d\sin(x)$$

$$+ \frac{f(x)}{2}$$

$$\cos(x)$$

$$\sin^2$$

## 27. Introduction to Forces

Mechanics has been studied for thousands of years. It enables us to understand what makes objects move and predict how they will behave. Today we use mechanics for a wide variety of engineering problems, including to design buildings and engines, to make aeroplanes fly, to understand how humans stand and walk, and to launch rockets and satellites.

For example, an aeroplane in flight is subject to four forces: weight, lift, thrust, and drag, as shown in Figure 27.1.

Figure 27.1: The forces on an aeroplane in flight.

In order to keep an aeroplane airborne and moving forwards at a constant height and velocity, the lift force must balance the weight force and the thrust force from the engines must balance the resistive force of drag. When the aeroplane takes off the thrust force must be greater than the drag force, and the lift force must be greater than the weight force. When the aeroplane lands the thrust must be less than the drag, and the lift must be less than the weight.

Consequently, engineers designing aeroplanes use mechanics to ensure that aeroplanes will fly, and can be controlled by the pilot.

Developing theories of mechanics that match behaviour observed in the real world has been the subject of much mathematical work. Aristotle (384BC - 322BC) theorised that a force was required in order to keep an object in motion. He believed that the greater the force, the greater the velocity. This theory was widely accepted and remained largely undisputed for around 2000 years.

Galileo (1564 - 1642) revolutionised thinking about mechanics. He believed that a body could be in horizontal motion at a constant velocity without any overall force acting on it. This leap in thinking was fundamental to our understanding of how forces work and objects move.

In 1687 Newton published Principia in which he stated his three laws of motion on which modern mechanics is based. This chapter covers the three laws of motion and explores how to use them to describe and predict motion.

## 27.1  Forces and Motion

In this section we will introduce key forces and discuss how they relate to the motion of the objects that they act on. We begin by defining the concept of a force.

> **Definition 27.1 — Force**
> A force is a push or pull upon an object that arises from the interaction between two objects. It has both magnitude and direction, so in order to fully describe a force we need to know both its size and the direction in which it is acting.

### 27.1.1  Types of Forces

There are many different types of forces, all of which act in different ways to change the motion of an object. For example, a person pushing a broken down car is exerting a pushing force on the car. In contrast, the force exerted on the car by a tow-line is a pulling force.

> **Remark**
> Objects in the real world are typically subject to a number of different forces at any instant. We will introduce these forces throughout the following section. Figures in this chapter only show the forces acting on an object that we have introduced up to that point and may therefore omit some of the more complex forces that we will investigate later.

We already know that if we drop an object it will fall downwards, rather than upwards or sideways. The following definition explains why this happens.

> **Definition 27.2 — Weight**
> The Earth exerts a gravitational force on every object. This force is called *Weight* and it is always directed from the centre of mass of an object to the centre of the Earth.

> **Remark**
> It is a popular story that Newton discovered gravity because an apple fell on his head when he was sitting under a tree at Woolsthorpe Manor in Lincolnshire. Whilst this story has almost certainly been embellished, it is believed to be true that Newton's early

thoughts about gravity originated from observing apples falling, and asking himself why they fell straight down, and not sideways or upwards.

---

**Example 27.1**

A mass is dropped from the roof of a skyscraper and falls to the ground. What type of force is indicated by the blue arrow?

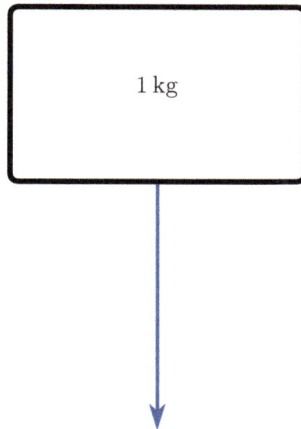

1 kg

**Solution:**
The force indicated by the blue arrow is a weight force arising from the mass of the object under gravity.

---

We now look at the types of forces that we can use to make an object move.

**Definition 27.3 — Pushing force**
A *pushing* force arises when one object (or person) exerts pressure on another object directly.

**Definition 27.4 — Driving force**
A *driving* force is exerted on an object by an engine operating to push the object.

---

**Example 27.2**

A car engine causes the car to move forward. What type of force is that? What type of force causes the motion if the car is broken down and the car is pushed by a person?

**Solution:**
The engine applies a driving force to the car. In contrast, the person pushing the car applies a pushing force.

**27.1.2   Newton's First Law of Motion**

In order to understand the concept of a force completely, we need to think about the relationship between forces and motion using Newton's laws of motion.

---

**Law 27.1 — Newton's First Law of Motion**

An object will remain in its state of motion unless it is acted upon by an external force. This means that an object moving at a constant speed in a straight line will continue moving at the same speed and in the same direction unless it is acted upon by a force outside of the object. This is also the case if the object is at rest (i.e. if the speed is $0\,\text{ms}^{-1}$).

---

**Remark**

This theorem dispels a common misconception that an object that is moving must be subject to a force in order to continue moving.

Considering many real-world scenarios may suggest that this is not true, because we see objects slowing as they roll across a surface, for example. However, this change in speed occurs because of an invisible resistance force acting upon the object. We will examine these forces in more detail later.

---

Newton's laws are natural physical laws and, as such, cannot be proven mathematically (although they can be derived using other mathematical statements of physical laws). Instead, the laws are confirmed by observations from scientific experiments. The following activity is a simple way to observe Newton's first law of motion in action.

---

**Activity 27.1 — Demonstrating Newton's first law of motion**

In this activity, we try to drop a bean bag on a stationary target whilst running past it. First, we predict the outcome of the experiment.

What is likely to happen? Where must the ball be released to ensure that it hits the target? Why?

Now try the experiment and observe what happens. Does the reality agree with the predicted outcomes? What effect does the speed of the runner have on the result? How does this practical experiment demonstrate Newton's First Law?

---

**27.1.3   More Types of Forces**

Consider the Figure 27.2 showing mass hanging from a ceiling by a rope. We know that the mass is subject to a weight force vertically down due to gravity. However, it is not falling because the weight force is balanced by an identical but opposite force vertically upwards in the rope. This force is tension.

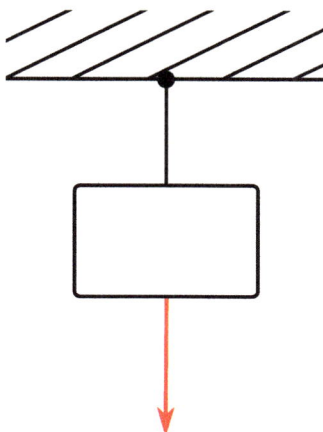

Figure 27.2: A mass hanging from a ceiling by a rope.

**Definition 27.5 — Tension**

Tension is a pulling force transmitted by a taut connector (eg a rope, chain, cable, wire or rod) and exerted on two objects, one at either end of the connector. The tension force is exerted on the objects as two equal and opposite forces. The figure below demonstrates this for the mass attached to the ceiling.

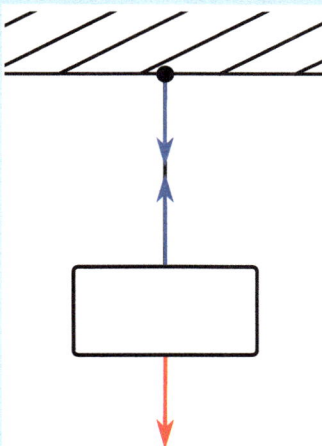

In this case, the two objects are the mass and the ceiling. The tension force exerted on the mass balances out the weight force and prevents the mass from falling. There is also a tension force exerted on the ceiling. This is also balanced by a force vertically upwards - a reaction force - and we will examine this in later sections.

Now consider the example of a person squeezing a drinks can between their hands to squash it ready for recycling. Both hands are pushing the can in equal and opposite directions. The can changes shape due to the movement of each end of the can. The force in action in this example is compression.

**Definition 27.6 — Compression**
A compression force is comprised of two equal and opposite pushing forces exerted on one object. The force acts to squeeze, or compress, the object.

---

**Example 27.3**
A photograph is hung in a frame by a single vertical string that is attached to a nail in a wall. Draw a picture of the photo-frame string with arrows showing the direction of the tension force.

**Solution:**

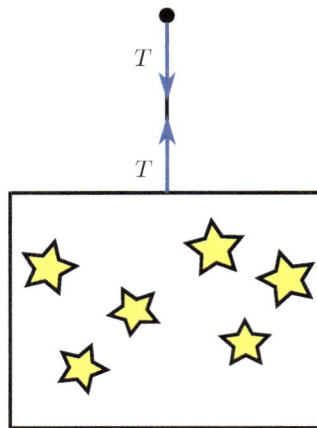

---

**Example 27.4**
Consider again the example of a person squashing a drinks can between their hands. Draw a picture of the can with arrows showing the direction of the compression force.

**Solution:**

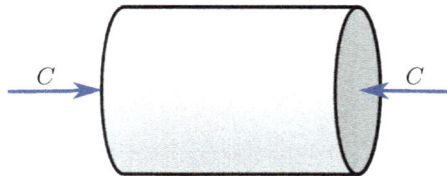

---

## 27.1.4 Newton's Second Law of Motion

We now move onto examining how the force acting on an object relates to its acceleration or deceleration. This will enable us to calculate the acceleration produced by a known force.

**Law 27.2 — Newton's Second Law of Motion**

The acceleration of an object is directly proportional to the magnitude of the force exerted on this object, and is in the same direction as the force. The mass of the object must remain constant.

This can be stated algebraically as

$$F = ma, \tag{27.1}$$

where $F$ represents the magnitude of the force, and $a$ represents the magnitude of the acceleration.

**Remark**

We can see that Newton's first law of motion is actually a special case of Newton's second law. In Newton's first law, the object moves with a constant velocity; that is, the acceleration is zero. If we set $a = 0$ in Newton's second law we obtain $F = 0$ as required by Newton's first law.

**Definition 27.7 — Newtons**

We can use Newton's second law to derive the units of measurement for a force. Recall that the standard unit for mass is kg and the standard unit for acceleration is $\text{ms}^{-2}$. Substituting into $F = ma$ we have that

$$\text{Unit for force} = \text{kg} \times \text{ms}^{-2} = \text{kgms}^{-2}.$$

This unit is called the *Newton*, represented by N. Therefore, $1\,\text{N} = 1\,\text{kgms}^{-2}$.

**Activity 27.2 — Demonstrating Newton's second law of motion**

In this activity we are going to throw a bean bag horizontally forwards (overarm) and observe the path that it follows.

Firstly, we think about what is likely to happen. Which forces will play a role in the motion? Which direction will they operate in? When will they start to affect the motion and when will they stop? How will they affect the motion of the ball? Predict and draw the path that the bean bag will follow.

Now try it and observe what happens. Does the reality agree with the predictions we made? How can we use Newton's second law to explain why?

## 27.1.5  Acceleration Due to Gravity

We have already discussed that a weight force is due to a gravitational force exerted on an object by the Earth. This force is associated with an acceleration that arises from this gravitational attraction.

**Definition 27.8 — Acceleration Due to Gravity and Calculating Magnitude of the Weight Force**

The magnitude of the weight force can be found from Newton's second law of motion,

$F = ma$. In this case, $m$ is the mass of the object (not that of the Earth) and $a = g$ is the *acceleration due to gravity*. Therefore,

$$W = mg.$$

Acceleration due to gravity is not a universal constant. It depends on the location of an object in the universe. In the vicinity of the Earth's surface it varies between $g = 9.76\,\text{ms}^{-2}$ and $g = 9.83\,\text{ms}^{-2}$, however, it is often modelled as a constant. The standard acceleration due to gravity is internationally agreed to be $g = 9.806\,65\,\text{ms}^{-2}$.

**Tip**
Exam questions may give a value for $g$ which should then be used throughout that question. This value will often be $g = 9.81\,\text{ms}^{-2}$ or $g = 10\,\text{ms}^{-2}$. If no value is given then use $g = 9.8\,\text{ms}^{-2}$.

---

**Example 27.5**
A mass of $1\,\text{kg}$ is dropped from the roof of a skyscraper and falls to the ground. What type of force is indicated by the blue arrow? What is the magnitude of this force?

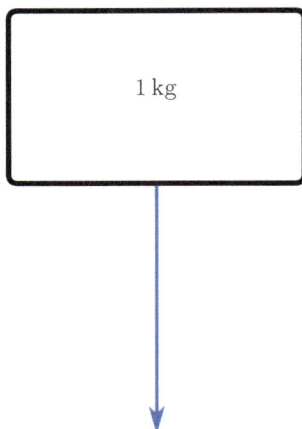

**Solution:**
As we found in Example 27.2, the force indicated by the blue arrow is a weight force arising from the mass of the object under gravity.
We are not given a specific value for acceleration due to gravity, therefore we use $g = 9.8\,\text{ms}^{-2}$.

$$W = mg$$
$$= 1\,\text{kg} \times 9.8\,\text{ms}^{-2}$$
$$= 9.8\,\text{N}.$$

### 27.1.6  Force Diagrams

A force diagram is a diagram that represents the direction of all the forces acting on an object. The only objects shown are those that we are considering (*i.e.* the mass but not the ceiling), simplified to a dot or a box. Each force is represented by an arrow that points in the direction of the force.

> **Tip**
> These diagrams enable us to simplify complex problems so that we may focus on analysing the forces applied to a single object and how they will affect its motion.

> **Remark**
> Forces diagrams can also be used to represent the relative magnitude of the forces. To do this, the length of each arrow should correspond to the magnitude of the force it represents. It is important to be aware of whether the exam question you are answering requires this.

---

**Example 27.6**

Identify the types of forces acting on a 3 kg mass attached to a ceiling by a rope. Draw the force diagram.

**Solution:**

Two forces will act on the mass: a weight force vertically downwards and a tension force in the rope acting vertically upwards.

The force diagram shows only the object in question (in this case, the mass) and the forces acting on it.

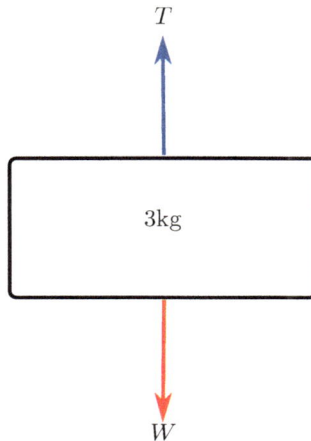

$T$

3kg

$W$

---

**Activity 27.3 — Force Diagrams**

The figure below shows a block of mass 0.3 kg resting on a plane inclined at 30° to the horizontal. The block is attached to a wall by a rope parallel to the plane. The tension in the rope is 6 N.

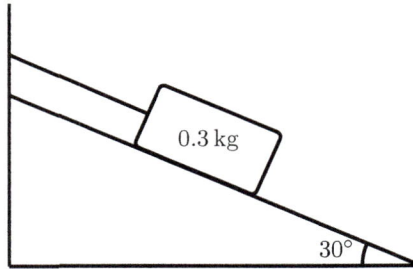

Draw the forces diagram for the block, assuming that $g = 10\,\mathrm{ms}^{-2}$ and neglecting any reaction forces. Find the weight of the block.

**Exercise 27.1**

Q1. Identify the forces described in the following. Neglect resistance and reaction forces.

(a) The force causing a ball to fall to the ground.

(b) The force exerted by a hammer on a nail.

(c) The force exerted on the disc of a disc brake as it is squeezed by a pair of brake pads.

(d) The force in a rope during a tug-of-war between two teams.

(e) The force exerted on a motorized bicycle by its engine.

(f) The force exerted on a bicycle pedal by a cyclist.

Q2. Identify the forces shown with blue arrows in the figure below which shows joists supporting a roof.

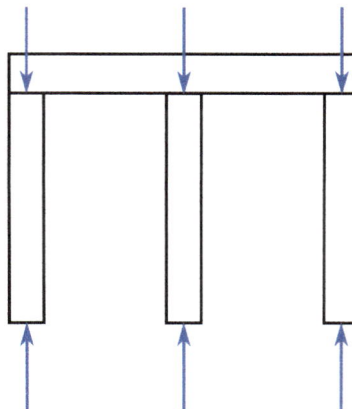

Q3. What is the weight of a 75 kg man when he is:
   (a) standing on Earth where $g = 9.8 \, \text{ms}^{-2}$?
   (b) standing on Mars where $g = 3.73 \, \text{ms}^{-2}$?
Q4. Find the magnitudes of the following forces. Neglect resistance forces.
   (a) The pushing force on a box of mass 3 kg that is accelerating horizontally at a rate of $2 \, \text{ms}^{-2}$.
   (b) The weight force on a ball of mass 20 g.
   (c) The driving force on a car of mass 1300 kg, given that the car accelerates from $0 \, \text{ms}^{-1}$ to $60 \, \text{ms}^{-1}$ in 10 s.
Q5. A box of mass 3 kg is subject to a pushing force of 12 N. Given that the initial velocity of the box is $2 \, \text{ms}^{-1}$, how far does the box travel in 10 s?
Q6. Draw force diagrams for the following scenarios. Neglect resistance and reaction forces.
   (a) A box hanging from a crane by a rope, where there is no wind force.
   (b) A moving car towing a trailer.
   (c) An empty elevator car which is supported by a single cable above the car.

## 27.2 From Forces to Motion

Our analysis so far has considered each force acting on an object independently from the others. We have focussed on the magnitude of the forces and, where we have related this to motion, we have only done so qualitatively. In this section we explore how Newton's laws enable us to analyse and predict the motion of an object precisely.

### 27.2.1 Multiple Collinear Forces

Consider the forces acting on a box that is being lifted by a crane, as shown in Figure 27.3. The box is subject to a weight force due to gravity and a tension force in the chain that attaches it to the crane.

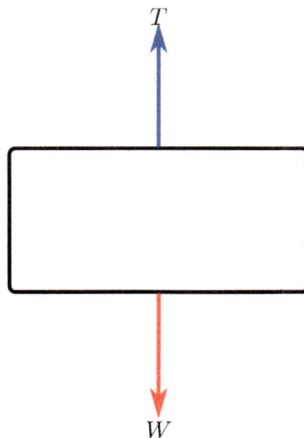

Figure 27.3: The forces on a box as it is lifted by a crane.

The forces are acting in opposite directions but are acting along the same line. We say they are *collinear*. From Newton's Second Law we know that $F = ma$, where $F$ is the overall force acting on the box. Therefore, in order to calculate the acceleration of the box as it is being lifted we need to find the sum of the forces acting on it.

---

**Definition 27.9 — Resultant force**
The resultant force on an object is the sum of all forces acting on that object. It has both magnitude and direction.

---

**Tip**
Recall from our work on vectors that every force has both magnitude and direction. Therefore, in our calculation of the resultant, some forces will be represented by a positive value for $F$ whilst others will be negative. It is important to ensure that the direction of each force is considered carefully in order to obtain the correct value for the resultant force.

---

**Example 27.7**
A box is moved along a flat, horizontal surface by one person pushing the box and one person pulling the box. The pushing force is equal to 2N and the pulling force is equal to 3N. Calculate the resultant force that is accelerating the box along the surface. Neglect resistance forces.

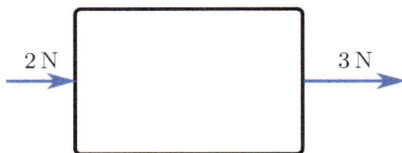

**Solution:**
The only forces acting on the box are the pushing and the pulling forces. Both forces are acting in the same direction and along the same line. Therefore, both are positive and the resultant is

$$F = 2 + 3 = 5\,\text{N}.$$

---

**Example 27.8**
A stationary crane is lifting a box of mass 0.5 kg. The tension in the chain connecting the box to the crane is 6 N. Draw the forces diagram for the box and calculate the resultant force. Neglect resistance forces and take acceleration due to gravity to be $g = 10\,\text{ms}^{-2}$.

**Solution:**
First we need to calculate the weight force acting on the box using the given value for acceleration due to gravity, $g = 10\,\text{ms}^{-2}$.

$$W = mg = 0.5 \times 10 = 5\,\text{N}.$$

The weight force is acting vertically downwards with magnitude $5\,\text{N}$, and the tension force is acting vertically upwards with magnitude $6\,\text{N}$. Therefore, we can now draw the forces diagram:

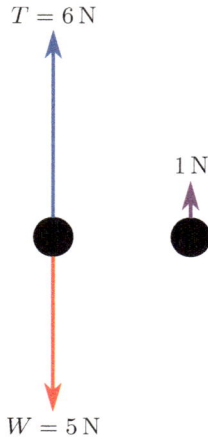

$$T = 6\,\text{N}$$

$$1\,\text{N}$$

$$W = 5\,\text{N}$$

The tension force is acting in the direction of motion (we are told that the box is being lifted), therefore, it is positive. The weight force is acting in the opposite direction and is therefore negative. The resultant is:

$$F = \text{Tension} + \text{Weight} = 6 + (-5) = 6 - 5 = 1\,\text{N}.$$

### 27.2.2 Multiple Perpendicular Forces

Let us revisit the example of the crane lifting a box. Now, in addition to the weight and tension forces, there is a strong wind blowing horizontally. These forces are shown in Figure 27.4.

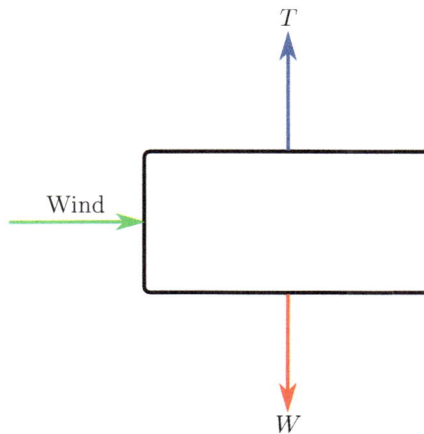

$$T$$

$$\text{Wind}$$

$$W$$

Figure 27.4: The forces acting on a box lifted by a crane with wind.

In our previous example, we found that the resultant force was vertically upwards, however, now the box is being pushed to the right as well. The following example shows how both the magnitude and the direction of the resultant force changes due to the effect of the wind force.

---

**Example 27.9**

Consider the box in Example 27.8. Now suppose that the box is also subject to a pushing force to the right of magnitude $2\,\mathrm{N}$ due to the wind. Find the resultant force and draw the forces diagram.

**Solution:**

We know from our calculations in Example 27.8 that the vertical component of the forces acting on the box is $1\,\mathrm{N}$. We are told that the horizontal component is $2\,\mathrm{N}$. Therefore, we know from our work on vectors that the resultant force will look like this:

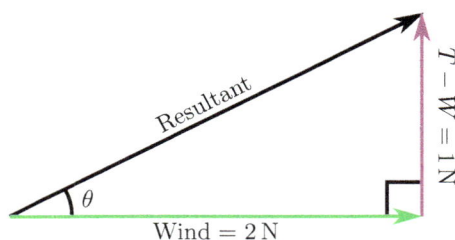

The magnitude of the resultant force is the hypotenuse of the right-angled triangle formed by the perpendicular forces. The direction of the resultant force is the angle between the hypotenuse and one of the sides (provided that we state which angle we are using, then either is acceptable). We can therefore use Pythagoras' Theorem to calculate the magnitude of the resultant:

$$R = \sqrt{1^2 + 2^2}$$
$$= \sqrt{1 + 4}$$
$$= \sqrt{5}$$
$$= 2.24\,\mathrm{N} \text{ to 2 decimal places.}$$

The angle between the resultant and the horizontal is found using:

$$\tan(\theta) = \frac{\text{opposite}}{\text{adjacent}} = \frac{1}{2}.$$

Therefore,

$$\theta \approx 26.57°.$$

The resultant force has magnitude $2.24\,\mathrm{N}$ and direction $26.57°$ to the horizontal.

---

Where an object is acted upon by perpendicular forces, the forces form a right-angled triangle and:
1. the *magnitude* of the *resultant force* is the *hypotenuse*; and
2. the *direction* of the *resultant force* is the angle between the hypotenuse and one of the sides.

In order to find the resultant force of multiple perpendicular forces we use the following procedure.

---

**Algorithm 27.10 — Finding the Resultant of Multiple Perpendicular Forces**
1. find $F_x$, the overall force in the $x$-direction, by summing all collinear forces, taking account of whether they are positive or negative;
2. find $F_y$, the overall force in the $y$-direction, by summing all collinear forces, taking account of whether they are positive or negative;
3. draw a right-angled triangle showing the directions of $F_x$, $F_y$, the resultant, and the angle that we will use to define the direction of the resultant;
4. use Pythagoras' theorem to find the magnitude of the resultant force: $F = \sqrt{F_x^2 + F_y^2}$; and
5. use trigonometry to find the direction of the resultant force, for example, $\tan \theta = \|F_y\|/\|F_x\|$.

---

**Example 27.10**

A particle is subject to the forces shown in the forces diagram below.

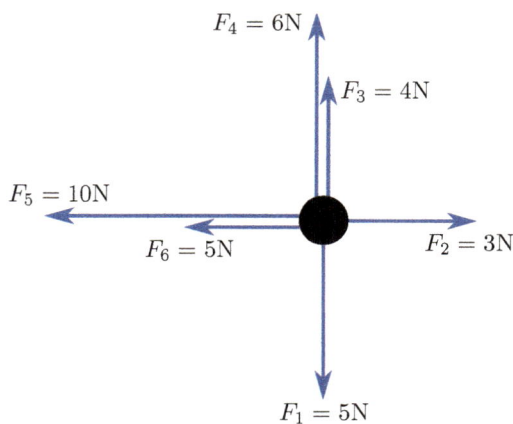

Find the resultant force.

**Solution:**

We use Algorithm 27.10.

First we find $F_x$, the overall horizontal force. We will treat forces to the right as positive and forces to the left as negative. We could choose either left or right as positive.

$$F_x = F_2 - F_5 - F_6 = 3 - 10 - 5 = -12\,\text{N}.$$

Now we find $F_y$, the overall vertical force. We will treat upwards forces as positive and downwards forces as negative.

$$F_y = F_3 + F_4 - F_1 = 4 + 6 - 5 = 5\,\text{N}.$$

Next we draw a triangle showing the directions of $F_x$, $F_y$, the resultant and the angle $\theta$.

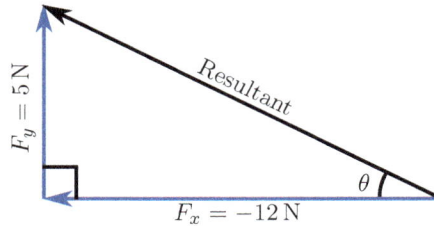

Notice that as we treated forces to the right as positive, and calculated $F_x$ as negative, the arrow for $F_x$ goes to the left and the resultant is directed up and to the left.

We are now able to use Pythagoras' Theorem to calculate the magnitude of the resultant force.

$$\begin{aligned}
R &= \sqrt{F_x^2 + F_y^2} \\
&= \sqrt{(-12)^2 + 5^2} \\
&= \sqrt{144 + 25} \\
&= \sqrt{169} \\
&= 13\,\text{N}.
\end{aligned}$$

Next we find the angle $\theta$ that will give us the direction of the resultant force. From trigonometry we know that

$$\tan(\theta) = \frac{|F_y|}{|F_x|} = \frac{5}{12}.$$

Therefore,

$$\theta \approx 22.62°.$$

Our examples so far have considered objects that are subject to horizontal and vertical forces only. However, the only requirement of our method is that the forces be perpendicular. The following example demonstrates that we can also use this method to find the resultant of forces that are perpendicular, but not horizontal or vertical.

**Example 27.11**

A particle is subject to the forces shown in the forces diagram below.

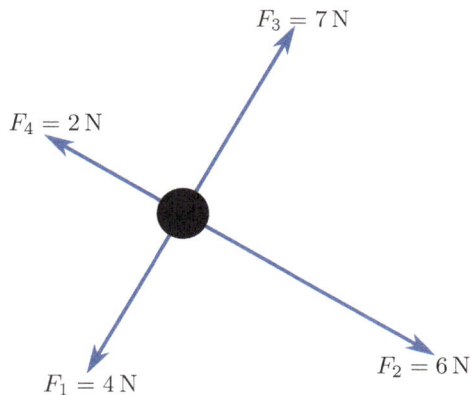

$F_3 = 7\,\text{N}$

$F_4 = 2\,\text{N}$

$F_2 = 6\,\text{N}$

$F_1 = 4\,\text{N}$

Find the resultant force on the particle.

**Solution:**

We again use Algorithm 27.10. Here, however, we will define the positive $x$-direction to be the direction of $F_2$ and the positive $y$-direction to be the direction of $F_3$.
First, we find $F_x$ by summing $F_2$ and $F_4$ as follows:

$$F_x = F_2 - F_4 = 6 - 2 = 4\,\text{N}.$$

Next we find $F_y$ by summing $F_1$ and $F_3$ as follows:

$$F_y = F_3 - F_1 = 7 - 4 = 3\,\text{N}.$$

Now we draw the triangle showing $F_x$, $F_y$, and the magnitude and direction of the resultant.

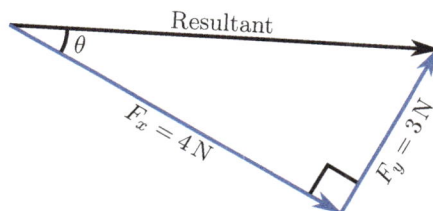

Resultant

$\theta$

$F_x = 4N$

$F_y = 3N$

This diagram shows us that the resultant is the hypotenuse of the right angled triangle formed by $F_x$ and $F_y$. Therefore, we calculate the magnitude of the resultant:

$$
\begin{aligned}
R &= \sqrt{F_x^2 + F_y^2} \\
&= \sqrt{4^2 + 3^2} \\
&= \sqrt{16 + 9} \\
&= \sqrt{25} \\
&= 5\,\text{N}.
\end{aligned}
$$

The direction of the resultant is given by $\theta$, which we calculate as follows:

$$
\tan(\theta) = \frac{F_y}{F_x} = \frac{3}{4}.
$$

Therefore,

$$
\theta \approx 36.87^\circ.
$$

**Tip**

Throughout this chapter, we have examined forces applied to a body, an object, a particle, a mass, and to real objects, such as boxes or cars. All of these may arise in exam questions. Whatever the name given to the object, our approach to the analysis is the same, provided that the object is a single unit.

### 27.2.3  Forces as Vectors

Our previous section enabled us to find the resultant of multiple forces operating in perpendicular directions in terms of its magnitude and direction (or angle $\theta$). As the force has both magnitude and direction it is a vector and can be expressed as a vector by using its horizontal and vertical components. Indeed, as part of our calculations we found $F_x$ and $F_y$ which are the components of the resultant force vector (provided that our perpendicular forces are horizontal and vertical).

**Example 27.12**

We revisit Example 27.10, however, here we will find the resultant force in vector form and use this to find its magnitude and direction.

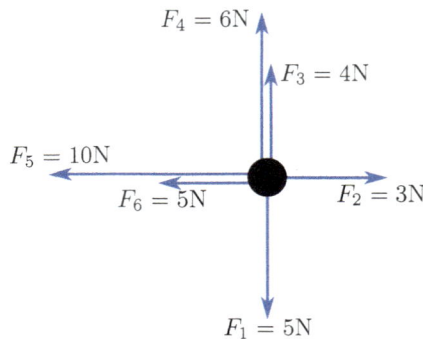

**Solution:**

In order to find the horizontal component of the vector resultant force we sum the horizontal components of the forces. This is equivalent to

$$\mathbf{R}_i = F_2 - F_5 - F_6 = 3 - 10 - 5 = -12\,\mathrm{N}.$$

The vertical component of the vector resultant force is the sum of the vertical components of the forces:

$$\mathbf{R}_j = F_4 + F_3 - F_1 = 6 + 4 - 5 = 5\,\mathrm{N}.$$

Therefore, the resultant force is

$$\mathbf{F} = \begin{pmatrix} -12 \\ 5 \end{pmatrix}\mathrm{N} = (-12\mathbf{i} + 5\mathbf{j})\mathrm{N}$$

in vector form.

We recall from our work on vectors that the magnitude of the vector $\mathbf{a} = a\mathbf{i} + b\mathbf{j}$ is $\sqrt{a^2 + b^2}$, and the direction is $\arctan(b/a)$. Therefore, the magnitude of the resultant force is

$$|\mathbf{R}| = \sqrt{(-12)^2 + 5^2} = \sqrt{144 + 25} = \sqrt{169} = 13\,\mathrm{N},$$

and its direction (angle $\theta$ from the *positive* horizontal) is given by

$$\theta = \arctan\left(\frac{5}{-12}\right) = -22.6198...$$

This equation gives us two possible answers for the direction: $\theta = -22.6198...°$ and $\theta = -22.6198... + 180 = 157.38...°$. Since the horizontal component of $\mathbf{R}$ is negative and the vertical component is positive we know that the resultant must go upwards and to the left. Therefore, $\theta = 157.38°$ (to 2 decimal places).

The magnitude of $\mathbf{R}$ is the same as we calculated in Example 27.10, however, $\theta$ is different. This is because we defined $\theta$ differently in our two examples.

**Remark**

In the previous section we could define our axes in whatever manner was most convenient for our calculations, however, when we are working with forces in vector form our axes must be horizontal (positive to the right) and vertical (positive upwards).

**Tip**

When we write forces as vectors it is important to remember to include the units of measurement just as we would for a scalar.

In real world situations it is unlikely that all the forces we are analysing will be perpendicular to one another. However, if we know all of the forces in vector component form then we are able to use our existing techniques to find the resultant force.

**Example 27.13**

A particle is subject to the forces shown in the diagram below. Find the resultant force in vector form and in magnitude and direction form.

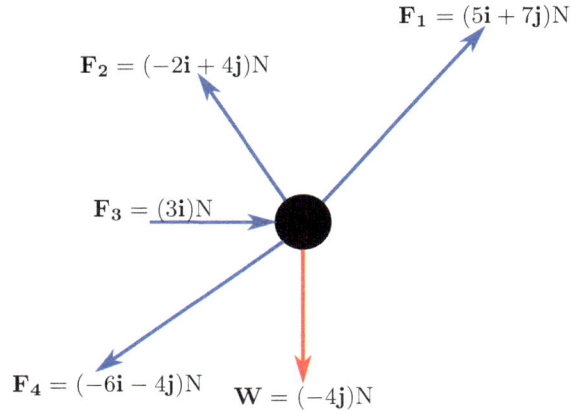

$$\mathbf{F_1} = (5\mathbf{i} + 7\mathbf{j})\text{N}$$

$$\mathbf{F_2} = (-2\mathbf{i} + 4\mathbf{j})\text{N}$$

$$\mathbf{F_3} = (3\mathbf{i})\text{N}$$

$$\mathbf{F_4} = (-6\mathbf{i} - 4\mathbf{j})\text{N} \quad \mathbf{W} = (-4\mathbf{j})\text{N}$$

**Solution:**

To find the resultant force in vector form we simply add all of the vector forces together:

$$\begin{aligned}
\mathbf{R} &= \mathbf{W} + \mathbf{F_1} + \mathbf{F_2} + \mathbf{F_3} + \mathbf{F_4}\\
&= (-4\mathbf{j}) + (5\mathbf{i} + 7\mathbf{j}) + (-2\mathbf{i} + 4\mathbf{j}) + (3\mathbf{i}) + (-6\mathbf{i} - 4\mathbf{j})\\
&= (5 - 2 + 3 - 6)\mathbf{i} + (-4 + 7 + 4 - 4)\mathbf{j}\\
&= (3\mathbf{j})\text{N}.
\end{aligned}$$

The magnitude of the resultant force is

$$|\mathbf{R}| = \sqrt{0^2 + 3^2} = 3\,\text{N}.$$

We would normally calculate the direction of the resultant by using

$$\tan\theta = \frac{\text{opposite}}{\text{adjacent}}.$$

However, in this example this calculation would require us to divide by zero which is not possible. Therefore, we will use another trigonometric identity instead:

$$\sin\theta = \frac{\text{opposite}}{\text{hypotenuse}} = \frac{3}{3} = 1.$$

Therefore, the direction of the resultant is

$$\theta = \arcsin(1) = 90°,$$

which is vertically upwards.

> **Remark**
> It is also possible to analyse vector forces using column vectors. We will explore this further in activity 27.4.

### 27.2.4 Analysing the Motion of an Object: Equilibrium

Our analysis so far in this section has focussed on describing the forces acting on an object. However, the power of Newton's laws lies in enabling us to relate the forces on an object to its motion and it is that we focus on now.

Consider a mass attached to a ceiling with string as shown in Figure 27.5. We assume that the string is inextensible (cannot stretch) and does not break. Therefore, the mass will not move and its acceleration will be zero.

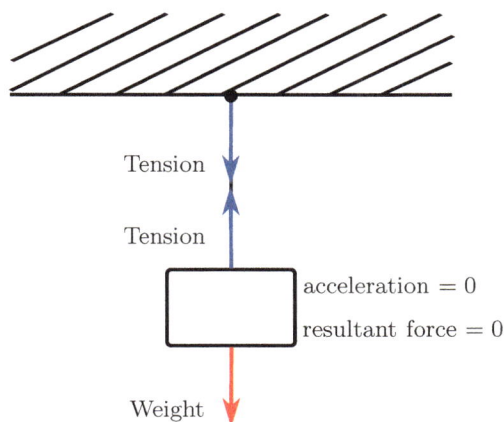

Figure 27.5: The forces on and acceleration of a mass hanging from a ceiling.

From Newton's Second Law we know that $F = ma$, therefore, since the acceleration is zero we know that the resultant force must also be zero. When the forces on an object are balanced so that the resultant force is zero, we say that the object is in a state of equilibrium.

> **Definition 27.11 — Equilibrium**
> A particle is in equilibrium if and only if the resultant force is zero.

> **Remark**
> We know from Newton's First Law that, if the resultant force on a particle is zero, it must have constant velocity. Therefore, a particle that is in equilibrium has constant (or zero) velocity and its acceleration is equal to zero.

**Example 27.14**

A 4kg mass is attached to a ceiling by an inextensible rope. The tension in the rope is 40 N. Is the mass in equilibrium? Take acceleration due to gravity to be $g = 10\,\text{ms}^{-2}$.

**Solution:**

The mass is subject to two forces, tension in the rope vertically upwards and weight due to gravity acting vertically downwards. The magnitude of the tension force is given as 40 N. We must calculate the magnitude of the weight force, using $g = 10\,\text{ms}^{-2}$:

$$W = mg = 4 \times 10 = 40\,\text{N}.$$

The weight force and the tension force act in opposite directions, therefore the resultant force is

$$F = W - T = 40 - 40 = 0\,\text{N}.$$

As the resultant force is 0 N the mass is in equilibrium.

---

**Example 27.15**

A body is subject to forces as shown in the diagram below. Is the body in equilibrium?

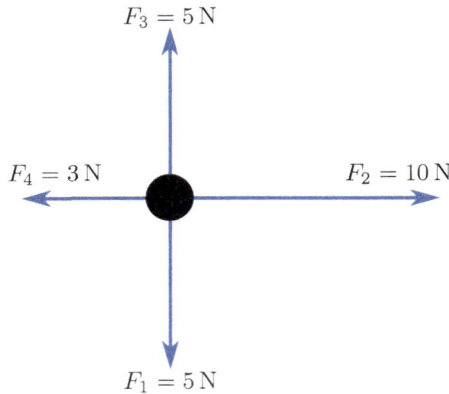

$$F_3 = 5\,\text{N}$$

$$F_4 = 3\,\text{N} \qquad F_2 = 10\,\text{N}$$

$$F_1 = 5\,\text{N}$$

**Solution:**

We will use Algorithm 27.10. First, we find $F_x$, the overall force in the $x$-direction. We will treat forces acting to the right as positive.

$$F_x = F_2 - F_4 = 10 - 3 = 7\,\text{N}.$$

Next, we find $F_y$, the overall force in the $y$-direction. We will treat forces that are vertically upwards as positive.

$$F_y = F_3 - F_1 = 5 - 5 = 0\,\text{N}.$$

As $F_y = 0$ we cannot draw the right-angled triangle that our algorithm requires as the vertical side would have zero length. However, this means we can deduce that our resultant force will be horizontal. The magnitude of the resultant is

$$F = \sqrt{F_x^2 + F_y^2} = \sqrt{7^2 + 0^2} = 7\,\text{N}.$$

The resultant force is non-zero, therefore, the particle is *not* in equilibrium.

**Remark**

Although overall the particle in Example 27.15 is not in equilibrium, as $F_y = 0$ we can say that the particle is in vertical equilibrium. However, as $F_x \neq 0$ the particle is not in horizontal equilibrium.

**Activity 27.4 — Equilibrium**

Which of the following objects are in equilibrium?

(a) A ball rolling across a smooth surface at constant velocity $v$ where the resultant force on the ball is $F = 0\,\text{N}$.

(b) A mass that is subject to the following forces:

$$\mathbf{F_1} = \begin{pmatrix} 0 \\ 3 \end{pmatrix}\,\text{N}, \quad \mathbf{F_2} = \begin{pmatrix} 0 \\ -3 \end{pmatrix}\,\text{N}, \quad \mathbf{F_3} = \begin{pmatrix} 5 \\ -2 \end{pmatrix}\,\text{N},$$

$$\mathbf{F_4} = \begin{pmatrix} 4 \\ 4 \end{pmatrix}\,\text{N}, \quad \mathbf{F_5} = \begin{pmatrix} -9 \\ -2 \end{pmatrix}\,\text{N}.$$

The following activity gives a practical demonstration of perpendicular components of forces and equilibrium.

**Activity 27.5 — Perpendicular forces and equilibrium**

In this activity we are going to hang an object in the middle of a length of string that is held horizontally and observe what happens as we try to straighten the string.

Firstly, we think about what is likely to happen. Which forces are balancing out the weight force in order to hold the object in equilibrium? Which directions are they acting in? What will happen as we try to straighten the string horizontally?

Now try it and observe what happens. Does the reality agree with the predictions we made? How can we use equilibrium and perpendicular components of forces to explain why?

If we know that an object is in equilibrium and, therefore, that the resultant force on the object is zero, we can find individual forces on that object that are unknown. We will explore this in the following example.

**Example 27.16**

The particle in the diagram below is in equilibrium. Find the magnitude and direction of force $F_1$.

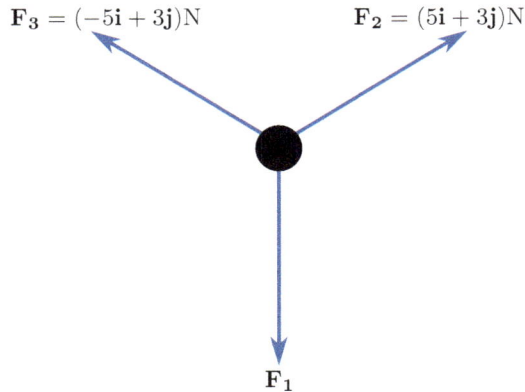

$$\mathbf{F_3} = (-5\mathbf{i} + 3\mathbf{j})\mathrm{N} \qquad \mathbf{F_2} = (5\mathbf{i} + 3\mathbf{j})\mathrm{N}$$

$$F_1$$

**Solution:**

We know that

$$\mathbf{F} = \mathbf{F_1} + \mathbf{F_2} + \mathbf{F_3}.$$

Therefore, we can find $\mathbf{F_1}$ from

$$\mathbf{F_1} = \mathbf{R} - \mathbf{F_2} - \mathbf{F_3}.$$

We are told that the particle is in equilibrium. This means that $\mathbf{F} = 0\mathbf{i} + 0\mathbf{j}$. Therefore,

$$\mathbf{F_1} = 0\mathbf{i} + 0\mathbf{j} - (5\mathbf{i} + 3\mathbf{j}) - (-5\mathbf{i} + 3\mathbf{j})$$
$$= (0\mathbf{i} - 6\mathbf{j})\mathrm{N}.$$

The magnitude of $\mathbf{F_1} = 6\,\mathrm{N}$ and its direction is vertically downwards.

### 27.2.5  Analysing the Motion of an Object: Using Newton's Second Law

When an object is not in equilibrium, it is either accelerating or decelerating (negative acceleration). If we know the resultant force, then we can use Newton's Second Law, $F = ma$, to calculate the acceleration, or, conversely, if we know the acceleration we can calculate the resultant force.

**Tip**

Some of the exam boards refer to $F = ma$ as the equation of motion and expect students to understand this terminology. Check what is expected from the relevant exam board.

**Example 27.17**

A crane is lifting a box of mass 40 kg. The tension in the chain attached to the box is 450 N. Find the acceleration of the box.

**Solution:**

In order to calculate the acceleration we will use $F = ma$. We therefore need to find the resultant force on the box. We start by finding the weight force acting on the box, using $g = 9.8\,\mathrm{ms}^{-2}$ as no value for $g$ is specified.

$$W = mg = 40 \times 9.8 = 392\,\mathrm{N}.$$

In order to formulate the equation for our resultant, it is helpful to draw a forces diagram for the box.

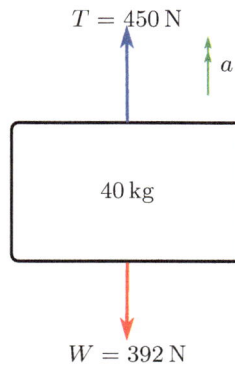

$T = 450\,\mathrm{N}$

$a$

40 kg

$W = 392\,\mathrm{N}$

Taking vertically upwards to be positive, the resultant is

$$F = T - W = 450 - 392 = 58\,\mathrm{N}.$$

Rearranging $F = ma$ for acceleration we have

$$a = \frac{F}{m} = \frac{58}{40} = 1.45\,\mathrm{ms}^{-2}.$$

The orientation for the acceleration is the same as we have assumed for forces. We assumed that a force vertically upwards is positive, therefore, a positive acceleration will also be upwards. Here, our box is accelerating upwards at $1.45\,\mathrm{ms}^{-}2$.

**Law 27.3 — Newton's Second Law for Vectors**

So far we have only applied Newton's Second Law to the magnitude of the resultant. We have worked out the direction of the resultant from diagrams or based on our understanding of the problem. However, it is also possible to write Newton's Second Law in terms of vectors:

$$\mathbf{F} = m\mathbf{a}.$$

Here, both $\mathbf{F}$ and $\mathbf{a}$ are vectors with the same direction, and mass, $m$, is always a scalar.

**Example 27.18**

A mass of 0.4 kg is subject to the forces shown in the diagram below. The acceleration is $\mathbf{a} = 4\mathbf{i} + 2\mathbf{j}$. Find force $\mathbf{F_1}$. Take $\mathbf{g} = (-10\mathbf{j})\mathrm{ms}^{-2}$.

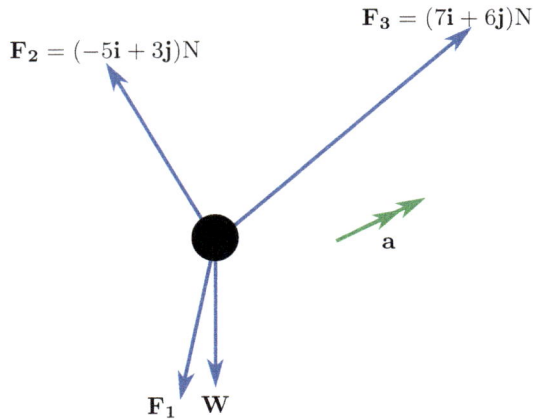

$\mathbf{F_3} = (7\mathbf{i} + 6\mathbf{j})\mathrm{N}$

$\mathbf{F_2} = (-5\mathbf{i} + 3\mathbf{j})\mathrm{N}$

a

$\mathbf{F_1}$  $\mathbf{W}$

**Solution:**

Firstly, we use Newton's Second Law to find the resultant force on the mass:

$$\mathbf{F} = m\mathbf{a} = 0.4 \times (4\mathbf{i} + 2\mathbf{j}) = 1\mathbf{i} + 0.5\mathbf{j}$$

Next, we find the weight force:

$$\mathbf{W} = m\mathbf{g} = 0.4 \times -10\mathbf{j} = -4\mathbf{j}\mathrm{N}.$$

We know that

$$\mathbf{F} = \mathbf{W} + \mathbf{F_1} + \mathbf{F_2} + \mathbf{F_3}.$$

Rearranging gives

$$\begin{aligned}
\mathbf{F_1} &= \mathbf{R} - \mathbf{W} - \mathbf{F_2} - \mathbf{F_3} \\
&= (1\mathbf{i} + 0.5\mathbf{j}) - (-4\mathbf{j}) - (-5\mathbf{i} + 3\mathbf{j}) - (7\mathbf{i} + 6\mathbf{j}) \\
&= -1\mathbf{i} - 4.5\mathbf{j}.
\end{aligned}$$

**Exercise 27.2**

Q1. Which of the following objects is in equilibrium? Neglect resistance forces.
- (a) A ball that is dropped from a height of 0.5 m above the ground;
- (b) A car travelling up a hill at a constant speed of 50 kms$^{-1}$;
- (c) A box that is lifted from the ground with an acceleration of 0.2 ms$^{-2}$; or
- (d) A box that is sliding down a slope at an increasing speed.

Q2. A mass of $500\,\text{g}$ is accelerating at $4\,\text{ms}^{-2}$. What is the resultant force acting on the mass?

    (a) $2000\,\text{N}$

    (b) $2\,\text{ms}^{-2}$;

    (c) $8\,\text{N}$; or

    (d) $2\,\text{N}$.

Q3. A box of mass $0.4\,\text{kg}$ is subject to a resultant force of $25\,\text{N}$. What is the acceleration of the box?

    (a) $62.5\,\text{ms}^{-2}$;

    (b) $0.016\,\text{ms}^{-2}$;

    (c) $10\,\text{ms}^{-2}$; or

    (d) $25\,\text{ms}^{-2}$.

Q4. Find the resultant force acting on the following objects.

    (a) the centre of a rope being used in a tug of war where one team is pulling with a force of $632\,\text{N}$ and the other team is pulling with a force of $785\,\text{N}$. Assume that the rope has zero mass.

    (b) the particle in the figure below

$$F_4 = 6\,\text{N}$$

$$F_5 = 8\,\text{N} \qquad\qquad F_3 = 3\,\text{N}$$

$$F_2 = 4\,\text{N}$$

$$F_1 = 4\,\text{N}$$

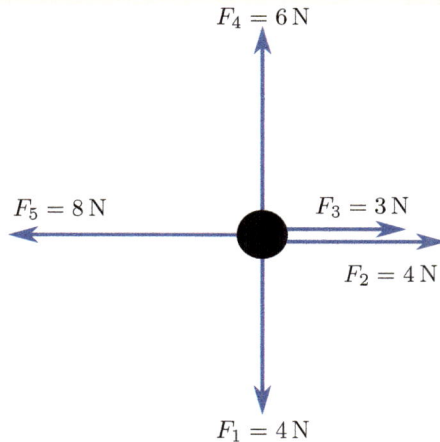

    (c) an object that is subject to the following vector forces:

$$\mathbf{F_1} = (-5\mathbf{j})\text{N};$$
$$\mathbf{F_2} = (-1\mathbf{i} + 3\mathbf{j})\text{N};$$
$$\mathbf{F_3} = (10\mathbf{i})\text{N}; \text{ and}$$
$$\mathbf{F_4} = (-6\mathbf{i} - 1\mathbf{j})\text{N}.$$

    (d) a particle that is subject to the following vector forces:

$$\mathbf{F_1} = \begin{pmatrix} 0 \\ 5 \end{pmatrix}\text{N}; \qquad \mathbf{F_2} = \begin{pmatrix} 0 \\ -5 \end{pmatrix}\text{N}; \qquad \mathbf{F_3} = \begin{pmatrix} 5 \\ 0 \end{pmatrix}\text{N};$$

$$\mathbf{F_4} = \begin{pmatrix} 1 \\ 7 \end{pmatrix}\text{N}; \qquad \mathbf{F_5} = \begin{pmatrix} -6 \\ -5 \end{pmatrix}\text{N}; \qquad \mathbf{F_6} = \begin{pmatrix} -3 \\ -1 \end{pmatrix}\text{N}.$$

Q5. A particle is subject to the forces shown below. Find the magnitude and direction of the acceleration of the particle.

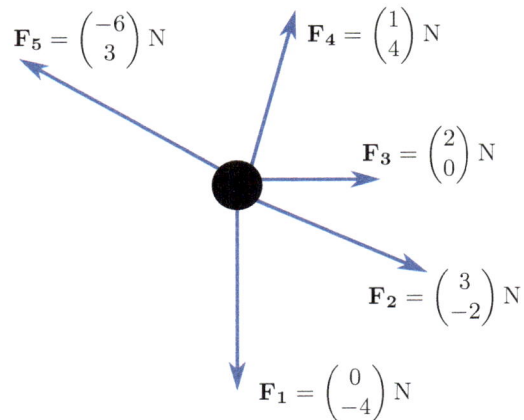

$\mathbf{F_5} = \begin{pmatrix} -6 \\ 3 \end{pmatrix}$ N　　　$\mathbf{F_4} = \begin{pmatrix} 1 \\ 4 \end{pmatrix}$ N

$\mathbf{F_3} = \begin{pmatrix} 2 \\ 0 \end{pmatrix}$ N

$\mathbf{F_2} = \begin{pmatrix} 3 \\ -2 \end{pmatrix}$ N

$\mathbf{F_1} = \begin{pmatrix} 0 \\ -4 \end{pmatrix}$ N

Q6. A car is accelerating up a hill at $\mathbf{a} = (-3\mathbf{i} + 0.7\mathbf{j})\text{ms}^{-2}$. The car is subject to the forces shown below. Find the driving force, shown on the diagram as $\mathbf{D}$. Assume that acceleration due to gravity is $g = 10\text{ ms}^{-2}$.

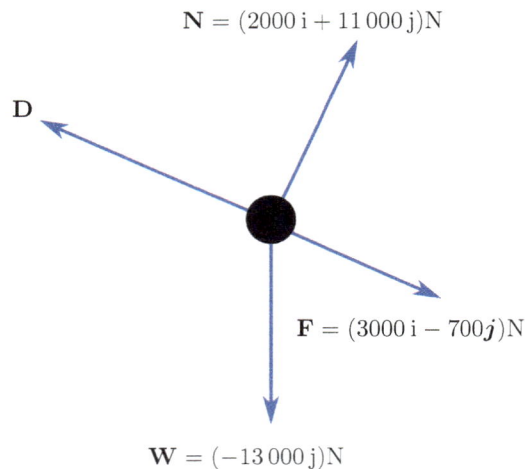

$\mathbf{N} = (2000\,\mathbf{i} + 11\,000\,\mathbf{j})\text{N}$

$\mathbf{D}$

$\mathbf{F} = (3000\,\mathbf{i} - 700\mathbf{j})\text{N}$

$\mathbf{W} = (-13\,000\,\mathbf{j})\text{N}$

Q7. A crane is lifting a box of unknown mass at constant speed. The tension in the rope is 60N. The rope breaks. What is the acceleration of the box after the rope is cut?

## Key Facts — Introduction to Forces

- A force is a push or pull upon an object that arises from the interaction between two objects. Forces are vector quantities.
- Important forces are:
  - Weight (the force due to gravity);
  - Pushing/Driving Forces;
  - Tension;
  - Compression.
- Newton's first law of motion: An object will remain in its state of motion unless it is acted upon by an external force.
- Newton's second law of motion: The acceleration of an object is directly proportional to the magnitude of the force exerted on this object, and is in the same direction as the force. The mass of the object must remain constant.
  This can be stated algebraically as

  $$F = ma,$$

  where $F$ represents the magnitude of the force, and $a$ represents the magnitude of the acceleration.
- The unit of force is the Newton. This is a derived S.I. unit with $1\,\mathrm{N} = 1\,\mathrm{kgms}^{-2}$.
- The force due to gravity on an object is $mg$, where $m$ is the mass of the object in kg and $g$ is acceleration due to gravity, which is $10\,\mathrm{ms}^{-2}$, $9.8\,\mathrm{ms}^{-2}$ or $9.81\,\mathrm{ms}^{-2}$ on the Earth's surface, depending on the question.
- The resultant force on an object is the sum of all forces acting on that object. It has both magnitude and direction.
- Force diagrams make understanding the action of all forces and finding the resultant force simpler.
- A particle is in equilibrium if and only if the resultant force is zero. By Newton's first law, this *does not mean* that the particle is at rest.

## Chapter Assessment — Introduction to Forces

Download and sit the 30 minute assessment for this chapter from the digital book.

$$a_1 = \frac{F_1}{m_1} = \frac{6}{0.3} = 20\,\mathrm{ms}^{-2}.$$

# 28. Forces to Model the Real World

From Chapter 27 we now understand forces such as weight, tension, pushing, driving, and compression, and we know how to calculate the resultant force on an object and assess whether it is in equilibrium. However, many of the problems we have considered have either been artificial (*e.g.* a particle subject to a list of unnamed forces) or we have been required to neglect resistance forces. This is because there are other important sets of forces that we need to understand in order to model problems that arise in the real-world. Figure 28.1 shows the Saturn V rocket, which sent the first humans to the moon in 1969 using a *reaction force*.

Figure 28.1: The Saturn V rocket used Newton's Third Law to send astronauts to the moon.

This chapter introduces some of these additional forces and explores how we can analyse forces in systems comprised of multiple objects.

## 28.1 Reaction Forces

Consider the box shown below, resting on a flat surface, such as a desk. We know that the box is subject to a weight force vertically downwards. Our previous analysis would suggest that this force would accelerate the box down, however, we know that the box does not move through the desk but instead remains stationary on the desk, in equilibrium. This means that there must be another force that balances the weight force. This force is called a normal reaction force.

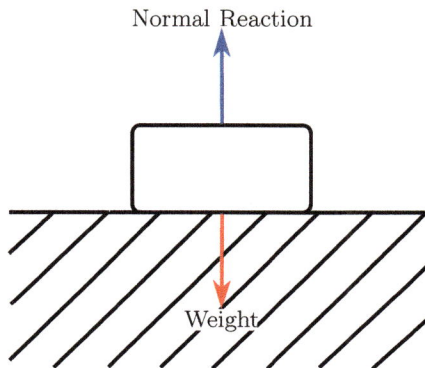

---

**Definition 28.1 — Normal Reaction Force**

A normal reaction force, or normal contact force, is a force that is exerted on an object by a surface that is in contact with the object. Its direction is perpendicular to the surface, and its magnitude is equal to the force that the object exerts on the surface in a direction perpendicular to the surface.

When the surface is horizontal, the magnitude of the normal reaction force is equal to the weight of the object.

The normal reaction force is typically denoted by $R$ and it arises due to the contact between the object and the surface. When contact is lost, there is no longer any normal reaction force.

---

The following activity enables us to visualise how a normal reaction force works.

---

**Activity 28.1 — Normal Reaction Force Demonstration**

In this activity we will observe the motion of a coin and use our observations to make deductions about the forces that are acting on the coin.

Firstly, we will place a piece of card on top of a cup, and rest a coin in the centre of the card.

What forces are acting on the coin? Is the coin in equilibrium? Explain why in terms of the forces acting on the coin.

Now quickly pull the card away. What happened? Explain why in terms of the forces acting on the coin.

---

**Example 28.1**

A box of mass 5 kg is resting on a horizontal surface. What is the magnitude and direction of the normal reaction force?

**Solution:**

The weight of the box is given by

$$W = mg = 5 \times 9.8 = 49\,\text{N}$$

The surface is horizontal, and therefore, the normal reaction force is equal and opposite to the weight of the box. The normal reaction force has magnitude $R = 49\,\text{N}$ and its direction is vertically upwards.

**Remark**

Although the direction of the normal reaction force is governed by the surface, its magnitude depends on other forces acting on the object. As we proceed through this section we will see that this is also the case for resistance forces.

### 28.1.1  Newton's Third Law

We have seen that the normal reaction force on an object arises from the contact between the object and the surface. Physical laws do not distinguish between an object and a surface; rather, a surface is simply considered to be another object. Therefore, it is reasonable to suppose that the contact also causes a force to be exerted on the surface.

Consider, for example, a heavy box resting on a weak table. The top of the table may begin to sag. This shows that the box exerts a force on the table. As a consequence, the table exerts a normal reaction force on the box. This pair of forces is explained by Newton's Third Law.

**Law 28.1 — Newton's Third Law of Motion**

All forces in the Universe occur in equal but oppositely directed pairs.

**Remark**

One force in the pair is typically called the *action* force and the other is the *reaction* force.

As with Newton's First and Second Laws, we cannot prove them, so instead we use a practical activity to demonstrate and observe it in action.

**Activity 28.2 — Demonstrating Newton's Third Law**

In this activity we are going to roll one object (*e.g.* a marble or a ball) into the centre of another identical object that is stationary. It is important that the direction of motion of the moving object is along the line passing through the centre of masses of the objects. First, predict what will happen to the motion of the objects when they collide. Which forces play a role in the changes to motion? What direction do they act in? Are they equal or different?

Now try it. What happened? Were our predictions correct? If not, why not?

**Remark**

The notion of equal and opposite forces might be surprising. If every force has an equal and opposite force, then how can any resultant force be non-zero? The answer is that the *reaction* force is not necessarily acting on the same object as the *action* force. Consider the example of a person pushing a box. The person exerts an *action* force on the box and the box exerts an equal and opposite *reaction* force on the person. One force acts on the box, and the other force acts on the person. Whether or not the box is in equilibrium depends solely on the forces that act on the box and not on the reaction force that the box is exerting on the person.

**Example 28.2**

Consider the figure below of a box at rest on a simple table which itself is stationary on a horizontal floor. Identify the equal and opposite force pairs.

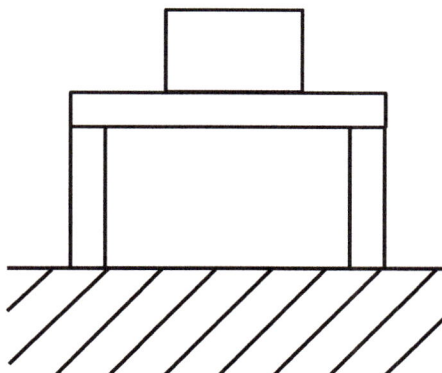

**Solution:**

The box is subject to a weight force. This weight force is exerted on the table as a pressure force. The corresponding equal and opposite force is the normal reaction force on the box.

The top of the table is subject to a weight force which is, in turn, exerted on the table legs as a pressure force. This results in an equal and opposite normal reaction force on the top of the table at each end.

Each table leg exerts a pressure force on the floor. The corresopnding equal and opposite forces are the normal reaction forces exerted on each leg.

This is summarised in the figure below, where action forces are shown in red and reaction forces are shown in blue.

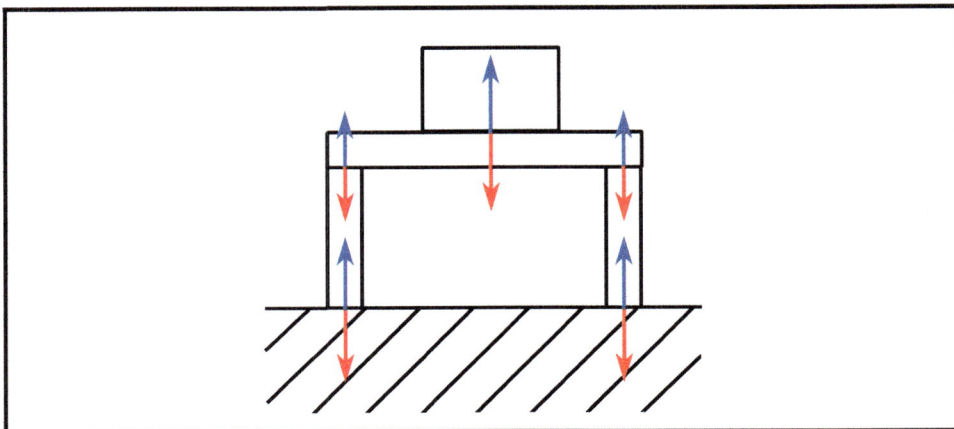

---

**Remark**

In Example 28.2, the pressure force acting downwards on the top of the table leg and the normal reaction force acting upwards on the bottom of the table leg combine to result in a compression force on the table leg.

---

**Example 28.3**

A block of mass $0.3\,\text{kg}$ is at rest on a smooth surface. Another block of mass $0.5\,\text{kg}$ slides across the surface and collides with the first block, hitting it with a force equal to $6\,\text{N}$. What is the acceleration of each block resulting from the collision?

**Solution:**

First we sketch the forces that act on the first block at the point of the collision.

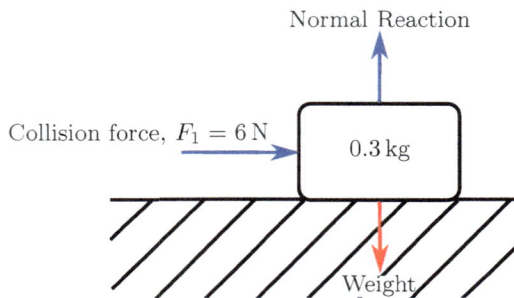

During the collision, the first block is hit with a force of $F_1 = 6\,\text{N}$. The acceleration that this force results in is given by

$$a_1 = \frac{F_1}{m_1} = \frac{6}{0.3} = 20\,\text{ms}^{-2}.$$

Now we sketch the forces that act on the second block at the point of the collision.

Normal Reaction

Collision force, $F_1 = 6\,\text{N}$

0.5 kg

Weight

From Newton's Third Law, we know that the collision results in an equal and opposite reaction force on the second block. Therefore, the force on the second block is $6\,\text{N}$ in the opposite direction. The acceleration of the second block is given by

$$a_2 = \frac{F_2}{m_2} = \frac{-6}{0.5} = -12\,\text{ms}^{-2}.$$

This acceleration will be in the opposite direction to that of the first block.

**Remark**

In Example 28.3, we only look at the forces that act at the point of collision because the forces arise from contact between the two blocks. For example, when the first block makes contact with the second block it exerts a pushing force on the second block. However, the blocks have different velocities after the collision, and, therefore, are no longer in contact with each other. This means that the pushing force, and its corresponding reaction force, cease to act from the moment that the blocks lose contact, immediately after the collision.

We can take advantage of equal and opposite force pairs to drive an object forward using another type of force, thrust.

**Definition 28.2 — Thrust**

Thrust is a type of driving force that comprises an equal and opposite force pair. A system forces a mass in one direction. The accelerated mass then causes a force of equal magnitude, but opposite direction, on that system.

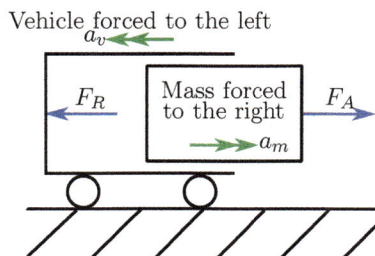

Vehicle forced to the left
$a_v$

$F_R$   Mass forced to the right   $F_A$
$a_m$

Figure 28.2: A vehicle using thrust as a means of propulsion.

Figure 28.2 shows a vehicle which has an engine that uses thrust. The engine forces a mass to the right with an action force $F_A$. This mass accelerates to the right at $a_m$. The action force $F_A$ results in an equal and opposite reaction force, $F_R$, on the vehicle. This reaction force causes the vehicle to accelerate to the left at $a_v$.

We use an activity to explore and understand thrust further.

---

**Activity 28.3 — Demonstrating Thrust - The Balloon Rocket**

We are going to make a simple rocket using a balloon, some string and a straw.

1. Tie one end of a piece of string to a chair.
2. Thread the string through the straw, and pull the straw all the way to the chair.
3. Blow up the balloon, but do not tie it off.
4. Stick the balloon to the straw in two places - one near the nozzle and the other near the top of the balloon. The nozzle end should be nearest to the chair.
5. Pull the string taut.
6. Let go of the balloon and watch what happens.

What happened? Discuss the motion of the balloon. Which action and reaction forces caused the motion? Which directions did they operate in?

---

**Example 28.4**

A rocket car of mass 10 kg has an engine that forces air out of the back of the car with force $F_{\text{air}} = 200$ N. What is the acceleration of the car and in which direction?

**Solution:**

As the air is forced out, an equal and opposite force is exerted on the car pushing it forwards. Therefore, the force pushing the car forwards is $F_{\text{car}} = -F_{\text{air}} = -200$ N. We rearrange $F = ma$ to find the acceleration:

$$a = \frac{F_{\text{car}}}{m} = -\frac{200}{10} = -20\,\text{ms}^{-2}.$$

The car will accelerate forwards, in the opposite direction to which the air travelled.

---

## 28.2  Resistance Forces

Newton's first law of motion tells us that an object that is in motion will continue at the same velocity unless it is acted upon by an external force. However, consider a box that is pushed along a horizontal surface. Our current model tells us that there are three forces acting on the box: a pushing force, weight and a normal reaction force. If we stop pushing the box, then, if our current model were complete, we would expect the box to continue to move at the same speed and never stop. However, we know that if we were to observe a box in motion it would slow down and eventually stop. This is because the box is also subject to a resistance force that opposes the motion.

---

**Definition 28.3 — Friction**

Friction is a force that one surface exerts on another when the two surfaces are sliding against each other, or when another force is trying to cause them to slide against each other. It always acts to oppose motion, and, therefore, its direction will always be opposite to the direction of motion or the direction of the force that is trying to cause motion.

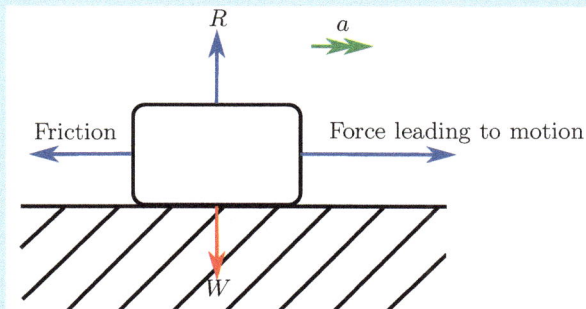

### Example 28.5

A person is trying to pull a box of 60 kg along horizontal, rough ground. The box is not moving. Given that the friction force is 350 N, what is the magnitude of the force the person pulling with?

**Solution:**

The box is in equilibrium, therefore, the resultant force must be zero; that is, $F = P - F_r = 0$, where $P$ is the pulling force. Rearranging, we obtain

$$P = F_r = 350 \, \text{N}.$$

**Tip**

If an object is not moving, or is moving at a constant speed, then it is in equilibrium.

Now we consider an example of friction on an object that is moving.

### Example 28.6

A box of 10 kg is being pushed from left to right along a horizontal surface. The magnitude of the friction force is 50 N and the acceleration of the box is $a = 2 \, \text{ms}^{-2}$. What is the pushing force?

**Solution:**

First, we use $F = ma$ to find the resultant force:

$$F = 10 \times 2 = 20 \, \text{N}.$$

We know that the resultant force is in the direction of motion, which is also the direction of the pushing force, $P$, whereas, the direction of the friction force is opposite to the direction of motion. Therefore,

$$F = P - F_r$$

and rearranging gives us the magnitude of the pushing force

$$P = F + F_r = 20 + 50 = 70 \, \text{N}.$$

We can generalise our definition of friction to include other resistance forces.

**Definition 28.4 — Resistance Force**
A resistance force is a force that opposes motion. If an object is subject to a resistance force then it can only be accelerated by a force that is sufficiently large to overcome the resistance force.
A resistance force always acts in the opposite direction from motion, or attempted motion.

**Example 28.7**
Two vehicles are alongside each other on a road and start moving at the same time, with the same driving force, $D = 400 \, \text{N}$. Each vehicle has mass $m = 1000 \, \text{kg}$. However, the vehicles are different shapes: Vehicle 1 is curved and of low height, while Vehicle 2 is square and tall. The air resistance force on Vehicle 1 is $100 \, \text{N}$ and the air resistance force on Vehicle 2 is $300 \, \text{N}$. Find the acceleration of each vehicle.

**Solution:**
As air resistance is a resistance force, it opposes motion and acts in the opposite direction to the driving force. Therefore, we can find the resultant forces for each vehicle. The resultant force acting on Vehicle 1 is

$$F_{V1} = 400 - 100 = 300 \, \text{N}$$

and the resultant force acting on Vehicle 2 is

$$F_{V2} = 400 - 300 = 100 \, \text{N}.$$

We then rearrange $F = ma$ to $a = F/m$ in order to find the acceleration of each vehicle. For Vehicle 1, this is

$$a_{V1} = \frac{F_{V1}}{m} = \frac{300}{1000} = 0.3 \, \text{ms}^{-2}.$$

For Vehicle 2, we obtain

$$a_{V2} = \frac{F_{V2}}{m} = \frac{100}{1000} = 0.1 \, \text{ms}^{-2}.$$

**Remark**
Air resistance is a resistance force that arises due to the movement (or attempted movement) of an object in air. It is caused by the friction between the air and the object. As the example above shows, the magnitude of the air resistance force on an object depends on the shape of an object. This is because the air molecules follow a path around the object as it moves through them. The smoother and flatter the object is

(a streamlined object), the smoother and flatter the path followed by the air molecules is, and the smaller the magnitude of the air resistance. The flow of molecules around objects is a major field of study in mathematics, so this description is necessarily very simplified.

**Example 28.8**

A block of mass $10\,\text{kg}$ is at rest on an inclined plane, as shown in the figure below. Find the frictional force required for the block to be in equilibrium given that $\mathbf{W} = \begin{pmatrix} 0 \\ -98 \end{pmatrix}\,\text{N}$ and $\mathbf{R} = \begin{pmatrix} -49 \\ 49 \end{pmatrix}\,\text{N}$.

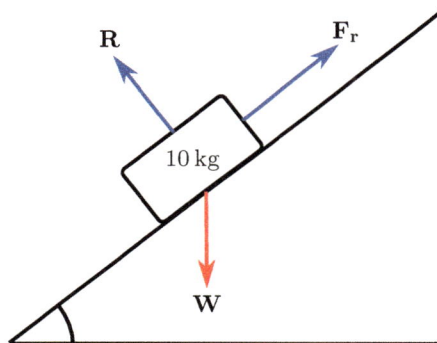

**Solution:**

In order for the block to be in equilibrium, the resultant force must be zero. This means that

$$\mathbf{F} = \mathbf{F_r} + \mathbf{R} + \mathbf{W} = \begin{pmatrix} 0 \\ 0 \end{pmatrix}.$$

Rearranging for $\mathbf{F_r}$ we obtain

$$\mathbf{F_r} = \mathbf{F} - \mathbf{R} - \mathbf{W}$$
$$= \begin{pmatrix} 0 \\ 0 \end{pmatrix} - \begin{pmatrix} -49 \\ 49 \end{pmatrix} - \begin{pmatrix} 0 \\ -98 \end{pmatrix}$$
$$= \begin{pmatrix} 49 \\ 49 \end{pmatrix}\,\text{N}.$$

Therefore, the frictional force required to keep the block in equilibrium is $\mathbf{F_r} = \begin{pmatrix} 49 \\ 49 \end{pmatrix}\,\text{N}$.

In the absence of friction, the block would slide down the plane due to the weight and normal reaction forces, however, the friction force opposes the motion and keeps the block in equilibrium.

**Remark**
We might wonder how the weight and normal reaction forces in Example 28.8 can cause the block to slide down the plane. Rather than considering each force in terms of its vertical and horizontal components, we could consider each force to have a component that is parallel to the plane and a component that is perpendicular to the plane. Then we would see that both the weight and normal reaction forces have a component that acts parallel to the plane, and down the plane, thus causing the block to slide down the plane. We will consider how we analyse systems such as this in more detail in Year 2.

**Definition 28.5 — Modelling Assumptions**
When we describe scenarios that we go on to analyse, or when we are setting up models, we often make assumptions that are reasonable for the situation and enable us to model the objects and environment more readily. We make many assumptions when we are modelling a situation involving forces and motion, indeed, we have already used some of these assumptions in our previous work. The main assumptions we might make are formalised below.

*Smooth:* A smooth surface is frictionless. That means that when an object slides, or attempts to slide, along a smooth surface there is no friction force opposing the motion.

*Vacuum:* If an object is in a vacuum then there are no air particles surrounding it to slide against or impact the object. Therefore, an assumption that an object is in a vacuum means that there is no air resistance.

*Light:* Situations will often involve ropes or rods connecting objects or connecting an object to a surface. Assuming that the connector is light means that we can neglect the weight of the connector.

*Inextensible or Inelastic:* Assuming that a connecting rope or string is inextensible, or inelastic, means that it cannot stretch (or compress) and that the magnitude of the tension force is the same at every point.

*Uniform:* Most real-world objects and environments have varying properties throughout. For example, temperature can vary depending on location in a room, or density may vary throughout an object. Assuming that an object or environment is uniform means that we can neglect these differences and any effects that may normally arise from them.

*Particle:* We have often used the term particle throughout our discussion of forces. If we model an object as a particle we assume that all forces are applied at a single, central point on the object. This means that we can ignore any rotational motion. We will always make this assumption at A-Level.

*Rigid:* If we assume that an object is rigid, then we may ignore bending and stretching forces. We will always make this assumption at A-Level.

**Activity 28.4 — Modelling Assumptions**
A uniform block of mass 4 kg is at rest on a smooth, inclined plane. The block is attached to a wall by a light, inextensible rod. The plane is located in a vacuum. Explain how the modelling assumptions are used here.

## 28.3 Internal and External Forces

Consider the example of a car moving along a road. We have already examined the forces that are acting on the car: a driving force, a friction force, a weight force and a normal reaction force. However, we know that inside the car there is an engine where components are exerting forces on one another. For example, the pistons compress the fuel and gases push the pistons back again. In turn, the pistons push the crankshaft. These forces are called internal forces because they act between objects that are part of the car. The driving, friction, weight and reaction forces are called external forces.

**Definition 28.6 — Internal Forces**
Internal forces act between objects that form part of a system.

**Definition 28.7 — External Forces**
External forces act on a system from outside of the system.

Internal forces cannot accelerate the overall system, although they can cause one part of the system to move relative to another. External forces will result in an acceleration to the whole system unless they are balanced out by other external forces.

**Remark**
This is the definition required for A-Level Mathematics, however, students who are also studying A-Level Physics or Further Mathematics may encounter a more formal definition that internal forces are those that cannot do work on the system. This means that internal forces do not change the total mechanical energy. The difference between this definition and ours above is the treatment of weight. For example, if an object falls under its own weight, then potential energy is converted into kinetic energy and the total mechanical energy does not change. So, using the strict definition, weight is an internal force. However, using our definition, weight is an external force because it arises due to gravity which is outside of the system.

In order to decide whether a force is an internal or external force, it is important first to define what the system is in any particular problem. Usually it will make our analysis easier to define the system in a particular way.

**Activity 28.5 — Demonstrating Internal and External Forces**
We are going to explore the concepts of internal and external forces by using a Newton's Cradle. Newton's cradle demonstrates many principles of mechanics (although it was not invented by Newton). To see Newton's Cradle working we pull one (or more) of the balls to one side and then let go, as shown in the figure below.

For demonstrating internal and external forces, we first define the system to be the entire cradle. With a system defined, we ask what are the internal and external forces. Are they accelerating the overall system? Are they accelerating only part of the system relative to another part of the system?

Now, we push the entire cradle along the desk slowly. Keeping our definition of the system the same, we again ask what are the internal and external forces. Are they accelerating the overall system or only part of the system?

---

**Example 28.9**

A crane has lifted a box and is driving forwards. Define the system that we would use to find the acceleration resulting from the driving force and identify the internal and external forces.

**Solution:**

The acceleration resulting from the driving force is the acceleration of the entire crane, box, and connecting chain. Therefore, we define the system to be the entire crane, box, and connecting chain.

The crane is moving due to a driving force. This force accelerates the whole system and so it is an external force.

The crane's motion is resisted by a friction force and air resistance. These forces arise from the interaction between the system and its environment, so they are external forces. The system is subject to a weight force and a normal reaction force which are balancing external forces. The mass used to calculate these forces would be the overall weight of the crane, box and the connecting chain.

The box has its own weight which is balanced by tension in the chain. These forces are acting between parts of the system only and so are internal forces.

The engine within the crane will have many forces acting between its components. These will all be internal forces.

---

## 28.4　Connected Particles

Consider a train moving along a track. The train is made up of an engine and several carriages that are all connected. Each part of the train is free to move relative to the other parts of the train, however, the parts of the train exert forces on one another indirectly through the connections. If we wish to understand the motion of the individual carriages, then we model the train as a series of connected objects. Our studies so far have focussed on individual objects, or on analysing the motion of a whole system. We need additional techniques in order to analyse the carriages. This section examines how to model the forces and motion of connected objects.

---

**Tip**

Many problems involving connected particles will use a pulley. A pulley is a wheel around which a string or cord passes. The wheel can spin, thus moving the cord around the wheel as well. Each end of the cord will be connected to something (typically a mass, or ceiling). Pulleys act to reduce the magnitude of the force required to lift an object and to reverse its direction.

---

**Activity 28.6 — Pulleys**

In this activity we are going investigate the effect of pulleys on the forces required to lift a mass. The remark below describes a method for making a homemade pulley, if one is not available.

Firstly, we experiment with a fixed pulley. Attach each end of the string to different masses and pass the string over the pulley. Pull on the smaller mass and observe the force required to lift the larger mass. Is the force required to lift the mass the same as its weight? Why?

Next, we use a free pulley; that is, a pulley that is free to move vertically. Fix one end of the string to a surface (*e.g.* a table, wall, or door frame), and attach one mass to the pulley. Now rest the pulley on the string and pull on the free end of the string to lift the mass. Is the force required to lift the mass the same as its weight? Why?

The pulley systems should be similar to those in the figure below.

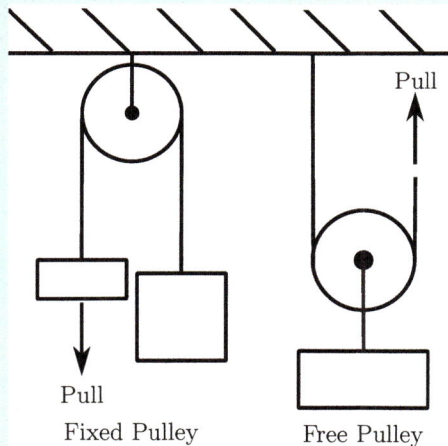

Pull

Pull

Fixed Pulley                  Free Pulley

---

**Remark**

If a manufactured pulley is not available for use with Activity 28.6, then the following method may be used to make a pulley in the classroom or at home.

**Materials Required:**
 (a)  A rod, for example, a piece of dowelling or a knitting needle;
 (b)  A cotton reel;
 (c)  Sticky tape; and
 (d)  A long piece of string.
**Method:** To make the fixed pulley, pass the rod through the centre of the cotton reel, and stretch the string across the top of the reel. Attach each end of the rod firmly to separate chairs or tables which are sufficiently wide apart to carry out the experiment. For the mobile pulley, fix one end of the string to a surface (a wall, table, or door frame will suffice). Pass the rod through the centre of the cotton reel and rest the cotton reel on top of the string.

---

**Example 28.10**

The figure below shows two masses, $m_1 = 0.5\,\text{kg}$ and $m_2 = 0.8\,\text{kg}$ connected by a light, inextensible string and a smooth, light pulley. The masses are initially held in place and then are released. Find the acceleration of each mass when they are released and the corresponding resultant forces.

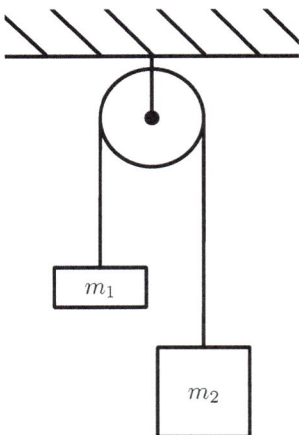

**Solution:**

First we draw the forces diagrams for each mass. The modelling assumptions that we are given tell us that the magnitude of the tension force is constant along the length of the string, that we can ignore the weight of the string and pulley, and that there is no friction between the pulley and the string. As the two masses are connected, their accelerations will be of the same magnitude but in opposite directions. We need to select axes systems for the masses that mean that $a$ is positive for each mass (recall that a negative value for $a$ represents a deceleration). Therefore, the force diagrams are:

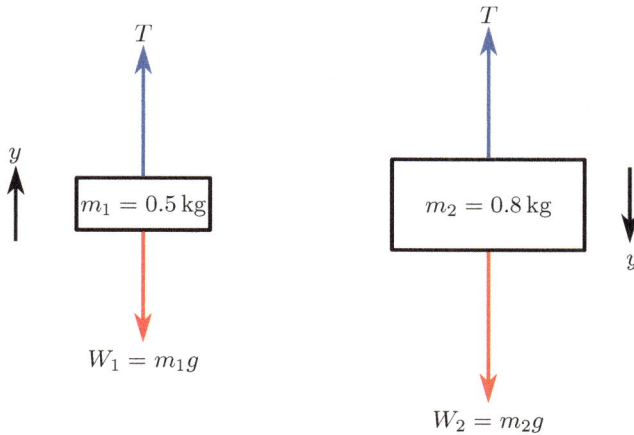

We now set up the equations for the resultant force for each mass, using the axes systems shown on the diagrams to decide which forces are positive and which are negative. For the first mass:

$$F_1 = T - W_1,$$

$$F_2 = W_2 - T.$$

We can find $W_1$ and $W_2$ by:

$$W_1 = m_1 g = 0.5 \times 9.8 = 4.9\,\text{N}$$

and

$$W_2 = m_2 g = 0.8 \times 9.8 = 7.84\,\text{N}.$$

So, the equations for the resultant forces become

$$F_1 = T - 4.9 \tag{28.1}$$

and

$$F_2 = 7.84 - T. \tag{28.2}$$

As the accelerations of the two objects are the same, we know that $F_1 = m_1 a = 0.5a$ and $F_2 = -m_2 a = -0.8a$. Substituting into equations (28.1) and (28.2) we obtain

$$0.5a = T - 4.9 \tag{28.3}$$

and

$$0.8a = 7.84 - T. \tag{28.4}$$

We now have two equations with two unknowns and can solve these simultaneously to find $a$ and $T$. Firstly, we notice that, if we add the two equations together, $T$ will cancel, and we will be able to find $a$.

$$0.5a + 0.8a = T - 4.9 + 7.84 - T,$$
$$\Rightarrow \qquad 1.3a = 2.94,$$
$$\Rightarrow \qquad a = \frac{2.94}{1.3},$$
$$= 2.26\,\text{ms}^{-2} \text{ to 2 d.p.}.$$

We now need to find the resultant forces. There are two possible methods, set out separately below.

**Method 1:**
We substitute the full value for $a$ that we have calculated into $F = ma$ to find the resultant force for each mass. Firstly, for mass 1:

$$F_1 = m_1 a = 0.5 \times \frac{2.94}{1.3} = 1.13\,\text{N to 2 d.p.}$$

Now, for mass 2:

$$F_2 = m_2 a = 0.8 \times \frac{2.94}{1.3} = 1.81\,\text{N to 2 d.p.}$$

**Method 2:**
We now find $T$ and use this to find the resultant forces. Substituting our full value for $a$ back into equation (28.3) we obtain

$$0.5 \times \frac{2.94}{1.3} = T - 4.9,$$
$$\Rightarrow \qquad T = 0.5 \times \frac{2.94}{1.3} + 4.9 = 6.03\,\text{N to 2 d.p.}$$

Next we substitute the full value for $T$ back into equations (28.1) and (28.2) in order to find the resultant forces on each mass:

$$F_1 = 6.03... - 4.9 = 1.13\,\text{N to 2 d.p.}$$

and

$$F_2 = 7.84 - 6.03... = 1.81\,\text{N to 2 d.p.}$$

Clearly, Method 1 is quicker than Method 2, however, this is because the question did not ask us to find $T$. If the question had asked for $T$ as well, then the second method would have involved no additional effort.

**Remark**
Instead of the two separate forces diagrams for each mass, it is acceptable to draw a single forces diagram on the pulley system. We would need to set up our axes systems in

the same way as shown in the example above. Conceptually, for the single diagram, this represents the positive $y$-direction going up, around the pulley and down the other side, as this may be easier to visualise. However, it is often easier to analyse the resultant force on each object if we use separate diagrams. Use Activity 28.7 to try each method and choose a preference.

**Activity 28.7 — Forces Diagrams for Pulley Systems**
A mass $m_1 = 1.2\,\text{kg}$ is attached to a mass $m_2 = 0.3\,\text{kg}$ by a light, inextensible string that passes over a smooth, light pulley. The masses are initially held stationary and are then released. Formulate the equations that we would use to find the acceleration and tension upon release:
  (a) using a single forces diagram; and
  (b) using separate forces diagrams for each mass.

The examples above are one-dimensional only; that is, all of the motion and forces are vertical. We now consider a two-dimensional pulley system.

**Example 28.11**
A mass of $1\,\text{kg}$ is held at rest on a smooth table. It is connected to another mass of $0.5\,\text{kg}$ by a light, inextensible string that is running over a light, smooth pulley at the edge of the table, as shown in the figure below. The masses are released. Find the tension in the string, the resultant force and the acceleration. Assume that acceleration due to gravity is $g = 10\,\text{ms}^{-2}$.

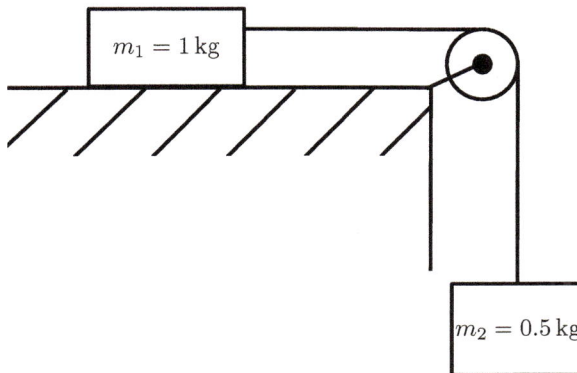

**Solution:**
First we draw the forces diagram for the mass and pulley system. The modelling assumptions that we have been given tell us that we can ignore the weight of the string and the pulley, that the tension force is constant along the length of the string, that there is no friction between the pulley and the string, and that there is no friction between the table and mass 1. As the table is frictionless, we would expect that, when the masses are released, mass 1 will move to the right and mass 2 will move down. The acceleration of mass 1 and mass 2 will have the same magnitude. The weight of mass

1 will be balanced by an equal and opposite normal reaction force. Therefore, we have the following forces diagram with axes as shown.

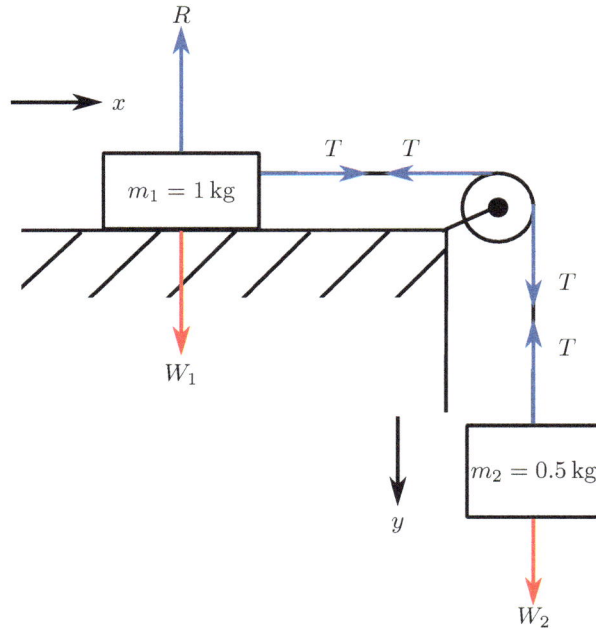

Notice that we have shown tension forces coming from the pulley. This is to show a complete representation of how the forces act, however, the only forces that are relevant to our analysis are those that act directly on each of the masses.

The separate forces diagrams will be as follows:

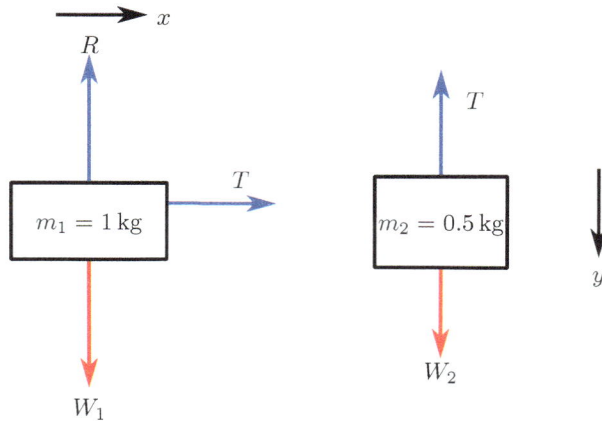

We have only indicated one axis dimension for each mass because each mass only moves in one dimension, therefore, we only need to know the positive direction for one dimension for each mass.

Now we write our equations for the resultant forces in the direction of motion. This means that $W_1$ and $R$ do not play a part.

$$F_1 = T, \tag{28.5}$$
$$F_2 = W_2 - T. \tag{28.6}$$

We simplify the second equation by calculating $W_2$, using $g = 10\,\text{ms}^{-2}$:

$$W_2 = m_2 g = 0.5 \times 10 = 5\,\text{N}.$$

As we know that both masses will have the same acceleration, $a$, we can use $F = ma$ to write

$$m_1 a = T,$$
$$m_2 a = 5 - T.$$

Substituting the values we know for the masses this becomes

$$a = T, \tag{28.7}$$
$$0.5a = 5 - T. \tag{28.8}$$

We now have two equations with two unknowns, which we can solve simultaneously to find $a$ and $T$. First, we substitute Equation (28.7) into Equation (28.8):

$$0.5T = 5 - T.$$

Rearranging we obtain

$$T = \frac{10}{3} = 3.33\,\text{N to 2 d.p.}$$

From Equation (28.7), we immediately know that the acceleration is also:

$$a = \frac{10}{3} = 3.33\,\text{ms}^{-2} \text{ to 2 d.p.}$$

We also need to find the resultant forces $F_1$ and $F_2$.

**Method 1:**
We substitute the full value we have calculated for $a$ into $F = ma$ as follows.

$$F_1 = m_1 a = 1 \times \frac{10}{3} = 3.33\,\text{N to 2 d.p.},$$
$$F_2 = m_2 a = 0.5 \times \frac{10}{3} = 1.67\,\text{N to 2 d.p..}$$

**Method 2:**
We substitute the full value we have calculated for $T$ into Equations (28.5) and (28.6)

$$F_1 = T = 3.33\,\text{N to 2 d.p.},$$
$$F_2 = W_2 - T = 5 - \frac{10}{3} = 1.67\,\text{N to 2 d.p.}$$

The following example considers how we might solve this problem if the table were not smooth.

---

**Example 28.12**

A mass of $0.8\,\text{kg}$ is held at rest on a table. It is connected to another mass of $1\,\text{kg}$ by a light, inextensible string that is running over a light, smooth pulley at the edge of the table, as shown in the figure below. The masses are released.

(a) Find the tension in the string, the resultant force and the acceleration, given that the friction force between the mass and the table is $1.6\,\text{N}$.

(b) Use the answer to (a) to find the distance travelled by the masses in $2\,\text{s}$.

Assume that acceleration due to gravity is $g = 10\,\text{ms}^{-2}$.

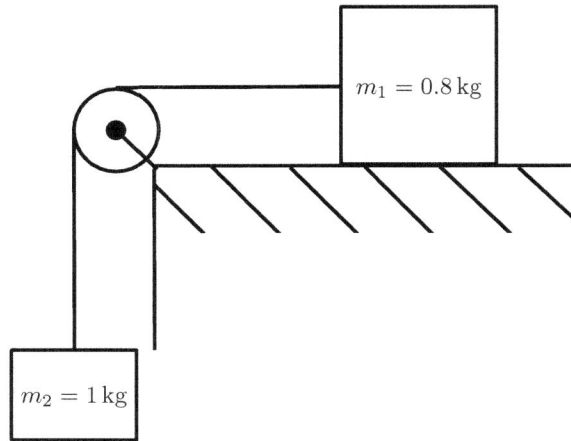

$m_1 = 0.8\,\text{kg}$

$m_2 = 1\,\text{kg}$

**Solution:**

(a) Firstly, we draw the forces diagrams. The modelling assumptions tell us that we can neglect the weight of the string and the pulley, that tension is constant along the length of the string, and that there is no friction between the string and the pulley. The magnitude of the friction force is small relative to the weight of the hanging mass, therefore, we would expect that mass 1 will move to the left and mass 2 will move down when the masses are released. As the masses are connected they will have the same acceleration. Therefore, the forces diagram is as follows.

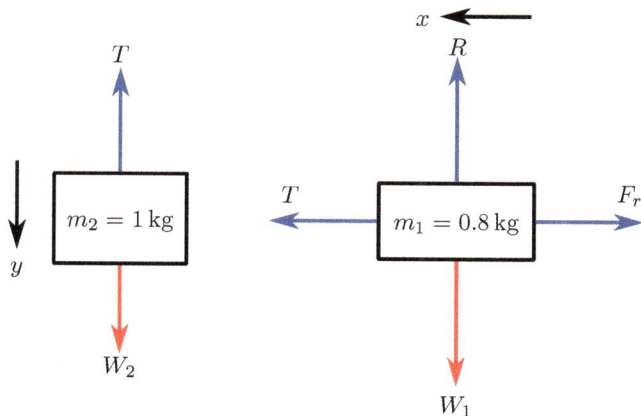

Next, we write our equations for the resultant forces on the masses in the direction of motion.

$$F_1 = T - F_r, \tag{28.9}$$
$$F_2 = W_2 - T. \tag{28.10}$$

We know that $F_r = 1.6\,\text{N}$, and we can calculate $W_2$:

$$W_2 = m_2 g = 1 \times 10 = 10\,\text{N}.$$

Therefore, Equations (28.9) and (28.10) become

$$F_1 = T - 1.6, \tag{28.11}$$
$$F_2 = 10 - T. \tag{28.12}$$

Substituting the resultant forces for $F = ma$, we have

$$0.8a = T - 1.6, \tag{28.13}$$
$$a = 10 - T. \tag{28.14}$$

We now have two equations with two unknowns and we may solve simultaneously. We add Equation (28.13) to Equation (28.14) to find the acceleration.

$$0.8a + a = T - 1.6 + 10 - T,$$
$$\Rightarrow \quad 1.8a = 8.4,$$
$$\Rightarrow \quad a = \frac{8.4}{1.8},$$
$$\Rightarrow \quad a = 4.67\,\text{ms}^{-2} \text{ to 2 d.p.}$$

Now, we use Equation (28.14) to find the tension, using the full value that we have calculated for the acceleration.

$$a = 10 - T,$$
$$\Rightarrow \quad T = 10 - a,$$
$$\Rightarrow \quad T = 10 - \frac{8.4}{1.8},$$
$$\Rightarrow \quad T = 5.33 \,\text{N to 2 d.p.}$$

Next, we find the resultants, again showing both possible methods.

**Method 1:**
We substitute the full value we have calculated for acceleration into $F = ma$ for each mass.

$$F_1 = m_1 a = 0.8 \times \frac{8.4}{1.8} = 3.73 \,\text{N to 2 d.p.},$$
$$F_2 = m_2 a = 1 \times \frac{8.4}{1.8} = 4.67 \,\text{N to 2 d.p.}$$

**Method 2:**
We substitute the full value that we have calculated for tension into Equations (28.11) and (28.12).

$$F_1 = T - 1.6 = 5.33... - 1.6 = 3.73 \,\text{N to 2 d.p.},$$
$$F_2 = 10 - T = 10 - 5.33... = 4.67 \,\text{N to 2 d.p.}$$

(b) We now need to find the distance travelled by the masses in $2\,\text{s}$. To do this, we use the equations of constant acceleration, with the full value we have calculated for acceleration. We know that the masses were initially at rest, therefore, $u = 0\,\text{ms}^{-2}$.

$$s = ut + \frac{1}{2}at^2$$
$$= 0 \times 2 + \frac{1}{2} \times 4.67... \times 2^2$$
$$= 9.33 \,\text{m to 2 d.p.}$$

In Activity 28.6 we considered a pulley that was free to move. The following example shows how we analyse a system involving a free pulley.

**Example 28.13**
Two masses are connected by the pulley system shown in the figure below and are held at rest. The pulley connected to the ceiling is fixed in place, whereas the pulley connected to mass $m_1$ is free to move vertically. The pulley and string system is light, smooth, and inextensible. The masses are released. Find the tension in the string, the resultant forces, and the acceleration, given that $m_1 = 2\,\text{kg}$ and $m_2 = 8\,\text{kg}$.

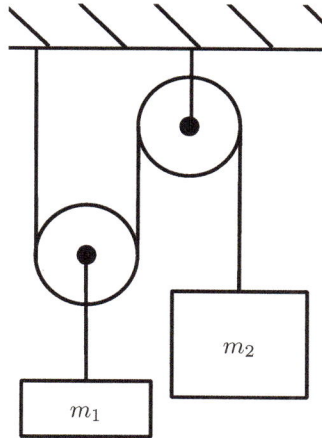

**Solution:**

Firstly, we draw the forces diagrams. The modelling assumptions mean that we can ignore the weight of the pulleys and the string, and that there is no friction between the string and the pulleys. As mass 2 is much larger than mass 1 we would expect that mass 2 will move down and mass 1 will move up. It is helpful in visualising this problem to draw first a single diagram on the pulley system, and then to draw the separate diagrams for each mass. Firstly, the single diagram is

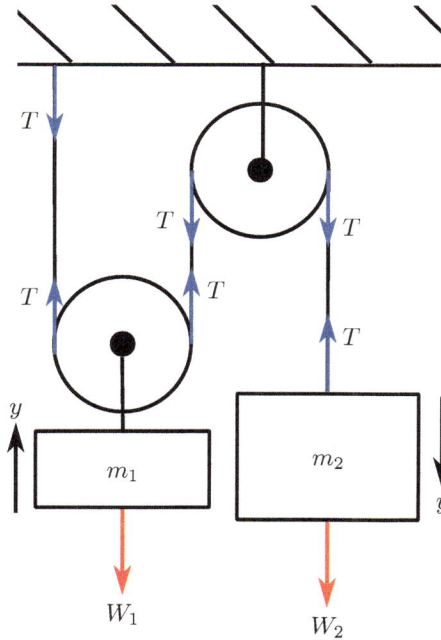

Now we draw the separate forces diagrams, so that we can see how these forces act on each mass.

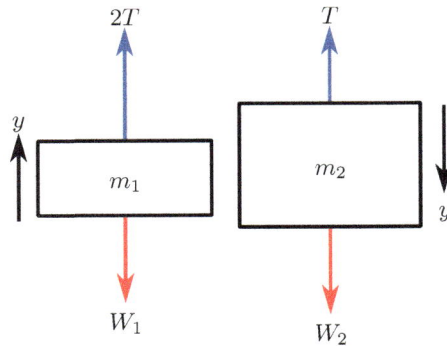

The tension force acting on mass 1 is twice that acting on mass 2. This is because mass 1 is, via the pulley, connected to two sections of the string, whereas, mass 2 is only connected to one.

We now compose our equations for the resultant forces on each mass.

$$F_1 = 2T - W_1,$$ (28.15)

$$F_2 = W_2 - T.$$ (28.16)

We can find $W_1$ and $W_2$ as follows.

$$W_1 = m_1 g = 2 \times 9.8 = 19.6\,\text{N},$$

$$W_2 = m_2 g = 8 \times 9.8 = 78.4\,\text{N}.$$

Therefore, Equations (28.15) and (28.16) become

$$F_1 = 2T - 19.6,$$ (28.17)

$$F_2 = 78.4 - T.$$ (28.18)

In contrast to our previous examples, the acceleration of each mass will be different. This is because the string above mass 1 is essentially doubled up, compared with the string above mass 2. The pulleys must pass the same length of string for each mass during motion, however, the string above mass 1 travels down one side and up the other side of the pulley, whilst the string above mass 2 only travels down, directly above the mass. This means that the distance travelled by the masses is different and, therefore, the accelerations are different. We substitute $F = ma$ into Equations (28.17) and (28.18) as follows.

$$2a_1 = 2T - 19.6,$$ (28.19)

$$8a_2 = 78.4 - T.$$ (28.20)

We now have two equations with three unknowns: $T$, $a_1$, and $a_2$. We need to find a separate relation between two of these unknowns in order to solve the problem. Using our explanation above, we know that, as mass 2 descends a distance $s$, the distance travelled by mass 1 is only $s/2$. From our equations of constant acceleration we know that $s \propto a$. Therefore, we arrive at a relation between the accelerations of each mass:

$$a_1 = \frac{1}{2}a_2.$$

We substitute this into equations (28.19) and (28.20) to yield two equations of two unknowns.

$$2a_1 = 2T - 19.6, \tag{28.21}$$
$$16a_1 = 78.4 - T. \tag{28.22}$$

To solve these equations we first seek to cancel $T$. We can do this by multiplying Equation (28.22) by 2 and adding to Equation (28.21).

$$2a_1 + 32a_1 = 2T - 19.6 + 156.8 - 2T,$$
$$\Rightarrow \quad 34a_1 = 137.2,$$
$$\Rightarrow \quad a_1 = \frac{137.2}{34},$$
$$\Rightarrow \quad a_1 = 4.04\,\text{ms}^{-2} \text{ to 2 d.p.}$$

We substitute (the unrounded) $a_1$ into the equation relating $a_1$ and $a_2$ to find $a_2$.

$$\frac{1}{2}a_2 = 4.03...,$$
$$\Rightarrow \quad a_2 = 2 \times 4.03...,$$
$$\Rightarrow \quad a_2 = 8.07\,\text{ms}^{-2} \text{ to 2 d.p.}$$

Next, we substitute into Equation (28.22) to find $T$.

$$16 \times 4.03... = 78.4 - T,$$
$$\Rightarrow \quad T = 78.4 - 16 \times 4.03...,$$
$$\Rightarrow \quad T = 13.84\,\text{N to 2 d.p.}$$

Now, we again show two different methods to find the resultant forces.

**Method 1:**
Firstly, we use $F = ma$ with our two calculated accelerations:

$$F_1 = m_1 a_1 = 2 \times 4.04... = 8.07\,\text{N to 2 d.p.},$$
$$F_2 = m_2 a_2 = 8 \times 8.07... = 64.56\,\text{N to 2 d.p.}$$

**Method 2:**
Now, we substitute the value we have calculated for $T$ into Equations (28.17) and (28.18).

$$F_1 = 2T - 19.6 = 2 \times 13.84... - 19.6 = 8.07\,\text{N to 2 d.p.},$$
$$F_2 = 78.4 - T = 78.4 - 13.84... = 64.56\,\text{N to 2 d.p.}$$

We now explore how we might solve a problem involving three masses.

---

**Example 28.14**

Three masses are connected by two pulley systems, as shown in the figure below, and are initially held at rest. Both pulley and string systems are light, smooth, and inextensible. The surface on which mass $m_2$ rests is smooth. The masses are released. Find the tension in each string, the resultant forces on each mass, and the acceleration.

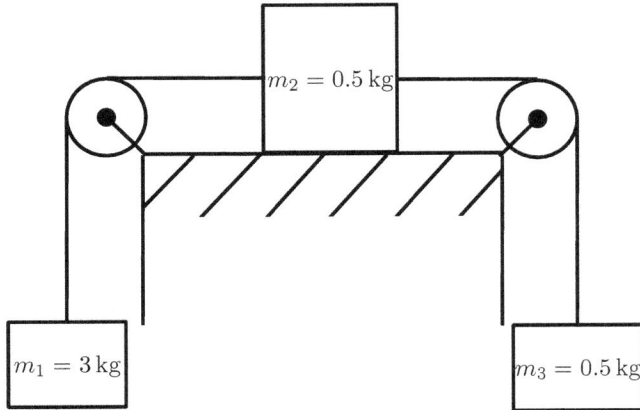

**Solution:**

We begin by drawing the forces diagrams. The modelling assumptions mean that we can neglect the weight of the pulleys and strings, that there is no friction between the pulleys and strings, or between mass 2 and the surface, and that tension is constant along the length of each string. However, as the strings are separate from one another, the tension in each string may be different. The magnitude of the acceleration of each mass will be the same. We would expect mass 1 to move downwards, mass 2 to move to the left, and mass 3 to move upwards. Therefore, the forces diagrams are as below. We now write the equations for the resultant forces on the masses in the direction of motion. As there are three masses there will be three equations.

$$F_1 = W_1 - T_{12}, \tag{28.23}$$
$$F_2 = T_{12} - T_{23}, \tag{28.24}$$
$$F_3 = T_{23} - W_3. \tag{28.25}$$

We can find $W_1$ and $W_3$ as follows.

$$W_1 = m_1 g = 3 \times 9.8 = 29.4 \, \text{N},$$
$$W_3 = m_3 g = 0.5 \times 9.8 = 4.9 \, \text{N}.$$

Substituting into Equations (28.23), (28.24), and (28.25) we obtain

$$F_1 = 29.4 - T_{12}, \tag{28.26}$$
$$F_2 = T_{12} - T_{23}, \tag{28.27}$$
$$F_3 = T_{23} - 4.9. \tag{28.28}$$

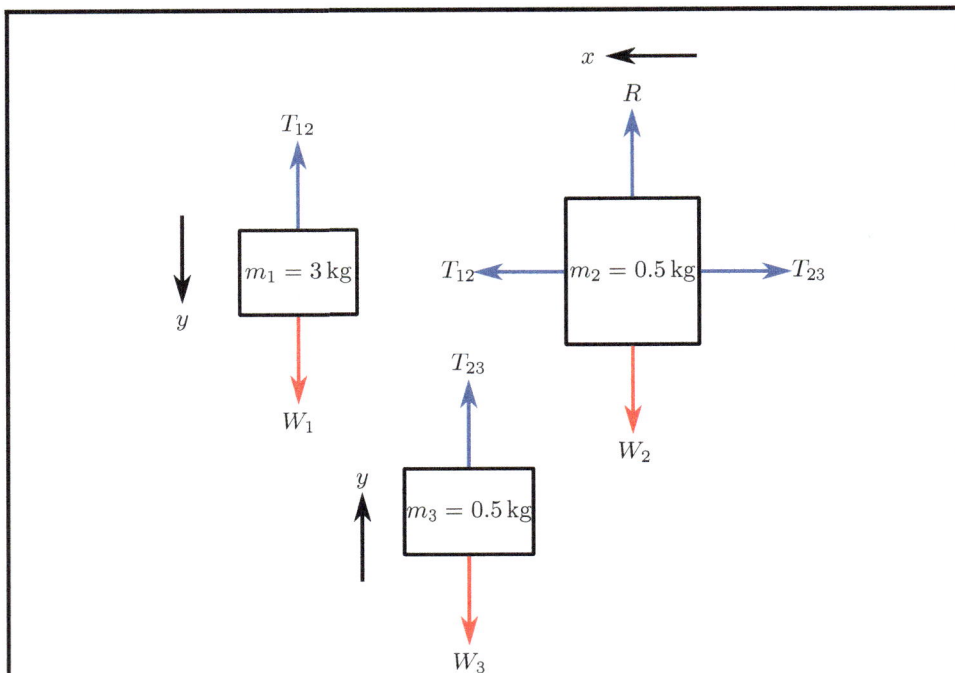

Substituting for $F = ma$, these equations become

$$3a = 29.4 - T_{12}, \tag{28.29}$$
$$0.5a = T_{12} - T_{23}, \tag{28.30}$$
$$0.5a = T_{23} - 4.9. \tag{28.31}$$

We now have three equations with three unknowns. We solve by adding Equations (28.29) and (28.30) together to eliminate $T_{12}$, giving us a system of two equations with two unknowns, which we know how to solve. We then substitute the value we find for $a$ into Equation (28.29) to find $T_{12}$.

Firstly, adding Equations (28.29) and (28.30) together.

$$3a + 0.5a = 29.4 - T_{12} + T_{12} - T_{23},$$
$$\Rightarrow \qquad 3.5a = 29.4 - T_{23}.$$

Therefore, our new system of two equations that we must solve is

$$3.5a = 29.4 - T_{23}, \tag{28.32}$$
$$0.5a = T_{23} - 4.9. \tag{28.33}$$

We now add Equations (28.32) and (28.33) together to eliminate $T_{23}$, and rearrange to find $a$.

$$3.5a + 0.5a = 29.4 - T_{23} + T_{23} - 4.9,$$
$$\Rightarrow \qquad 4a = 24.5,$$
$$\Rightarrow \qquad a = \frac{24.5}{4},$$
$$\Rightarrow \qquad a = 6.125 \, \text{N}.$$

Substituting into Equation (28.32) we have

$$3.5 \times 6.125 = 29.4 - T_{23},$$
$$\Rightarrow \qquad T_{23} = 29.4 - 3.5 \times 6.125,$$
$$\Rightarrow \qquad T_{23} = 7.9625 \, \text{N}.$$

Next we substitute our calculated value for $a$ into equation (28.29) in order to find $T_{12}$.

$$3 \times 6.125 = 29.4 - T_{12},$$
$$\Rightarrow \qquad T_{12} = 29.4 - 3 \times 6.125,$$
$$\Rightarrow \qquad T_{12} = 11.025 \, \text{N}.$$

If we wish, we can check that our calculated values are consistent, by checking that the left and right hand sides are equal in the equation which has all three unknowns, (28.30).

$$0.5a = 0.5 \times 6.125 = 3.0625,$$
$$T_{12} - T_{23} = 11.025 - 7.9625 = 3.0625.$$

Now, we need to find the resultant forces on each mass. As with previous examples, we show two possible methods.

**Method 1:**
Firstly, we substitute our calculated value for $a$ into $F = ma$ for each mass.

$$F_1 = m_1 a = 3 \times 6.125 = 18.375 \, \text{N},$$
$$F_2 = m_2 a = 0.5 \times 6.125 = 3.0625 \, \text{N},$$
$$F_3 = m_3 a = 0.5 \times 6.125 = 3.0625 \, \text{N}.$$

**Method 2:**
Now, we substitute our calculated values for the tension in each string into Equations (28.26), (28.27), and (28.28).

$$F_1 = 29.4 - T_{12} = 29.4 - 11.025 = 18.375 \, \text{N},$$
$$F_2 = T_{12} - T_{23} = 11.025 - 7.9625 = 3.0625 \, \text{N},$$
$$F_3 = T_{23} - 4.9 = 7.9625 - 4.9 = 3.0625 \, \text{N}.$$

We now use our understanding of pulleys to make deductions about a simple pulley system.

**Example 28.15**

Two masses of $3\,\mathrm{kg}$ are connected by the simple pulley shown in the figure below, and are held at rest. The pulley and string system is light, smooth, and inextensible. The masses are released. Deduce what the acceleration of the masses will be.

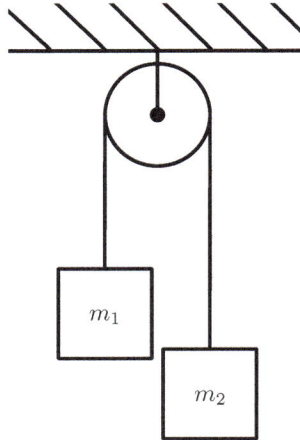

**Solution:**

Our modelling assumptions enable us to ignore the weight of the pulley and string, and to assume that there is no friction between the pulley and the string, and that tension is constant along the length of the string. As the masses are connected, they will have the same acceleration. The two objects have the same mass, and, therefore, are subject to the same weight force. We can deduce that the weight forces will balance out and that neither mass will move. The acceleration will be $a = 0\,\mathrm{ms}^{-2}$.

We will now check that our deductions are correct. We have the following forces diagrams.

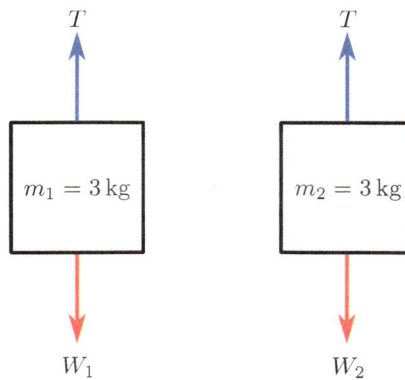

We can write down the equations of the resultant forces as follows. For now, we will assume that mass 1 will move up, and mass 2 will move down.

$$F_1 = T - W_1,$$
$$F_2 = W_2 - T.$$

Since $m_1 = m_2$, we know that $W_1 = W_2 = W$, and these equations become

$$F_1 = T - W,$$
$$F_2 = W - T.$$

We now substitute in $F = ma$ to the left hand side of each equation, using $m_1 = m_2 = 3\,\text{kg}$.

$$3a = T - W,$$
$$3a = W - T.$$

Adding these equations together we obtain

$$3a + 3a = T - W + W - T,$$
$$\Rightarrow \qquad\qquad\qquad 6a = 0,$$
$$\Rightarrow \qquad\qquad\qquad a = 0\,\text{ms}^{-2}.$$

Consider the example of a train, moving along a straight, level track. The carriages of the train are all moving in the same direction, with the same acceleration. Since we have assumed that the track is straight and level, there is no lateral or vertical motion (see remark). Therefore, there is no relative motion between the carriages and they can be modelled as a single particle with the mass of the overall system. We show this in the following example.

**Remark**
If the track on which a train is moving is straight, then the carriages' motion will be collinear. In contrast, if the track is curved, then the carriages will be at different angles from one another and their directions of motion at any point in time will be different. In this case, we would need to model the carriages as separate particles, in 2-dimensions. Similarly, if the track is not level (*i.e.* there are bumps, hills, or dips), then the carriages will rise and fall in turn. They will, therefore, have different directions of motion from one another in the vertical plane. Again, we would need to model the carriages as separate particles, in 2-dimensions.

**Example 28.16**
A toy train consisting of two carriages is pulled along a smooth, straight, level track with a force of 0.15 N. Each carriage has mass 30 g and they are connected by a light, rigid, inextensible rod. Find the acceleration of each carriage, modelling the carriages as a single particle.

**Solution:**
Our modelling assumptions mean that we may neglect friction and the weight of the rod, the rod will not compress, and tension is the same for the whole length of the rod. As the carriages are modelled as a single particle, we can assume that their weight force acts through the centre of the train. Therefore, we have the following forces diagram.

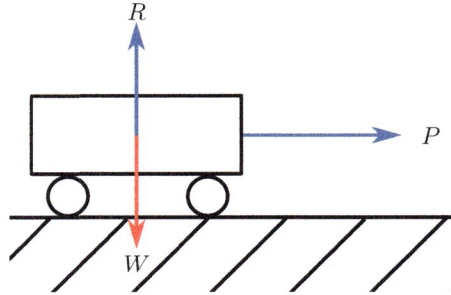

The carriages will have the same acceleration in the same direction, given by rearranging $F = ma$:

$$a = \frac{m_1 + m_2}{F} = \frac{0.03 + 0.03}{0.15} = 0.4\,\text{ms}^{-2}.$$

**Tip**
Particles that are connected and in contact can almost always be modelled as a single particle.

From the examples in this section we can see that there is a general methodology that we follow in solving problems involving connected particles. This is set out in the following algorithm.

**Algorithm 28.8 — Solving Problems Involving Connected Particles**
1. Identify whether the connected particles have any relative motion. If not, the particles should be modelled as a single particle.
2. Identify all of the modelling assumptions and their impact on the problem.
3. Consider the likely motion and, from this, determine axes for each mass.
4. Draw the forces diagram.
5. Write the equations for the resultant forces.
6. Use $F = ma$ to express the equations for the resultant forces in terms of $a$ instead of $F_i$, where $i$ is the index of an object.
7. Solve the two resulting equations simultaneously for $a$ and $T$.
8. If required, substitute back into the original equations, or use $F = ma$, to find the resultant forces $F_i$.

**Example 28.17**

A box of mass 0.3 kg is lifted by a crane. The box contains an air conditioning unit of mass 90 kg. Find the tension in the chain required to lift the box.

**Solution:**

From Algorithm 28.8, we first consider whether the box and the air conditioning unit have any relative motion. As the air conditioning unit is within the box, there is no relative motion between the unit and the box. Therefore, we model the box and the unit as a single particle of mass 90.3 kg.

The particle is subject to two forces: weight and tension in the rope. In order for the crane to lift the particle, the tension force in the rope must overcome the weight. Algebraically, this means

$$T > W.$$

The weight of the particle is given by

$$W = mg = 90.3 \times 9.8 = 884.94 \, \text{N}.$$

Therefore, the tension in the chain required to lift the particle is

$$T > 884.94 \, \text{N}.$$

**Exercise 28.1**

Q1. A box of mass 50 g is at rest on a horizontal surface. What is the normal reaction force on the box?
   (a) 490 N vertically upwards;
   (b) 0.49 N vertically upwards;
   (c) 490 N vertically downwards; or
   (d) 0.49 N vertically downwards.

Q2. A box of mass 0.5 kg is pushed along a horizontal surface with a force of 10 N. The friction force is 2.5 N. What is the resultant force on the box?
   (a) 12.5 N in the direction of the pushing force;
   (b) 5 N in the direction of the pushing force;
   (c) 7.5 N in the direction of the pushing force; or
   (d) 2.5 N in the direction of the pushing force.

Q3. A test rocket of mass 15 kg is ready to lift-off into the air. It forces air out of its base with $F_{\text{air}} = 1000$ N. What is the resultant force on the rocket once it has lost contact with the ground?
   (a) 853 N vertically upwards;
   (b) 1000 N vertically upwards;
   (c) 0 N; or
   (d) 1147 N vertically upwards.

Q4. A box of mass 5 kg is at rest on a horizontal surface. The box has a rocket inside it which forces air out of the back of the box with $F_{\text{air}} = 35$ N. Given that friction is $F_r = 10$ N, what is the resultant force on the box?

Q5. A box of mass $3\,\mathrm{kg}$ is at rest on a horizontal surface. What is the normal reaction force on the box?

Q6. A box of mass $10\,\mathrm{kg}$ is at rest on a smooth slope (inclined plane). The component of the weight of the box that is perpendicular to the slope is $5\sqrt{2}$ and the component of the weight of the box that is parallel to the slope is $5\sqrt{2}$. The magnitude of the weight force is $100\,\mathrm{N}$. What is the normal reaction force?

Q7. A box of mass $5\,\mathrm{kg}$ is in equilibrium on an inclined plane. The weight on the box is $\mathbf{W} = -50\mathbf{j}\mathrm{N}$ and the normal reaction force is $\mathbf{R} = (-21.7\mathbf{i} + 12.5\mathbf{j})\mathrm{N}$. The box is subjected to a pushing force up the slope of $\mathbf{P} = (29.295\mathbf{i} + 33.125\mathbf{j})\mathrm{N}$. What is the friction force on the box?

Q8. A box of mass $3\,\mathrm{kg}$ is moving along a horizontal floor with an initial velocity of $3\,\mathrm{ms}^{-2}$. It is subject to a pushing force in the direction of motion of $3\,\mathrm{N}$ and a friction force of $9\,\mathrm{N}$. What is the velocity of the box after it has travelled $1.25\,\mathrm{m}$?

Q9. A box of mass $2\,\mathrm{kg}$ is in equilbrium on a horizontal surface. The box is subjected to a pushing force of $6\,\mathrm{N}$. What friction force is required to maintain equilibrium?

Q10. Find the following forces.
    (a) A box of mass $120\,\mathrm{kg}$ is sliding on a smooth plane. What is the friction force?
    (b) A ball of mass $20\,\mathrm{g}$ is falling through a vacuum. What is the force due to air resistance?
    (c) A light tow rope connects two cars of mass $1200\,\mathrm{kg}$. What is the weight of the tow rope?

Q11. A mass of $5\,\mathrm{kg}$ is held at rest and is connected to another mass of $8\,\mathrm{kg}$ by a light, inextensible rope that runs over a smooth, light pulley. The mass is released. Find
    (a) the tension in the rope;
    (b) the acceleration of each mass; and
    (c) the resultant force on each mass.

Q12. A box of mass $25\,\mathrm{kg}$ is held at rest on a desk. The box is connected to a hanging mass of $2\,\mathrm{kg}$ by a smooth, light, inextensible rope and pulley system. The friction force between the box and the desk is $F_r = 10\,\mathrm{N}$. Find
    (a) the tension in the rope;
    (b) the acceleration of each mass; and
    (c) the resultant force on each mass.

Q13. Three masses are connected by a single pulley as shown. The rope and pulley system is smooth, light, and inextensible.

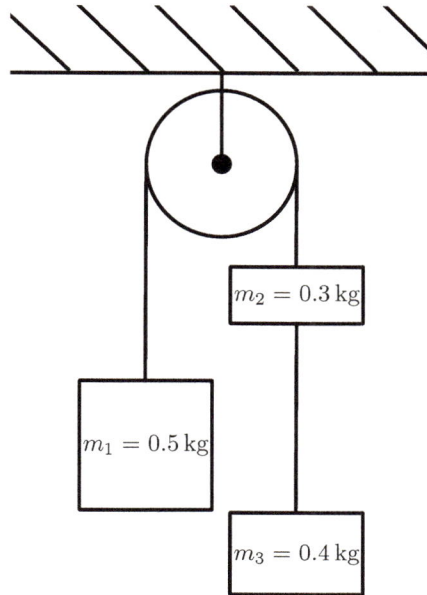

Find
  (a) the tension in the rope; and
  (b) the acceleration of each mass.
Q14. A person of mass $75\,\text{kg}$ is standing within an elevator car of mass $300\,\text{kg}$. Find the tension force in the cable connected to the lift, assuming that the car is moving at a constant velocity.

## Key Facts — Forces to Model the Real World

- A normal reaction force, or normal contact force, is a force that is exerted on an object by a surface that is in contact with the object. Its direction is perpendicular to the surface, and its magnitude is equal to the force that the object exerts on the surface in a direction perpendicular to the surface.
- Newtons third law: All forces in the Universe occur in equal but oppositely directed pairs.
- Thrust is a type of driving force that comprises an equal and opposite force pair.
- Friction is a force that one surface exerts on another when the two surfaces are sliding against each other, or when another force is trying to cause them to slide against each other.
- A resistance force is a force that opposes motion. Friction is an example of a resistance force.
- When modelling the real world, we need to make a number of *modelling assumptions* to simplify the problem.

**Chapter Assessment — Forces to Model the Real World**
Download and sit the 30 minute assessment for this chapter from the digital book.

# 29. Introduction to Probability

Games of chance have been played since the dawn of civilisation, and the mathematical study of such games has been important since the 17th Century. The desire to understand games of chance (no doubt influenced in part by their sponsors) led many famous mathematicians to investigate games and gambling. Amongst these, Pascal, Fermat, Huygens, Jakob Bernouli, de Moivre and Laplace made major contributions to the field of probability.

(a)

(b)

Figure 29.1: Christian Huygens (a) and Pierre-Simon Laplace (b)

Probability plays a central role in many aspects of modern life, for example; calculating insurance premiums, understanding the National Lottery and deciding if a patient should be tested for a medical condition. As Pierre-Simon Laplace wrote in 1814 "the most important problems in life are, for the most part, only problems in probability".

In this chapter we begin to look at the mathematical theory of probability and explore the application of this theory to simple situations.

## 29.1 Events and the Sample Space

We begin by making a number of definitions that will be required throughout this and subsequent chapters.

> **Definition 29.1 — Simple Probability Terms**
> Here we define some common terminology.
>
> **Experiment:** A procedure which can be repeated and has defined outcomes.
> **Outcomes:** The possible results of an experiment.
> **Sample space:** The sample space is the collection of all possible outcomes from the experiment. For example, if we were to roll a single six-sided die the sample space would be $\{1,2,3,4,5,6\}$, since these are all the possible outcomes. The sample space is sometimes referred to as $\Omega$.
> **Event:** An event is a subset of the sample space and is a collection of outcomes from the experiment.
> **Equally Likely:** Events/outcomes are said to be equally likely if they all have the same theoretical probability of occurring.

> **Remark**
> Familiarity with a standard pack of cards is expected. In such a pack (or deck) there are 52 cards. These are divided up into four suits: hearts, diamonds, spades and clubs. Of these suits, hearts and diamonds are represented as red cards, whilst spades and clubs are black cards. Each suit's 13 cards are: Ace, 2, 3, 4, 5, 6, 7, 8, 9, 10, Jack, Queen, King. The Jack, Queen and King are known as picture cards.

> **Example 29.1**
> If we randomly choose a card from a regular pack of cards, how many possible outcomes are there?
>
> **Solution:**
> Since every card in the deck is unique (*i.e.* a number/picture and suit) there are 52 possible outcomes.

> **Remark**
> When we *randomly choose* an outcome (or event), it means that picking any of the possible outcomes (or events) is equally likely.

**Example 29.2**

In a bag containing 5 red balls and 9 blue balls, let $E_1$ be the event "We pick a red ball" and $E_2$ be the event "We pick a blue ball". Are $E_1$ and $E_2$ equally likely?

**Solution:**

Whilst the probability of picking any particular ball will be $\frac{1}{14}$, $E_1$ and $E_2$ are not equally likely, as the probability of selecting a red ball or a blue ball are $\frac{5}{14}$ and $\frac{9}{14}$, respectively.

**Activity 29.1 — Listing Outcomes**

Find the possible outcomes for the following situations.

(a) An experiment where a fair six-sided die is rolled and a fair coin is flipped.

(b)   i. If two coins are flipped simultaneously, how many distinct outcomes are there?

      ii. Imagine that we now flip one coin after the other and we are concerned about the order - does the number of outcomes change?

(c) A football team is to play two matches. What are the possible outcomes from the two games?

**Definition 29.2 — Probability of an Event**

The probability of an event $A$ occurring is defined to be the number of outcomes in $A$ divided by the total number of equally likely outcomes in the sample space (including $A$).

The probability of an event $A$ is usually written as $P(A)$.

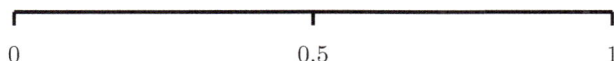

$$0 \qquad\qquad\qquad 0.5 \qquad\qquad\qquad 1$$

Figure 29.2: A probability scale.

Probabilities can only take values between 0 to 1, as shown in Figure 29.2. An event which is impossible (*e.g.* rolling a 7 on a normal six-sided dice) will have a probability of 0. An event which is certain (*e.g.* rolling a number between 1 and 6 on a normal six-sided dice) will have a probability of 1. This means that, for an event $A$, $0 \leq P(A) \leq 1$. It also indicates that the sample space, $\Omega$, must always have a probability equal to 1 since it covers all possible outcomes. Thus $P(\Omega) = 1$.

**Activity 29.2 — Probability - True or False?**

Consider the statements below - are they true or false? Justify the answer.

(a) If we get a tail with ten flips of a coin, then we will get a tail on the next flip of the coin.

(b) When we roll two fair six-sided dice, since there are six outcomes for each then there are twelve outcomes in total. The probability of rolling two sixes is therefore $\frac{2}{12}$ (or $\frac{1}{6}$ as a simplified fraction).

(c) If we get a tail when we flip a coin, then when we next flip the coin we will get a head.
(d) In a game of bingo, we are less likely to win if we choose the number 13.
(e) The probability of it raining in London is less than the probability of it raining in Edinburgh.
(f) The probability of getting a head with two flips of a coin is $\frac{1}{2}$.
(g) The probability that we roll a six is equal to the probability that we roll a four with a fair six-sided die.

**Definition 29.3 — Probability of an Event not Occurring**
$A'$ is defined to be the event that $A$ does not occur and is often referred to as the complement of $A$. Since the sample space has probability equal to 1, we define $P(A')$, the probability of event $A$ not occurring, to be

$$P(A') = 1 - P(A).$$

**Example 29.3**
A fair spinner, as shown below, is spun.
(a) What is the probability of the spinner landing on an even number?
(b) Hence, or otherwise, what is the probability of the spinner not landing on an even number?

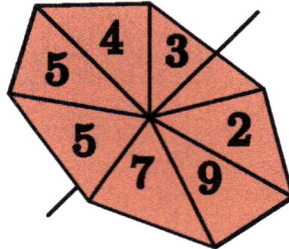

**Solution:**
(a) The spinner has seven equally divided sections, therefore, there are seven possible outcomes (although some are the same). Let $E$ denote the event "the spinner lands on an even number". We now consider which of the possible outcomes are even. Out of the seven possible outcomes, only two (2 and 4) are even numbers, hence

$$P(E) = \frac{2}{7}.$$

(b) $E'$ is the event "the spinner lands on a number which is not even" (equivalently "the spinner lands on a number which is odd"). Since the probability of the sample space (*i.e.* all possible outcomes) is equal to 1, then

$$P(E') = 1 - P(E).$$

Thus,

$$P(E') = 1 - \frac{2}{7}$$
$$= \frac{5}{7}.$$

**Remark**

Note, in the above example we formally state the event "the spinner lands on an even number" and let $E$ denote this event. This should always be done when defining events since, for example, saying P(even) is meaningless.

**Exercise 29.1**

Q1. Which probability below might be assigned to the event that it will rain in England during the month of April? Explain this selection.

    0.16        0.57        0.68        0.85

Q2. Three fair coins are flipped.
    (a) List all the possible outcomes when these coins are flipped.
    (b) What is the probability of getting exactly two heads?

Q3. Consider a fair, standard cubical die.
    (a) What is the probability of rolling a four?
    (b) What is the probability of *not* rolling a four?

Q4. Consider a regular pack of playing cards.
    (a) What is the probability of drawing a spade?
    (b) What is the probability of *not* drawing a spade?
    (c) What is the probability of drawing a seven?
    (d) What is the probability of drawing a seven that is *not* a spade?

Q5. A customer in an ice cream parlour is choosing a two-scoop ice cream sundae.
    (a) Given the choice of vanilla, chocolate or strawberry ice cream, how many different combinations are there?
    Assuming the customer chooses at random, and his choice of second scoop is independent of the previous scoop,
    (a) what is the probability that a given ice cream has a least one scoop of chocolate?
    (b) what is the probability that a given ice cream has two scoops of the same flavour?

Q6. There are two blue counters and three yellow counters in a box. If two counters are removed at random, what is the probability that both counters are blue?

Q7. A fair, four-sided spinner is spun. Each section is equally likely and is coloured in either blue, red or green. There are two red sections of the spinner.
    (a) What is the probability of the spinner *not* landing on blue?
    (b) What is the probability the spinner lands on red or green?

Q8. Given the word PROBABILITY, what is the probability that, in two randomly selected letters, there will be:
(a) the letter B;
(b) the letter B or I;
(c) *neither* a B *nor* an I.

## 29.2  Venn Diagrams and Set Notation

Events occurring within a sample space can be represented using mathematician John Venn's 1880 invention, 'Eulerian Circles', more commonly known as Venn diagrams.

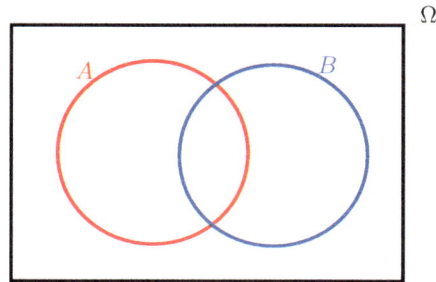

Figure 29.3: A Venn diagram with two sets $A$ and $B$.

The enclosing box (often a rectangle) in a Venn diagram represents the sample space, $\Omega$, which is the set of all possible outcomes (see Section 29.1). A closed curve (usually a circle or ellipse) is used to represent each event within the sample space.

The events $A$ and $A'$ can be represented using Venn diagrams as shown in Figures 29.4 and 29.5 respectively.

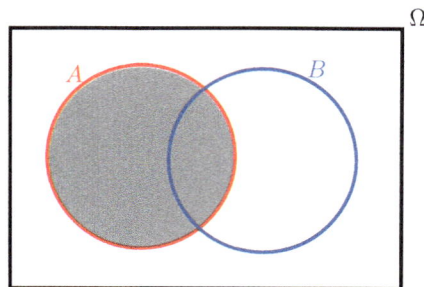

Figure 29.4: The event $A$ represented using a Venn diagram.

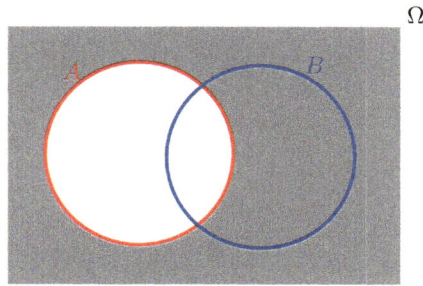

Figure 29.5: The event $A'$ represented using a Venn diagram.

Since an event is a set of outcomes (see Section 29.1), we are able to use set notation to identify different events which correspond to parts on the Venn diagram.

---

**Definition 29.4 — Intersection**

The intersection of two events $A$ and $B$ is the event where both $A$ *and* $B$ occur (see below). The intersection of two events (or sets), $A$ and $B$, is usually written as $A \cap B$.

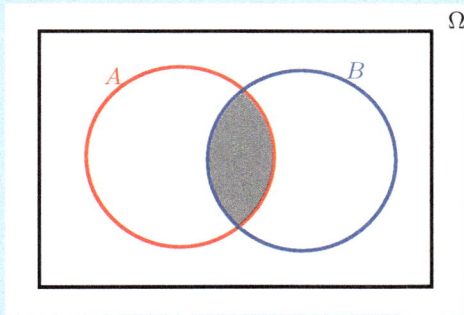

---

**Definition 29.5 — Union**

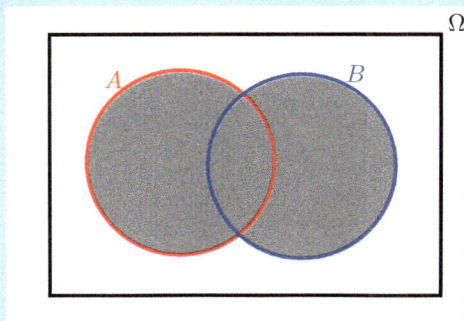

The union of two events $A$ and $B$ is the event where *either* $A$ or $B$ occur, as shown above. The union of two events (or sets), $A$ and $B$, is usually written as $A \cup B$.

**Example 29.4**

There are 60 people at an activity centre. In total, 31 go sailing, 24 go both sailing and climbing, 12 do neither activity. How many only go climbing?

**Solution:**

In order to answer this question it is beneficial to construct a Venn diagram representing the outcomes. Let the event $S$ represent those that go sailing and the event $C$ represent those that go climbing. As usual, $\Omega$ will represent the sample space. Using the information provided we can draw the Venn diagram below.

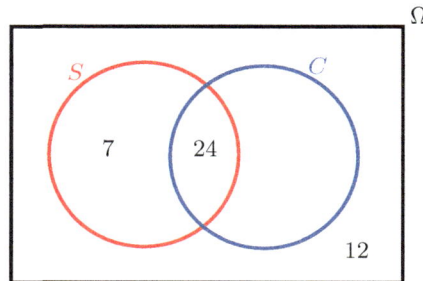

Since we know that 24 people go sailing *and* climbing, we can deduce that 7 people just go sailing, since 31 people go sailing in total. We conclude the missing value (those that only go climbing) to be 17 people. The completed Venn diagram is shown below.

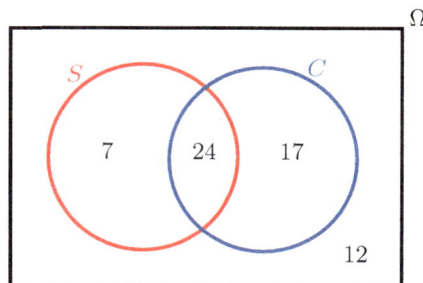

This means that 41 people go climbing in total.

**Activity 29.3 — Three Event Venn Diagrams**

A given sample space $\Omega$ has three events, $A$, $B$ and $C$.

(a) Represent the sample space $\Omega$ and shade $A \cap B$, $B \cup C$ and $A \cap B \cap C$ on different Venn diagrams.

(b) Represent $\Omega$ as a Venn diagram and label each section using set notation.

---

**Example 29.5**

John is considering the probabilities of drawing, at random, a seven from a standard pack of cards (event $A$) or a red card (event $B$). Construct a Venn diagram of these probabilities .

**Solution:**

Since there are four different suits in a normal deck of cards, there are four sevens in the deck and so, $P(A) = \frac{4}{52} = \frac{1}{13}$. Additionally, two of these suits are red, so half of the deck will be red and $P(B) = \frac{26}{52} = \frac{1}{2}$.

Thus, as two of the sevens must be red, $P(A \cap B) = \frac{2}{52} = \frac{1}{26}$, as can be seen in the intersection in the figure below. From this, the other probabilities can be calculated, since the probabilities in each closed curve must sum to the total probability for that event.

$$P(A \cap B') = \frac{1}{13} - \frac{1}{26} = \frac{1}{26},$$

$$P(B \cap A') = \frac{1}{2} - \frac{1}{26} = \frac{12}{26} = \frac{6}{13}$$

Finally, since $P(A \cup B) = \frac{28}{52}$, the rest of the possibilities (to be included in the sample space) must have a probability of $\frac{24}{52}$ or $\frac{6}{13}$ when simplified. The final Venn diagram is shown below.

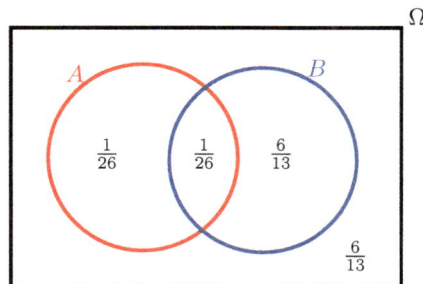

---

**Tip**

As demonstrated in Examples 29.4 and 29.5, the sections of a Venn diagrams may either state a quantity, or the probability of an event.

**Exercise 29.2**

Draw a Venn diagram when answering the following questions.

Q1. During one hour at a coffee shop, 100 people are served a drink. 72 people order coffee, 41 order tea and 15 order both tea and coffee. Find the probability that a randomly selected customer

(a) bought tea;

(b) bought tea and coffee;

(c) bought coffee but not tea.

Q2. In a class of 30 students, there are 17 studying French and 15 studying Spanish. Given that 5 students do not study either language, find the probability that a randomly selected student

(a) studies French and Spanish;

(b) studies French but not Spanish;

(c) studies Spanish but not French.

Q3. Amy is at an archery range practising her technique. Her instructor tells her that the probability she hits the target is 0.57 and the probability she misses the board altogether is 0.22.

(a) Draw a Venn diagram of the events $T$, that she hits the target, and $B$, that she hits the board.

(b) What is the probability that Amy will hit the board but not the target?

Q4. A card is drawn at random from a standard deck of cards.

(a) Construct a Venn diagram of probabilities for the event that a picture card is drawn ($A$) or a prime number card is drawn ($B$).

(b) Find

    i. $\text{P}(A')$;      ii. $\text{P}(A \cup B)$;      iii. $\text{P}(A \cap B)$;      iv. $\text{P}(A' \cap B)$.

Q5. Zac asks people to randomly choose a number between 1 and 10. The events being considered are $P$, the number is prime, and $E$ the number is even.

(a) Construct a Venn diagram of probabilities for Zac's experiment.

(b) Find

    i. $\text{P}(P \cap E)$;      ii. $\text{P}(E')$;      iii. $\text{P}(P \cup E')$;      iv. $\text{P}(P' \cap E')$.

Q6. A survey is conducted at a local leisure centre about the sports and activities that people take part in. Two of the most popular results were football ($F$) and swimming ($S$). The probability that a randomly selected person from the survey played football is 0.76 and the probability that they swim is 0.69. Given that the probability that someone does not do either event is 0.14, find

(a) $\text{P}(F \cup S)$;

(b) $\text{P}(F \cap S)$;

(c) $\text{P}(F')$;

(d) $\text{P}(F' \cup S)$.

## 29.3 Finding Probabilities Using the Addition Rule

Venn diagrams are an extremely useful tool when studying probability as they help us visualise the theoretical probabilities we have been asked to find.

If we reconsider John's card experiment (Example 29.5) it is clear that $P(A \cup B) \neq P(A) + P(B)$.

Take two events, $A$ and $B$, in the sample space $\Omega$. Let $P(A) = p$, $P(B) = q$ and $P(A \cap B) = r$. Then we can construct the Venn diagram given in Figure 29.6.

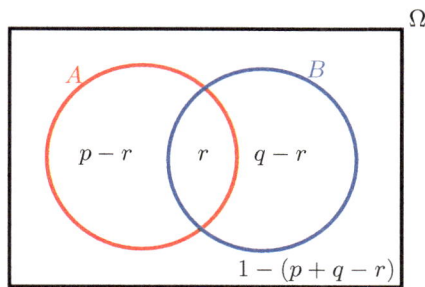

Figure 29.6: The addition rule for Union and Intersection.

This means that the probability of the union of $A$ and $B$ can be calculated as

$$P(A \cup B) = (p - r) + r + (q - r)$$
$$= p + q - r$$
$$= P(A) + P(B) - P(A \cap B).$$

---

**Formula 29.1 — The Addition Rule**
For two events, $A$ and $B$, the probability of their union is found by

$$P(A \cup B) = P(A) + P(B) - P(A \cap B).$$

---

**Tip**
The addition rule can be rearranged to give a formula for the intersection

$$P(A \cap B) = P(A) + P(B) - P(A \cup B).$$

---

**Example 29.6**
Two events, $A$ and $B$, are such that $P(A) = 0.7$, $P(B) = 0.5$ and $P(A \cap B) = 0.4$. Find
(a) $P(A \cup B)$;
(b) $P(A')$;
(c) $P(A' \cap B)$.

**Solution:**

(a) We can use the addition rule (Formula 29.1) and the given information to answer this:

$$P(A \cup B) = 0.7 + 0.5 - 0.4$$
$$= 0.8.$$

We are now able to construct a Venn diagram of the information. The completed diagram is shown below.

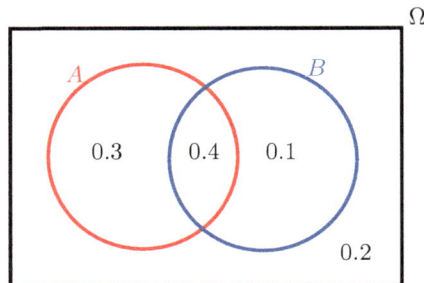

(b) From the above figure, $P(A') = 0.3$.

(c) The probability of an outcome being in both $A'$ and $B$ is 0.1. Thus, $P(A' \cap B) = 0.1$.

**Exercise 29.3**

Q1. Given that $P(A) = 0.4$, $P(B) = 0.5$ and $P(A \cap B) = 0.2$, find
   (a) $P(A \cup B)$;
   (b) $P(A' \cap B)$;
   (c) $P(A \cup B')$.

Q2. Given that $P(M) = 0.3$, $P(N) = 0.5$ and $P(M \cup N) = 0.6$, find
   (a) $P(M \cap N)$;
   (b) $P(M \cap N')$;
   (c) $P(M')$.

Q3. Given that $P(X') = 0.6$, $P(X \cap Y) = 0.1$ and $P(X' \cap Y') = 0.2$, find
   (a) $P(X)$;
   (b) $P(X \cap Y')$;
   (c) $P(X' \cup Y)$.

Q4. The users of a fitness tracking app are randomly selected. The probability that a selected user cycles is 0.77, whereas the probability that they run is 0.58. The probability that a selected user runs or cycles or does both is 1. Find the probability that a randomly selected user
   (a) runs and cycles;
   (b) does not run or cycle;
   (c) runs but does not cycle.

Q5. Aliens from the Sentinal Moon are called Necromorphs and have different anatomies to the average human being. There is a probability of 0.46 that a Necromorph will have three eyes and a probability of 0.53 that a Necromorph will have three arms. Given that the probability a randomly selected Necromorph will have three eyes or three arms or both is 0.76, find the probability that the Necromorph
   (a) will have three eyes and three arms;
   (b) neither three eyes nor three arms;
   (c) three eyes but not three arms.

Q6. Aysha goes to buy a new car from a second hand dealer. From the range of cars available, the probability the car is silver is 0.62 and the probability it has two passenger doors is 0.44. The probability that Aysha selects a silver car with two passenger doors is 0.28. Find the probability that Aysha randomly chooses
   (a) a car that is either silver or has two passenger doors or both;
   (b) a car that is silver that does not have two passenger doors;
   (c) a car which is not silver.

## 29.4 Mutually Exclusive Events

**Definition 29.6 — Mutually Exclusive Events**
Two events, $A$ and $B$, are said to be mutually exclusive if only one of these events can occur at the same time, *i.e.* if $A$ occurs then $B$ does not, and vice versa.

Consider the set of elements $\{1, 2, 3, 4, 5, 6, 7, 8, 9, 10\}$. Let event $A$ be selection of an even number and event $B$ be selection of an odd number. These events are *mutually exclusive*, since we cannot choose an even number which is also odd.

**Example 29.7**
Consider the set of numbers $\{2, 4, 5, 6, 7, 9, 13, 16, 20, 23, 24, 32, 35\}$. Let $A$ be the event "number is a multiple of 4" and $B$ be the event "number is prime". Are $A$ and $B$ mutually exclusive?

**Solution:**
From the definitions of events $A$ and $B$

$$A = \{4, 16, 20, 24, 32\},$$
$$B = \{2, 5, 7, 13, 23\}.$$

As there are no elements common to the set representations of $A$ and $B$, the events $A$ and $B$ are mutually exclusive. Indeed, it is not possible for a positive integer to be a multiple of 4 and a prime number.

**Activity 29.4 — Mutually Exclusive or Not?**
Alice rolls two dice and adds together the numbers shown. Explain, with reasons, which pairs of the following events are mutually exclusive.
  A. The total is a prime number.
  B. The total is odd.

  C.  One die shows the number 4.
  D.  The total is 8.
  E.  The total is even.
  F.  Both numbers on the dice are odd.
  G.  The total is more than 10.
  H.  The total is less than 6.

Two events, $A$ and $B$, which are mutually exclusive can be shown on a Venn diagram as in Figure 29.7.

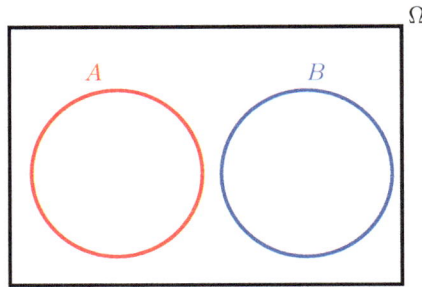

Figure 29.7: Two mutually exclusive events.

---

**Formula 29.2 — Empty Intersection for Mutually Exclusive Events**
If events $A$ and $B$ are mutually exclusive ,then

$$P(A \cap B) = 0.$$

---

As a result of Formula 29.7, we can adapt the addition rule (Formula 29.1) when applied to mutually exclusive events. Since the intersection of two mutually exclusive events is zero, their union is simply the probability of each event added together.

---

**Formula 29.3 — Addition Rule for Mutually Exclusive Events**
If events $A$ and $B$ are mutually exclusive, then

$$P(A \cup B) = P(A) + P(B).$$

---

**Example 29.8**
The events $A$ and $B$ are mutually exclusive. Given that $P(A) = 0.6$ and $P(B) = 0.2$, find

  (a) $P(A \cup B)$;        (b) $P(A')$;        (c) $P(A' \cap B)$;        (d) $P(A' \cap B')$.

**Solution:**
As events $A$ and $B$ are mutually exclusive, $P(A \cap B) = 0$, from Formula 29.3, and so the events can be represented on a Venn diagram as shown below.

$\Omega$

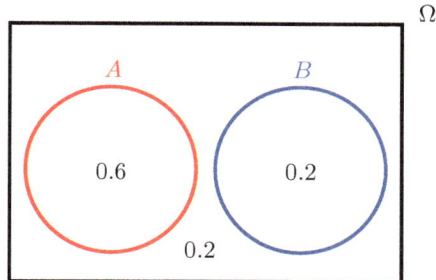

(a) From Formula 29.3, $P(A \cup B) = P(A) + P(B)$, hence,

$$P(A \cup B) = 0.6 + 0.2$$
$$= 0.8.$$

(b) Using Definition 29.3

$$P(A') = 1 - P(A)$$
$$= 1 - 0.6$$
$$= 0.4.$$

(c) From the above figure, we can see that $P(A' \cap B) = P(B)$. Therefore $P(A' \cap B) = 0.2$.

(d) Given that $A$ and $B$ are mutually exclusive, we can see that $P(A' \cap B')$ will give us the space that is outside of both events. Using the Venn diagram, we can identify the probability to be 0.2, therefore, $P(A' \cap B') = 0.2$.

**Example 29.9**
Arsenal and Tottenham Hotspur are due to play a football match. The probability that Arsenal win is 0.31 and the probability that Tottenham win is 0.36.
   (a) Draw a Venn diagram of probabilities for the football match.
   (b) Find the probability that Tottenham do not win the game.
   (c) Find the probability that the result is a draw.

**Solution:**

(a) The Venn diagram is shown below.

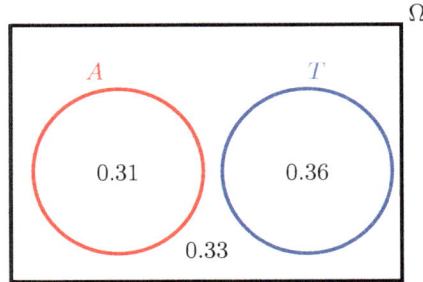

(b) From Definition 29.3

$$P(T') = 1 - P(T)$$
$$= 1 - 0.36$$
$$= 0.64.$$

(c) A draw indicates that neither Tottenham nor Arsenal win the game, therefore we consider $P(T' \cap A')$. Since $T$ and $A$ are mutually exclusive, and using the above figure, we can identify this space as the area outside of both event $T$ and $A$. Thus, $P(T' \cap A') = 0.33$.

---

**Exercise 29.4**

Q1. Which of the following is the probability of the intersection of $A$ and $B$ where A and B are mutually exclusive?

0.5              1              0              0.25

Explain this selection.

Q2. Given that $P(A) = 0.3$, $P(B') = 0.43$ and that the events $A$ and $B$ are mutually exclusive.

   (a) Draw a Venn diagram of probabilities to represent these events.
   (b) Find
       i. $P(B)$;
       ii. $P(A')$.

Q3. If $P(A) = 0.5$ and $P(B) = 0.5$, are events $A$ and $B$ mutually exclusive? If not, why not?

Q4. Dan likes his eggs done only one way for breakfast. The probability that he has poached eggs is 0.47 and the probability that he has scrambled eggs is 0.25.

   (a) Find the probability he does not have either poached or scrambled egg.
   (b) Find the probability that he does not have scrambled egg.

Q5. The probability that a new born baby, whose parents have brown eyes, also has brown eyes is 0.75. The probability that the baby has blue eyes is 0.06.

   (a) Find the probability that the baby does not have blue eyes.
   (b) Find the probability that the baby does not have blue or brown eyes.

Q6. Andy and Novak are playing a tennis match. The probability that Andy wins is 0.29. Draw a Venn diagram of probabilities to represent the possible outcomes.

Q7. A bag contains 7 red balls, 7 yellow balls and 1 black ball. A ball is randomly chosen from the bag. Considering the events of selecting either a red ball or a black ball,

  (a) represent this information in a Venn diagram of probabilities;

  (b) find the probability that

    i. a black ball is not chosen;

    ii. a yellow ball is not chosen.

Q8. Two rugby teams, Sale Sharks and Wasps, are playing a friendly match. The probability that Sale Sharks win is 0.37 and the probability that Wasps win is 0.59.

  (a) Find the probability that neither team win.

  (b) Find the probability that Wasps do not win.

## 29.5  Independent Events

**Definition 29.7 — Independent Events**

Two events, $A$ and $B$, are said to be statistically independent if $A$ happening, or not, has no influence upon the probability of $B$ happening.

**Example 29.10**

If we flip a fair coin, then roll a fair six-sided die and record what is shown on each, then the result from the coin will not influence the result from the die. For example, the probability of rolling a 5 is $1/6$ no matter whether a head or a tail was thrown. Throwing a head and rolling a 5 are *independent* events.

**Formula 29.4 — Independent Events**

If $A$ and $B$ are independent events then

$$P(A \cap B) = P(A) \times P(B).$$

**Example 29.11**

Given that the events $A$ and $B$ are independent, $P(A) = 0.36$ and $P(B) = 0.45$, find

  (a) $P(A \cap B)$;

  (b) $P(A \cup B)$.

**Solution:**

  (a) Since $A$ and $B$ are independent, using Formula 29.4 gives

$$P(A \cap B) = 0.36 \times 0.45$$
$$= 0.162.$$

(b) Using the addition rule (Formula 29.1),

$$P(A \cup B) = 0.36 + 0.45 - 0.162$$
$$= 0.648.$$

### Exercise 29.5

Q1. Determine whether events $A$ and $B$ are independent given:
   (a) $P(A) = 0.4,$     $P(B) = 0.3,$     $P(A \cap B) = 0.12.$
   (b) $P(A) = 0.1,$     $P(B) = 0.1,$     $P(A \cap B) = 0.1.$
   (c) $P(A) = 0.5,$     $P(B) = 0.49,$     $P(A \cap B) = 0.244.$
   (d) $P(A) = 0.44,$     $P(B) = 0.8,$     $P(A \cap B) = 0.352.$
   (e) $P(A) = 0.17,$     $P(B) = 0.62,$     $P(A \cap B) = 0.1054.$
   (f) $P(A) = 0.78,$     $P(B) = 0.22,$     $P(A \cap B) = 0.1617.$
Q2. Given that the events $A$ and $B$ are independent, find (i) $P(A \cap B)$ and (ii) $P(A \cup B)$ when:
   (a) $P(A) = 0.4,$     $P(B) = 0.5;$
   (b) $P(A) = 0.72,$     $P(B) = 0.4;$
   (c) $P(A) = 0.57,$     $P(B) = 0.68.$

### Example 29.12

An all year round ice cream van is selling ice creams. The probability that the van sells at least five ice creams during each hour of trade is 0.92 during the month of July $(J)$. The probability that the vans sells at least five ice creams during each hour of trade is 0.23 during the month of November $(N)$.
   (a) Are these events statistically independent?
   (b) Find $P(J \cap N)$ and state, with context, what this represents.
   (c) Find $P(J \cup N)$.

**Solution:**
   (a) Since the sales during July do not directly affect the sales made during November, it is reasonable to suggest that these events *are* statistically independent.
   (b) Given the suggested answer to part (a), we can use Formula 29.4 to determine that

$$P(J \cap N) = 0.92 \times 0.23$$
$$= 0.2116.$$

   This represents the probability that the ice cream van sells at least five ice creams during each hour of trade in both July and November.
   (c) Using the addition rule (Formula 29.1)

$$P(J \cup N) = 0.92 + 0.23 - 0.2116$$
$$= 0.9384.$$

**Exercise 29.6**

Q1. Kate has two identical bags containing four blue counters and seven green counters. Let $A$ be the event that Kate draws a blue counter from the first bag and let $B$ be the event that Kate draws a blue counter from the second bag.

(a) Determine whether or not the events $A$ and $B$ are independent, with justification.

(b) Find $P(A \cap B)$.

(c) Find $P(A' \cup B)$ and state, with context, what this represents.

Q2. A library's automated booking system determines that the probability of someone withdrawing fiction books ($A$) is 0.87, whereas the probability of someone withdrawing non-fiction books ($B$) is 0.34.

(a) Discuss the arguments for and against these events being statistically independent.

(b) Assuming that events $A$ and $b$ are independent, find

i. $P(A \cap B)$;

ii. $P(A' \cap B)$;

iii. $P(A' \cup B)$.

Q3. Three fair coins, $a$, $b$ and $c$, are flipped at the same time. The events that each coin lands on a head are represented by $H$, $I$ and $J$, respectively. The outcome from one coin has no influence on the outcomes of the others and, therefore, $H$, $I$ and $J$ are all independent events.

(a) Find $P(H \cap I)$;

(b) Find $P(H \cap I \cap J)$ and state what this represents;

(c) Find $P(I \cup J')$.

## 29.6  Tree Diagrams

Tree diagrams are another way of representing the probabilities of outcomes and events of experiments and are particularly helpful when an experiment is conducted more than once.

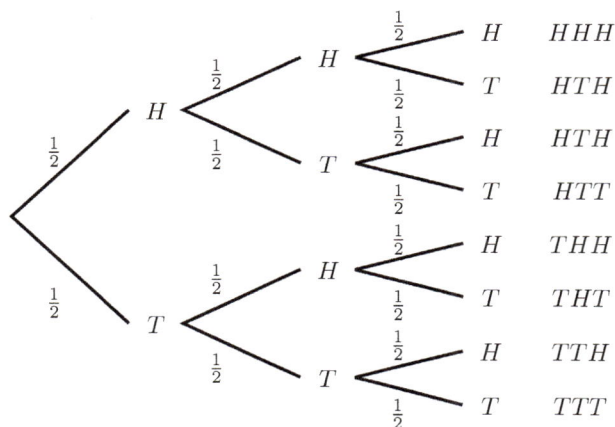

Figure 29.8: Example of a tree diagram, where a coin is flipped three times.

For example, if we were to flip a coin and then repeat the process twice, we would be able to draw the tree diagram in Figure 29.8, where the probability of each event is denoted on each branch.

---

**Example 29.13**

A bag contains five blue balls and seven yellow balls. Shannon draws a ball from the bag and then replaces it. She then draws another ball and replaces it. The events in the experiment are $Y$: "Shannon takes a yellow ball", and $B$: "Shannon takes a blue ball".
  (a) Draw a tree diagram to represent the outcomes of the experiment.
  (b) Find the probability that Shannon draws two blue balls from the bag.
  (c) Find the probability that Shannon draws at least one yellow ball from the bag.

**Solution:**
  (a) We know that $P(Y)$ is $\frac{7}{12}$ and $P(B)$ is $\frac{5}{12}$. These will be the probabilities on each of our branches, as shown below.

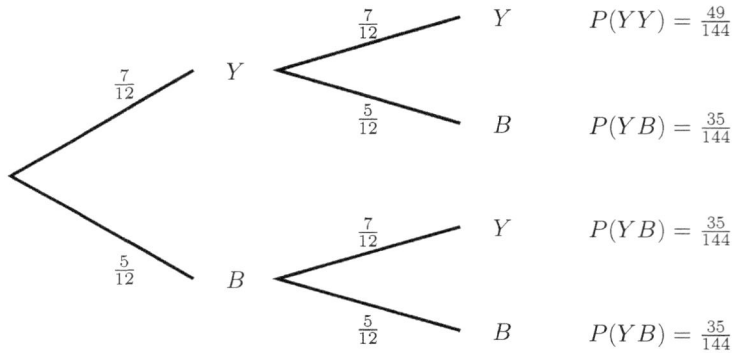

Tree diagram branches:
- $\frac{7}{12}$ to $Y$, then $\frac{7}{12}$ to $Y$: $P(YY) = \frac{49}{144}$
- then $\frac{5}{12}$ to $B$: $P(YB) = \frac{35}{144}$
- $\frac{5}{12}$ to $B$, then $\frac{7}{12}$ to $Y$: $P(YB) = \frac{35}{144}$
- then $\frac{5}{12}$ to $B$: $P(YB) = \frac{35}{144}$

  (b) Since the events are independent, we can find the probabilities of each possible outcome by multiplying along each branch.
  There is only one branch that results in two blue balls being drawn from the bag and the probability of this outcome will be $\frac{5}{144} \times \frac{5}{12} = \frac{25}{144}$.

  (c) From the tree diagram, we can see that there are three different outcomes which result in at least one yellow ball being drawn from the bag. This problem (given the previous parts) can be solved in two ways.
    • Using part (b) we can see that we have already considered the only possibility where a yellow ball is not drawn from the bag. Therefore, since $\Omega = 1$, we can deduce that the probability of drawing at least one yellow ball from the bag is $1 - \frac{25}{144} = \frac{119}{114}$.
    • Consider each outcome individually $(YY, YB, BY)$.
    Selecting two yellow balls has the probability $\frac{7}{12} \times \frac{7}{12} = \frac{49}{144}$
    Selecting a yellow ball and then a blue ball has probability $\frac{7}{12} \times \frac{5}{12} = \frac{35}{144}$.
    Selecting a blue ball and then a yellow ball has probability $\frac{5}{12} \times \frac{7}{12} = \frac{35}{144}$.

> Adding these probabilities together gives the probability of drawing at least one yellow blue from the bag to be $\frac{119}{144}$, as before.

**Exercise 29.7**

Q1. Tracey draws and then replaces a card from a standard deck of cards. Lisa then draws a card from the same pack.
   (a) Draw a tree diagram for the event that Tracey and Lisa both draw a card from the suit of spades;
   (b) find the probability that exactly one card drawn is a spade;
   (c) find the probability that both cards drawn are not from the suit of spaces.

Q2. The probability that Tim wins a squash match against Richard is 0.38. The probability that Tim wins against Jack is 0.71.
   (a) Draw a tree diagram showing the outcomes when Tim plays Jack followed by Richard.
   (b) Find the probability that Tim wins both games.
   (c) Find the probability that Tim wins at least one game.

Q3. Tom is travelling to work in his car. On his way he passes through two sets of traffic lights. Based on his previous journeys, the probability that the first set are green is 0.72 and the probability that the second set are green is 0.54.
   (a) Draw a tree diagram of the outcomes.
   (b) Find the probability that Tom is able to pass through both sets of lights without stopping.
   (c) Find the probability that Tom stops exactly once on his journey.

**Learning Resource 29.1 — Revising Venn Diagrams**

The activity available to download below, in the digital book, tests understanding of Venn diagrams and set notation.

**Exercise 29.8**

Q1. We are given that $P(A) = 0.27$, $P(B) = 0.41$ and $P(A \cap B) = 0.15$. Find
   (a) $P(A \cup B)$;
   (b) $P(A' \cup B)$.
   Given that $P(C) = 0.21$, events $A$ and $C$ are independent and events $B$ and $C$ are mutually exclusive,
   (c) find $P(A \cap C)$;
   (d) draw the Venn diagram of the sample space and events $A$, $B$ and $C$;
   (e) find (i) $P(B \cup C)$; (ii) $P(B' \cap C)$.

Q2. During a school sports day, the track and field competitors were recorded and placed into the table below.

| | Track | Field | Total |
|---|---|---|---|
| Year 7 | 37 | 41 | 78 |
| Year 8 | 43 | 38 | 81 |
| Year 9 | 22 | 19 | 41 |
| Total | 102 | 98 | 200 |

(a) Find the probability that a randomly selected student from the school
   i. did a field event;
   ii. was a year 8 track competitor.
(b) Show that the event that a student was 'in year 7' and the event that a student 'completed a track event' are not statistically independent.

Q3. A survey of the way 100 people travel to work was made. It was found that
   59 people travel by car,
   18 people walked,
   10 people used the bus and walked,
   12 people used a different form of transport.

(a) Draw a Venn diagram of probabilities to display the information from the survey.
(b) What is the probability that a randomly selected person only used the bus to travel to work?
(c) Find the probability that a randomly selected person used one of the stated motorised transports.
(d) State, giving a reason, whether or not the event that a person travelled by bus and the event that a person walked are statistically independent.
(e) State, giving a reason, whether or not any events from the survey are mutually exclusive.

Q4. The members of a semi-professional jazz band are asked which instruments they are able to play. The probabilities of the instrument a randomly selected player plays are displayed in the Venn diagram below, where $b$ and $w$ are probabilities.

$W$ is the event a player plays a woodwind instrument.
$P$ is the event a player plays the piano.
$B$ is the event a player plays a brass instrument.

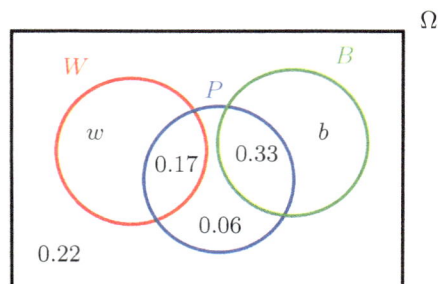

(a) Given that the probability that someone plays only a woodwind instrument and the probability that someone plays only a brass instrument are equal,

find the probabilities $b$ and $w$.
  (b) Show that the events that someone plays a woodwind instrument and that someone plays a brass instrument are mutually exclusive.
  (c) State, giving a reason, why the events that someone playing the piano and someone playing a brass instrument are not statistically independent.
  (d) Given that there are 18 players in the band, suggest how many players play only the piano.

Q5. The Venn diagram below shows the probabilities of lunchtime purchases in a local shop. $a$, $b$ and $c$ are also probabilities.

$M$ is the event that a customer bought a meal (*e.g.* a sandwich, salad, wrap, *etc.*)
$S$ is the event that a customer bought a snack (*e.g.* fruit, crisps, chocolate, *etc.*)
$D$ is the event that a customer bought a drink.

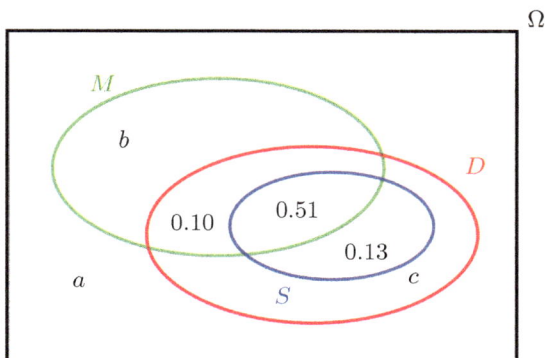

  (a) Given that every customer purchases at least one item, write down the value of probability $a$.
Given also that events $M$ and $S$ are independent,
  (b) find the value of $b$;
  (c) hence, find the value of $c$;
  (d) find
      i. $P((M \cup S)')$;
      ii. $P((D \cap S']')$.

---

**Key Facts — Introduction to Probability**

- A experiment procedure which can be repeated and has defined outcomes.
- The sample space is the collection of all possible outcomes from the experiment.
- An event is a subset of the sample space and is a collection of outcomes from the experiment.

- Events/outcomes are said to be equally likely if they all have the same theoretical probability of occurring.
- When we randomly choose an outcome (or event), it means that picking any of the possible outcomes (or events) is equally likely.
- The probability of an event $A$ occurring is defined to be the number of outcomes in $A$ divided by the total number of equally likely outcomes in the sample space (including $A$). The probability of an event $A$ is usually written as $P(A)$.
- $A'$ is defined to be the event that $A$ does not occur and is often referred to as the complement of $A$. Since the sample space has probability equal to 1, we define $P(A')$, the probability of event $A$ not occurring, to be

$$P(A') = 1 - P(A).$$

- The intersection of two events $A$ and $B$ is the event where both $A$ *and* $B$ occur. The intersection of two events (or sets), $A$ and $B$, is usually written as $A \cap B$ and is shown below.

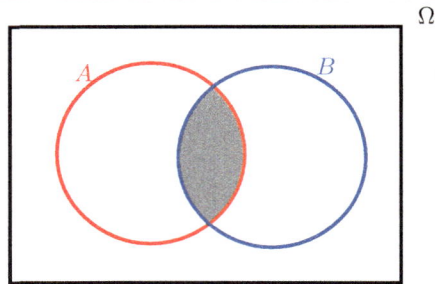

- The union of two events $A$ and $B$ is the event where *either A or B* occur. The union of two events (or sets), $A$ and $B$, is usually written as $A \cup B$ and is shown below.

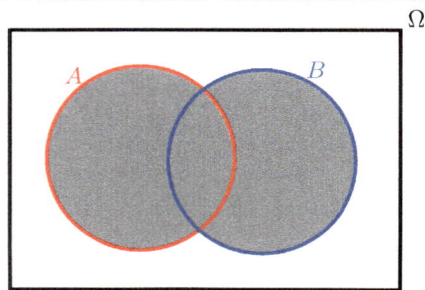

- For two events, $A$ and $B$, the probability of their union is found by

$$P(A \cup B) = P(A) + P(B) - P(A \cap B).$$

This is known as the addition rule.

- Two events, $A$ and $B$, are said to be mutually exclusive if only one of these events can occur at the same time, *i.e.* if $A$ occurs then $B$ does not, and vice versa.
- If events $A$ and $B$ are mutually exclusive ,then

  $P(A \cap B) = 0.$

- If events $A$ and $B$ are mutually exclusive, then

  $P(A \cup B) = P(A) + P(B).$

- Two events, $A$ and $B$, are said to be statistically independent if $A$ happening, or not, has no influence upon the probability of $B$ happening.
- Tree diagrams can be used to represent the probabilities of one event happening followed by another.

**Chapter Assessment — Introduction to Probability**

Download and sit the 30 minute assessment for this chapter from the digital book.

$$E(X) = \sum_x x\mathbb{P}(X = x).$$

# 30. Discrete Probability Distributions

The ideas and techniques of probability and statistics are in use all around us. Taking out a loan or life insurance, being prescribed medicine or given a vaccination, using a secure website to make a purchase: these, and many other activities rely on them. When modern probability theory developed in the seventeenth century, things were rather different. Isaac Newton, Samuel Pepys and many famous mathematicians wrote on the subject, but their discussions typically concerned games of dice. One common question they considered involved a game of chance which, for some reason, is not played to the end. How should the money be divided between the players?

(a)                                              (b)

Figure 30.1: Luca Pacioli (a) and Samuel Pepys (b).

Two centuries earlier, in 1494, the Italian friar Luca Pacioli proposed a solution for the problem of dividing the stakes. He suggested that the pot should be shared in proportion to the number of rounds each player had won so far, and gave a convoluted technique for calculating this division. His method was unsatisfactory, since he ignored the number of rounds remaining. If the target score is one hundred and player one has not yet scored, then player two takes the entire pot whether they have scored one point or ninety-nine. More importantly, his treatment of the problem did not distinguish between games of chance and games of skill: he did not have a clear idea of a random event. In 1693, Samuel Pepys wrote to Isaac Newton with a subtle variation of this problem. Pepys wanted to compare the chances of winning different games. He asked who was most likely to succeed: player A who must throw one six with six dice, player B who is to throw two sixes with twelve dice, or player C with a target of three sixes with eighteen dice. Investigating this question turned out to be remarkably fruitful.

It took another three centuries before there was a firm mathematical foundation for the answer, but along the way a whole new branch of mathematics was created: one which we could hardly imagine doing without today. When probabilistic techniques are applied correctly, they can be powerful tools for tackling real-world as well as mathematical problems. However, history shows that the application is not always obvious and can be counter-intuitive.

## 30.1 Games of Chance

Isaac and Sam are playing a game of Heads and Tails. The first to win three rounds scores ten points. They have won two rounds each when the coin is lost. They disagree about how they should divide the points. Isaac thinks they should have five points each. Sam, who lost the first two rounds, thinks his luck has changed and that he should have more of the points.

> **Question**
> How should the points be divided between Sam and Isaac? Why?

Isaac is making certain claims about the game.

(a) Every round is independent of the others. We cannot guess who will win the next round by looking at who won the earlier rounds, or by looking at any other information.

(b) Every round is the same as the others. The chance of winning a round does not change during the game.

(c) They each have an equal chance of winning a round.

But what is "an equal chance"? A more mathematical description is that as they play more rounds, Isaac will win roughly half of them, and Sam will win roughly half. The more rounds they play, the closer these ratios will get to exactly one half, so that

$$\frac{\text{rounds won by Sam}}{\text{total rounds played}} \to \frac{1}{2} \quad \text{and} \quad \frac{\text{rounds won by Isaac}}{\text{total rounds played}} \to \frac{1}{2}.$$

Isaac has created a mathematical model of the game and used that model to calculate a fair division of the points. This model is called a *probability distribution*. It can be used to make calculations about a sequence of events where each event is unpredictable, but the behaviour of many events is known or can be estimated.

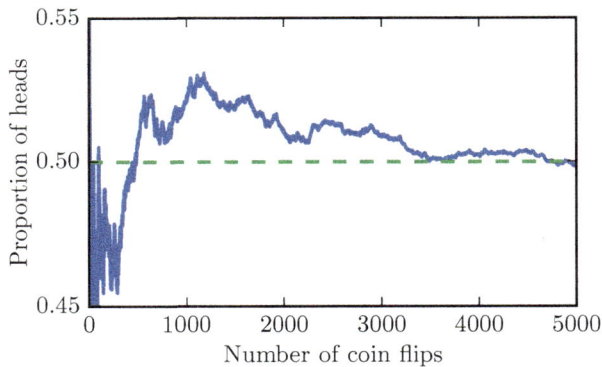

Figure 30.2: An example of the implied probability when flipping a fair coin.

## 30.2  Discrete Probability Distributions

Probability distributions can be *discrete* or *continuous*. A distribution is discrete if the set of possible outcomes can be written down in a list, such as {heads, tails} or $\{1, 2, 3, \ldots\}$. [1] Suppose that we have an experiment or observation with a random outcome and that we can perform the experiment, or make the observation, as many times as we like. We would like to model this using a probability distribution. First, we must check that it satisfies the first two properties of the Heads and Tails game: that the observations are independent and that we expect the probabilities to be constant between observations. If these two conditions hold, then we say that the observations are *independent* and *identically distributed*.

Next, for every possible outcome $x$m we must state what proportion of the time we expect to see that outcome. For example, if we roll a die we expect to roll a three one sixth of the time. Writing this more formally, we must choose a probability $p_x$ such that in the long run the ratio of occurrences of $x$ to the total number of observations is very close to $p_x$:

$$\frac{\text{occurrences of x}}{\text{number of observations}} \to p_x.$$

---

**Activity 30.1 — Calculating Probabilities**

If a fair coin is flipped twice, what is the probability that exactly two heads are obtained? What is the probability if it is flipped three times, or ten times?

---

The unknown outcome of a single observation is often denoted by a capital letter, such as $X$. This is called a *random variable*. The probability that we see the value $y$ for a given observation is written as $\mathbb{P}(X = y)$, which is read as "the probability that $X$ equals (or takes the value) $y$". The set of values that $X$ can take is called the *range* or the *support* of $X$.

---

[1] Continuous probability distributions can be used to model data such as the speed of a passing car, or a person's height.

In the long run, we expect each value of $X$ to occur in proportion to its probability. This means that we also expect the mean value of the observations to tend to a fixed value. If we make a series of observations $\{x_1, x_2, x_3, \ldots\}$ then the mean value is given by

$$\frac{1}{N} \sum_{i=0}^{N} x_i = \sum_{x} \frac{x \times (\text{occurrences of } x)}{\text{number of observations}} \to \sum_{x} x \mathbb{P}(X = x).$$

This value is called $\mathbb{E}(X)$, the *expected value* or *expectation* of $X$. It is also called the *mean* of the distribution.

> **Formulae 30.1 — Expectation of a Random Variable**
> The expectation of a random variable $X$ is given by
>
> $$\mathbb{E}(X) = \sum_{x} x \mathbb{P}(X = x).$$

Note that the expectation of a random variable is derived in a similar way to its distribution. We start with a finite set of observations and calculate the frequency of each outcome (as a proportion of the total) and the mean value. As the set of observations becomes larger and larger, the frequency ratios tend to the probability distribution, and the mean value tends to the expectation of the distribution.

---

**Example 30.1**
A fair coin is flipped twice. Write down the probability distribution for the number of times the coin lands showing heads. What is the expected number of heads?

**Solution:**
For each observation, we see one of the following:
- heads then heads (two heads),
- heads then tails (one head),
- tails then heads (one head) or
- tails then tails (no heads).

If we performed many observations then, because the coin is fair, we would expect to see each of these possibilities roughly an equal number of times. There are four possibilities, so each should be seen about a quarter of the time.
From this, we can write down the probability distribution as

$$\mathbb{P}(0 \text{ heads}) = \frac{1}{4}, \qquad \mathbb{P}(1 \text{ head}) = \frac{1}{2}, \qquad \mathbb{P}(2 \text{ heads}) = \frac{1}{4}.$$

This is shown in table 30.1.

| Number of heads | Probability |
|:---------------:|:-----------:|
| 0 | 0.25 |
| 1 | 0.5 |
| 2 | 0.25 |

Table 30.1: Probability distribution for tossing two coins.

Let $X$ be the number of heads. The expected value of $X$ is given by

$$\sum_{i=0}^{2} i \times \mathbb{P}(X = i) = 0.\mathbb{P}(X = 0) + 1.\mathbb{P}(X = 1) + 2.\mathbb{P}(X = 2) = 1.$$

This definition of a probability distribution has been developed from our intuitive ideas about probabilities. Fortunately, it gives us a definition of probability which is the same as the theoretical one we have already seen. In particular, we can deduce from the definition that:

(a) $0 \leq \mathbb{P}(X = x) \leq 1$.
(b) If $A$ is the event that $(X = a)$ and $B$ is the event that $(X = b)$ where $a \neq b$, then $A$ and $B$ are mutually exclusive, $\mathbb{P}(A \cap B) = 0$ and $\mathbb{P}(X = a \text{ or } b) = \mathbb{P}(A \cup B) = \mathbb{P}(A) + \mathbb{P}(B)$.
(c) The probabilities sum to one: $\sum_{x} \mathbb{P}(X = x) = 1$ where we take the sum over all possible values of $X$.

**Activity 30.2 — Proving Properties of a Probability Distribution**
How can we convince ourselves that these statements are necessarily true, given our definition of the probability of a random event? Choose one of the statements and write out a proof.

**Example 30.2**
A random variable $X$ can take the values $\{-2, -1, 0, 1, 2\}$. Its probability distribution is given by $\mathbb{P}(X = x) = \gamma(x^2 + 1)$ for some constant value $\gamma$.
Q1. Calculate $\gamma$.
Q2. For one observation of $X$ we are told that the value is non-negative. Calculate $\mathbb{P}(X = 1)$.

**Solution:**
Q1. The sum of the probabilities over the range of $X$ must equal one. Therefore,

$$\sum_{x=-2}^{2} \mathbb{P}(X = x) = \sum_{x=-2}^{2} \gamma(x^2 + 1) = \gamma(5 + 2 + 1 + 2 + 5) = 15\gamma = 1$$

and so $\gamma = \frac{1}{15}$.
Q2. This is a question about conditional probabilities so we will use the identity

$$\mathbb{P}(A|B) = \frac{\mathbb{P}(A \cap B)}{\mathbb{P}(B)}.$$

Here, $A$ is the event that $X = 1$ and $B$ is the event that $X$ is non-negative. Therefore, $A \cap B$ is the event that $X = 1$ *and* $X$ is non-negative: in other words, $A \cap B = A$.

The probability distribution tells us that $\mathbb{P}(A) = \mathbb{P}(X = 1) = \frac{2}{15}$ and that

$$\mathbb{P}(B) = \mathbb{P}(X = 0) + \mathbb{P}(X = 1) + \mathbb{P}(X = 2) = \frac{1}{15} + \frac{2}{15} + \frac{5}{15} = \frac{8}{15}.$$

Therefore, $\mathbb{P}(A|B) = \frac{2}{15}/\frac{8}{15} = \frac{1}{4}$.

**Exercise 30.1**

Q1. A fair coin is flipped four times.
   (a) Write down the set of possible outcomes.
   (b) What is the probability of getting at least two heads in a row?
   (c) The random variable $X$ is equal to the length of the longest run of heads. Write down the probability distribution of $X$.

Q2. We throw a pair of fair dice.
   (a) Write down a set of outcomes which all have the same probability.
   (b) What is the probability that at least one 5 or 6 is rolled?
   (c) After rolling the dice, we are given a score. If we have rolled a double, we score one point. Otherwise, we score the higher of the two dice. Write down the probability distribution for our score. Is its expectation higher or lower than the expectation when rolling a single die?

Q3. We are looking for our calculator. We know that it is in the top, the middle, or the bottom drawer of our desk.
   (a) Write down the probability distribution for the location of our calculator.
   (b) We open the top drawer, and the calculator is not in it. How does this change the probability distribution?

Q4. The probability distribution of a random variable $X$ is given by $\mathbb{P}(X = x) = \alpha x(x + 3)$ for $x \in \{1, 2, 3\}$ and for some constant value $\alpha$.
   (a) Calculate $\alpha$.
   (b) Given that $X$ is odd, calculate $\mathbb{P}(X = 1)$.
   (c) Another random variable $Y$ can take values in the set $\{3, 4, 5\}$. The probability that it takes the value $y$ is proportional to $y$. Calculate the probability distribution of $Y$.
   (d) The random variables $X$ and $Y$ are independent. Calculate $\mathbb{P}(Y - X = 3)$.

Q5. A bag contains 8 red balls and 4 green balls. A ball is drawn from the bag, its colour is noted, and the ball is discarded. A second ball is drawn and its colour noted.
   (a) What is the probability distribution for the first ball drawn?
   (b) What is the probability distribution for the second ball, if the first ball was red?
   (c) What is the probability distribution for the second ball, if the colour of the first ball is unknown?

## 30.3 The Binomial Distribution

Suppose we want to decide whether a coin is fair. One possibility is to flip the coin a very large number of times and check that the ratio of heads to the total number of tosses gets closer and closer to one half. This raises the question of what value to choose to be "a very large number".

In theory, if we saw a thousand heads in a row, this would not prove that the coin was unfair, since that is a possible (but unlikely) outcome, even for a fair coin. All we can say is that the coin is fair (or unfair) with a certain probability.

We know what the long term behaviour of the coin should be, and if we see the coin behaving in a very different way, we will say the coin is probably unfair. The more the behaviour we see differs from the behaviour we expect, the more certain we will be that the coin is unfair.

In order to do this, we first have to work out what behaviour we expect to see from more than one observation. For example, if we tossed a coin 50 times and we saw 20 heads, what is the chance that the coin is fair? To answer this, we need to know the probability of seeing 20 heads with a fair coin.

We start with an experiment or observation which satisfies the usual conditions:

- We can make a large number of observations.
- The observations are independent and identically distributed.

We will consider an observation which has only two possible outcomes, such as:

- whether a coin lands showing heads;
- whether a car passing a certain point exceeds the speed limit;
- whether a manufactured item is faulty.

There is some fixed probability $p$ that the event will occur for a given observation. This very simple probability distribution is called the Bernoulli distribution and we will use it to build up more interesting distributions.

We record the outcome of $n$ observations. Our random variable $X$ will be defined as the number of times that the event occurs. The possible outcomes are $\{0, 1, 2, \ldots, n\}$.

---

**Exercise 30.2**

Q1. A fair coin is tossed 3 times. Find the probability that we see:
  (a) three heads;
  (b) the sequence $\{heads, tails, tails\}$;
  (c) the sequence $\{tails, heads, tails\}$;
  (d) the sequence $\{tails, tails, heads\}$;
  (e) exactly one head.

Q2. Four fair dice are thrown. Find the probability that:
  (a) no fives or sixes are thrown;
  (b) every dice shows a five or a six;
  (c) exactly one die shows a five or a six.

---

**Learning Resource 30.1 — Managing Flight Bookings**

The airline industry has a notoriously low profit margin, often quoted as being around 1%. With such a low profit margin flying routed with planes that are not at capacity would be a serious weakness. Given that on average 5% of passengers fail to turn up to a given flight, airlines often overbook tickets and rely on statistics to enable them to do this without upsetting a lot of customers. An activity investigating the modelling of this can be downloaded from the digital book.

This probability distribution is called the *binomial distribution*. It is used to model observations of a fixed number of independent events, each of which has the same probability of success. If $X$ has a binomial distribution then we write $X \sim B(n, p)$ where $n$ is the number of events observed, and $p$ is the probability of success for each event.

The probability of seeing any particular sequence of successes and failures during one set of observations depends only on the number of successes. If we know the number of successes, then we can calculate the number of failures, since the total number of observations is fixed at $n$.

For example, if a biased coin has probability $p$ of showing heads and $n = 5$, then the probability of seeing the sequence {heads, heads, tails, heads, tails} is

$$p.p.(1 - p).p.(1 - p) = p^3(1 - p)^2$$

and we will get the same result for any sequence with three heads and two tails.

In general, the probability of seeing a particular sequence with $k$ heads is $p^k(1 - p)^{n-k}$. To calculate the probability of seeing *any* sequence with $k$ heads in it, we must multiply this by the number of ways that $k$ heads can appear in a sequence of length $n$.

Using techniques from Chapter 8 the number of ways that $k$ heads can appear in a sequence of length $n$ is the number of ways we can choose $k$ items out of a list of length $n$. It is given by

$$\binom{n}{k} = \frac{n(n-1)\ldots(n-k+1)}{k(k-1)\ldots 1} = \frac{n!}{k!(n-k)!}.$$

This is usually read as "$n$ choose $k$" and is sometimes written as $^nC_k$.

Multiplying these two results together gives us the probabilities for the binomial distribution.

---

**Formulae 30.2 — Binomial Probability Distribution**

If $X \sim B(n, p)$ then

$$\mathbb{P}(X = k) = \binom{n}{k} p^k (1 - p)^{n-k} \qquad k \in \{0, 1, \ldots, n\}.$$

The mean of $X$ is $np$ and the variance is $np(1 - p)$.

---

The binomial distribution for $n = 10$ and $p = 0.3$ is shown in Figure 30.3

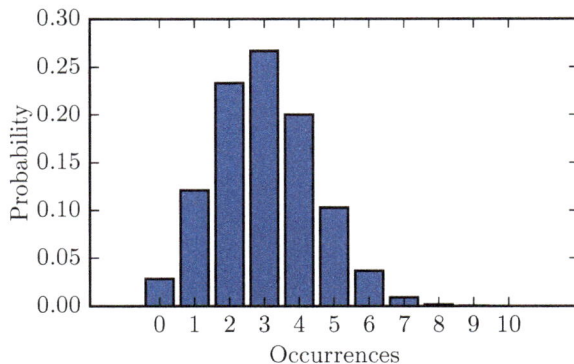

Figure 30.3: The Binomial Distribution $B(10, 0.3)$.

---

**Interactive Activity 30.1 — The Shape of the Binomial Distribution**

The Geogebra interactive applet below can be used to examine what happens qualitatively to the distribution of probabilities as $n$ and $p$ are varied.

---

**Remark**

Notice the similarities between the binomial distribution and the binomial theorem studied in Chapter 8.

---

**Interactive Activity 30.2 — Calculating Binomial Probabilities**

Use the interactive Geogebra applet below to calculate the following probabilities.

Q1. Let $X \sim B(10, 0.5)$, find
    (a) $\mathbb{P}(X = 6)$,
    (b) $\mathbb{P}(X \geq 7)$,
    (c) $\mathbb{P}(X \leq 4)$.

Q2. Let $X \sim B(15, 0.3)$, find
    (a) $\mathbb{P}(X > 2)$,
    (b) $\mathbb{P}(X = 11)$,
    (c) $\mathbb{P}(X \leq 5)$,
    (d) $(\mathbb{P}(X < 2) \text{ or } \mathbb{P}(X > 8))$.

**Remark**

Returning to the problem discussed in the introduction to this chapter which had been proposed by Pepys. He asked who was most likely to succeed: player A who must throw one six with six dice, player B who is to throw two sixes with twelve dice, or player C with a target of three sixes with eighteen dice. Newton immediately picked up the ambiguity in the question, and it was agreed that the intended task was to throw "at least" the stated number of sixes.

The solution is fairly simple given modern techniques and a calculator. If $X_n \sim B(6n, \frac{1}{6})$ then we simply compare

$$\mathbb{P}(X_1 \geq 1) \simeq 0.665, \tag{T30.1}$$
$$\mathbb{P}(X_2 \geq 2) \simeq 0.619 \text{ and} \tag{T30.2}$$
$$\mathbb{P}(X_3 \geq 3) \simeq 0.597, \tag{T30.3}$$

to see that player A is most likely to succeed. Newton calculated the first two of these values exactly and stated that the third answer would be smaller than the second. He was thus able to answer Pepys' question correctly.

Pepys could not understand why this should be the correct answer and Newton attempted an intuitive explanation. He said that for player B to meet their target, they must meet player A's target with six dice, and then meet it again with the remaining six dice. This must surely be harder than player A's task, who must meet this target only once.

In fact, Pepys' intuition served him well, since Newton's explanation was, to say the least, incomplete. It ignores the occasions when players throw multiple sixes with the six dice. The more likely this is, the greater the advantage of throwing more dice.

Suppose that the game is played with biased dice, where each die lands on a six 90% of the time. If player A is very unlucky on his turn and fails to throw a six, he misses his target. If player B has the same level of bad luck, and does not throw a single six with the first six dice, they still have another six dice and are very likely to throw two sixes with them, meeting their target. We can see that player B is much less likely to fail than player A. With these dice, the original situation has been reversed and player C has the easiest task.

Therefore, the answer to Pepys' question depends on the probability of throwing a six. But Newton's explanation works just as well for biased dice, so it cannot be correct. The question is really whether a probability of $\frac{1}{6}$ is sufficiently high for player C to have the advantage. Newton had proved arithmetically that it is not, but he did not supply Pepys with a correct intuitive argument for the result.

---

**Activity 30.3 — The Mean and Variance of the Binomial Distribution**

Prove that the mean and variance of the binomial distribution is given by $np$ and $np(1-p)$ respectively.

The variance of a random variable $X$ is given by

$$\text{Var}(X) = \mathbb{E}(X^2) - [\mathbb{E}(X)]^2.$$

Use the results that for two independent random variables $X$ and $Y$, $\mathbb{E}(X + Y) =$

$\mathbb{E}(X) + \mathbb{E}(Y)$ and $\mathbb{E}(XY) = \mathbb{E}(X)\mathbb{E}(Y)$.
*Hint:* If $X_n \sim B(n,p)$ and $X_1 \sim B(1,p)$, what is the distribution of $X_n + X_1$?

**Interactive Activity 30.3 — The Binomial Distribution - True or False**
In the interactive activity below, decide which statements are true and which are false.

**Exercise 30.3**
Q1. The random variable $X$ has a binomial distribution with parameters $n = 8$ and $p = 0.3$. Calculate, to three decimal places:
  (a) $\mathbb{P}(X = 0)$,
  (b) $\mathbb{P}(X = 4)$,
  (c) $\mathbb{P}(X \geq 2)$,
  (d) $\mathbb{P}(X$ is odd $|X < 4)$.
Q2. A coin is flipped seven times. Let $X$ be the number of heads, and $Y$ be the number of tails. If $X$ has the distribution $B(7, 0.5)$, what distribution does $Y$ have? Calculate:
  (a) $\mathbb{P}(X = 1)$,
  (b) $\mathbb{P}(X = 2)$,
  (c) $\mathbb{P}(Y = 5)$,
  (d) $\mathbb{P}(Y \leq 3)$.
Q3. We have a set of six fair dice. At each throw, we roll all six dice.
  (a) Which is most likely to succeed: rolling at least one six in a single throw, at least two sixes in two throws, or at least three sixes in three throws?
  (b) The dice are switched for biased ones. Each one rolls a six 25% of the time. Does this change the answer to the question?
Q4. A factory makes widgets and sells them in boxes of ten. One out of every 20 widgets is defective.
  (a) What is the probability that a box contains exactly one defective widget?
  (b) What is the probability that a box contains at least one defective widget?
  (c) We receive two boxes of ten widgets each. We have been told that the first box contains at least one defective widget. What is the probability that the first box contains at least two defective widgets?
  (d) We pick a widget at random from the second box. It is defective. What is the probability that the second box contains at least two defective widgets?

## 30.4 Other Discrete Distributions

Whilst being applicable to many situations, the binomial distribution is not suitable for everything. In this section we discuss two other discrete distributions that have wide applicability. It should be noted that these are not formally on the syllabus for A-Level Mathematics (and are studied in more detail as part of the Further Mathematics A-Level), however, an awareness of them enhances understanding of the statistical modelling cycle which is discussed in Year 2.

### 30.4.1 The Poisson Distribution

Suppose that a flood monitoring station has an automatic sensor which sends an alert whenever the river level exceeds a certain height. If we want to model the number of days on which an alert is sent, we might use the binomial distribution.

> **Activity 30.4 — Flood Warnings**
> Suppose that the probability that an alert is sent on any given day is 0.2. Over the course of a week, the days on which an alert is sent are recorded. Write down the probability distribution of the total number of days on which at least one alert is sent. What is the expected value for this distribution? Why might we expect the mean number of alerts sent to be higher than this?

Suppose we want to model the total number of alerts sent in a week. If we used our model for the number of days on which an alert is sent, we will underestimate the total number. This is because whenever multiple alerts are sent on a single day, we will only record one of them.

Suppose we were told that the mean number of alerts in a week was equal to 2. We could increase our value of $p$ to compensate for the missed alerts. The distribution $B(7, 0.286)$ gives us the correct mean number of alerts each week. On the other hand, it will underestimate the number of days when no alert is sent and, since it still assumes a maximum of one alert per day, will also underestimate the probability of a high number of alerts in a week.

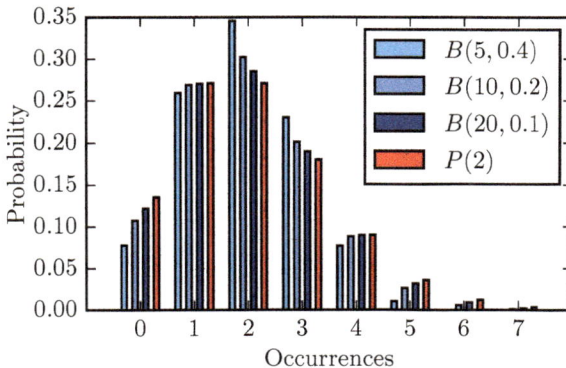

Figure 30.4: $B(5, 0.4)$, $B(10, 0.2)$ and $B(20, 0.1)$ approximating a Poisson distribution with $\lambda = 2$.

However, if we let $n$ tend to infinity, the binomial distribution tends towards a distribution which is much simpler to use, as seen in Figure 30.4. As $n$ increases, we reduce $p$ so that the mean number of events, $np$, stays the same. The constant value of $np$ is often denoted by $\lambda$, the Greek letter lambda. The distribution we reach in the limit is called the Poisson distribution. It can take any value in $\{0, 1, 2, \ldots\}$ and depends only on the value of $\lambda$.

---

**Formulae 30.3 — Poisson Probability Distribution**

If $X$ has a Poisson distribution with parameter $\lambda$ then

$$\mathbb{P}(X = k) = \frac{\lambda^k e^{-\lambda}}{k!} \qquad k \in \{0, 1, 2, \ldots\}.$$

The mean and variance of $X$ are both $\lambda$.

---

The Poisson distribution can be used to model events which happen randomly and independently of each other. If the mean number of events in a given time interval is $\lambda$, then the total number of events in that interval has a Poisson distribution with parameter $\lambda$.

We derived the Poisson distribution by taking the limit of the binomial distribution when $n$ is large and $np$ is a fixed value. This means that for certain ranges of $n$ and $p$, the Poisson distribution is a good approximation to the binomial distribution. This is useful, because the binomial distribution can be difficult to use. If tables of values are being used, they do not usually have entries for large values of $n$. If the values are being calculated by computer then the answers can become inaccurate, since the expression includes factorial terms which become very large, and power terms which become very small. Mathematical techniques have to be used to ensure the answers have the required accuracy. In this case, the Poisson distribution can give more reliable answers.

A Poisson distribution can be used to approximate a binomial distribution $B(n, p)$ for large $n$ and small $p$. As a guideline, consider using a Poisson approximation if:

$$n \geq 20 \qquad \text{and} \qquad np \leq 10.$$

---

**Interactive Activity 30.4 — The Poisson Distribution**

The Geogebra applet below (in the digital book) can be used to investigate the Poisson distribution in more detail. By moving the sliders appropriately, answer the following.

Q1. A random variable $X$ has a Poisson distribution with parameter $\lambda = 3.8$.
   (a) Calculate $\mathbb{P}(X = 0)$;
   (b) Calculate $\mathbb{P}(X = 4)$;
   (c) Calculate $\mathbb{P}(X \geq 4)$;
   (d) Calculate $\mathbb{P}(X = 6 | X \geq 5)$.

**Exercise 30.4**

Q1. A random variable $X$ has the probability distribution

$$\mathbb{P}(X = k) = \frac{3^k e^{-3}}{k!} \qquad k \in \{0, 1, 2, \ldots\}.$$

    (a) Calculate $\mathbb{P}(X = 0)$.
    (b) Calculate $\mathbb{P}(X \geq 2)$.
    (c) Calculate $\mathbb{P}(X = 3 | X \geq 3)$.

Q2. A data centre has a large number of disk drives. Drives fail independently of each other and the mean number of failed drives per hour is 2.
    (a) What is the probability that more than two drives fail in an hour?
    (b) What is the mean number of failed drives per day?
    (c) What is the probability distribution for the number of failed drives per day.
    (d) The next delivery of spare drives is due in two hours and there are three drives left in stock. What is the probability that the next shipment of drives will be needed before they arrive?

Q3. The mushroom delivery mechanism on a frozen pizza production line drops an average of 32 pieces of mushroom, evenly spread, on each pizza. If a pizza is divided into eight equal slices, what is the probability that the pizza has at least one slice with no mushroom on it?

Q4. Approximately 8% of men are colour blind. Use the Poisson approximation to the binomial to estimate the probability that, out of one hundred men, fewer than four are colour blind.

Q5. The exponential function can be written as an infinite series:

$$e^x = \sum_{k=0}^{\infty} \frac{x^k}{k!}.$$

Prove that the expected value of the Poisson distribution is equal to its parameter $\lambda$.

### 30.4.2 The Geometric Distribution

We derived the binomial distribution by considering a sequence of independent trials with a constant probability of success. We made a fixed number of observations and asked how many successes we might see.

Suppose that for the same sequence of observations, we asked a different question. We now want to model the number of failures we might see before the first success. Like the Poisson distribution, this distribution does not have a finite range. It can take any non-negative integer value, that is any value in the set $\{0, 1, 2, \ldots\}$.

**Activity 30.5 — Waiting for Success**

Given a series of random events, each with probability $p$ of success, what is the probability that success is achieved at the first attempt? What is the probability of one failure followed by a success? What is the probability that there are exactly two failures before a success? In general, what is the probability there are exactly $k$ failures, followed by a success?

The probability distribution given by this model is called the geometric distribution, because the probabilities form a geometric progression. Each probability is equal to the previous probability multiplied by $(1-p)$.

**Formulae 30.4 — Geometric Probability Distribution**
For a sequence of independent trials, each with probability $p$ of success, the probability of observing exactly $k$ failures before the first success is given by

$$\mathbb{P}(X=k) = p(1-p)^k.$$

**Remark**
We can use the geometric distribution as an example of the power of probabilistic methods to solve problems very simply. Consider the distribution $\mathbb{P}(X=k) = p(1-p)^k$ for a value $0 \le p \le 1$. We know that this describes the number of failures before a success is seen and since the chance that we never see a success is zero (we expect every sequence of trials to end in a success eventually), we can say that

$$\sum_{k=0}^{\infty} p(1-p)^k = 1$$

and therefore

$$\sum_{k=0}^{\infty} (1-p)^k = \frac{1}{p}.$$

By letting $r = 1-p$, this immediately gives us the result that the sum of a geometric progression for $0 \le r \le 1$ (and with initial starting value $a=1$) is given by

$$\sum_{k=0}^{\infty} r^k = \frac{1}{1-r}.$$

We can also calculate $\sum kp^k$ similarly. We can easily prove that

$$\mathbb{P}(X=k+1|X \ge 1) = \mathbb{P}(X=k)$$

and then we can calculate the expectation of $X$ by considering the outcome of the first trial. If the first trial is a success, $X$ takes the value zero, so

$$\mathbb{E}(X|X=0) = 0.$$

If the first trial is a failure, we use the result above to show that

$$\mathbb{E}(X|X \geq 1) = \sum_{k=0}^{\infty}(k+1)\mathbb{P}(X = k+1|X \geq 1)$$

$$= \sum_{k=0}^{\infty}(k+1)\mathbb{P}(X = k)$$

$$= \sum_{k=0}^{\infty}k\mathbb{P}(X = k) + \sum_{k=0}^{\infty}\mathbb{P}(X = k) = \mathbb{E}(X) + 1.$$

Putting these together, we have that

$$\mathbb{E}(X) = \mathbb{P}(X = 0)\mathbb{E}(X|X = 0) + \mathbb{P}(X \geq 1)\mathbb{E}(X|X \geq 1) \qquad \text{(T30.4)}$$
$$= p.0 + (1-p)\,(\mathbb{E}(X) + 1) \qquad \text{(T30.5)}$$
$$= (1-p)\,(\mathbb{E}(X) + 1) \qquad \text{(T30.6)}$$

which can be rearranged to give $\mathbb{E}(X) = (1-p)/p$.
Then we can use the standard definition of expectation to show that

$$\mathbb{E}(X) = \sum_{k=0}^{\infty}k\mathbb{P}(X = k) = \sum_{k=0}^{\infty}kp(1-p)^k = p\sum_{k=0}^{\infty}k(1-p)^k$$

and putting these two expressions for $\mathbb{E}(X)$ equal to each other, we have that

$$\sum_{k=0}^{\infty}k(1-p)^k = \frac{1-p}{p^2}$$

or, equivalently,

$$\sum_{k=0}^{\infty}kp^k = \frac{p}{(1-p)^2}.$$

Suppose that the random variable $X$ has a geometric distribution then, calculating probabilities, we can notice that $\mathbb{P}(X = 4|X \geq 1) = \mathbb{P}(X = 3)$. More generally, for $k, m \geq 0$,

$$\mathbb{P}(X = m + k|X \geq m) = \mathbb{P}(X = k).$$

This is known as the *memoryless property*.

**Remark**
To prove that this property holds in general for a Geometric distribution we note that the probability that $X \geq m$ is the probability that the first $m$ trials are all failures.

That is, $\mathbb{P}(X \geq m) = (1 - p)^m$.

$$\mathbb{P}(X = m + k | X \geq m) = \frac{\mathbb{P}(X = m + k)}{\mathbb{P}(X \geq m)}$$

$$= \frac{p(1-p)^{m+k}}{(1-p)^m} = p(1-p)^k = \mathbb{P}(X = k).$$

When we use the geometric distribution to model the number of failures before the first success, this result has a simple interpretation. It tells us that the probability of seeing $k$ more failures is always the same, however many failures we have already seen. The number of failures we expect to see in the future always has the same geometric distribution.

**Interactive Activity 30.5 — The Geometric Distribution**

The Geogebra applet below (in the digital book) can be used to investigate the geometric distribution in more detail. By moving the sliders appropriately, answer the following.

Q1. Given a random variable $X$ with distribution

$$\mathbb{P}(X = k) = \frac{1}{4}\left(\frac{3}{4}\right)^k \qquad k \in \{0, 1, 2, \ldots\},$$

calculate
(a) $\mathbb{P}(X = 0)$;
(b) $\mathbb{P}(X = 3)$;
(c) $\mathbb{P}(X \geq 7)$;
(d) $\mathbb{P}(X \geq 10 | X \geq 2)$.

**Exercise 30.5**

Q1. Given a random variable $X$ with distribution

$$\mathbb{P}(X = k) = \frac{2}{5}\left(\frac{3}{5}\right)^k \qquad k \in \{0, 1, 2, \ldots\},$$

calculate
(a) $\mathbb{P}(X = 0)$;
(b) $\mathbb{P}(X \geq 2)$;
(c) $\mathbb{P}(X \geq 4)$;
(d) $\mathbb{P}(X \geq 4 | X \geq 2)$.

Q2. A food packing plant has an automated system to separate large strawberries from smaller ones. As they pass the system, the strawberries are scanned. If they are too small, the system diverts them off the main packing line. It is known that 10% of the strawberries are classed as small.
(a) Write down the probability distribution for the number of large strawberries scanned before the first small strawberry is found.
(b) It is discovered that as the system diverts one strawberry, it fails to scan the

next two. What is the probability that it misses a small strawberry while it diverts another one?

(c) What is the probability that we see exactly one large strawberry before the first small one?

(d) If we have already seen three large strawberries, what is the probability that we see exactly one more before the first small one?

---

**Key Facts — Discrete Probability Distributions**

- The outcomes of a discrete probability distribution can be written in a list, and probabilities assigned to each possible outcome.
- Random variables are usually denoted by a capital letter.
- The expectation of a random variable $X$ is given by

$$\mathbb{E}(X) = \sum_x x \mathbb{P}(X = x).$$

- Properties of discrete probability distributions.
    (a) $0 \le \mathbb{P}(X = x) \le 1$.
    (b) If $A$ is the event that $(X = a)$ and $B$ is the event that $(X = b)$ where $a \ne b$, then $A$ and $B$ are mutually exclusive, $\mathbb{P}(A \cap B) = 0$ and $\mathbb{P}(X = a \text{ or } b) = \mathbb{P}(A \cup B) = \mathbb{P}(A) + \mathbb{P}(B)$.
    (c) The probabilities sum to one: $\sum_x \mathbb{P}(X = x) = 1$ where we take the sum over all possible values of $X$.
- The binomial distribution can be used if the following conditions are satisfied:
    (a) The number of trials, $n$, is fixed.
    (b) Each observation is independent.
    (c) Each observation is either a "success" or a "failure".
    (d) The probability of "success", $p$, is the same for each trial.
- If $X \sim B(n, p)$ then

$$\mathbb{P}(X = k) = \binom{n}{k} p^k (1 - p)^{n-k} \qquad k \in \{0, 1, \ldots, n\}.$$

- The mean of the binomial distribution $X \sim B(n, p)$ is $np$ and the variance is $np(1 - p)$.
- The Poisson distribution can be used if:
    (a) The events occur independently.
    (b) The events occur at random.
    (c) The probability of an event occurring in a given time interval does not change.
- If $X$ has a Poisson distribution with parameter $\lambda$ then

$$\mathbb{P}(X = k) = \frac{\lambda^k e^{-\lambda}}{k!} \qquad k \in \{0, 1, 2, \ldots\}.$$

- For the Poisson distribution $X \sim B(n, p)$, the mean and variance of $X$ are both $\lambda$.
- The geometric distribution can be used if:

    (a) The number of trials is not fixed.

    (b) Trials continue until the first success.

    (c) The probability of success, $p$, is the same for every trial.

- If $X \sim \text{Geo}(p)$ then

$$\mathbb{P}(X = k) = p(1 - p)^k.$$

**Chapter Assessment — Discrete Probability Distributions**

Download and sit the 30 minute assessment for this chapter from the digital book.

$H_0: \quad p = 0.79,$

$H_1: \quad p \leq 0.79.$

$+ \dfrac{f(-x)}{2}$

# 31. Hypothesis Testing

Real-world decisions are often made based on quantities that are estimated from a large population. In practice, we cannot collect data from every individual in a given population because it is simply too large. Instead, we collect data from a representative sample of the population, then use sample statistics to estimate quantities such as the mean and variance. In order to make decisions regarding the population based on sample data, we use *hypothesis testing*.

Examples of situations in which hypothesis testing is appropriate are:

- A medical researcher may wish to decide on the basis of experimental evidence whether a certain vaccine is superior to the one presently being used.
- An engineer may need to decide whether an apparent increase in the strength of steel wire made by a new process is strong enough evidence to justify changing to the new process for all future production.
- A cosmetics company may need to assess the impact on customer satisfaction of a change to the formulation of a moisturiser.
- A food technologist wishes to determine if the mean height of oil seed rape seedlings is increased by the addition of a treatment agent.

There are several different types of hypothesis test. Our choice of test is determined by the underlying distribution of our data and the statistic that we wish to test. Throughout this chapter, we will focus on the hypothesis test for the *proportion in the binomial distribution*. To conduct a hypothesis test for the proportion in the binomial distribution, our sample data must have the form of a finite number of independent trials. Each trial is sampled randomly and is regarded as either a *success* or *failure*.

## 31.1 Null and Alternative Hypotheses

When conducting a hypothesis test, we test an assumption about the population against sample data. The initial assumption is the *null hypothesis*, which we test against the *alternative hypothesis*.

---

**Definition 31.1 — Null Hypothesis, $H_0$**
The *null hypothesis*, denoted $H_0$, is the hypothesis that we wish to test. This is the current situation, or presumed value of a particular quantity.

---

**Example 31.1**
In the examples above, the null hypotheses are:
- that the new vaccine is equally effective as the existing vaccine,
- that there is no difference in the strength of wire made by the new and existing industrial process,
- that customer satisfaction is unchanged by the use of the new formulation,
- that there is no difference in the mean height of the seedlings.

---

Essentially, the null hypothesis is the outcome if our experimental data provides no new information, *i.e.* nothing has changed as a result of the experiment.

---

**Definition 31.2 — Alternative Hypothesis, $H_1$**
The *alternative hypothesis*, denoted $H_1$, is the alternative result, or outcome to the null hypothesis.

---

**Example 31.2**
In the examples above, suitable alternative hypotheses are:
- the new vaccine is more effective than the existing vaccine,
- the new industrial process produces stronger wire than the existing process,
- the measure of customer satisfaction is affected by the new formulation,
- the addition of the treatment leads to an increase in mean height of the seedlings.

---

## 31.2  One and Two-Tailed Tests

The null hypothesis, $H_0$, always involves equality. When conducting a test for the proportion, $p$, of a binomial distribution, the null hypothesis is

$$H_0: \quad p = p_0,$$

where $p_0$ is the proportion that we wish to test.
There are three options for the alternative hypothesis, $H_1$:
- One-tailed test:
  − we are only interested in *increases* in $p$, so

  $$H_1: \quad p > p_0,$$

  − we are only interested in *decreases* in $p$, so

  $$H_1: \quad p < p_0.$$

- Two-tailed test: we are interested in *any change* in $p$, so

  $$H_1: \quad p \neq p_0.$$

> **Remark**
> Often, one-tailed and two-tailed tests are referred to as one-sided and two-sided respectively. This is to reflect that the hypothesis test has a one- or two-sided alternative in each case.

**Decision rule:** In general, $H_0$ and $H_1$ are not treated equally. We decide that $H_0$ is true unless there is sufficient evidence against $H_0$ in favour of $H_1$, *i.e.* "innocent until proven guilty". We must specify a criterion to determine how much evidence is "sufficient" to reject $H_0$.

> **Remark**
> We must be specific with our vocabulary when writing the outcome of a hypothesis test. We should *not* write "we accept $H_1$", since the hypothesis test provides no evidence that $H_1$ is true. Instead, we write "we reject $H_0$ in favour of $H_1$". However, we are allowed to write "we accept $H_0$" if there is insufficient evidence against it.

> **Activity 31.1 — Choice of Test**
> For each of the following situations,
>  - Decide whether a one- or two-tailed test is appropriate and justify your answer. *Hint:* there may be more than one answer.
>  - Write down the null and alternative hypotheses.
>  - Write down the possible conclusions from the hypothesis test.
>
> Situation 1
>
> A university has introduced a new timetabling system to minimise clashes where each student has the option to choose from several combinations of modules. The university conducted a hypothesis test to check their claim that only 10% of students have a clash in their timetable.
>
> Situation 2
>
> A car salesman claims that he sells a car to 60% of customers that he gives his business card to. To verify this claim, a hypothesis test is conducted by comparing how many business cards are given out against the total number of successful sales in a given time period.
>
> Situation 3
>
> In an experiment to assess the reliability of First Class mail, $n$ letters were posted First Class to a particular destination, and $k$ (with $k < n$) were found to arrive the next day. A hypothesis test was conducted to substantiate the postage company's claim that 80% of First Class mail arrives the next day.

## 31.3 The Significance Level

There are two ways in which we can come to the wrong conclusion having conducted a hypothesis test.
 - We reject $H_0$ in favour of $H_1$ when $H_0$ is in fact true.
 - We do not reject $H_0$ when $H_0$ is in fact false.

Incorrect conclusions arise either because we have not conducted our test appropriately, or because we do not have enough data to be representative of the population.

We can control the probability of rejecting $H_0$ given that $H_0$ is true by prescribing a *significance level* $\alpha$, where

$$\alpha = P(\text{reject } H_0 | H_0 \text{ true}).$$

By choosing a significance level, we choose the likelihood of incorrectly concluding that we reject $H_0$. Of course, we could argue that the best choice is $\alpha = 0$. However, this would lead to us *always* accepting $H_0$, even when $H_0$ is false, so we must have $\alpha > 0$.

---

**Remark**

The two errors above are often referred to as *type I* and *type II* errors, where
- Type I error: we reject $H_0$ in favour of $H_1$ when $H_0$ is in fact true.
- Type II error: We do not reject $H_0$ when $H_0$ is in fact false. This error occurs when there is insufficient evidence in our data to reject the null hypothesis, even though it is false.

The four possible outcomes of a hypothesis test are summarised in the table below.

| Decision | $H_0$ True | $H_0$ False |
|---|---|---|
| Accept $H_0$ | Correct Decision | Type II Error |
| Reject $H_0$ | Type I Error | Correct Decision |

It should be noted that Type 1 and Type II errors are not on the A-Level syllabus, but appear in some Further Mathematics courses. However, an understanding that mistakes can be made is important.

---

**Example 31.3**

An analogy of a hypothesis test is the story of "the boy who cried wolf". In this story, a boy is known for repeatedly calling for help, claiming that he is being attacked by a wolf. On all occasions so far, on running to help him, people have found the boy perfectly safe, with no wolf to be seen.

Each occasion the boy cries for help is an independent trial. The null hypothesis is that there is no wolf (nothing has changed) and the alternative hypothesis is that there is a wolf (the situation has changed). We could make the following incorrect conclusions.
- We believe the boy, running to save him, only to find out that he was lying again.
- We ignore the boy's call when there really was a wolf and leaving him to fend for himself.

---

**Activity 31.2 — Murder trial**

The decision made by the jury on a murder trial may be considered as a hypothesis test. For this analogy, answer the following questions:
  (a) What is the null hypothesis?
  (b) What is the alternative hypothesis?
  (c) Which two incorrect conclusions could we make?

## 31.4  Conducting a Hypothesis Test

There are five steps in conducting a hypothesis test. We explain each of them in turn.

**Step 1: State the null and alternative hypotheses.**
We clearly state the statistic that we wish to test in the null hypothesis $H_0$, followed by the one- or two-tailed alternative, $H_1$.

**Step 2: Determine the appropriate test statistic.**
The choice of test statistic is determined by the underlying distribution of the data sampled, along with the value that we wish to test. In this section, we consider the hypothesis test for the proportion in the binomial distribution. For our null hypothesis, $H_0$, we assume that each member of a sample of size $n$ has the same probability $p_0$ of having some attribute. Our test statistic is the binomial random variable random variable $X$, the number of individuals having the attribute ($n - X$ individuals do not have the attribute). Assuming $H_0$ is true,

$$X \sim \text{Bin}(n, p_0).$$

We wish to test whether the value of $X$ obtained from the sample is sufficiently close to its expected value, given that $p_0$ is the correct proportion, so that we accept $H_0$. Currently, we are performing a test for the proportion of the binomial distribution, but in Year 2 we will also consider the hypothesis test for the mean of a normal distribution.

**Step 3: Choose the significance level.**
We choose the probability of incorrectly rejecting $H_0$. The significance level also quantifies how much evidence is "sufficient" to reject $H_0$ in favour of $H_1$. The smaller the significance level, the more evidence we require against $H_0$. The significance level defines a *critical region*, examples of which are in a remark below. The remaining region is the *acceptance region* since, if the test statistic is in this region, we accept $H_0$. We locate the edge of the critical region using a *critical value for a one-tailed test*, or *two critical values for a two-tailed test*.

If we choose a significance level of $\alpha$ for a one-tailed test, the corresponding critical value of $X$, denoted $X^{\text{crit}}$, depends on the alternative hypothesis $H_1$.

- For $H_1:$   $p < p_0$, $X^{\text{crit}}$ is the greatest integer such that $P(x \leq X^{\text{crit}}) \leq \alpha$.
- For $H_1:$   $p > p_0$, $X^{\text{crit}}$ is the least integer such that $P(x \geq X^{\text{crit}}) \leq \alpha$.

For a two-tailed test, we obtain two critical values, $X_1^{\text{crit}}$ and $X_2^{\text{crit}}$, which define the lower and upper boundaries of the critical region respectively. The critical values satisfy the two conditions above, where $X_1^{\text{crit}}$ is the greatest integer and $X_2^{\text{crit}}$ is the least integer such that

$$P(x \leq X_1^{\text{crit}}) \leq \frac{\alpha}{2}, \ P(x \geq X_2^{\text{crit}}) \leq \frac{\alpha}{2},$$
$$\Rightarrow P(X_1^{\text{crit}} \leq x \leq X_2^{\text{crit}}) \geq 1 - \alpha.$$

Common choices of $\alpha$ are 0.1, 0.05 and 0.01. We often say that we test at the "$100\alpha\%$ level", so if $\alpha = 0.05$, then we test at the 5% level.

**Step 4: Compute the value of the test statistic.**
In the case of conducting a test for the proportion in the binomial situation, the test statistic is simply $X$, the number of individuals with the given attribute in the sample.

**Step 5: Make the decision and interpret in the context of the hypothesis**
Comparing the test statistic with the critical value, we check whether it lies in the critical region or not. If $z$ is in the critical region, then we have sufficient evidence to reject $H_0$ in favour of $H_1$. Otherwise, we conclude that there is insufficient evidence to reject $H_0$. We then interpret this decision in terms of the original data, deciding whether our data has provided sufficient evidence for a change in our previous belief regarding the statistic being tested.

**Remark**

The critical and acceptance regions for both one and two sided hypothesis tests of differing significance levels in the case that the distribution, $X \sim \text{Bin}(20, 0.6)$, are shown below.

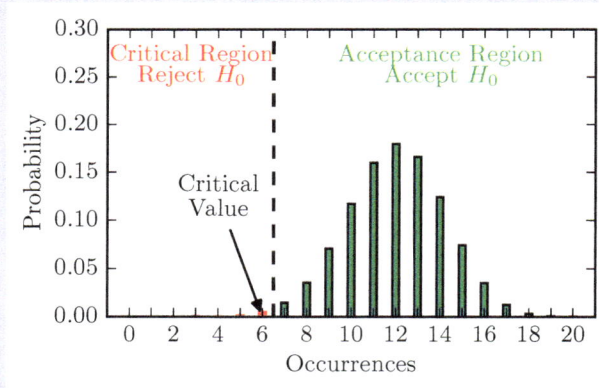

Figure 31.1: Critical and acceptance regions for a one-tailed test (where $H_1 : p < p_0$) at the 1% level, with $X \sim \text{Bin}(20, 0.6)$.

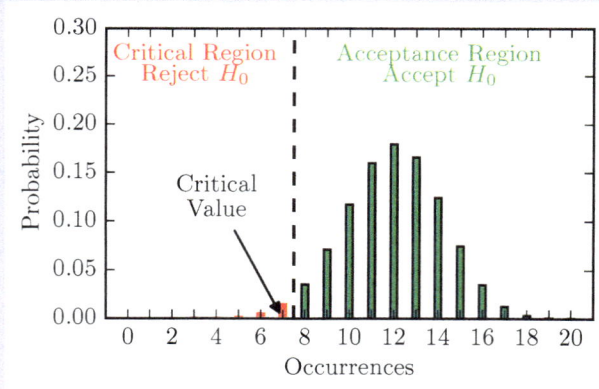

Figure 31.2: Critical and acceptance regions for a one-tailed test (where $H_1 : p < p_0$) at the 5% level, with $X \sim \text{Bin}(20, 0.6)$.

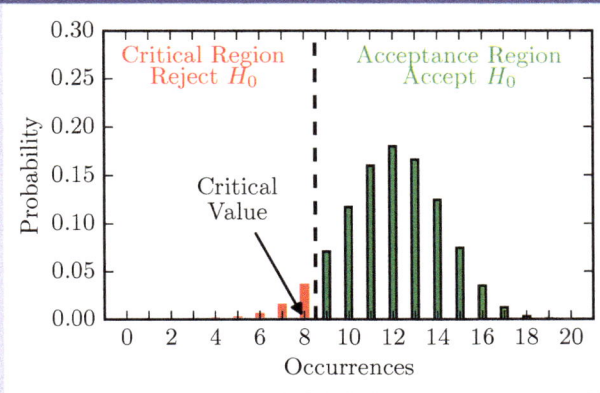

Figure 31.3: Critical and acceptance regions for a one-tailed test (where $H_1 : p < p_0$) at the 10% level, with $X \sim \text{Bin}(20, 0.6)$.

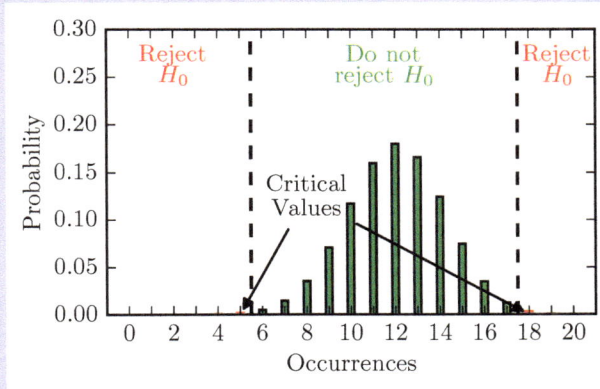

Figure 31.4: Critical and acceptance regions for a two-tailed test (where $H_1 : p \neq p_0$) at the 1% level, with $X \sim \text{Bin}(20, 0.6)$.

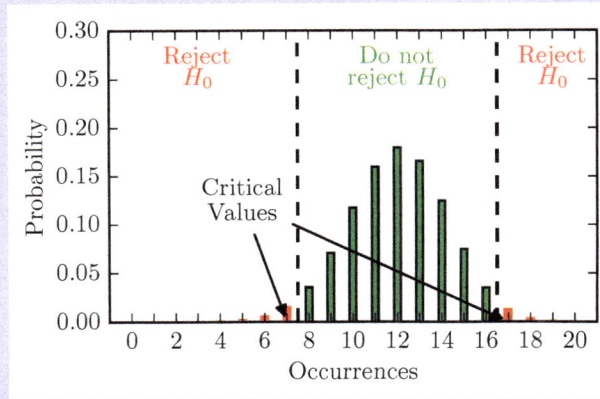

Figure 31.5: Critical and acceptance regions for a two-tailed test (where $H_1 : p \neq p_0$) at the 5% level, with $X \sim \text{Bin}(20, 0.6)$.

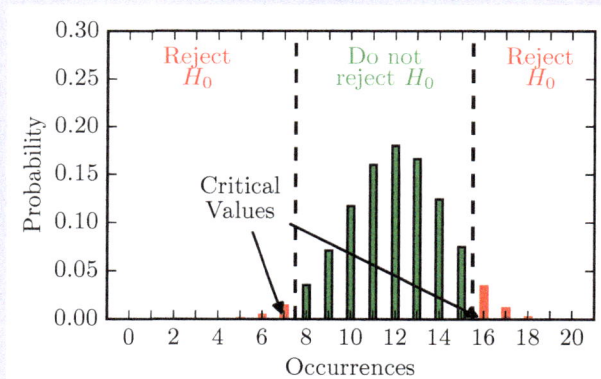

Figure 31.6: Critical and acceptance regions for a two-tailed test (where $H_1 : p \neq p_0$) at the 10% level, with $X \sim \text{Bin}(20, 0.6)$.

---

**Use of Technology 31.1 — Finding Binomial Probabilities**

Any probabilities required when conducting a hypothesis test can be computed using a calculator.

---

**Example 31.4**

In a Sixth Form College, the probability that a student passes an assessed physics practical is 0.79. A newly employed teacher is taking their first A-Level class. Out of a group of 20 students, 11 people pass the practical. Should the Head of Science be concerned?

**Solution:**

The information in the question gives $n = 20$, $p_0 = 0.79$ and $X = 11$. The Head of Science should be concerned if it is statistically significant that the pass rate has decreased.

1) We state the null and alternative hypotheses:

$$H_0 : \quad p = 0.79,$$
$$H_1 : \quad p \leq 0.79.$$

We choose a one-tailed test since we are interested in whether the pass rate *has* decreased.

2) We regard the student passing as a success and not passing as a failure. Hence, we are sampling from an underlying binomial distribution, so the appropriate test statistic is $X$, where $X \sim \text{Bin}(20, 0.79)$.

3) We choose a significance level of $\alpha = 0.5$. Following the procedure above, $X^{\text{crit}}$ is the greatest integer such that $P(x \leq X^{\text{crit}}) \leq 0.05$. Since $P(x \leq 12) = 0.0419$

and $P(x \leq 13) = 0.1071$, $X^{\mathrm{crit}} = 12$. The acceptance and critical regions are illustrated below.

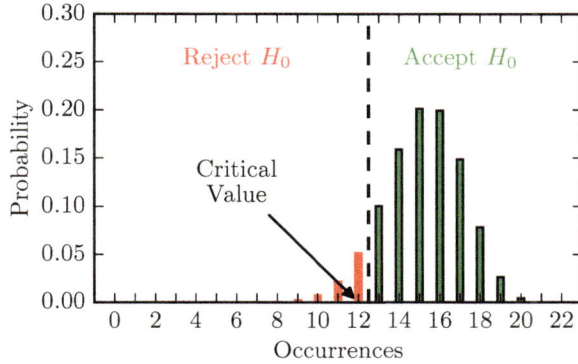

4) We are given that $X = 11$.
5) Since $11 < 12 = X^{\mathrm{crit}}$, the test statistic lies in the *critical region*. Hence, we have sufficient evidence to reject $H_0$ in favour of $H_1$. We conclude that there is sufficient evidence for the Head of Science to be concerned about the quality of teaching.

**Example 31.5**

A darts player hits a treble 20 with probability 0.15. In his first match after getting a new coach, the darts player throws 50 darts, 11 of which hit treble 20. Conduct a hypothesis test at the 5% level to decide whether the new coach has improved the player's performance.

**Solution:**
The information in the question gives $n = 50$, $p_0 = 0.15$ and $X = 11$.
1) We state the null and alternative hypotheses:

$$H_0 : \quad p = 0.15,$$
$$H_1 : \quad p > 0.15.$$

We choose a one-tailed test since we are interested in whether the darts player's performance has improved or not.
2) We regard the darts player hitting a treble 20 as a success, and hitting anything else as a failure. Hence, we are sampling from an underlying binomial distribution, so the appropriate test statistic is $X$, where $X \sim \mathrm{Bin}(50, 0.15)$.
3) As instructed, we choose a significance level of $\alpha = 0.05$. Following the procedure above, $X^{\mathrm{crit}}$ is the least integer such that $P(x \geq X^{\mathrm{crit}}) \leq 0.05$. Using a calculator, we find that $P(x \geq 12) = 0.0628$, and $P(x \geq 13) = 0.0301$, so $X^{\mathrm{crit}} = 12$. The acceptance and critical regions are illustrated below.

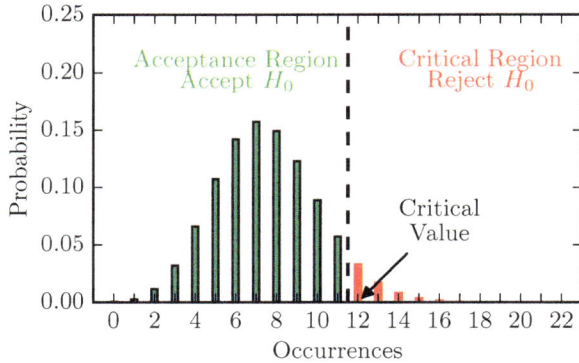

4) We are given that $X = 11$.

5) Since $11 < 12 = X^{\text{crit}}$, the test statistic lies in the *acceptance region*. Hence, we have insufficient evidence to reject $H_0$ in favour of $H_1$. There is insufficient evidence that the new coach has improved the darts player's performance.

---

**Example 31.6**

A tennis player serves an ace with probability 0.08. In the most recent match, the player serves 40 times and 9 are aces. Conduct a hypothesis test at the 10% significance level to verify the claim that the tennis player's success rate of serving aces has changed.

**Solution:**

The information in the question gives $n = 40$, $p_0 = 0.08$ and $X = 9$.

1) We state the null and alternative hypotheses:

$$H_0 : \quad p = 0.08,$$
$$H_1 : \quad p \neq 0.08.$$

We choose a two-tailed test since we are only interested in whether the success rate *has* changed.

2) We regard the tennis player serving an ace as a success, and not serving an ace as a failure. Hence, we are sampling from an underlying binomial distribution, so the appropriate test statistic is $X$, where $X \sim \text{Bin}(40, 0.08)$.

3) We choose a significance level of $\alpha = 0.1$. Following the procedure above, $X_1^{\text{crit}}$ is the greatest integer such that $P(x \leq X_1^{\text{crit}}) \leq 0.05$ and $X_2^{\text{crit}}$ is the least integer such that $P(x \geq X_2^{\text{crit}}) \leq 0.05$. Since $P(x = 0) = 0.0356$ and $P(x \leq 1) = 0.1594$, $X_1^{\text{crit}} = 0$. Furthermore, $P(x \geq 6) = 0.0967$, $P(x \geq 7) = 0.0376$, $X_2^{\text{crit}} = 7$. The acceptance and critical regions are illustrated below.

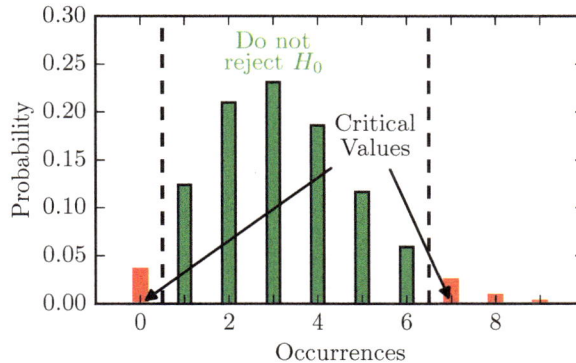

4) We are given that $X = 9$.

5) Since $9 > 7 = X_2^{\text{crit}}$, the test statistic lies in the *critical region*. Hence, we have sufficient evidence to reject $H_0$ in favour of $H_1$. We conclude that there is sufficient evidence that the tennis player's success rate of serving aces is different to 0.08.

---

**Interactive Activity 31.1 — Finding the Critical Region**

Use the interactive below (in the digital book) to explore how the critical region changes for different distributions, significance levels and types of tests.

---

**Exercise 31.1**

Q1. For each of the following tests, state the critical and acceptance regions and whether to accept or reject the null hypothesis.

(a) $H_0 : p = 0.2$, $H_1 : p < 0.2$. The critical value is 10 and the value of $X$ from the sample is 7.

(b) $H_0 : p = 0.6$, $H_1 : p \neq 0.6$. The critical values are 6 and 14 and the value of $X$ from the sample is 12.

(c) $H_0 : p = 0.5$,d $H_1 : p > 0.5$. The critical value is 35 and the value of $X$ obtained from the sample is 45.

Q2. For each of the following tests, state the null hypothesis $H_0$ and the alternative hypothesis $H_1$.

(a) A mobile phone company claims that 10% of their customers have made a complaint in the last 2 years. A survey of 50 customers is conducted to test whether the company has underestimated the number of complaints made against them.

(b) A confectionery company produces sweets in 5 flavours. The company claims that there is an equal number of a given flavour in each bag of 20 sweets. A bag is chosen at random, then the number of raspberry flavoured sweets is counted to conduct a hypothesis test to verify the company's claim that each flavour is represented equally.

(c) A keen gardener grows tomato plants every year. Each year, 70% of the seeds planted produce a usable crop of tomatoes. This year, the gardener has bought an expensive greenhouse that claims to improve the likelihood that seeds planted grow into healthy tomato plants. At the end of the season, the gardener conducts a hypothesis test on her 80 planted seeds to see whether her purchase was justified.

Q3. For each of the hypothesis tests in Q2, find the critical region for the 1%, 5% and 10% significance levels.

Q4. A hypothesis test is conducted using a sample of size 30, where $H_0 : p = 0.3$, $H_1 : p < 0.3$. The critical value at the 5% level is 4.

(a) Find the probability of incorrectly rejecting $H_0$. Explain why the value obtained is not equal to 5%.

(b) If we conduct the same test at the 10% level, is the new critical value less than or more than 4?

(c) If the value from $X$ obtained from the sample is 3, state whether we accept or reject $H_0$ at the 5% and 10% levels.

Q5. A glass factory produces orders on time if 85% of glassware that comes off the production line has no defects. An inspector chooses a sample of 100 products and conducts a hypothesis test at the 5% level to determine whether the factory is running at the required rate of quality.

(a) Write down the null and alternative hypotheses and find the critical region.

(b) Decide whether to accept or reject the null hypothesis if the inspector finds
   i. 79 products with defects;
   ii. 75 products with defects.

Q6. In an opinion poll, a sample of $n = 1235$ people were interviewed and 495 said they supported the Government. Conduct a hypothesis test at the 1% level to investigate whether or not the true proportion is 0.45.

## 31.5 Providing p-values

In our tests so far, once we have chosen a significance level, we conclude our test by either accepting or rejecting the null hypothesis. However, sometimes our test statistic is close to the critical value, meaning that a small change to the significance level may change the outcome of the test. We require a measure of the strength of evidence against $H_0$ to distinguish cases that are close to the edge of the critical region from those that are far into it.

### Definition 31.3 — p-value
The p-value for a hypothesis test is the probability of obtaining a test statistic, $X$, at least as extreme as that obtained from the sample given that the null hypothesis is true. The smaller the p-value, the more evidence we have *against* $H_0$. In particular, a p-value of $\alpha$ corresponds to being on the edge of the critical region, so if we are conducting a test at the $100\alpha\%$ level, we require a p-value less than $\alpha$ to reject $H_0$.

The calculation of the p-value for the hypothesis test for the proportion in the binomial distribution depends on the alternative hypothesis, $H_1$. Assuming $X \sim \text{Bin}(n, p_0)$, then the most likely value for $X$ is $np$. If $x$ is our observed value of $X$, then the calculation of the p-value is summarised in the table below.

| $H_1$ | p-value |
|---|---|
| $p < p_0$ | $P(X \leq x \,|\, p = p_0)$ |
| $p > p_0$ | $P(X \geq x \,|\, p = p_0)$ |
| $p \neq p_0$ | If $x < np$, $P(X \leq x \,|\, p = p_0)$<br>If $x > np$, $P(X \geq x \,|\, p = p_0)$ |

**Example 31.7**

A hypothesis test is conducted where $X \sim \text{Bin}(80, p)$, and $H_0 : p = 0.4$, $H_1 : p \neq 0.4$. For each of the following values of the test statistic, find the p-value to 4 decimal places and state whether $H_0$ is rejected in favour of $H_1$ at the 5% level.

(a) $X = 30$;

(b) $X = 46$.

**Solution:**

Recall that, for a two-tailed test, the critical region is divided equally on each side of the acceptance region. Hence, we seek p-values less that 0.025 to reject $H_0$. The most likely value of $X$ is $np = 32$, so we consider values much higher and lower than 32.

(a) Since $30 < np$, the p-value is given by

$$P(X \leq 30) = 0.3687.$$

Since $0.3687 > 0.025$, we have insufficient evidence to reject $H_0$ at the 5% level.

(b) Since $46 > np$, the p-value is given by

$$P(X \geq 46) = 1 - P(X \leq 45) = 0.0012.$$

Since $0.0012 < 0.025$, we have sufficient evidence to reject $H_0$ in favour of $H_1$ at the 5% level.

**Remark**

Although hypothesis testing is widely used in statistical analysis, there is some controversy in the science community over its use. Some common criticisms of hypothesis testing are:

- Hypothesis tests provide the opposite of the desired conclusion. A p-value is the probability of obtaining a result at least as *extreme* as that obtained in the sample, given that $H_0$ is true. We really want to know whether $H_0$ is *true* given the sample data. In practice, we can only conclude whether to reject, or not reject, $H_0$ without ever discovering whether $H_0$ is true or not.
- The null hypothesis is *always* false, since it is practically impossible for us to find a sample mean that is precisely equal to the population mean.
- The binary nature of the conclusion of a hypothesis test is restrictive. If we obtain a p-value of 0.051, we reach the exact opposite conclusion compared with a p-value of 0.049 at the 5% level. In reality, there is little difference between each of these results.
- Hypothesis tests are open to abuse by poor experimental design. By adjusting the significance level, or carefully choosing a one- or two-tailed test, we can ma-

nipulate the same set of data to provide evidence for opposing conclusions. This is potentially very dangerous, for example when analysing data from a medical trial for an experimental drug.

There are many other arguments against the use of hypothesis testing. On balance, hypothesis testing is a useful tool for making basic inferences on data obtained, while the analysis can be supplemented with other statistical techniques, such as parameter estimation, to make further inferences from a data set.

**Exercise 31.2**

Q1. A hypothesis test is conducted where $X \sim \text{Bin}(200, p)$, and $H_0 : \quad p = 0.75$, $H_1 : \quad p \neq 0.75$. Find the p-value to 4 decimal places for the following values of the test statistic.
    (a) $X = 145$;
    (b) $X = 160$;
    (c) $X = 151$;
    (d) $X = 130$.

Q2. A hypothesis test is conducted where $X \sim \text{Bin}(50, p)$, and $H_0 : \quad p = 0.05$, $H_1 : p > 0.05$. Find the p-value to 4 decimal places for the following values of the test statistic and state whether $H_0$ is rejected in favour of $H_1$ at the 10% level.
    (a) $X = 5$;
    (b) $X = 8$;
    (c) $X = 10$.

Q3. A hypothesis test is conducted to conclude whether a normal six-sided die is fair when 1 six is thrown in 24 throws.
    (a) Find the p-value of this hypothesis test.
    (b) What is the conclusion of the hypothesis test at the 5% level?
    (c) Does the conclusion at the 5% level change if no sixes were thrown in 24 throws?
    (d) What if one six is thrown in 50 throws?

Q4. The maternity unit of a hospital claims that only 1 in 4 women must wait for more than 20 minutes for their antenatal scan. A midwife believes that the unit is understaffed and, as a result, the true proportion of women that must wait for more than 20 minutes is larger than 1 in 4. The midwife conducts a hypothesis test at the 5% level with a sample of 30 women that visit the unit, 15 of whom waited more than 20 minutes for their scan.
    (a) State a condition on the method of choosing the women so that a binomial probability model can be used to test the claim.
    (b) State a reason for assuming that the sample is drawn from a binomial distribution in this case.
    (c) State the null and alternative hypotheses of the midwife's test.
    (d) What is the conclusion of the midwife's test?

Q5. A cereal company runs a competition in which it claims that 1 in 10 cereal boxes contain a prize. A disgruntled customer believes that this proportion decreases over time and conducts a hypothesis test at the 5% level using a sample of 60 boxes of cereal that she has collected.
    (a) State $H_0$ and $H_1$ for this hypothesis test.

The customer initially thought that $k$ of the cereal boxes contained prizes, which led her to reject $H_0$ in favour of $H_1$. She then found one more box containing a prize, which led her to accept $H_0$.

(b) Find the value of $k$.

Q6. The marketing manager of a toy company claims that 60% of customers that visit the company's website and add at least one item to their basket make a purchase. A member of the sales team disputes this claim, stating that the true proportion of successful sales is less than 60%. They conduct a hypothesis test at the 1% level with a sample of 40 customers who visited the website and added an item to their basket, 21 of whom completed a purchase. What is the conclusion of their test?

Q7. A pharmaceutical company claims that their new drug for reducing blood pressure is successful in 85% of patients.

(a) A junior doctor believes that the drug has a lower success rate than that claimed by the pharmaceutical company, so conducts a hypothesis test at the 5% level using a sample of 50 patients. If the drug is successful for 38 patients, what is the outcome of the doctor's test?

(b) A rival company claims that their new drug has a higher success rate than the drug tested in **(a)**. Their research team conduct a hypothesis test at the 10% level using a sample of 25 patients to verify the claim. If the drug is successful for 24 patients, what is the outcome of the company's test?

(c) Comment on the reliability of the conclusions from the hypothesis tests in (a) and (b).

Q8. A bus company with a fleet of 300 buses found that the emissions systems of 7 out of 22 they randomly chose and tested *failed* to meet the pollution control guidelines. To plan for the next costing period, the company would like to test if this is strong evidence that more than 20% of the fleet is out of compliance. Use this information to conduct an appropriate hypothesis test and conclude whether there is significant evidence that more than 20% the fleet is out of emissions compliance.

Q9. A supermarket is proud of the fact that 60% of their branded products are cheaper than the same product sold by their competitors. In a brand matching exercise, a market researcher picks 65 branded products at random and finds that 40 of them are cheaper in the supermarket than their competitors. Conduct a hypothesis test at the 10% level to substantiate the supermarket's claimed proportion of cheaper branded products.

Q10. The coin used in a tennis tournament to determine the order of service is suspected to be biased towards heads. In an experiment by an independent adjudicator, the coin is found to land on heads more than twice as often as tails.

(a) Find the p-value to test, at the 5% level, whether the coin is biased towards heads if

    i. the coin is tossed 9 times;

    ii. the coin is tossed 21 times.

(b) If there is sufficient evidence for us to conclude that the coin is biased in favour of heads, find the minimum number of times the coin was tossed.

**Key Facts — Hypothesis Testing**

- The *null hypothesis*, denoted $H_0$, is the hypothesis that we wish to test. This is the current situation, or presumed value of a particular quantity.
- The *alternative hypothesis*, denoted $H_1$, is the alternative result, or outcome to the null hypothesis.
- The significance level of a test, $\alpha$ is defined to be $\alpha = P(\text{reject } H_0 | H_0 \text{ true})$.
- There are 5 steps to completing an hypothesis test:
    1. State the null and alternative hypotheses.
    2. Determine the appropriate test statistic.
    3. Choose the significance level.
    4. Compute the value of the test statistic.
    5. Make the decision and interpret in the context of the hypothesis.
- The critical value, acceptance region, and the critical (rejection) region for a one sided hypothesis test for $X \sim \text{Bin}(20, 0.6)$, with $\alpha = 0.1$ are shown below.

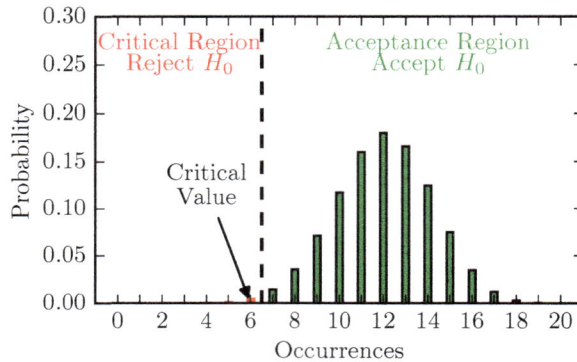

- The conclusion to a hypothesis test must always be given in terms of the situation.
- The p-value for a hypothesis test is the probability of obtaining a test statistic, $X$, at least as extreme as that obtained from the sample given that the null hypothesis is true.
- The smaller the p-value, the more evidence we have *against* $H_0$.
- A p-value of $\alpha$ corresponds to being on the edge of the critical region, so if we are conducting a test at the $100\alpha\%$ level, we require a p-value less than $\alpha$ to reject $H_0$.
- Calculation of the $p$-value is summarised below:

| $H_1$ | p-value |
|---|---|
| $p < p_0$ | $P(X \leq x \,|\, p = p_0)$ |
| $p > p_0$ | $P(X \geq x \,|\, p = p_0)$ |
| $p \neq p_0$ | If $x < np$, $P(X \leq x \,|\, p = p_0)$<br>If $x > np$, $P(X \geq x \,|\, p = p_0)$ |

**Chapter Assessment — Hypothesis Testing**
Download and sit the 30 minute assessment for this chapter from the digital book.

$$\frac{\sin(x)}{}$$

$$f'(x) = \lim_{h \to 0} \frac{f(x+h) - f(x)}{h} + f($$

$$d \sin($$

# 32. Answers

**Answers — Exercise 2.1**

Q1.  (a)  256
     (b)  1
     (c)  7
     (d)  $\frac{1}{64}$
     (e)  $\frac{1}{5}$
     (f)  16
     (g)  $\frac{125}{27}$
     (h)  216
     (i)  $\frac{81}{16}$
     (j)  $\frac{27}{8}$

Q2.  (a)  $x^8$
     (b)  $48y^6$
     (c)  $\frac{2x^6}{3}$
     (d)  $18x^4y^2$
     (e)  $\frac{x^4}{4}$
     (f)  $\frac{y^2}{49x^4}$
     (g)  $x^3$
     (h)  $4x^{\frac{5}{2}}$
     (i)  $x^{\frac{3}{4}}$
     (j)  $x^{\frac{11}{6}}$

Q3.  (a)  $x = -2$
     (b)  $x = 2$

(c) $x = \frac{3}{11}$

(d) $x = \frac{13}{7}$

(e) $x = \frac{16}{3}$

(f) $x = -\frac{5}{2}$

(g) $x = 1$

## Answers — Exercise 2.2

Q1. (a) 5

(b) 4

(c) 42

(d) 6

(e) $\sqrt{6}$

(f) $2\sqrt{2}$

(g) 6

(h) $\frac{1}{2}$

Q2. (a) $3\sqrt{2}$

(b) $2\sqrt{30}$

(c) $6\sqrt{2}$

(d) $2\sqrt{6}$

(e) $3\sqrt{5}$

(f) $4\sqrt{3}$

(g) $7\sqrt{2}$

(h) $13\sqrt{3}$

Q3. (a) 1

(b) $21 - 4\sqrt{5}$

(c) $18 + 7\sqrt{6}$

(d) $3 + \sqrt{21} + \sqrt{15} + \sqrt{35}$

Q4. (a) $\frac{\sqrt{7}}{7}$

(b) $4\sqrt{5}$

(c) $\frac{\sqrt{14}}{2}$

(d) $\frac{\sqrt{6}}{2}$

(e) $\frac{4 - \sqrt{2}}{14}$

(f) $3\sqrt{5} + 3$

(g) $-3 - 2\sqrt{2}$

(h) $2 + \sqrt{3}$

## Answers — Exercise 3.1

Q1. (a) $\mathbf{a} - \mathbf{b}$;

(b) $\mathbf{a} - \mathbf{b}$;

(c) $\mathbf{c} - \mathbf{b}$;

(d) $\mathbf{c} - \mathbf{a}$.

Q2. $\frac{1}{2}(\mathbf{a} + \mathbf{b})$.

Q3. (a)

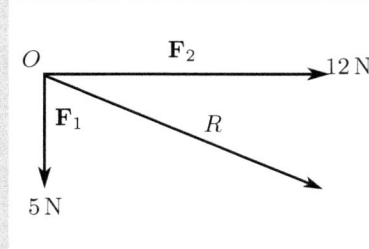

    (b) 13 N;

    (c) $\approx 0.395$ radians.

Q4. $\frac{1}{5}\mathbf{a} + \frac{4}{5}\mathbf{b}$.

Q5. Perpendicular height of the parallelogram.

Q8. $DF : FC = 4 : 3$.

---

**Answers — Exercise 3.2**

Q1.  (a) $(1, 1)$, $\begin{pmatrix} 1 \\ 1 \end{pmatrix}$;

    (b) $(2, 0)$, $\begin{pmatrix} 2 \\ 0 \end{pmatrix}$;

    (c) $(0, -3)$, $\begin{pmatrix} 0 \\ -3 \end{pmatrix}$;

    (d) $(3, 2)$, $\begin{pmatrix} 3 \\ 2 \end{pmatrix}$;

    (e) $(-4, 1)$, $\begin{pmatrix} -4 \\ 1 \end{pmatrix}$;

    (f) $\left( \frac{1}{2}, -\frac{3}{2} \right)$;

    (g) $\begin{pmatrix} 1/2 \\ -3/2 \end{pmatrix}$.

Q2.  (a) $|\mathbf{a}| = 5$, $\hat{\mathbf{a}} = \frac{3}{5}\mathbf{i} + \frac{4}{5}\mathbf{j}$;

    (b) $|\mathbf{a}| = \sqrt{5}$, $\hat{\mathbf{a}} = \frac{1}{\sqrt{5}}(1, -2)$;

    (c) $|\mathbf{a}| = 2\sqrt{5}$, $\hat{\mathbf{a}} = \frac{1}{\sqrt{5}}\begin{pmatrix} -1 \\ 2 \end{pmatrix}$;

    (d) $|\mathbf{a}| = \sqrt{2}$, $\hat{\mathbf{a}} = \frac{1}{\sqrt{2}}(-1, 1)$.

Q3.  (a) $\overrightarrow{AB} = (-1, 0)$, $|\overrightarrow{AB}| = 1$;

    (b) $\overrightarrow{AB} = (-2, -4)$, $|\overrightarrow{AB}| = 2\sqrt{5}$;

    (c) $\overrightarrow{AB} = (4, -4)$, $|\overrightarrow{AB}| = 4\sqrt{2}$;

    (d) $\overrightarrow{AB} = (-7, -1)$, $|\overrightarrow{AB}| = 5\sqrt{2}$.

Q4. (a) $\dfrac{9}{5}(3,4)$;

(b) $\dfrac{3}{\sqrt{5}}(1,-2)$;

(c) $\dfrac{1}{3\sqrt{5}}(-2,1)$;

(d) $(1,1)$.

Q5. (a) $74.05°$.

(b) $26.57°$.

(c) $53.13°$.

## Answers — Exercise 4.1

Q1. (a) $|AB| = 2\sqrt{2}$, gradient of $AB = 1$, midpoint of $AB = (3,3)$.

(b) $|AB| = 3\sqrt{2}$, gradient of $AB = -1$, midpoint of $AB = \left(\frac{19}{2}, \frac{7}{2}\right)$.

(c) $|AB| = 6$, gradient of $AB$ is undefined, midpoint of $AB = (6,0)$.

(d) $|AB| = 2\sqrt{26}$, gradient of $AB = \frac{1}{5}$, midpoint of $AB = (7,6)$.

(e) $|AB| = 3\sqrt{13}$, gradient of $AB = \frac{3}{4}$, midpoint of $AB = \left(0, \frac{3}{2}\right)$.

(f) $|AB| = 4\sqrt{6}$, gradient of $AB = -\frac{3}{2}$, midpoint of $AB = (13,2)$.

(g) $|AB| = 5$, gradient of $AB = -1$, midpoint of $AB = \left(1, \frac{13}{2}\right)$.

(h) $|AB| = 3\sqrt{2}$, gradient of $AB = 2$, midpoint of $AB = \left(-\frac{1}{2}, \frac{1}{2}\right)$.

Q2. (a) $\sqrt{2}a$.

(b) $5\sqrt{5}a$.

(c) $\sqrt{16t^2 + t^4}$.

Q3. $p = 5$.

Q4. $z = 11$.

Q5. The gradient of $AB$ is equal to that of $BC$ and so the three points can be joined by a straight line.

Q6. The midpoint of both line segments is $(11,4)$.

Q7. $B(12,7)$.

Q8. $y_A = -1$, $x_B = 12$.

Q9. $AB$, $CD$ and $CE$ are all parallel. $CD$, $AB$, $CE$ are all perpendicular to $EF$.

## Answers — Exercise 4.2

Q1. (a) $2$.

(b) $0$.

(c) $-4$.

(d) $-2$.

(e) $\frac{1}{3}$;

(f) $\frac{-2}{3}$;

(g) $\frac{1}{3}$;

(h) $\frac{13}{6}$.

Q2. (a) $3x - y - 2 = 0$;

(b) $-x + 2y + 12 = 0$;

(c) $3x + y + 7 = 0$;

(d) $12y - 4x - 15 = 0$.

Q3. $y = -\frac{1}{2}x + 4$.

Q4. $-7x - 12y + 84 = 0$.

Q5. 27.

Q6. $m = 4$.

Q7. $a = -1$ and $b = 2$.

Q8. (a) $y = 2x$;

(b) $y = 3x - 7$;

(c) $y = -x + 5$;

(d) $y = -2x - 13$;

(e) $y = \frac{3}{2}x + 6$;

(f) $y = -\frac{2}{3}x - 5$;

(g) $y = 6x + 13$;

(h) $y = \frac{2}{13}x + \frac{1}{26}$.

Q9. $y = \frac{5}{2}x - a$.

Q10. (a) $-x + y - 2 = 0$;

(b) $x + y - 5 = 0$;

(c) $-4x + 3y + 9 = 0$;

(d) $x + 5y - 42 = 0$;

(e) $-7x + 3y + 11 = 0$;

(f) $5x + 6y - 29 = 0$;

(g) $-x + y + 11 = 0$;

(h) $-6x + 13y + 77 = 0$.

Q11. $-2x - 3y + 12 = 0$.

Q12. $P(4, 5)$ and $2x - 5y + 17 = 0$.

Q13. (a) $-x - 3y + 14 = 0$;

(b) $-7x - 5y + 42 = 0$;

(c) $8x + 3y + 31 = 0$;

(d) $11x + 6y - 76$.

---

**Answers — Exercise 4.3**

Q1. (a) $3x + 2y = 10$.

(b) $2x + 3y = 15$.

Q2. $(b)$ and $(d)$ could be true.

Q3.

Q4. $\alpha > \gamma$ and $\beta = \delta$ are true.

Q5. (a) $|AB| = \sqrt{113}$, $|BC| = 2\sqrt{17}$ and $|CA| = 9$ so the triangle is scalene.

(b) 36 square units.

(c)   i. $AB : -8x - 7y + 90 = 0$.

ii. $BC : 4x - y - 18 = 0$.

iii. $CA : y = -2$.

Q6. (a) $A(3, 5)$.

(b) $B(9, 0)$.

(c) $C\left(\frac{1}{2}, 0\right)$

Q7. (a) $c = 15n + 30$.

(b) £375.

(c) 8900 business cards.

Q8. (a) $225x - y - 15 = 0$.

(b) You cannot have a negative number of people.

(c) 915.

(d) The school should reach a steady state population after 5 years due to Year 11 leaving the school.

Q9. The coordinates are as follows: $A(3,2)$, $B(6,3)$, $C(5,6)$, $D(2,5)$ and $E(4,4)$. The equations of the lines are given below:

$$\text{Line joining } A \text{ to } B : -x + 3y - 3 = 0,$$
$$\text{Line joining } B \text{ to } C: -3x + y - 21 = 0,$$
$$\text{Line joining } C \text{ to } D : x - 3y - 13 = 0,$$
$$\text{Line joining } D \text{ to } A : 3x + y - 11 = 0,$$
$$\text{Line joining } A \text{ to } C : 2x + 2y - 12 = 0.$$

The area of the square is 10 square units.

## Answers — Exercise 5.1

Q1. (a) Yes, (b) No, (c) Yes, (d) Yes, (e) Yes, (f) No

## Answers — Exercise 5.2

Q1. The functions a, b and c are quadratic.

## Answers — Exercise 5.3

Q1. (a) $x = 0$, $(0,1)$
(b) $x = -1$, $(-1,-1)$
(c) $x = -1$, $(-1,0)$
(d) $x = -1$, $(-1,4)$
(e) $x = -2$, $(-2,0)$
(f) $x = -2$, $(2,11)$
(g) $x = 2$, $(2,0)$
(h) $y = -1/2$, $(7/4,-1/2)$

Q2. The plot is symmetrical.

## Answers — Exercise 5.4

Q2. (a) $(x+1)(x+11)$.
(b) $(x+1)(x+17)$.
(c) $(x+1)(x+41)$.
(d) $(t+1)(t+23)$.
(e) $(x+1)(x-13)$.
(f) $(x-1)(x-23)$.

Q3.  (a) $(x+1)(x+24)$.
     (b) $(x+4)(x+7)$.
     (c) $(x+3)^2$.
     (d) $(x+5)((x+4)$.
     (e) $(x+3)(x+7)$.
     (f) $(x+2)(x+4)$.
     (g) $(x-2)(x+3)$.
     (h) $(x+7)(x-5)$.
     (i) $(x+4)(x-1)$.
     (j) $(x+3)(x-5)$.
     (k) $(x+4)(x-2)$.

     (l) $(x+2)(x-4)$.
     (m) $(x+1)(x-5)$.
     (n) $(x-3)(x+4)$.
     (o) $(x-2)(x-3)$.
     (p) $(x-1)(x-6)$.
     (q) $(x-3)(x-8)$.
     (r) $(x-7)(x-5)$.
     (s) $(x-2)(x-12)$.
     (t) $(x-6)(x-7)$.
     (u) $(x+6)(x-7)$.
     (v) $(x+7)(x-3)$.

## Answers — Exercise 5.5

Q1. See worked solutions.

Q2.  (a) $(2x+1)(x+1)$.
     (b) $(2x+3)(x+1)$.
     (c) $(3x+2)(x+3)$.
     (d) $(5x+3)(x+1)$.

     (e) $(5x+4)(x+2)$.
     (f) $(7x-1)(x+3)$.
     (g) $(7x+2)(x-5)$.
     (h) $(13x-3)(x-2)$.

Q3.  (a) $(2x+2)^2$.
     (b) $(2x+2)(2x+3)$.
     (c) $(2x+2)(3x+1)$.
     (d) $(x+1)(4x+3)$.
     (e) $(2x-3)(4x+5)$.

     (f) $(3x-1)(4x-2)$.
     (g) $(5x+2)(2x-1)$.
     (h) $(10x+2)(x+3)$.
     (i) $(2x+5)(6x+1)$.
     (j) $(5x+2)(6x+7)$.

## Answers — Exercise 5.6

Q1. See worked solutions.

Q2.  (a) $(x+2)^2$.
     (b) $(x+7)^2$.
     (c) $(x+8)^2$.

     (d) $(x+3)^2$.
     (e) $(x-1)^2$.
     (f) $(x-5)^2$.

Q3.  (a) $\left(x+\frac{1}{3}\right)^2$.
     (b) $\left(x-\frac{1}{3}\right)^2$.
     (c) $\left(x+\frac{4}{5}\right)^2$.

     (d) $\left(x-\frac{3}{7}\right)^2$.
     (e) $\left(x-\frac{1}{5}\right)^2$.
     (f) $\left(x+\frac{4}{9}\right)^2$.

## Answers — Exercise 5.7

Q1.   (a) $y = (x+1)^2 + 1.$
  (b) $y = (x+1)^2 + 4.$
  (c) $y = (x+2)^2 + 5.$
  (d) $y = (x+3)^2 + 2.$
  (e) $y = (x+3)^2 - 5.$

  (f) $y = (x+2)^2 - 8.$
  (g) $y = (x+1)^2 - 6.$
  (h) $y = (x-2)^2 + 5.$
  (i) $y = (x-4)^2 + 6.$
  (j) $y = (x-1)^2 + 15.$

Q2.   (a) $y = \left(x+\frac{3}{2}\right)^2 + \frac{7}{4}.$
  (b) $y = \left(x+\frac{3}{2}\right)^2 + \frac{19}{4}.$
  (c) $y = \left(x+\frac{1}{2}\right)^2 + \frac{15}{4}.$
  (d) $y = \left(x+\frac{5}{2}\right)^2 - \frac{53}{4}.$
  (e) $y = \left(x+\frac{5}{2}\right)^2 + \frac{11}{4}.$

  (f) $y = \left(x-\frac{3}{2}\right)^2 - \frac{21}{4}.$
  (g) $y = \left(x-\frac{11}{2}\right)^2 - \frac{105}{4}.$
  (h) $y = \left(x-\frac{5}{2}\right)^2 - \frac{37}{4}.$
  (i) $y = \left(x-\frac{1}{2}\right)^2 - \frac{5}{4}.$
  (j) $y = \left(x-\frac{5}{2}\right)^2 - \frac{1}{4}.$

Q3.   (a) $y = -(x+3)^2 + 7.$
  (b) $y = -(x-1)^2 + 2.$
  (c) $y = -(x-2)^2 + 1.$
  (d) $y = -(x+4)^2 + 20.$

  (e) $y = -\left(x-\frac{5}{2}\right)^2 + \frac{17}{4}.$
  (f) $y = -\left(x+\frac{3}{2}\right)^2 + \frac{19}{4}.$
  (g) $y = -(x+7)^2.$
  (h) $y = -\left(x-\frac{7}{2}\right)^2 + \frac{33}{4}.$

Q4.   (a) $y = 2(x+1)^2 + 4.$
  (b) $y = 2(x-2)^2 - 6.$
  (c) $y = 3(x+2)^2 - 11.$
  (d) $y = 5(x-2)^2 - 22.$

  (e) $y = 2\left(x-\frac{3}{4}\right)^2 - \frac{1}{8}.$
  (f) $y = 3\left(x-\frac{2}{3}\right)^2 + \frac{11}{3}.$
  (g) $y = 7\left(x-\frac{3}{14}\right)^2 - \frac{383}{28}.$
  (h) $y = 2\left(x-\frac{6}{5}\right)^2 - \frac{21}{5}.$

Q5.   (a) $(-1, -6).$
  (b) $(-3, -6).$
  (c) $(3, 2).$
  (d) $\left(\frac{1}{2}, \frac{43}{4}\right).$
  (e) $\left(\frac{3}{2}, -\frac{25}{4}\right).$

  (f) $\left(\frac{1}{2}, -\frac{19}{4}\right).$
  (g) $\left(-\frac{5}{8}, \frac{441}{16}\right).$
  (h) $\left(\frac{1}{4}, \frac{39}{8}\right).$
  (i) $\left(\frac{7}{10}, -\frac{29}{10}\right).$
  (j) $(-4, -39).$

## Answers — Exercise 5.8

Q1.   (a) $x_1 \approx -0.2$ and $x_2 \approx 4.2.$
  (b) $x_1 \approx -1.2$ and $x_2 \approx 3.2.$
  (c) $x_1 \approx -3.2$ and $x_2 \approx 1.2.$
  (d) $x_1 \approx -5.8$ and $x_2 \approx -0.2.$

  (e) $x_1 = -1$ and $x_2 = -5.$
  (f) $x_1 \approx -1.4$ and $x_2 \approx 3.4.$
  (g) $x_1 \approx -4.7$ and $x_2 \approx 1.7.$
  (h) $x_1 \approx -3.6$ and $x_2 \approx 0.6.$

Q2.  (a) $x_1 = 3$ and $x_2 = -1$.
     (b) $x_1 = -4$ and $x_2 = -7$.
     (c) $x_1 = -3$ and $x_2 = 4$.
     (d) $x_1 = \frac{-1}{2}$ and $x_2 = -3$.

     (e) $x_1 = \frac{3}{2}$ and $x_2 = -1$.
     (f) $x_1 = \frac{3}{2}$ and $x_2 = \frac{1}{3}$.
     (g) $x_1 = \frac{-5}{3}$ and $x_2 = \frac{1}{2}$.
     (h) $x_1 = \frac{7}{4}$ and $x_2 = \frac{1}{3}$.

Q3.  (a) $x = -2 + \sqrt{5}$ and $x = -2 - \sqrt{5}$.
     (b) $x = 4 + \sqrt{7}$ and $x = 4 - \sqrt{7}$.
     (c) $x = -4 + \sqrt{20}$ and $x = -4 - \sqrt{20}$.
     (d) $x = -1 + \sqrt{2}$ and $x = -2 - \sqrt{2}$.

     (e) $x = \frac{3}{2} + \sqrt{\frac{21}{4}}$ and $x = \frac{3}{2} - \sqrt{\frac{21}{4}}$.
     (f) $x = -3$ and $x = -1$.
     (g) $x = 1 + \sqrt{3}$ and $1 - \sqrt{3}$.
     (h) $x = -3 + \sqrt{6}$ and $x - -3 - \sqrt{6}$.

Q4.  (a) $x = \frac{-3-\sqrt{5}}{2}$ and $x = \frac{-3+\sqrt{5}}{2}$.
     (b) $x = \frac{-5+\sqrt{33}}{2}$ and $x = \frac{-5-\sqrt{33}}{2}$.
     (c) $x = \frac{3+\sqrt{5}}{2}$ and $x = \frac{3-\sqrt{5}}{2}$.
     (d) $x = -1$, a repeated root.

     (e) $x = \frac{-3+\sqrt{41}}{4}$ and $x = \frac{-3-\sqrt{41}}{4}$.
     (f) $x = \frac{-5+\sqrt{65}}{10}$ and $x = \frac{-5-\sqrt{65}}{10}$.
     (g) $x = \frac{-4+2\sqrt{7}}{3}$ and $x = \frac{-4-2\sqrt{7}}{3}$.
     (h) $x = -\frac{1}{3}$ and $x = 2$.

## Answers — Exercise 5.9

Q1.  (a) 17
     (b) $-19$
     (c) 12
     (d) $-20$
     (e) 24

     (f) 40
     (g) 48
     (h) $-7$
     (i) 0
     (j) 24

Q2.  (a) Two real and distinct roots.
     (b) Two real and distinct roots.
     (c) No real roots.
     (d) One real, repeated root.

     (e) Two real and distinct roots.
     (f) Two real and distinct roots.
     (g) No real roots.
     (h) One real, repeated root.

Q3. The quadratic equation has real and distinct roots for all $k$.

Q4. $p = \frac{9}{4}$.

Q5. $k > 0$ or $k < -1$.

## Answers — Exercise 5.10

Q1. This question can be checked with any graph plotting software.

Q2. This question can be checked with any graph plotting software.

## Answers — Exercise 5.11

Q1. $b = -9$.

Q2. $a = \pm 4$. $x^2 - 8x + 20$ or $x^2 + 8x + 20$.

Q3. The common difference is 2. $b = -6$.

Q4. The quadratics are $y = x^2 - 4x + 10$ and $y = x^2 + 4x + 6$.

Q5. $k = \frac{1}{4}$ and the minimal point is $\left(-\frac{1}{2}, \frac{1}{4}\right)$.

Q6. $x = -3$, $x = -4$, $x = 2$ or $x = 1$.

Q7. (a) 5 and 7.

(b) $25, -25, 14, -14, 11, -11, 10$ and $-10$.

(c) Depending on assumptions made there are either an infinite number of possibilities or the square can be $9, 16, 21, 24$ or $25$.

## Answers — Exercise 6.1

Q1. (a)

$$f(-1) = (-1)^3 - 4 \times (-1)^2 - (-1) + 4 = -1 - 4 + 1 + 4 = 0,$$
$$f(1) = (1)^3 - 4 \times (1)^2 - 1 + 4 = 1 - 4 - 1 + 4 = 0,$$
$$f(4) = (4)^3 - 4 \times (4)^2 - 4 + 4 = 64 - 64 - 4 + 4 = 0.$$

(b) Using this information we can sketch the graph.

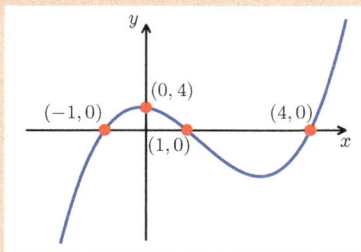

Q2. For large values of $x$, the expression is dominated by $-3x^5$. For large, negative values of $x$, $-3x^5$ will be a large positive number. For large, positive values of $x$, $-3x^5$ will be a large negative number.

Since the graph goes from positive to negative, and the polynomial expression is continuous, it must cross the $x-$axis at least once.

Q3. (a) $a = -3$, $b = 0$.

(b) We complete the table as follows.

| $x$ | -2 | -1 | 0 | 1 | 2 | 3 | 4 |
|---|---|---|---|---|---|---|---|
| $f(x)$ | -16 | 0 | 4 | 2 | 0 | 4 | 20 |

(c) From the table, we can sketch the following polynomial.

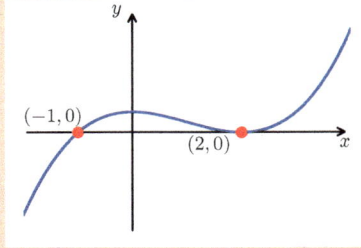

Q4. (a) For very large lawns, the cost of the feed will eventually be larger.
(b) Let $x$ be the width of the lawn. The total cost is $2x^2 + 12x$.
(c) The area of the lawn is $72\,\text{m}^2$.

## Answers — Exercise 6.2

Q1. (a) $x^3 + x^2 + 2x - 10$.
(b) $-2x^3 + 5ax^2 - 3x - 6$.
(c) $x^2 - 4$.
(d) $2x^3 - 6x - 4$.
(e) $-9x^3 + 2x^2 - 3x - 1$.
Q2. (a) $x^2 + (c - 3)x + (2 - 3c)$.
(b) $c = -1$, $f(0) = 5$.

## Answers — Exercise 6.3

Q1. (a) $(2x^2 + 2x - 12) \div (x + 3) = (2x - 4)$.
(b) $(x^3 + x^2 - 5x + 3) \div (x - 1) = (x^2 + 2x - 3)$.
(c) $(2x^4 - x^3 - 6x^2 + 8x + 12) \div (2x + 3) = (x^3 - 2x^2 + 4)$.
(d) $(-2x^4 + 7x^3 - 14x^2 + 19x - 6) \div (-x + 2) = (2x^3 - 3x^2 + 8x - 3)$.
Q2. $(x^3 + 11x^2 + 4x - 60) \div (x + 10) = (x^2 + x - 6)$.
$f(x) = (x + 10)(x + 3)(x - 2)$.
Q3. (a) The remainder is equal to 8.
(b) $f(x) = (x - 1)(6x^2 + x - 2) + 8$.
Q4. (a) $p = 5$.
(b) The remainder 30.
Q5. (a) $a = 6$, $b = 5$.
(b) The remainder is 140.
Q6. (a) $a = 9$.
(b) The roots of $4x^2 - 9x + 2$ are 2 and 1/4.
(c) $t = \frac{1}{2}$ and $t = -1$.
Q7. (a) $f(x) = x^2 + (3k - 2)x - 8k$.
(b)

$$f(x) = \left(x - 1 + \frac{3k}{2}\right)^2 - \left(\frac{9}{4}k^2 + 5k + 1\right).$$

(c) $g(x) = 9x^2 + 20x + 4$.

(d)

$$g(x) = 9x^2 + 20x + 4 = (x+2)(9x+2).$$

The repeated root of $f(x)$ is equal to 4.

## Answers — Exercise 7.2

Q1. (a) $S$ is a rhombus $\Leftarrow$ $S$ is a square.

(b) $X$ is a rhombus $\Rightarrow$ $X$ is a parallelogram.

(c) $n^2 > 8$ and $n > 0 \Leftrightarrow n \geq 3$.

(d) $x^2$ is rational $\Leftarrow$ $x$ is rational.

Q2. (a) Valid argument.

(b) Invalid arguemnt.

(c) Valid argument.

(d) Invalid argument.

(e) Invalid arguement.

## Answers — Exercise 7.3

Q4. (a) If $n^2$ is odd, then $n$ is odd.

(b) If $mn$ is odd, then $m$ and $n$ are odd.

(c) If $mn$ is even, then $m$ and $n$ are even.

(d) If $mn$ is even, then $m$ is odd and $n$ is even.

(e) If $7n + 4$ is even, then $n$ is even.

Q9. There are no integers $a$, $b$ and $c$ for which $4a + 3 = b^2 + c^2$.

## Answers — Exercise 8.1

Q1. (a) 120.

(b) 720.

(c) 40 320.

(d) 42.

Q2. (a) 10;

(b) 2;

(c) 4;

(d) 21;

(e) 126;

Q3. (a) 4.

(b) 1.

(c) 10.

(d) 20.

(e) 1.

(f) $n$.

(g) 1.

Q4. (a) $\binom{2}{0}$  $\binom{2}{1}$  $\binom{2}{2}$,

(b) $\binom{4}{0}$  $\binom{4}{1}$  $\binom{4}{2}$  $\binom{4}{3}$  $\binom{4}{4}$.

Q5. (a) 43949268;

(b) 10400600;

(c) 10737573;

(d) 108043253365600;

(e) 29462227291176635718126;

## Answers — Exercise 8.2

Q1. (a) $1 + 11x + 55x^2 + 330x^3 + \ldots$

(b) $1 - 18x + 144x^2 - 672x^3 + \ldots$

(c) $32 + 80x + 80x^2 + 40x^3 + \ldots$

(d) $\dfrac{1}{4} - \dfrac{3}{4}x + \dfrac{27}{16}x^2 - \dfrac{27}{8}x^3 + \ldots$

(e) $1 - \dfrac{1}{2}x - \dfrac{1}{8}x^2 - \dfrac{1}{16}x^3 - \ldots$

(f) $\dfrac{1}{5\sqrt{5}} + \dfrac{3}{25\sqrt{5}}x + \dfrac{3}{50\sqrt{5}}x^2 + \dfrac{7}{250\sqrt{5}}x^3 + \ldots$

Q2. (a) $1 - 12x + 54x^2 - 108x^3 + 81x^4$.

(b) $32 - 80x + 80x^2 - 40x^3 + 10x^4 - x^5$.

(c) $1 - x + \dfrac{3}{8}x^2 - \dfrac{1}{16}x^3 + \dfrac{1}{256}x^4$.

(d) $32 + \dfrac{80}{x} + \dfrac{80}{x^2} + \dfrac{40}{x^3} + \dfrac{10}{x^4} + \dfrac{1}{x^5}$.

(e) $x^3 + 3x^2y + 3xy^2 + y^3$.

(f) $32x^5 - 80x^4y + 80x^3y^2 - 40x^2y^3 + 10xy^4 - y^5$.

(g) $x^4 + 4x^2 + 6 + \dfrac{4}{x^2} + \dfrac{1}{x^4}$.

(h) $x^5 - 5x^3 + 10x - \dfrac{10}{x} + \dfrac{5}{x^3} - \dfrac{1}{x^5}$.

Q3. (a) $1 - 36x + 594x^2 - 5940x^3 + 40\,095x^4 + \ldots$

(b) $1 + \dfrac{21}{2}x + \dfrac{189}{4}x^2 + \dfrac{945}{8}x^3 + \dfrac{2835}{16}x^4 + \ldots$

(c) $x^{15} + 30x^{14}y + 420x^{13}y^2 + 3640x^{12}y^3 + 21\,840x^{11}y^4 + \ldots$

(d) $2048x^{11} - 22\,528x^9 + 112\,640x^7 - 337\,920x^5 + 675\,840x^3 + \ldots$

Q4. (a) $x^2 + 4x^3 + 6x^4$.

(b) $64x - 768x^2 + 3840x^3 - \ldots$

(c) $2 + 4x - 5x^2 + \ldots$

(d) $2 - 5x^2 - 2x^4 - \ldots$

(e) $x + 3x^2 + 6x^3 + \ldots$

(f) $x^2 - \dfrac{1}{2}x^3 + \dfrac{3}{8}x^4 - \ldots$

(g) $\dfrac{1}{2} - \dfrac{3}{4}x + \dfrac{9}{8}x^2 - \ldots$

(h) $\dfrac{1}{4} + \dfrac{19}{4}x + \dfrac{615}{16}x^2 + \dots$

Q5. (a) $-7$ (b) $432$ (c) $-\dfrac{2}{25}$ (d) $-48$

Q6. (a) $x^{10} + 10x^8 + 45x^6 + 120x^4 + 210x^2 + 252 + 210x^{-2} + 120x^{-4} + 45x^{-6} + 10x^{-8} + x^{-10}$.

(b) $x^{10} - 10x^7 + 45x^4 - 120x + 210x^{-2} - 252x^{-5} + 210x^{-8} - 120x^{-11} + 45x^{-14} - 10x^{-17} + x^{-20}$.

Q7. (a) $15,504$.

(b) $7920$.

(c) $-\dfrac{945}{256}$.

(d) $-\dfrac{5}{1024\sqrt{2}}$.

(e) $-\dfrac{35}{1024}$.

(f) $3402$.

(g) $80x^2$.

(h) $20y^3$.

Q8. (a) $1 - 2nx + 2n(n-1)x^2 - \dfrac{4}{3}n(n-1)(n-2)x^3 + \dfrac{2}{3}n(n-1)(n-2)(n-3)x^4 - \dots$

(b) $2^n + 2^{n-1}.3nx + 2^{n-3}.9n(n-1)x^2 + 2^{n-4}.9n(n-1)(n-2)x^3 + 2^{n-7}.27n(n-1)(n-2)(n-3)x^4 + \dots$

Q9. (a) $7 + 5\sqrt{2}$.

(b) $1393 - 744\sqrt{3}$.

(c) $292 + 112\sqrt{6}$.

(d) $\dfrac{89}{108}\sqrt{2} + \dfrac{17}{24}\sqrt{3}$.

(e) $153 + 28\sqrt{6} - 48\sqrt{3}$.

(f) $-2545 + 920\sqrt{5}$.

Q10. (a) $1 + 12x + 60x^2 + 160x^3 + \dots$

(b) $x = 0.01$, $(1.02)^6 \approx 1.1262$.

(c) $x = -0.02$, $(0.96)^6 \approx 0.7347$.

Q11. (a) $243 - 1620x + 4320x^2 - 5760x^3 + \dots$

(b) $x = 0.01$, $(2.96)^5 \approx 227.2378$.

(c) $x = -0.02$, $(3.08)^5 \approx 273.6259$.

Q12. $p = 5$.

Q13. $a = 3$.

Q14. $a = 96$, $b = 272$ and $c = 320$.

Q15. $4347.98592$.

## Answers — Exercise 9.1

Q1. Axis intercepts

(a) $(0,0)$

(b) $(0,0)$

(c) $(0,0)$

(d) $(0,0)$ and $(3,0)$

(e) $(-7,0)$, $(2,0)$ and $(0,-14)$

(f) $(-3,0)$ and $(9,0)$

(g) $(-4, 0)$, $\left(\frac{1}{2}, 0\right)$ and $(0, -4)$
(h) $(-2, 0)$, $(0, 0)$ and $(2, 0)$
(i) $(-2, 0)$, $(1, 0)$ and $(0, 4)$
(j) $(-3, 0)$, $(5, 0)$ and $(0, 75)$
(k) $(-3, 0)$, $(-2, 0)$, $(-1, 0)$ and $(0, 12)$
(l) $(0, 0)$, $(2, 0)$ and $(4, 0)$
(m) $(-1, 0)$, $(1, 0)$, $(2, 0)$ and $(0, 2)$

Q2. (a) Intercepts at $(0, -1)$, asymptotes at $x = 1$ and $y = 0$
(b) Intercepts at $(-1, 0)$, asymptotes at $x = 0$ and $y = 1$
(c) Intercepts at $(-2, 0)$ and $(0, 1)$, asymptotes at $x = 2$ and $y = -1$
(d) Intercepts at $(0, 1)$, asymptotes at $x = 4$ and $y = 1$

## Answers — Exercise 9.2

Q1. Rearranging we get,

$$0 = 3x^3 + x^2 - 2$$
$$2 = 3x^3 + x^2$$
$$\frac{2}{x^2} = 3x + 1.$$

Q2. Sketch showing two parallel tangents.

Q3. Possible equations,
(a) $y = (x - 3)(x - 5)$
(b) $y = (x - 2)(x - 6)$
(c) $y = (x - 1)(x - 5)$

Q4. Possible equations,
(a) $y = x^3$
(b) $y = x(x - 3)(x + 3)$
(c) $y = (x - 2)(x - 4)(x + 2)$

## Answers — Exercise 9.3

Q1. $y = \frac{15}{x}$
Q2. $\frac{2}{3}\pi$
Q3. The ratio between Celsius and Fahrenheit is not constant.
Q4. $2.7 \, \mathrm{g/cm^3}$
Q5. 911925

## Answers — Exercise 10.1

Q1. (a) The blue graph in the figure below.
(b) The green graph in the figure below.

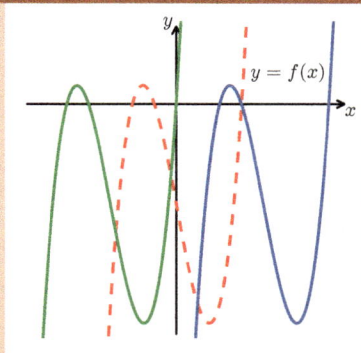

(c) The sketch is shown below.

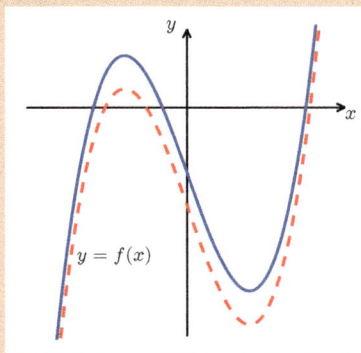

Q2. (a) $f(x) - 5$ is the blue graph in the figure below.

(b) $f(x) + 2$ is the green graph in the figure below.

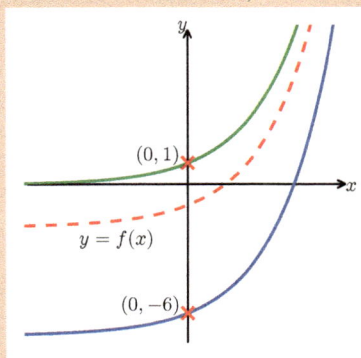

(c) $f(x - 3)$ is the blue graph in the figure below.

(d) $f(x + 4)$ is the green graph in the figure below.

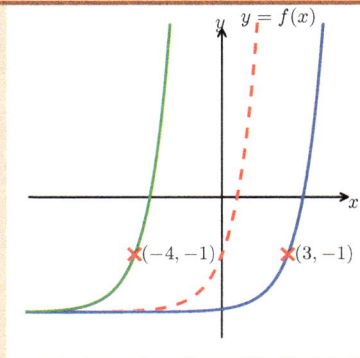

Q3. (a) or (d). The coordinates of the new point are $(1,1)$.
Q4.  (a) The sketch is shown below.

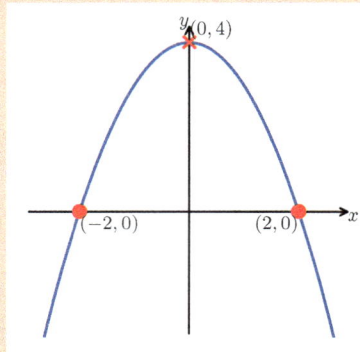

(b) $\begin{pmatrix} 0 \\ -3 \end{pmatrix}$
(c) $f(x-3) = -x^2$

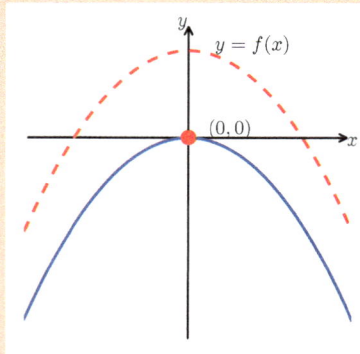

Q5.  (a) The sketch is shown below.

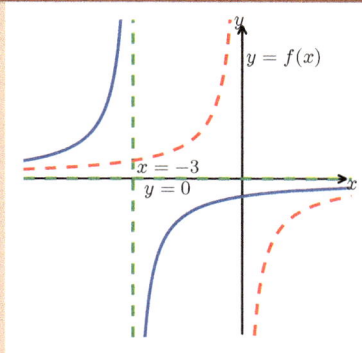

(b) The sketch is shown below.

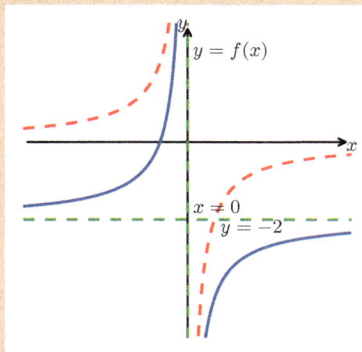

Q6. (a) The blue graph in the figure below.
(b) The green graph in the figure below.

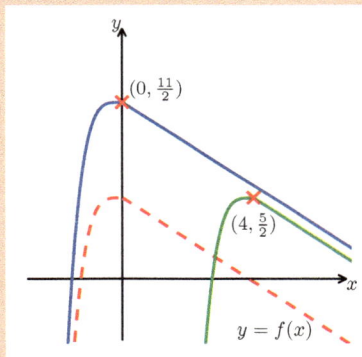

(c) The sketch is shown below.

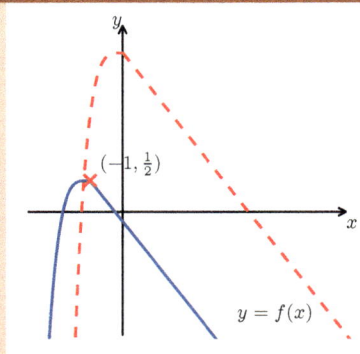

$f(x+1) - 2.$

**Answers — Exercise 10.2**

Q1.  (a) ii.
     (b) $y = -\frac{1}{3}g(x).$
Q2.  (a) The blue graph in the figure below.
     (b) The green graph in the figure below.

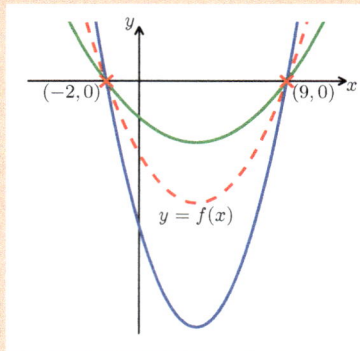

(c) The sketch is shown below.

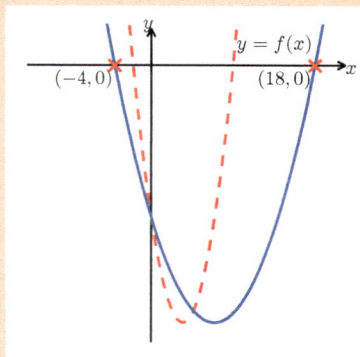

(d) The sketch is shown below.

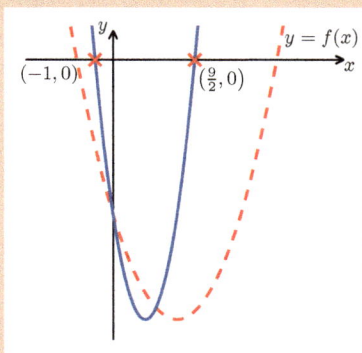

Q3. A stretch parallel to the $y$ axis by scale factor $-\frac{1}{2}$.

Q4. (a) The blue graph in the figure below.

(b) The green graph in the figure below.

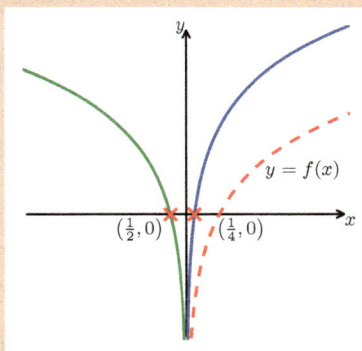

(c) The blue graph in the figure below.

(d) The green graph in the figure below.

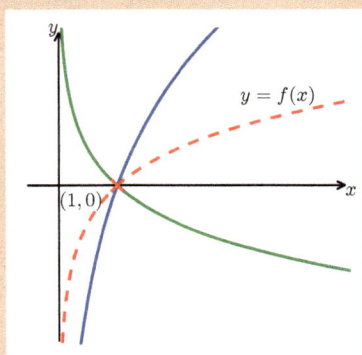

Q5. (a) $-2f(x) = -2(x-1)(x-2)(x-3)$

(b) The sketch is shown below.

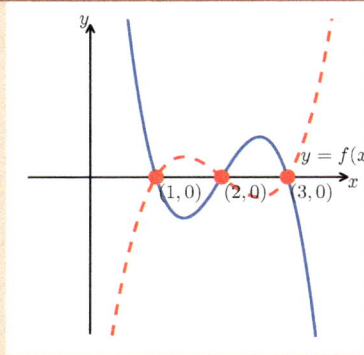

Q6.  (a)  The sketch is shown below.

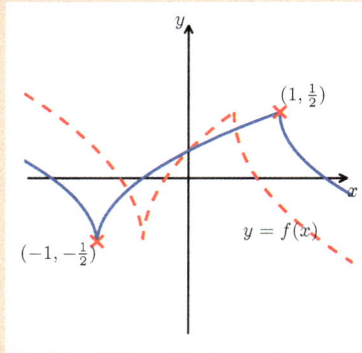

(b)  The sketch is shown below.

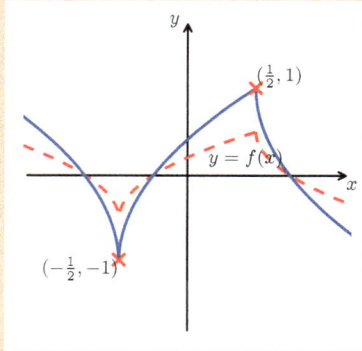

(c)  The sketch is shown below.

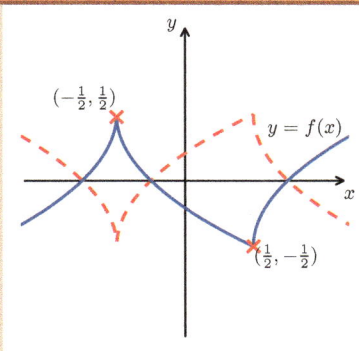

**Answers — Exercise 10.3**

Q1. (a)  i. $\frac{1}{4}f(x)$.
      ii. $f(x-4)$.
      iii. $-f(x)$.
      iv. $f(x)-3$.
      v. $f\left(-\frac{1}{2}x\right)$.

   (b)  i. Reflection in the $y$ axis.
      ii. Translation parallel to the $x$ axis by the column vector $\begin{pmatrix} 1 \\ 0 \end{pmatrix}$.
      iii. Translation parallel to the $y$ axis by the column vector $\begin{pmatrix} 0 \\ 2 \end{pmatrix}$.
      iv. Stretch parallel to the $y$ axis by scale factor $\frac{1}{2}$.
      v. Stretch parallel to the $x$ axis by scale factor $2$.

Q2. (a) $A$ transformed to the coordinates $\left(\frac{1}{2},0\right)$.
   (b) $A$ and $B$ transformed to coordinates $(1,0)$ and $(2,1)$ respectively.
   (c) $B$ transformed to the coordinates $(0,3)$.
   (d) $A$ and $B$ transformed to coordinates $(-1,1)$ and $(0,2)$ respectively.

Q3. (a) $y=f(x)$, $f(x)=3x-4$.

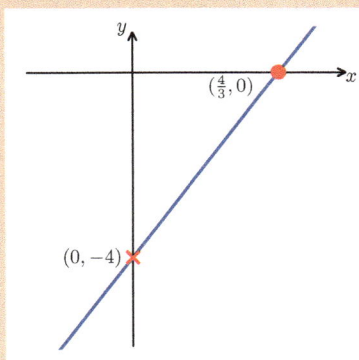

   (b) $y=-3x+4$.
   (c) $-f(x+2)$.

Q4. (a) $A$ transformed to coordinates $(6,-1)$.

    (b) $A$ transformed to coordinates $\left(3, -\frac{1}{3}\right)$.

    (c) $A$ transformed to coordinates $(0, -1)$.

Q5.  (a) $y = f(x + 4)$.

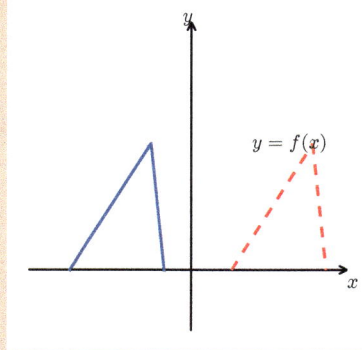

    (b) $y = 3f(x)$.

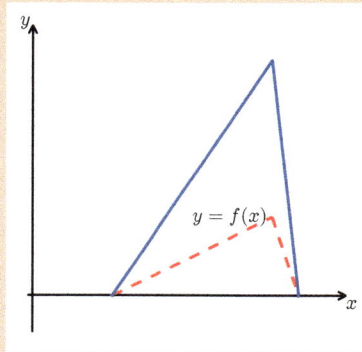

Q6.  (a) $3f(x)$.

    (b) $f(-x)$.

    (c) $f(x + 1) - 1$.

**Answers — Exercise 11.1**

Q1.  (a) $x = 3$, $y = 4$.

  (b) $x = \frac{-11}{4}$, $y = \frac{117}{16}$.

  (c) $x = \frac{12}{7}$, $y = 3$.

  (d) $t = 12$, $p = 5$.

Q2.  (a) $x = 5$, $y = 1$.

  (b) $x = 8$, $y = 4$.

  (c) $x = \frac{37}{8}$, $y = \frac{-9}{16}$.

  (d) $x = \frac{89}{7}$, $y = \frac{25}{7}$.

Q3.  (a) $(x, y) = (8, 2)$ and $(x, y) = (-8, -2)$.

  (b) $(x, y) = (10, 2)$ and $(x, y) = (-10, -2)$.

  (c) $(x, y) = \left(-2\sqrt{30}, \frac{-\sqrt{30}}{4}\right)$ and $\left(2\sqrt{30}, \frac{\sqrt{30}}{4}\right)$.

  (d) $(x, y) = \left(\sqrt{10}, \frac{\sqrt{10}}{4}\right)$ and $\left(-\sqrt{10}, -\frac{\sqrt{10}}{4}\right)$.

Q4.  (a) $(x, y) = (2, 3)$ and $(x, y) = (-2, 3)$.

  (b) $(x, y) = (5, 3)$ and $(x, y) = (5, -3)$.

  (c) $(x, y) = \left(\frac{8}{3}, \frac{3}{\sqrt{2}}\right)$ and $(x, y) = \left(\frac{-8}{3}, \frac{3}{\sqrt{2}}\right)$.

  (d) $(x, y) = (1, 2)$ and $(x, y) = (-1, -2)$.

Q5.  (a) $x = 4$, $y = 2$.

  (b) $p = 5$, $q = 3$.

  (c) $x = \frac{57}{14}$, $y = \frac{33}{14}$.

  (d) $x = \frac{183}{23}$, $y = \frac{100}{23}$.

Q6.  The cost of a cable is £7.50 and a calliper costs £22.50.

Q7.  (a) $(x, y) = \left(\frac{3}{\sqrt{2}}, \frac{3}{\sqrt{2}}\right)$ and $(x, y) = \left(-\frac{3}{\sqrt{2}}, -\frac{3}{\sqrt{2}}\right)$.

  (b) $(x, y) = \left(\frac{-3+\sqrt{39}}{5}, \frac{1+3\sqrt{39}}{5}\right)$ and $(x, y) = \left(\frac{-3-\sqrt{39}}{5}, \frac{1-3\sqrt{39}}{5}\right)$.

  (c) $(x, y) = \left(\frac{-1+\sqrt{17}}{2}, \frac{1+\sqrt{17}}{2}\right)$ and $(x, y) = \left(\frac{-1-\sqrt{17}}{2}, \frac{1-\sqrt{17}}{2}\right)$.

  (d) $(x, y) = \left(\frac{1+\sqrt{7}}{2}, \frac{-1+\sqrt{7}}{2}\right)$ and $(x, y) = \left(\frac{1-\sqrt{7}}{2}, \frac{-1-\sqrt{7}}{2}\right)$

Q8.  Side plates and dinner plates cost £7.10 and £14.20 respectively.

**Answers — Exercise 11.2**

Q1.  $x = 1$, $y = 2$.

Q2.  It appears that there is only one point of intersection. In fact there are two, namely, $(1, 8)$ and $(13/12, 53/6)$.

Q3.  There are three solutions: $(x, y) = (2, 2)$, $x, y = (\sqrt{3} - 1, -1 - \sqrt{3})$ and $x, y = (-1 - \sqrt{3}, \sqrt{3} - 1)$.

Q4.  $(l, w) = \left(4 + \sqrt{2}, 2 - \sqrt{2}\right)$ or $(l, w) = \left(2 - \sqrt{2}\right)$.

Q5.  $a = 4$ and $u = 2$.

Q6.  $\mu = \frac{3090}{23}$ and $\sigma = \frac{300}{23}$.

Q7.  $a \approx \mathrm{m\,s}^{-1}$, $T_1 \approx 69.2$N and $T_2 \approx 63.4$N.

Q8.  $A = \left(\frac{1}{3}, -\frac{4}{9}\right)$, $B = (-1, 4)$, $C = (5, -2)$, $D = (-1, 0)$, $E = (7, 4)$ and $F = (1, -2)$.

Q9.  $x = \frac{de - bf}{da - cb}$ and $y = \frac{fa - ce}{da - cb}$.

Q13.  $\frac{1}{x} + \frac{1}{y} = \frac{7}{2}$.

Q14. $A = \left(\frac{4}{3}\right)$, $B = \left(3, \frac{11}{2}\right)$, $C = \left(\sqrt{11} - 1, 7 - \sqrt{11}\right)$, $D = \left(\frac{11}{4}, \frac{13}{4}\right)$, $E = \left(7, \frac{15}{2}\right)$ and
$F = \left(1, \frac{3}{2}\right)$.

Q10. $d = \frac{29}{2000}v^2 + \frac{3}{8}v - \frac{13}{10}$.

## Answers — Exercise 12.1

Q1.  (a) $x \geq 5$;
     (b) $x > \frac{2}{3}$;
     (c) $x \geq 4$;
     (d) $x \leq 2$.

Q2.  (a) $x > \frac{20}{3}$;
     (b) $x > -1$;
     (c) $x \geq 2$;
     (d) $x > -14$.

Q3.  (a) $x > 8$;
     (b) $\leq 2$;
     (c) $x > \frac{1}{16}$;
     (d) $x > \frac{14}{9}$.

Q4.  (a) $0 < x \leq 24$;
     (b) $\frac{7}{3} \leq x < 7$;
     (c) $-\frac{13}{5} < x < \frac{1}{5}$.

## Answers — Exercise 12.2

Q1.  (a) $-5 \leq x \leq 5$;
     (b) $x < -3, x > 3$;
     (c) $-4 \leq x \leq 4$;
     (d) $-3 \leq x \leq 3$.

Q2.  (a) $-2 < x < 9$;
     (b) $-5 \leq x \leq 1$;
     (c) $x \leq -10, x \geq 1$;
     (d) $x < \frac{3}{4}, x > 1$.

## Answers — Exercise 12.3

Q1.  (a) $-1 \leq x < 2$;
     (b) $x \geq 2$;
     (c) $x > 6$.

Q2.  (a) $x < -1$;
     (b) $2 < x < 3$;
     (c) $0 < x < 2$.

**Answers — Exercise 12.4**

Q1.  (a) See the plot below.          (b) See the plot below.

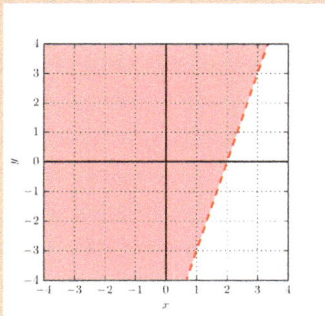
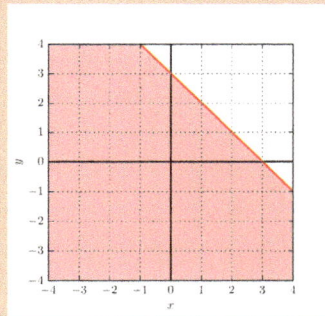

(c) See the plot below.

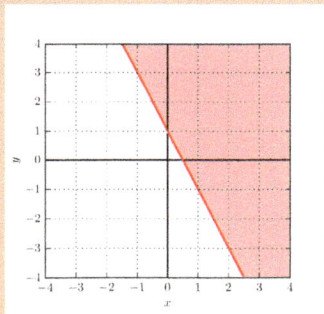

Q2.  (a) See the plot below.          (b) See the plot below.

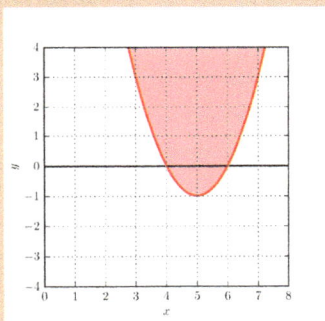
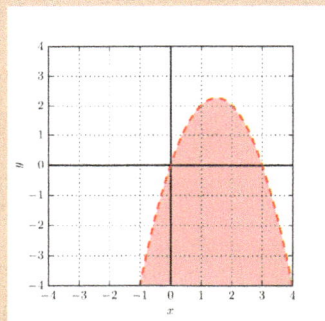

(c) See the plot below.

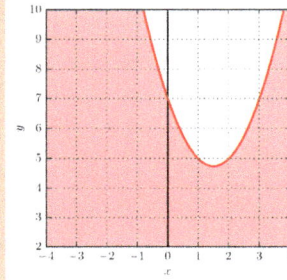

Q3.   (a) See the plot below.                (b) See the plot below.

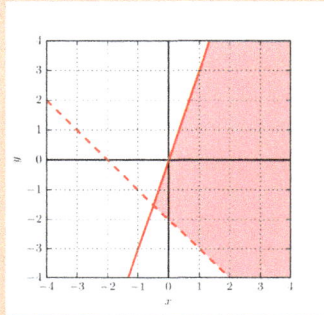                     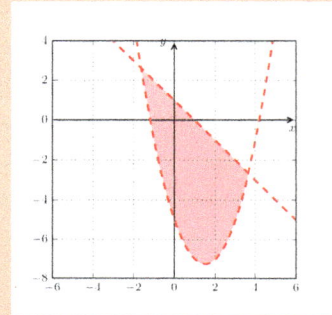

(c) See the plot below.

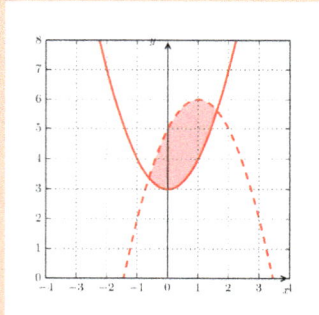

**Answers — Exercise 12.5**

Q1. Any reasonable suggestion.

Q2. (d).

Q3.   (a) $4x - 12 \leq 60 \longrightarrow x \leq 18$;

      (b) $x(x - 6) \geq 160 \longrightarrow x \geq 16$;

      (c) $16 \leq x \leq 18$.

**Answers — Exercise 13.1**

Q1. (a) $x^2 + y^2 = 4$;
(b) $x^2 + y^2 = 16$;
(c) $(x - 3)^2 + (y - 3)^2 = 1$;
(d) $(x + 1)^2 + (y - 2)^2 = 36$;
(e) $(x + 4)^2 + (y + 5)^2 = 9$;
(f) $(x - 6)^2 + (y + 4)^2 = 25$.

Q2. (a) Centre $(0,0)$, radius $2\sqrt{2}$;
(b) Centre $(0,0)$, radius 2;
(c) Centre $(0,0)$, radius $2\sqrt{3}$;
(d) This is not an equation of a circle;
(e) Centre $(4,2)$, radius 2;
(f) Centre $(3,3)$, radius 3;
(g) Centre $(-1,0)$, radius $\sqrt{5}$;
(h) Centre $(0,2)$, radius 6.

**Answers — Exercise 13.2**

Q1. (a) The centre is $(-4,0)$ and the radius is 6.
(b) The centre is $(0,2)$ and the radius is 4.
(c) The centre is $(-5,6)$ and the radius is 9.
(d) This is not an equation of a circle.
(e) The centre is $(1,-1)$, the radius is 1.
(f) This is not the equation of a circle.
(g) The centre is $(4,2)$, the radius is 2.
(h) The centre is $\left(\frac{4}{3}, \frac{2}{3}\right)$ and the radius is 2
(i) The centre is $(3,-2)$ and the radius is $\sqrt{20}$.
(j) This is not an equation of a circle.

Q2. (a) $\left(x - \frac{15}{2}\right)^2 + (y + 1)^2 = \frac{61}{4}$;
(b) $\left(x - \frac{27}{2}\right)^2 + (y - 3)^2 = \frac{29}{4}$;
(c) $(x + 3)^2 + (y + 1)^2 = 29$;
(d) $\left(x - \frac{11}{2}\right)^2 + \left(y - \frac{9}{2}\right)^2 = \frac{45}{2}$.

Q3. (a) $(x - 7)^2 + (y - 3)^2 = 16$;
(b) Inside;
(d) $s\sqrt{7}$.

Q4. No.

Q5. (a) $(5,4)$;
(c) $\sqrt{41} - 5$;
(d) 4.

**Answers — Exercise 13.3**

Q1. $a = \sqrt{r^2 - \frac{c^2}{4}}$

Q2. $c = 2\sqrt{r^2 - a^2}$

Q3. (a) $b = \frac{\sqrt{r^2 - 9r^2/4}}{2}$.

(b) $c = \frac{3r}{4}$.

Q4. (a) $\frac{2r}{\sqrt{3}}$.

(b) $\frac{\sqrt{7}r}{4}$.

Q5. (a) $8\,\text{cm}$;

(b) $6.4\,\text{cm}$.

Q6. (a) $k > -5$

(c) The equation represents a single point at $(2, -1)$

Q7. (a) $(x + 2)^2 + \left(y + \frac{5}{2}\right)^2 = \frac{205}{4}$;

Q8. (a) $r\sqrt{3}$;

(b) $2r$;

(c) $3r$.

---

## Answers — Exercise 14.1

Q1. Given in the order $\sin(A), \cos(A), \tan(A), \sin(B), \cos(B), \tan(B), AB, AC, BC$. The information given in the question is included in brackets.

(a) $\left[\frac{3}{5}\right], \frac{4}{5}, \frac{3}{4}, \frac{4}{5}, \frac{3}{5}, \frac{4}{3}, [5], 4, 3\text{cm}$.

(b) $\frac{4}{5}, \frac{3}{5}, \left[\frac{4}{3}\right], \frac{3}{5}, \frac{4}{5}, \frac{3}{4}, [0.1], 0.06, 0.08\text{m}$.

(c) $\frac{5}{13}, \frac{12}{13}, \frac{5}{12}, \frac{12}{13}, \left[\frac{5}{13}\right], \frac{12}{5}, 52, [48], 20\text{mm}$.

Q2. $x = 3$

Q3. $\cos(\theta) = \frac{\sqrt{5}}{3}, \tan(\theta) = \frac{2}{\sqrt{5}}$.

Q4. $\cos(\theta) = \frac{3}{\sqrt{10}}, \tan(\theta) = \frac{1}{\sqrt{10}}$.

Q5. (a) $\theta = 36.87°$;

(b) $\theta = 44.42°$;

(c) $\theta = 19.47°$;

(d) $\theta = 30°$.

Q6. $x = \frac{4}{\sqrt{3} - 1}$.

Q7. (a) $x = 1.40$;

(b) $x = 3.35$;

(c) $x = 9.81$;

(d) $x = 52.27$.

Q8. $24.5\,\text{m}$

---

## Answers — Exercise 14.2

Q4 $\theta = -300°, -240°, 60°, 120°, 420°, 480°$.

Q5 $\theta = -480°, -240°, -120°, 120°, 240°, 480°$.

Q6 $\theta = -300°, -120°, 60°, 240°$.

Q7 $\theta = -315°, 135°, 45°, 225°$.

Q8 (a) $y = \sqrt{3}$. $OP$ makes an angle of $150°$ with the positive $x$-axis

(b) $x = \frac{2}{\sqrt{3}}$. $OQ$ makes an angle of $30°$ with the positive-$x$ axis.

**Answers — Exercise 14.3**

Q1. (a) $7\sqrt{3}$ square units;
(b) 12.26 square units;
(c) 23.30 square units;
(d) 11.95 square units.

Q2. (a) Both methods give the area as $\frac{5\sqrt{11}}{2} \approx 8.29$ square units.
(b) There is a discussion to be had here.

Q3. $x = 8\,\text{cm}$.

Q4. $x \approx 6.7\,\text{cm}$.

Q5. $x = 7\,\text{cm}$.

Q6. $x \approx 51.06°$ or $x \approx 128.95°$.

**Answers — Exercise 14.4**

Q1. No.

Q2. (a) $|AB| \approx 6.53$;
(b) $|BC| =\approx 3.88$;
(c) $|CA| =\approx 266.95$;
(d) $|BC| =\approx 10.17$;
(e) $|AB| =\approx 6.09$.

Q3. (a) $\theta \approx 25.4°$;
(b) $\theta \approx 51.5°$;
(c) $\theta \approx 53.9°$;
(d) $\theta \approx 77°$;
(e) $\theta \approx 50.2\circ$;

Q4. (a) $65°$;
(b) 8 miles;
(c) 6.8 miles.

Q5. $8848\,\text{m}$.

Q6. $x = \frac{2}{3\sqrt{2}-2}$.

**Answers — Exercise 14.5**

Q1. (a) $11.31\,\text{cm}$
(b) $3.46\,\text{m}$
(c) $4.32\,\text{km}$

Q2. (a) $|BC| \approx 4.96$.
(b) Percentage error $= 0.84\%$.

Q3. $31.7\,\text{km}$.

Q4. $117\,\text{m}$.

Q5. (a) $92.9°$
(b) $25.2°$
(c) $48.5°$
(d) $106.6°$

Q6. 8.81 and 6.81.

Q8. The bearing is approximately $91.32°$.

**Answers — Exercise 15.1**

Q1.  (a) 6.5;

(b) 4.875;

(c) 3.5;

(d) 2.90625;

(e) 2.375;

(f) 1.90625;

(g) 1.655;

It is not yet clear that we have reached a limit, and so we cannot conclude anything about the value of the derivative.

Q2. $y' = 5x^4$.

4. (c) $m = 75$.

**Answers — Exercise 15.2**

Q1.  (a) $f'(x) = 6x^5$;

(b) $\frac{dy}{dx} = 8x^7$;

(c) $f'(x) = 9x^8$;

(d) $\frac{dy}{dx} = -4x^{-5}$;

(e) $f'(x) = -7x^{-8}$;

(f) $\frac{dy}{dx} = -x^{-2}$;

(g) $f'(x) = \frac{1}{2\sqrt{x}}$;

(h) $\frac{dy}{dx} = \frac{1}{5x^{\frac{4}{5}}}$;

(i) $f'(x) = \frac{3}{4}x^{-\frac{1}{4}}$;

(j) $\frac{dy}{dx} = 7x^6$;

(k) $f'(x) = 4x^3$;

(l) $\frac{dy}{dx} = \frac{9}{2}x^{\frac{7}{2}}$;

(m) $f'(x) = 3x^2$;

(n) $\frac{dy}{dx} = 6x^5$;

(o) $f'(x) = -x^{-2}$;

(p) $\frac{dy}{dx} = 6x^5$;

(q) $f'(x) = 9x^8$;

(r) $\frac{dy}{dx} = -6x^{-7}$;

(s) $f'(x) = -\frac{1}{4x^{\frac{5}{4}}}$;

(t) $\frac{dy}{dx} = \frac{5}{6x^{\frac{1}{6}}}$.

Q2.  (a) $\frac{dy}{dx} = 6$;

(b) $f'(x) = 21x^2$;

(c) $\frac{dy}{dx} = 28x^6$;

(d) $f'(x) = -12x^{-4}$;

(e) $\frac{dy}{dx} = -48x^{-5}$;

(f) $f'(x) = -35x^{-6}$;

(g) $\frac{dy}{dx} = 18x^{-4}$;

(h) $f'(x) = -2$;

(i) $\frac{dy}{dx} = 5x^{-2}$;

(j) $f'(x) = \frac{1}{8}x^{-\frac{3}{4}}$;

(k) $\frac{dy}{dx} = \frac{1}{2}x^{-3}$;

(l) $f'(x) = \frac{1}{10}x^{-\frac{9}{5}}$.

Q3. (a) 5120;

(b) $-9$;

(c) 1;

(d) $\frac{1310}{243}$.

Q4. (a) This function is increasing for all $x \in \mathbb{R}$.

(b) Increasing when $x \geq 1$, decreasing when $x \leq 1$;

(c) This function is increasing for all $x \in \mathbb{R}$;

(d) Increasing for $x \geq 0$, decreasing for $x \leq 0$.

Q5. (a) $\frac{dy}{dx} = 12x^3 + 6x^2 + 4$;

(b) $f'(x) = 12x^2 + 2x$;

(c) $\frac{dy}{dx} = 36x^5 - 20x^4 + 9x^2 + 8x + 1$;

(d) $f'(x) = \frac{3}{2}x^2 + \frac{4}{3}x - \frac{3}{4}$;

(e) $\frac{dy}{dx} = 3x^3 + \frac{4}{3}x - \frac{1}{5}$;

(f) $f'(x) = \frac{35}{3}x^4 + \frac{8}{3}x^3 + \frac{9}{4}x^2 - x + 2$;

(g) $\frac{dy}{dx} = \frac{3}{4}x^{-\frac{1}{2}} + 2x^{\frac{3}{2}}$;

(h) $f'(x) = \frac{28}{15}x^{\frac{4}{3}} + \frac{12}{25}x^{-\frac{1}{5}} - \frac{2}{9}x^{-\frac{1}{3}}$;

(i) $\frac{dy}{dx} = \frac{18}{25}x^{\frac{1}{6}} - \frac{1}{2}x^{-\frac{1}{4}} + \frac{1}{6}x^{-\frac{1}{2}}$;

(j) $f'(x) = -16x^{-5} + 6x^{-3} + 2x^{-4}$;

(k) $\frac{dy}{dx} = 3 + 2x^{-2} + 12x^{-3}$;

(l) $f'(x) = 8x + 3 + \frac{7}{3}x^{-2} - \frac{20}{7}x^{-3}$;

(m) $\frac{dy}{dx} = \frac{10}{12}x^{\frac{1}{4}} + \frac{4}{10}x^{-\frac{1}{2}} + \frac{1}{5x^{\frac{3}{2}}} + \frac{9}{16x^{\frac{7}{4}}}$;

(n) $f'(x) = \frac{8}{3}x - x^{-\frac{1}{2} - \frac{3}{2}x^{-\frac{3}{2}} - \frac{3}{4\sqrt{x}} + \frac{4}{9x^{\frac{4}{3}}}}$;

(o) $\frac{dy}{dx} = 7x^6 - \frac{10}{12}x^{\frac{1}{4}} + \frac{1}{3\sqrt{x}} + \frac{1}{x^{\frac{6}{5}}}$

Q6. (a) $\frac{dy}{dx} = 3x^2 - 6x + 2$;

(b) $f'(x) = 8(3x^2 + 3x - 2)$;

(c) $\frac{dy}{dx} = 6x^2(10x^2 + 3)$;

(d) $f'(x) = 12x - 5$;

(e) $\frac{dy}{dx} = 8x - 4$;

(f) $f'(x) = 24x + 19$;

(g) $\frac{dy}{dx} = 18x^2 + 4x + 15$;

(h) $f'(x) = 2(9x^2 - 15x + 5)$;

(i) $\frac{dy}{dx} = 6x(6x^2 - 1)$;

(j) $f'(x) = \frac{3}{x^2} + 10$;

(k) $\frac{dy}{dx} = \frac{13x - 12}{x^3}$;

(l) $f'(x) = -\frac{3}{x^4} - \frac{2}{x^2} + 16x$;

(m) $\frac{dy}{dx} = -\frac{4}{x^2}$;

(n) $f'(x) = 6x + \frac{7}{2}$;

(o) $\frac{dy}{dx} = \frac{9}{2}$;

(p) $f'(x) = -\frac{4}{x^2}$;

(q) $\frac{dy}{dx} = -\frac{2}{x^3}$;

(r) $f'(x) = 7x - \frac{3}{x^2}$;

(s) $\frac{dy}{dx} = \frac{5x^2+2}{\sqrt{x}}$;

(t) $f'(x) = \frac{2(10x^2-3x-1)}{3x^{\frac{4}{3}}}$;

(u) $\frac{dy}{dx} = \frac{21x^3+6x+4}{6\sqrt{x}}$;

(v) $f'(x) = \frac{3(4x^3-x-2)}{4x^{\frac{5}{2}}}$;

(w) $\frac{dy}{dx} = \frac{11x^7+16x^2+5x-2}{5x^{\frac{7}{5}}}$;

(x) $f'(x) = \frac{2(4x^3-3x-10)}{9x^{\frac{8}{3}}}$.

Q7. (a) $\frac{1}{2\sqrt{2}}$;

(b) $4 + \frac{1}{\sqrt{2}}$;

(c) 17.

## Answers — Exercise 16.1

Q1. (a) $\frac{1}{4}x^4 + c$

(b) $A = 4$

(c) $A \approx 6.25$

(d) 0.5625

Q2. The area calculated with each method is $A = 10$. The error of the rectangle method in this case is 0. This is because $f(x)$ is a straight line so each rectangle has the same area above $f(x)$ as below it.

## Answers — Exercise 16.2

Q1. (a) The indefinite integral is $x^4 + c$.

(b) The integrand is $4x^3$.

(c) The constant of integration is $c$.

(d) The integration variable is $x$.

Q2. (a)

$$\int 3x^2 \, dx = x^3 + c.$$

(b)

$$\int 6x^{3/2} \, dx = \frac{12}{5}x^{5/2} + c.$$

(c)

$$\int (5x+2)(x^2-3) \, dx = \frac{5}{4}x^4 + \frac{2}{3}x^3 - \frac{15}{2}x^2 - 6x + c.$$

(d)

$$F(x) = \frac{1}{4}(x^4 + 4x^2 - 1).$$

(e)

$$\int t^3 \, dt = \frac{1}{4}t^4 + c.$$

(f)

$$\int \frac{1}{x^3} \, dx = -\frac{1}{2x^2} + c.$$

(g)

$$\int x^4 \, dt = x^4(t + c).$$

(h)

$$F(x) = x^3 + \frac{1}{x} - 2x - \frac{1}{3}.$$

**Answers — Exercise 17.1**

Q1. The similarities of the graphs are:
  (a) Both graphs have values for all $x \in \mathbb{R}$.
  (b) Both graphs have a y-intercept at $(0,1)$.
  (c) Both graphs grow very large, very quickly.
  (d) There are no negative values for $f(x)$ or $g(x)$, and both graphs tend to zero
      as $x$ approaches $-\infty$
      The difference in the graphs is that for $g(x)$, the graph increases at a faster
      rate than $f(x)$.
Q2. The plot is shown below

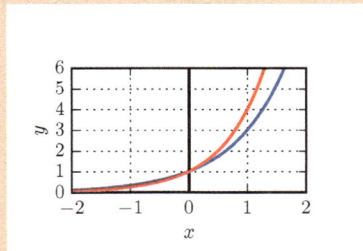

  (a) The red curve.
  (b) The blue curve.
Q3. Both the graphs for (a) and (c) will look like this:

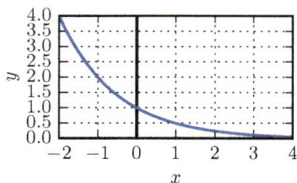

(b) This graph is a 'downhill' shape. It approaches zero as $x$ approaches $+\infty$.

(d) Both graphs are the same. This is because:

$$y = \frac{1}{2}^x$$
$$= (2^{-1})^x$$
$$= 2^{-x}$$

Q4. (a) The distinguishing features of an exponential graph are:
   - The graphs have a $y$-intercept at $(0,1)$
   - The domain is $x \in \mathbb{R}$
   - The range is $f(x) > 0$
   - The graph is increasing when $a > 1$ and decreasing when $0 < a < 1$
   - The graphs are asymptotic to the x-axis as $x$ approaches $-\infty$ when $a > 1$ and as $x$ approaches $\infty$ when $0 < a < 1$
   - The graphs are smooth and continuous.

   (b) $a > 0$ otherwise it could lead to calculations such as $-2^{\frac{1}{2}}$, which has no real answer. $a \neq 1$ because when $a = 1$, the answer will always be 1, and therefore the function will not be exponential.

---

**Answers — Exercise 17.2**

Q1. (b)

Q2. (d)

Q3. (d)

Q4. (a) £10,000

   (b) 2.4%

   (c) £10995.12

   (d) 30 years.

   (e) £6805.50

Q5. The growth factors between each piece of data are 1.20 to 3 s.f. and therefore the growth is approximately exponential.

   (a) 20% .

   (b) $V = 2700 \times 1.2^t$

   (c) £8629.73.

Q6. Any scenario where the initial value is 5.4 and the growth factor is 1.06. This could be also be any situation with a percentage increase of 6%.

Q7. Between 2033 and 2035.
Q8. The estimate of 2 million cells is not sensible.

**Answers — Exercise 17.3**
Q1. Answers below
    (a) 32
    (b) 0.32
    (c) 68%
Q2. Answers below
    (a) $I = 200 \times 0.7^t$
    (b) 248.02 mg.
    (c) 259.55 mg.
Q3. Answers below
    (a) £47334.40
    (b) See graph below

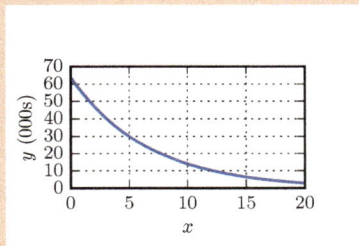

    (c) Between 4 and 5 years
Q4. (a) 30
    (b) 125 or 126
    (c) The whole population will become infected.

**Answers — Exercise 17.4**
Q1. Graphs shown below.
    (a) $y = e^{2x}$

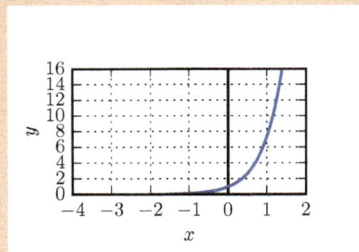

    (b) $y = e^{-3x}$

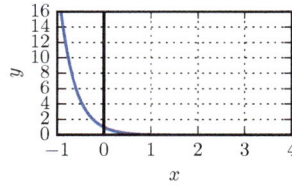

(c) $y = \frac{3}{2}e^{-2x}$

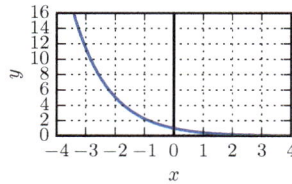

(d) $y = e^{\frac{x}{2}}$

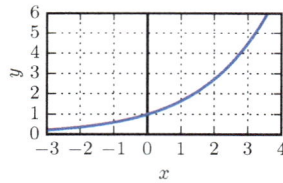

Q2. $V = 24\,000e^{-0.2t}$

    (a) £24 000

    (b) £13 171 (nearest £)

    (c) 4 years.

    (d) Graph drawn

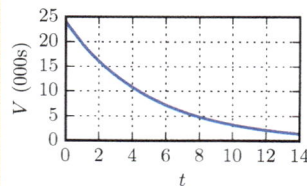

    (e) 7 years

Q3.  (a) 30

**Answers — 18.1**

Q1.  (a) 6;
     (b) $-1$;
     (c) $\frac{1}{2}$;
     (d) 0;
     (e) 1;
     (f) $-2$;
     (g) 4;
     (h) 2;
     (i) $-3$;
     (j) $-\frac{1}{3}$.

Q2.

  (a)

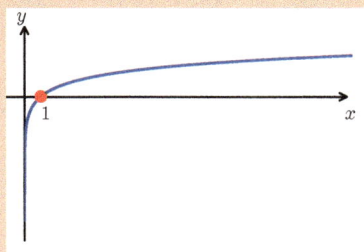

  (b)

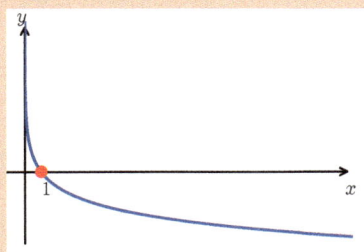

Q3.  (a) 10;
     (b) 3;
     (c) $16x$;
     (d) $12x$;
     (e) 3;
     (f) $\frac{1}{2}$.

**Answers — 18.2**

Q1.  (a) $\log_2(48)$;
     (b) $\log_3(14)$;
     (c) $\log_4(2)$;
     (d) $\ln(5e)$;
     (e) $\log_{10}(20)$;

(f) $\log_3(1) = \log_9(1) = 0$;

(g) $\log_4(216)$;

(h) $\log_3\left(\frac{1}{3}\right)$.

Q2. (a) $\frac{1}{\log_4(10)}$;

(b) $\log_2(\sqrt{13})$;

(c) $\log_{100}(x^2)$.

Q3. (a) $\log_4\left(3^5\right)$;

(b) $\log_3\left(7^{1/2}\right)$;

(c) $\ln\left(2 \times 8^4\right)$;

(d) $\log_{10}\left(\frac{1}{2000}\right)$.

Q4. (a) $\log_{10}((x+3)(x+2)) = \log_{10}(x^2 + 5x + 6)$;

(b) $\log_2(x+4)$;

(c) $\ln(x^2 + x + 1)$;

(d) $\log_{10}(x^3 y^4)$.

## Answers — 18.3

Q1. (a) $\pm 50^{1/4} = \pm 2.6591$ to 4 d.p.;

(b) $\frac{3}{7e^2} = 0.0580$ to 4 d.p.;

(c) $\sqrt{17} - 4 = 0.1231$ to 4 d.p.;

(d) $\sqrt[7]{\frac{1}{9}} = 0.7306$ to 4. d.p.

Q2. (a) $x = 4.9164$ to 4 d.p.;

(b) $x = -2.2619$ to 4 d.p.;

(c) $x = 1.4979$ to 4 d.p.;

(d) $x = \frac{2}{5}$;

(e) $x = 1\frac{1}{4}$;

(f) $x = 2.8226$ and $x = -0.8226$ to 4 d.p.;

(g) $x = -1$ and $x = 2$;

(h) $x = 1.6604$ to 4 d.p.;

(i) $x = 1.5579$ to 4 d.p.;

(j) $x = 2.7444$ and $x = 0.5775$ to 4 d.p.

## Answers — 18.4

Q1. The plots are shown below. Plots are identical because, if we take logs of both sides of $y = 100x^4$, we obtain $\log_{10}(y) = 4\log_{10} x + \log_{10}(100) = 4\log_{10} x + 2$.

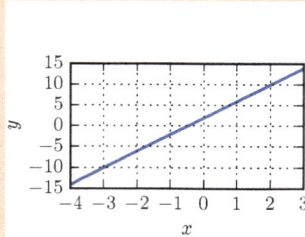

(a)

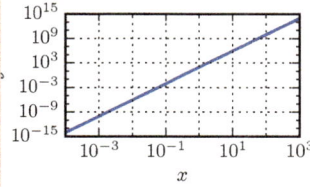

(b)

Q2.

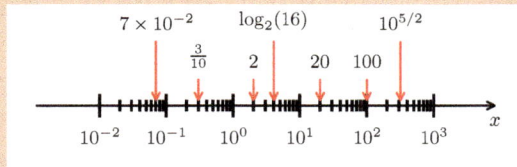

Q3.

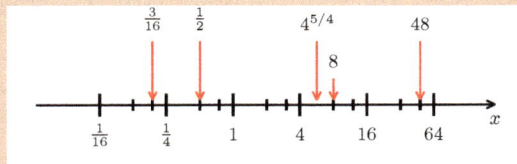

Q4. The plot is shown below and shows the model is reasonable with $y \approx 25x^{1.65}$.

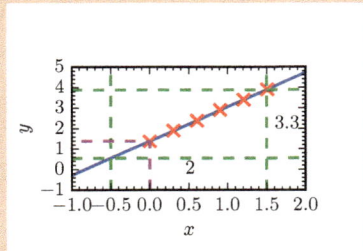

Q5. The plot is shown below. The data fits the law and the mass of the earth is approximately $6 \times 10^{24}$ kg.

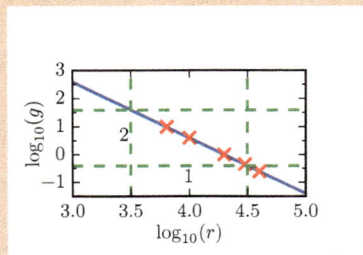

Q6. A possible model is $N = Ae^{kt}$, where $A$ and $k$. $A = 308$ and from the graph below, $k \approx 0.43$. The number of bacteria will double roughly every 1.6 hours.

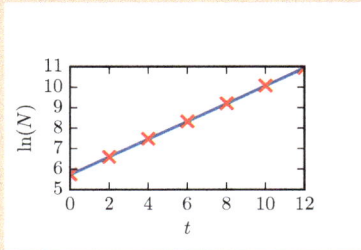

Q7. The plot is shown below. The model holds with $A \approx 65$ and $B \approx 150$.

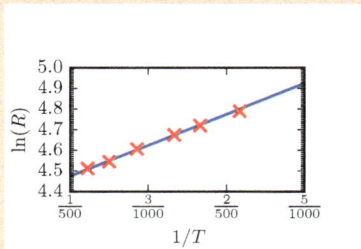

**Answers — Exercise 19.1**

Q1. (a) $\theta = 19.8°, 160.2°$.
   (b) $\theta = \pm 164.0°, \theta = \pm 16.0°$.
   (c) $\theta = 80°, \theta = -40°$.
   (d) $\theta = 280.3°, \theta = 460.3°$.
   (e) $\theta = 40.6°$.
   (f) $\theta = 14.3°, \theta = 75.7°$.

Q2. $a = 3, \theta = 9.38°, \theta = 93.96°$.

Q3. (a) $\theta \in \{-138.2°, -41.8°, 41.8°, 138.2°\}$.
   (b) $\theta = 19.5°, \theta = 160.5°$.
   (c) $\theta \in \{-161.6°, -63.44°, 18.4°, 116.57°\}$.
   (d) $\theta \in \{-180°, 0°, 180°\}$.
   (e) $\theta \in \{-138.2°, -41.8°, 30°, 150°\}$.
   (f) $\theta \in \{-120°, -109.47°, 109.47°, 120°\}$.
   (g) $\theta \in \{-148.3°, -125.8°, -58.3°, -35.8°, 31.7°, 54.2°, 121.7°, 144.2°\}$.

Q4. $\theta = -41.8°$.

Q5. $\theta \in \{9.2°, 99.2°, 189.2°, 279.2°\}$.

Q6. (a) Using the identity $\cos(\theta) = \sin(\theta + 90°)$,

$$\sin(45°) = \sin(-45° + 90°) = \cos(-45°) = \cos(45°).$$

Alternatively, since $\tan(45°) = 1$, we know that $\sin(45°) = \cos(45°)$. Therefore,

$$\sin^2(45°) + \cos^2(45°) = \sin^2(45°) + \sin^2(45°) = 2\sin^2(45°) = 1.$$

And therefore, $\sin(45°) = \frac{1}{\sqrt{2}}$.

(b) $\theta \in \{45°, 135°, 225°, 315°\}$.

---

**Answers — Exercise 19.2**

Q1. $1 - \sin^2(\theta)$.

Q2. $f(x) = \sin^2(x) - \sin^4(x)$.

Q3. $\theta = 60°$ and $\theta = 120°$.

Q4. $\theta = 45°$ and $\theta = 315°$.

Q5. $\theta \in \{35.26°, 215.26°, 395.26°\}$.

Q6. $f(x) = 3\cos(x) - 2\cos^3(x)$.

---

**Answers — Exercise 20.1**

Q1. (a) $-4x + y = 1$

(b) $2x + y = -3$

(c) $9x + 2y = -11$

(d) $-x + 2y = \sqrt{3} - \frac{\pi}{3}$

(e) $-3x + y = -4$

Q2. (a) $-x + 3y = -11$

(b) $-4x + 12y = -19$

(c) $4x + 9y = 34$

(d) $x + 2y = 2$

(e) $2x + 3y = -\frac{\pi}{6}$

Q3. $(-6, -24)$

Q4. $\frac{25}{4}$

Q5. $k = 12$

Q6. $x = -2$ and $x = 1$

Q7. $p = 4$

---

**Answers — Exercise 20.2**

Q1. (a) $x = -\frac{7}{4}$

(b) $x = -3$, $x = -\frac{1}{3}$

(c) $x = 2 \pm \sqrt{3}$

(d) $x = -\frac{2}{3}$, $x = \frac{5}{2}$

(e) $x = -2$, $x = 0$, $x = 1$

(f) $x = -1$, $x = 1$

(g) $x = \pm\frac{\sqrt{2}}{2}$

(h) $x = 1$
Q2. (a) $x = 3$.
(b) 8.

## Answers — Exercise 20.3

Q1. (a) $x = \frac{7}{4}$ is a *minimum*
(b) $x = -3$ is a *maximum* and $x = -\frac{1}{3}$ is a *minimum*
(c) $x = 2 - \sqrt{3}$ is a *maximum* and $x = 2 + \sqrt{3}$ is a *minimum*
(d) $x = -\frac{\sqrt{2}}{2}$ is a *maximum* and $x = \frac{\sqrt{2}}{2}$ is a *minimum*
(e) $x = 1$ is a *maximum*
Q2. (a) $x = -\frac{6}{5}$ is a *maximum*, $x = 0$ is an *inflection* and $x = 2$ is a *minimum*
(b) $x = \frac{1}{8}$ is a *minimum*
(c) $x = -\sqrt{3}$ is a *minimum* and $x = \sqrt{3}$ is a *maximum*
(d) $x = -\ln 2$ is a *minimum*
Q3. $-1 < x < 1$
Q4. $x < -1$, $0 < x < 1$
Q5. $914 \, \text{cm}^2$
Q6. $\frac{49}{32}$
Q7. $2000 \, \text{cm}^3$

## Answers — Exercise 21.1

Q1. c.
Q2. c.

Q3. (a) 15/2.  (c) −15/4.  (e) 77/10.
(b) −9.  (d) 119/12.

Q4. 1248.
Q5. The area between $y = \sqrt{x+1}$ and the $y$-axis is 4/3. The area to the $x$-axis is 14/3.
Q6. 135/4
Q7. £2.58.
Q8. 7/20.
Q9. 7.70 to 2 decimal places, or 45/64. Either decimals or fractions is acceptable as the question uses both.
Q10. The area under $f(x) = x\lfloor x \rfloor$ between $x = 0$ and $x = 4$ is 17.

$$\int_0^a f(x)\,\mathrm{d}x = \sum_{i=0}^{a-1} i^2 + \frac{1}{2}i.$$

**Answers — Exercise 22.1**

Q1. (a) The population is the set of all the cheeses which have reached the correct age. A sample is a randomly chosen, smaller subset of the cheeses.

 (b) The choice of cheeses is not random. He is always testing the cheeses at certain positions on the shelf.

 A better method would be to use stratified sampling, and choose three random cheeses from each shelf. This would not take significantly more time than his current method.

Q2. (a) The sample could be constructed using opportunity sampling. Since we cannot force healthy people to try the drug, the sample group can only be composed of those people willing to take part. The disadvantage is that the group is self-selecting, and so is unlikely to be particularly representative of the general population.

 (b) The first trials of a new drug are conducted on small sample groups because there is a small chance of very serious side effects being seen. Since the group is small and self-selected, we will have low confidence in the results being seen. If there is a 0.1% chance of a serious reaction, we are unlikely to see it during the trial. If the small trial does not show any significant problems with the drug, larger trials can be performed.

Q3. (a) The school could gather the names of all pupils taking the exam and list them in a random order. They could then take the sample to be the first twenty names on the list.

 (b) In simple random sampling, each choice should be independent. For the method described, after four pupils have been chosen from one class, no-one else in the class will be selected. Therefore, the choices are not independent.

Q4. (a) 16.95 pounds.

 (b) The values in the sample are fairly close together. This may be because the fish in the lake are similar sizes. In this case, our estimated mean could be reasonably accurate.

 (c) The sample is very small. A larger sample would increase the accuracy of the estimate.

 The sample has been constructed using opportunity sampling, with no attempt to measure a representative sample of the fish in the lake. This could be improved by capturing fish at different locations within the lake.

Q5. The researcher constructed two samples and obtained two different sample means. The different values may have come about by chance. Different samples will give different results. If the population variance is very high, then two sample means can have very different values.

Alternatively, the different values may have come about because he constructed the samples using opportunity sampling at two different times. It could be that one, or both, of the samples were not representative of the overall population.

He could improve his survey by improving his sampling technique. For example, he could find the proportion of people in different age categories in the overall population, and use quota sampling to ensure that his sample matches these proportions.

**Answers — Exercise 23.1**

Q1. The mode and the median are equal to 2. The mean is equal to 1.84 to two decimal places.

Q2. The bar chart is shown below.

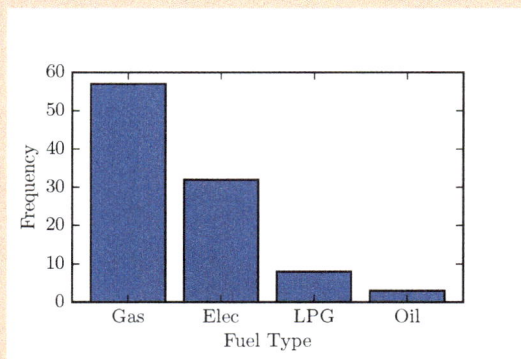

Q3. The bar chart is shown below.

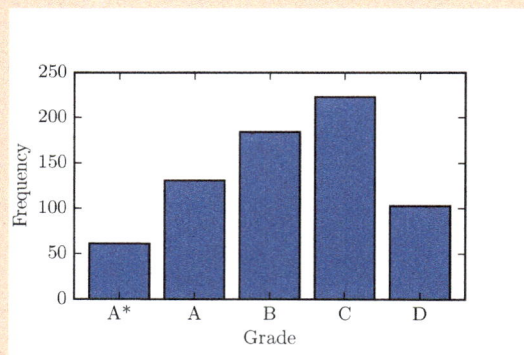

The median grade is B.

Q4. The bar chart is shown below.

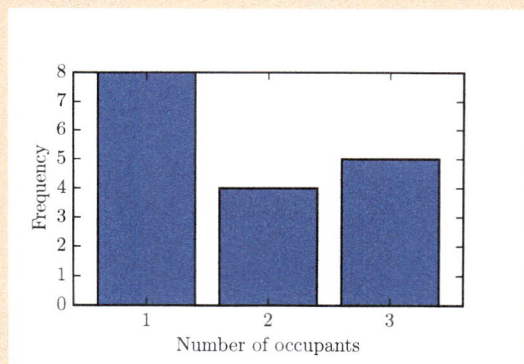

Q5.  A  The data has a mode at 3. The chart shows an outlier at value 9.
     B  The data has a mode at 6. This data has larger variation than Chart A. There are no outliers.
     C  The data does not have a clear modal value. It is roughly uniformly distributed over the range.
     D  The data is bimodal, with modes at 2 and 10.

## Answers — Exercise 23.2

Q1.  The bar chart data is shown in the following dot plot.

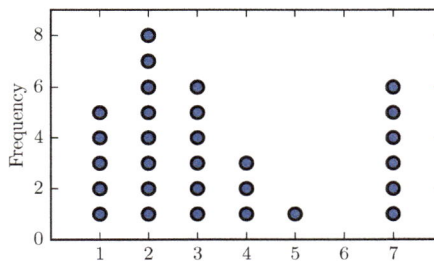

Q2.  (a)  Here are the results shown in a dot plot

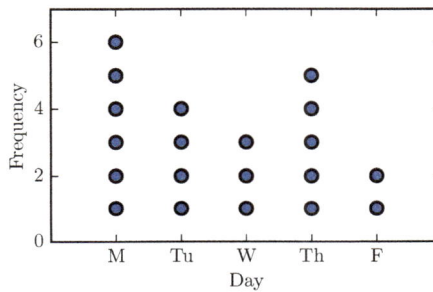

     (b)  4 people.
     (c)  Tuesday.

Q3.  (a)  In the following dot plot, the boys' results are shown as blue dots, and the girls' results in red.

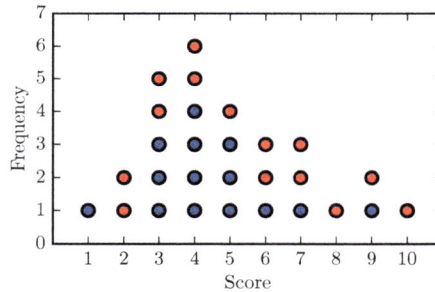

(b) 4.

(c) 5.5.

Q4. (a) The company's performance is shown in the following dot plot.

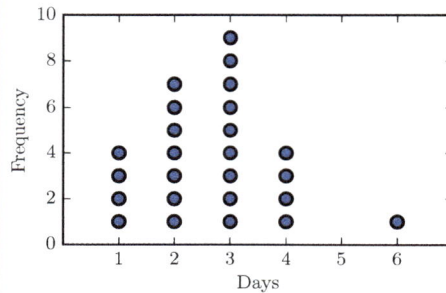

(b) 2.68 days.

(c) Of the 25 parcels in the sample, 20 were delivered within the target time. That is, 80% were delivered on time.

(d) A measurement of the company's performance is an attempt to measure customer satisfaction. The mean delivery time is not a good way to do this. If the sample is representative of the overall service, then one parcel out of every five is not delivered within the agreed time. This means that 20% of customers are likely to be unhappy with the service.

Parcels which are delivered early, lower the value of the mean. But those customers whose parcels are delivered late will not be reassured by being told that many other parcels were delivered early.

A better measurement would be to track the percentage of parcels delivered within the target time.

## Answers — Exercise 23.3

Q1. (a) Rounding to the nearest £1 000, we obtain the following plot, where 1|4 = £14 000.

```
1 | 4  5
2 | 0  3  5  7  7  7  8
3 | 0  0  0  1  2  3  5  5  6  7  7  8
4 | 1  1  4  4  6  8  8
5 | 2  8
6 |
7 |
8 | 5
9 | 5
```

(b)  £35 000

Q2. Weights of newborn babies, where 3.0|6 represents 3.06 kg

```
3.0 | 6  8
3.1 | 4  6  7  9
3.2 | 0  0  1  3  6  8
3.3 | 2  8  8
3.4 | 4  9
3.5 |
3.6 | 0
```

Q3. In this stem-and-leaf plot, 6|3 represents 63 $\mu g\, m^{-3}$.

```
 6 | 3
 7 |
 8 | 2  4
 9 | 1  1  8
10 | 0  3  4  8  9
11 | 2  3
12 | 4
```

The mean is 98.7 to one decimal place and the median is 101.5.

Q4. Results of men's 100 m semi-finals, where 9.8|6 represents 9.86 s.

```
 9.8 | 6
 9.9 | 2  4  5  7  8
10.0 | 1  1  1  3  5  5  7  8  8
10.1 | 1  2  3  3  6  7
10.2 | 3
```

The median is equal to 10.05. The median is equal to 10.05 to two decimal places.

Q5. Results of mental agility test, where 2|9|1 represents a score of 92 in the placebo group and 91 in the tonic group.

| Placebo group | | | | Tonic group | | | | | |
|---|---|---|---|---|---|---|---|---|---|
| 6 4 | 8 | | | | | | | |
| 6 1 0 | 9 | 1 | 4 | | | | | |
| 8 4 3 | 10 | 2 | 2 | 8 | | | | |
| 8 8 4 3 2 | 11 | 0 | 4 | 5 | 9 | | | |
| 7 4 2 | 12 | 2 | 2 | 3 | 4 | 7 | 8 | |
| 2 0 | 13 | 0 | 3 | 8 | 9 | | | |
| 7 3 | 14 | 0 | 1 | | | | | |
| 4 | 15 | | | | | | | |

Median of placebo group: 114
Median of tonic group: 122

**Answers — Exercise 23.4**

Q1. The histogram is shown below.

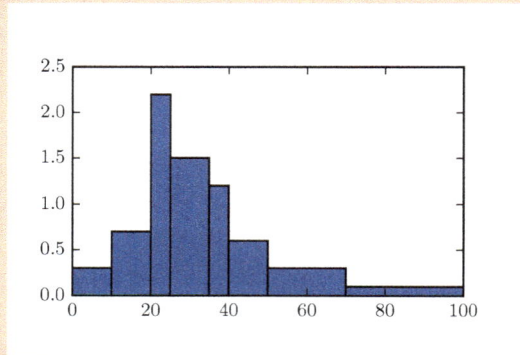

Q2.   (a) 14 cm
     (b) 24
     (c) 1 cm
     (d) 2 cm

Q3. The histogram is shown below.

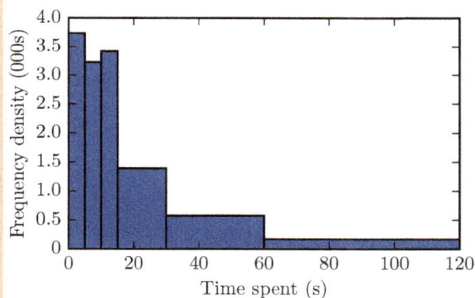

Q4. We assume that the greatest deviation from the standard weight, both above and below, is 0.567 grains.

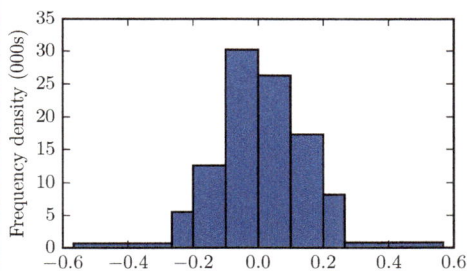

**Answers — Exercise 23.5**

Q1. The cumulative frequency values for the graph are shown below.

| Range | Cumulative Frequency A | Cumulative Frequency B |
|---|---|---|
| $0 \leq x < 25$ | 80 | 2435 |
| $25 \leq x < 50$ | 214 | 3789 |
| $50 \leq x < 75$ | 400 | 4632 |
| $75 \leq x < 100$ | 897 | 4876 |
| $100 \leq x < 125$ | 2431 | 5234 |
| $125 \leq x < 150$ | 6406 | 6592 |
| $150 \leq x < 175$ | 9271 | 8971 |
| $175 \leq x < 200$ | 10 000 | 10 000 |

This gives the following cumulative frequency graphs.

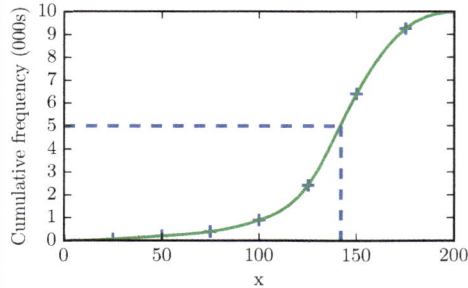

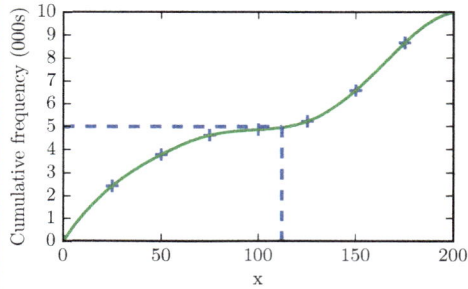

The estimated medians are 142 for data set A and 112 for data set B.
We expect the estimate for data set A to be more accurate, because the data points are denser around the median compared with data set B. A small change in the drawing of the graph will have less effect on the estimated median for data set B.

Q2. 290

Q3. 6.01 to two decimal places.

Q4. 123.2795 grains to four decimal places.

**Answers — Exercise 23.6**

Q1. (a) The quartiles for each class are given by:

|         | $Q_1$ | $Q_2$ | $Q_3$ |
|---------|-------|-------|-------|
| Class A | 51    | 64.5  | 75    |
| Class B | 46    | 55    | 82    |
| Class C | 60    | 76    | 85    |

(b) Class C has the highest values for all three of its quartiles, so we would say that overall it was the highest performing class.

Q2. The quartiles for each group of plants, in kg, are given by:

|  | $Q_1$ | $Q_2$ | $Q_3$ |
|---|---|---|---|
| Without fertiliser | 1.2 | 1.4 | 1.7 |
| With fertiliser | 1.0 | 1.2 | 1.6 |

The gardener saw mixed results from his tests with the fertiliser.
All three quartiles are lower for the fertilised plants, indicating that most plants had lower yields.
The results suggest that the fertiliser had a positive effect at the extremes: the entire range is higher when fertiliser is added. The highest yielding plants yield more, and the lowest yielding plants also do better, when the fertiliser is used.

Q3. The quartiles for washers created by the old and new machines, in mm, are given by:

|  | $Q_1$ | $Q_2$ | $Q_3$ |
|---|---|---|---|
| Old | 14.84 | 14.92 | 15.11 |
| New | 14.73 | 15.03 | 15.15 |

The variation, as measured by the interquartile range, of washers produced by the new machine is much larger than the old machine. For this reason, these results do not suggest that the new machine should be purchased.

Q4. (a) Our estimates of the quartiles, to one decimal place, are

$$Q_1 = 61.0, \qquad Q_2 = 69.3, \qquad Q_3 = 83.6.$$

(b) The estimate of the skew coefficient is 0.27 to two decimal places. The skew is positive.

**Answers — Exercise 23.7**

Q1. The mean is equal to 1371 and the standard deviation is 647.7 to one decimal place.

Q2. (a) We estimate the mean to be £33.80 an hour and the standard deviation to be £37.36 an hour.

(b) The mean is not a good measure to use in this case. The right hand tail of the data is very long and for highly asymmetric data sets, the mean is not a good description of the hourly rate. In this case, we can see that almost one third of the employees earn less than a third of the mean rate.

Q3. (a) For Nixon's speech, the mean and the standard deviation are:

$$\mu_N = 4.26 \text{ letters},$$
$$\sigma_N = 2.27 \text{ letters},$$

both to two decimal places. For Trump's speech, the values are:

$$\mu_T = 4.59 \text{ letters},$$
$$\sigma_T = 2.32 \text{ letters},$$

both to two decimal places.
(b) For Lincoln's speech, the mean and the standard deviation are:

$$\mu_L = 4.676 \text{ letters},$$
$$\sigma_L = 2.77 \text{ letters},$$

both to two decimal places.
(c) These results do not suggest that there is an overall trend to longer or shorter words in the speeches. The difference between the oldest and most recent speeches is less than 0.9 of a letter.

Q4. We estimate $\mu = 35.87$mph and $\sigma = 9.70$mph to two decimal places.

Q5. We calculate $a = 2$ and $b = 35$ to find the estimates

$$\mu_X = 40.03 \text{ GW},$$
$$\sigma_X = 7.20 \text{ GW},$$

to two decimal places.

## Answers — Exercise 23.8

Q1. (a) The median is equal to 86 and the mean is equal to 72.35.
(b) Data has a negative skew if its left hand tail is either longer or fatter than its right hand tail. We can classify a data set as having negative skew if its mean is less than its median. Alternatively, we can see the skew by plotting a histogram or box plot.
(c) Many pupils did very well in this test, with a majority scoring over 85%. This means that there was no possibility of having a long right hand tail, since the maximum possible score is 100%. The test is failing to differentiate between the most able students.

Q2. (a) To three decimal places,

$$\mu = 4.314,$$
$$\sigma = 0.863.$$

(b) To three decimal places, the skew coefficient is equal to 0.363.

Q3. (a) Our estimate of the median is 7.32 years to two decimal places.
(b) We estimate the skew to be $-0.33$ to two decimal places.

## Answers — Exercise 24.1

Q1. (a) Positive correlation. The test score is the response variable.
(b) No correlation.
(c) Negative correlation. The amount of sick leave taken is the response variable.

Q2. (a) The data has a weak positive correlation.

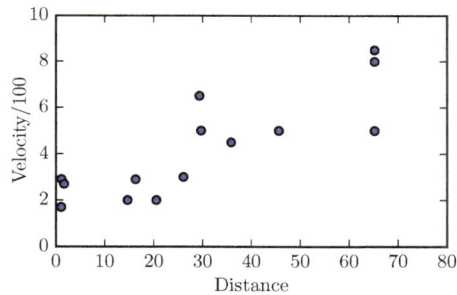

(b) The correlation is weaker than we would expect, which suggests that the figures that Hubble used were not very accurate.

(c) We would predict that a galaxy at a distance of 50 million light years would be travelling away from the earth at around $600\,\mathrm{km\,s^{-1}}$.

**Answers — Exercise 24.2**

Q1. (a) The scatter graph indicates a strong positive correlation between the length and the weight of the frogs.

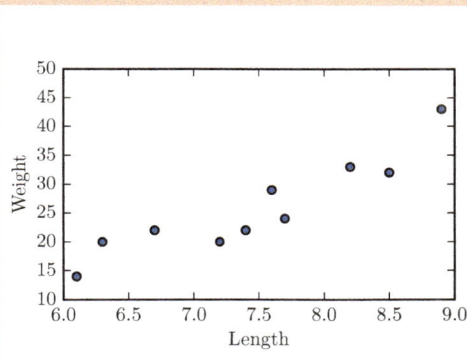

(b) We estimate a length of 7.9 cm to one decimal place.

(c) The scientist's estimate should be treated with caution, as it lies well outside the range of observed values.

Q2. (a) The price is the response variable.

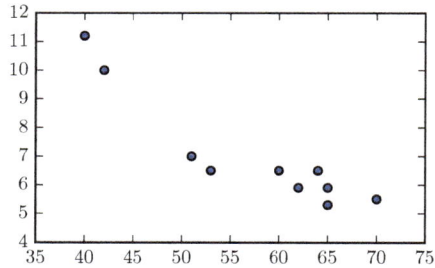

The graph shows a strong negative correlation.
(b) The covariance is equal to $\sigma_{AP} = -16.69$.
(c) The gradient of the line is given by $-0.1767$. This value indicates that the cars were losing around £177 in value each month.

Q3.  (a) The scatter graph suggests that there is no significant correlation between a film's title length and its box office takings.

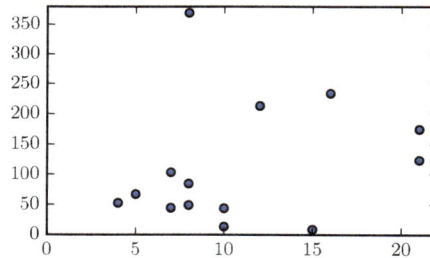

(b) The covariance is given by $\sigma_{lb} = 198.0$.
(c) The correlation coefficient is given by $\rho = 0.27$ to two decimal places. This is a low value for positively correlated data, much closer to zero than one, and so this confirms our observation that there is no significant correlation.

**Answers — Exercise 25.1**

Q1.  (a) The following graph shows the relationship between oil reserves (on the $x$-axis) and oil production (on the $y$-axis) for selected countries.

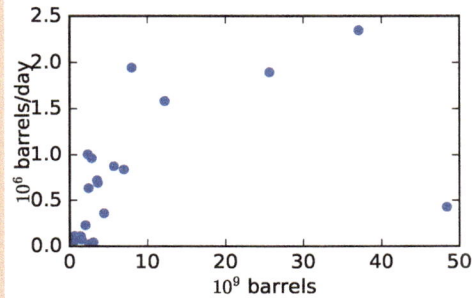

(b) There is a reasonably strong positive correlation between a country's oil reserves and oil production.

(c) The oil production from Libya in 2015 is significantly lower than would be expected. Possible reasons include:
- War, terrorism or other political instability.
- Sanctions imposed by other countries.
- Misreporting of reserves.

(d) The scatter diagram would suggest that the production in Oman was around 800 thousand barrels per day.

(e) Venezuela's production rate was only slightly higher than Nigeria's. However, its reserves are almost ten times the size of Nigeria's. If the correlation shown in the scatter graph were to continue, we would expect Venezuela's production rate to be much higher.

Q2. (a) $8.446 \times 10^6$ barrels per day.

(b) $3^{10}$ barrels.

(c) The median would be a better choice than the mean.

There is a sudden drop in Mexico's stated reserves, giving two outlier values in the data set. For asymmetric distributions like this, the outlying values tend to have an unreasonably large effect on the mean value. The median will be a more representative value.

(d) Possible reasons for such a sudden drop in the reported reserves include:
- A sharp drop in the price of oil. This could reduce the amount of oil which could be extracted profitably.
- A revision to incorrect or over-optimistic estimates.

Q3. (a) The blue line represents the production from the United Kingdom, and the red line represents Kuwait.
- Kuwait dropped its oil production significantly, by around 60%, during the 1980s. Reducing the supply would tend to push up the price.
- Following the Piper Alpha disaster in 1988, the UK's oil production fell by around half a million barrels of oil a day. It took about four years for the output to recover.
- The Iraqi invasion of Kuwait in 1990 stopped almost all oil production in the country. However, production recovered within a year or two.

(b) The oil production from the United Kingdom fell sharply after the peak around the year 2000. This could be due to falling oil prices, but we know that there have been periods of high oil prices since then. The more likely explanation is that production costs have risen. The most likely reason for rising costs, is that the remaining oil is much harder to extract.

(c) The graph shows that Kuwait's mean rate of production between 2004 and 2015 was over 2.5 million barrels per day. This means that the country produced over ten thousand million barrels in this time, or 10% of the stated reserves. It is not likely that additional reserves which were exactly equal to the oil produced were identified each year. Therefore, the stated values of the oil reserves are unlikely to be reliable.

Q4. (a) We must choose five points at random with replacement. For example, we might select $\{1991, 1999, 2002, 2003, 2003\}$. This would give us a sample mean of $1.9775 \times 10^6$ barrels per day.

(b) When we calculate the mean of five points selected at random, this sample mean will not be the same as the mean of the entire population. It could lie anywhere between the two extreme points, $1.8642 \times 10^6$ and $2.0302 \times 10^6$ barrels per day.

(c) To obtain a systematic sample, we can select a random starting point and then select every third point along. For example, if we start from the year 1994, we would have the data points $\{1864.2, 1936.3, 1994.1, 1951.5, 1894.6\}$. This gives a sample mean of $1.9281 \times 10^6$ barrels per day.

(d) The median value is 5.59%.

(e) Assuming that the consumption increases at 5.59% per year, we estimate that the consumption in 2015 will be $4.4489 \times 10^6$ barrels per day.

(f) Therefore, at this rate of increase, India's oil consumption will double every twelve to thirteen years.

Therefore, if India were to continue growing at this rate, by 2069 it would require the equivalent of the entire global output of 2005.

Q5. (a) This histogram is one way of presenting the data.

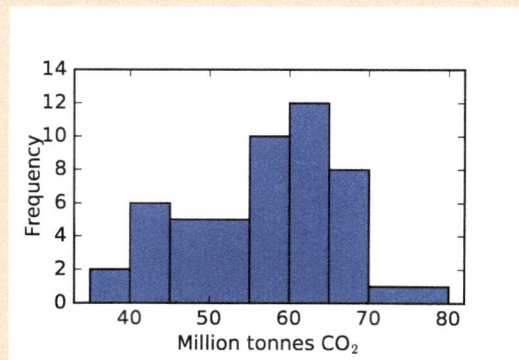

(b) We will identify outliers using the rule that the whiskers must not be more than 1.5 times the length of box.

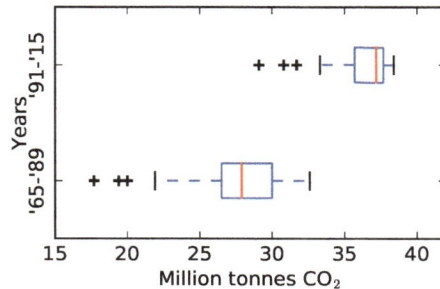

(c) Although both countries are oil producers, Norway produces around ten times as much as Denmark. Oil production was significantly higher oil in the years 1991 to 2015 compared with the years 1965 to 1989.

(d) The box plots show that Denmark's $CO_2$ emissions have dropped slightly, while Norway's have clearly increased. Therefore, the graphs tend to support the statement that oil production is positively correlated with carbon dioxide emissions. If we assume that carbon dioxide emissions are positively correlated with oil use, then the graph supports the statement.

(e) These graphs only look at two countries, and this sample size is not enough to draw a general conclusion. The difference we see could be due to other reasons, such as different government policies on emission reductions. Even if there is a link in this case, it does not necessarily hold in general.

Q6. (a) The data gives us the following graph.

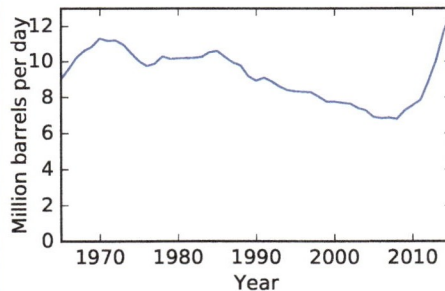

The graph is bimodal. Production reaches a (local) maximum output in 1970, which is followed by a gentle drop in production. In 2009, the output began to increase dramatically until production exceeded the previous maximum value.

(b) The most likely reason for the sudden increase in production is that the price of oil has increased. This can encourage investment to extract oil in areas which would not give a profit at lower prices.

Q7. (a) The United States was the largest emitter from 1965 to 2005. After 2006, China was the largest emitter. We can check the data directly to find this, or we can use a spreadsheet to find the answer automatically.

(b) We can plot the emission data for China against the size of its economy to obtain the following graph.

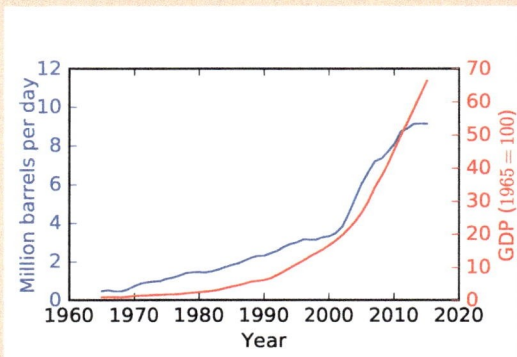

This graph suggests that the Chinese government has succeeded in curbing carbon emissions while the economy continued to grow.

## Answers — Exercise 26.1

Q1. (a) Travelling at $0.9\,\mathrm{m\,s^{-1}}$, it will take $10\,\mathrm{s}$ to cover $9\,\mathrm{m}$.

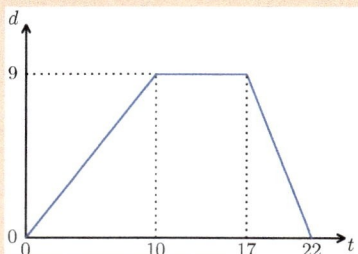

(b) $1.8\,\mathrm{m\,s^{-1}}$.

Q2. (a) $1.5\,\mathrm{m\,s^{-1}}$.

(b) $2\,\mathrm{m\,s^{-1}}$.

(c) The object is not moving.

(d) $-1.5\,\mathrm{m\,s^{-1}}$. The object is moving back towards point $A$ with a speed of $1.5\,\mathrm{m\,s^{-1}}$.

Q3. (a) At time $0\,\mathrm{s}$.

(b) At time $t = t_0$.

(c) The car hits the obstacle at time $t = t_1$. It decelerates quickly and bounces off the obstacle. It travels a short way backwards and comes to rest.

## Answers — Exercise 26.2

Q1. Acceleration is $-1\frac{5}{9}\,\mathrm{ms^{-2}}$. Speed is $3\frac{1}{3}\,\mathrm{ms^{-2}}$.

Q2. $\sqrt{85} + 5$ s.

Q3. 1200 m.

Q4. 4592 m to 4 s.f.

Q5. Acceleration is $4.761\,\text{ms}^{-2}$ to 3 d.p. Time to travel next 200 m is $2.507\,\text{s}$ to 3 d.p.

Q6. The initial speed is $20\,\text{ms}^{-1}$.

Q7. Dragster A will win. The time gap will be 0.26 s.

Q8. The ball will hit the floor for the second time at $t = \frac{2}{7}\left(5 + 2\sqrt{7}\right)$ s.

Q9. (a) $3.25\,\text{m s}^{-1}$.

   (b) At time $t = 10$s.

   (c) $\frac{2}{7}\text{ms}^{-2}$.

   (d) We need to show that the object has travelled the 16.25 m back to the starting point in the time interval $10 \leq t \leq 20.5$.
   First we calculate the velocity, $w$, at time $t_0 = 20.5$s.

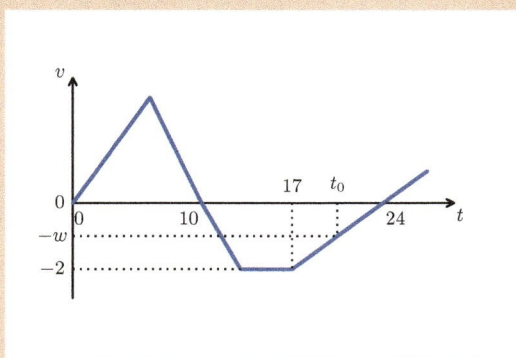

The acceleration in the interval $17 \leq t \leq 24$ is $\frac{2}{7}\text{ms}^{-2}$.

$$\frac{2}{7} = \frac{0 - w}{24 - 20.5} = \frac{-w}{3.5} \implies w = -1\text{ms}^{-1}.$$

From this, we can calculate the distance travelled in the interval $20.5 \leq t \leq 24$ by finding the area under the curve.

$$\text{distance} = \frac{1}{2} \times (24 - 20.5) \times 1 = 1.75\text{m}.$$

We know that the object travels 18 m in the interval $10 \leq t \leq 24$ and so the distance travelled in $10 \leq t \leq 20.5$ is $(18 - 1.75) = 16.25\text{m}$.
Therefore at time $t = 20.5$s, the object has travelled 16.25 m and has returned to the starting point.

   (e) 27.5 s.

Q10. (a) $t_0 = 28$.

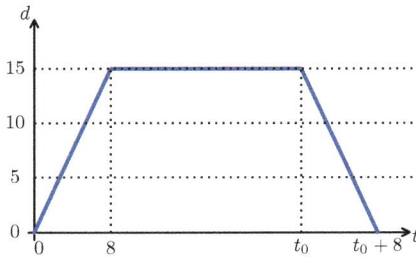

(b) $\frac{10}{3}$ ms$^{-2}$.
(c) The motorbike comes to rest at time $t_1 + 20 = 29$s.

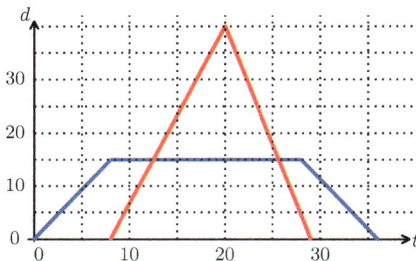

## Answers — Exercise 26.3

Q2. $s(t) = -\frac{t^4}{4} + t^3 - 10t$. At $t = 2$, the particle has diplacement $-20$.
Q3. $v(t) = 2t^2 + 2t$ and $s(t) = \frac{2}{3}t^3 + t^2$.
Q4. They will be $104.35$ m apart, to 2 d.p.
Q5. $11.02$ s.

## Answers — Exercise 27.1

Q1. (a) Weight.
    (b) Pushing.
    (c) Compression.
    (d) Tension.
    (e) Driving.
    (f) Pushing.
Q2. Compression.
Q3. (a) W=735 N.
    (b) W=279.75 N.

Notice that although the man's weight has changed by standing on Mars, his mass has remained constant.

Q4. (a) 6 N.

(b) $W = 0.196$ N.

(c) 7800 N.

Q5. 220 m.

Q6. (a)

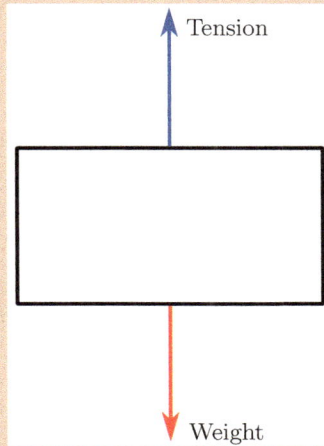

(b)

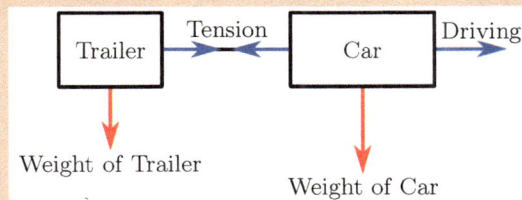

Or

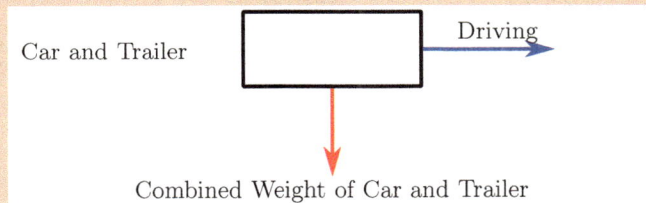

(c)

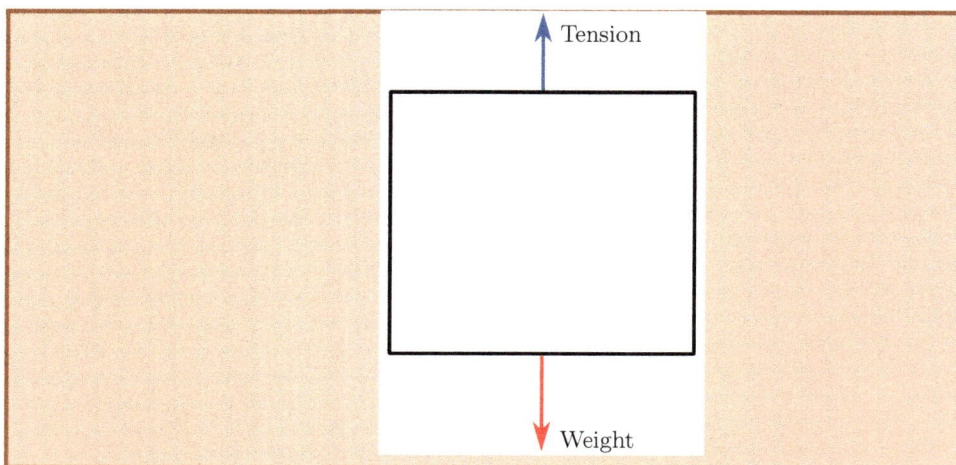

**Answers — Exercise 27.2**

Q1. (b).

Q2. (d).

Q3. (a).

Q4. (a) 153 N in the direction of the team pulling with 785 N.

(b) The magnitude of the resultant is 1.73 N to 2 decimal places and its direction is 63.34° from a left-pointing horizontal or 116.88° from a right-pointing horizontal, to 2 decimal places.

(c) $\mathbf{F} = 3\mathbf{i} - 3\mathbf{j}$ N.

(d) $\mathbf{F} = \begin{pmatrix} -3 \\ 1 \end{pmatrix}$ N.

Q5. The magnitude of the resultant is $|\mathbf{R}| = 1$ N and its direction is vertically upwards.

Q6. $\mathbf{D} = (-8,900\mathbf{i} + 3,610\mathbf{j})$ N.

Q7. $a = 9.8\,\text{ms}^{-2}$.

**Answers — Exercise 28.1**

Q1. (b).

Q2. (c).

Q3. (a).

Q4. $F = 25$ N.

Q5. $R = 29.4$ N vertically upwards.

Q6. $R = 5\sqrt{2}$ N perpendicular to the plane.

Q7. $F_r == (-7.595\mathbf{i} - 4.375\mathbf{j})$ N.

Q8. $v = 2\,\text{ms}^{-1}$.

Q9. $F_r = P = 6$ N.

Q10. (a) $F_r = 0$ N.

(b) $F_{\text{air}} = 0$ N.

(c) $W = 0$ N.

Q11. (a) $T = 60.31$ N to2 decimal places.
     (b) $a = 2.26\,\text{ms}^{-2}$ to 2 decimal places.
     (c)

$$F_1 = 11.31\,\text{N to 2 d.p.}$$
$$F_2 = 18.09\,\text{N to 2 d.p..}$$

Q12. (a) $T = 19.52$ N to 2 decimal places.
     (b) $a = 0.038$ N to 2 decimal places.
     (c)

$$F_1 = 9.52\,\text{N to 2 d.p.}$$
$$F_2 = 0.076\,\text{N to 2 d.p..}$$

Q13. (a) $T = 5.72$ N to 2 decimal places.
     (b) $a = 1.63$ N to 2 decimal places.
Q14. $T = 3675$ N.

**Answers — Exercise 29.1**

Q1. 0.68 or 0.85.
Q2. (a) HHH, HHT, HTH, THH, HTT, THT, TTH, TTT; (b) $\frac{3}{8}$.
Q3. (a) $\frac{1}{6}$; (b) $\frac{5}{6}$.
Q4. (a) $\frac{1}{4}$; (b) $\frac{3}{4}$; (c) $\frac{1}{13}$; (d) $\frac{3}{52}$.
Q5. (a) VV, VC, VS, CC, CS, SS; (b) $\frac{1}{2}$; (c) $\frac{1}{2}$.
Q6. $\frac{1}{10}$.
Q7. (a) $\frac{3}{4}$; (b) $\frac{3}{4}$.
Q8. (a) $\frac{2}{11}$; (b) $\frac{4}{11}$; (c) $\frac{7}{11}$.

**Answers — Exercise 29.2**

Q1. (i) $\frac{41}{100}$; (ii) $\frac{15}{100}$ (or equivalent); (iii) $\frac{57}{100}$.

Q2. (i) $\frac{7}{30}$; (ii) $\frac{10}{30}$ (or equivalent); (iii) $\frac{8}{30}$ (or equivalent).

Q3. (a) See below for the Venn diagram.

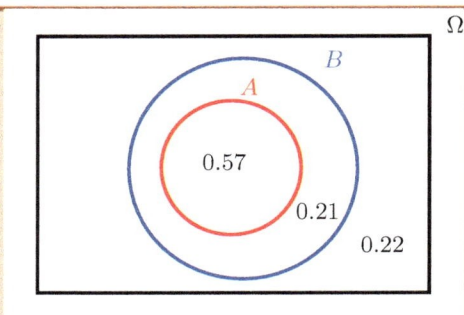

      (b) 0.43.

Q4.  (a) See below for the Venn diagram.

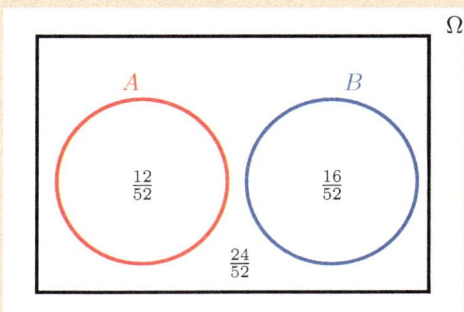

      (b)   i.  $\frac{40}{52}$ (or equivalent);

           ii.  $\frac{28}{52}$ (or equivalent);

           iii. 0;

           iv. $\frac{16}{52}$ (or equivalent).

Q5.  (a) See below for the Venn diagram.

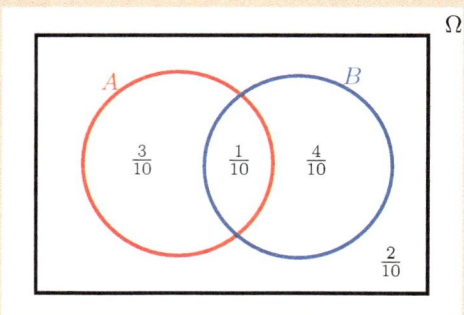

      (b)   i.  $\frac{1}{10}$;

           ii.  $\frac{1}{2}$ (or equivalent);

           iii. $\frac{6}{10}$;

           iv. $\frac{1}{5}$ (or equivalent).

Q6. (i) 0.86; (ii) 0.59; (iii) 0.24; (iv) 0.83.

**Answers — Exercise 29.3**

Q1. (i) 0.7; (ii) 0.3; (iii) 0.7.
Q2. (i) 0.2; (ii) 0.1; (iii) 0.7.
Q3. (i) 0.4; (ii) 0.3; (iii) 0.7.
Q4. (i) 0.35; (ii) 0; (iii) 0.23.
Q5. (i) 0.23; (ii) 0.24; (iii) 0.3.
Q6. (i) 0.78; (ii) 0.34; (iii) 0.38.

**Answers — Exercise 29.4**

Q1. 0.
Q2. (a) See below for Venn diagram.

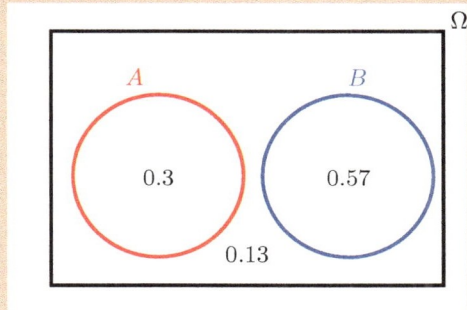

(b) (i) 0.57; (ii) 0.7.
Q3. (a) 0.28; (b) 0.75.
Q4. (a) 0.94; (b) 0.19.
Q5. See below for Venn diagram.

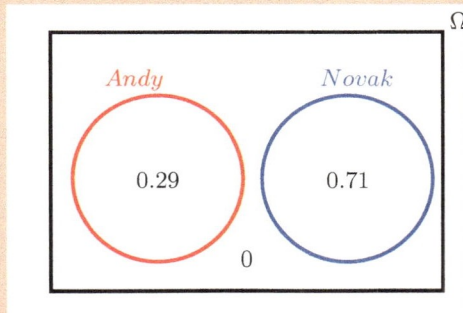

Q6. (a) See below for Venn diagram.

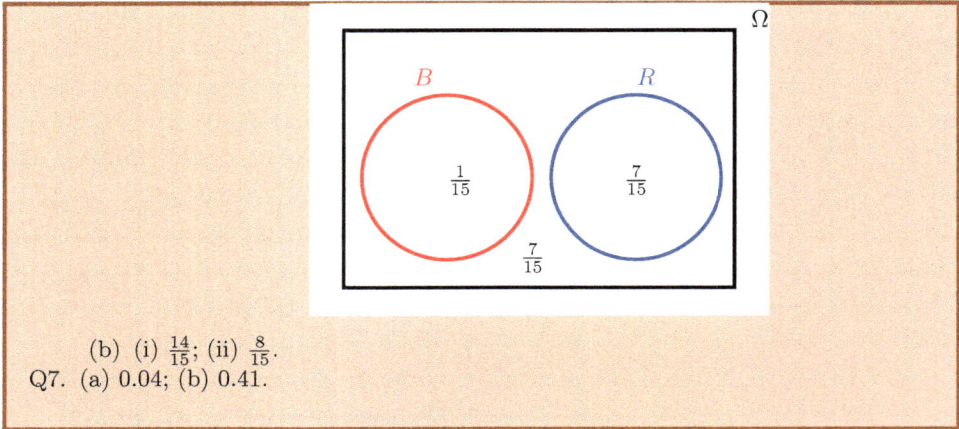

    (b) (i) $\frac{14}{15}$; (ii) $\frac{8}{15}$.

Q7. (a) 0.04; (b) 0.41.

### Answers — Exercise 29.5

Q1. Independent.

Q2. Not independent.

Q3. Not independent.

Q4. Independent.

Q5. Independent.

Q6. Not independent.

Q7. (i) 0.2; (ii) 0.7.

Q8. (i) 0.288; (ii) 0.832.

Q9. (i) 0.3876; (ii) 0.8624.

### Answers — Exercise 29.6

Q1. (a) The events $A$ and $B$ are independent since the probability of Kate drawing any coloured counter from the first bag *cannot* influence the outcome of which coloured counter she draws from the second bag.

    (b) $\frac{16}{121}$.

    (c) $\frac{44}{121}$.

Q2. (b) (i) 0.2958; (ii) 0.0442; (iii) 0.4258.

Q3. (a) $\frac{1}{8}$;

    (b) $\frac{1}{8}$;

    (c) $\frac{3}{4}$ (or equivalent).

**Answers — Exercise 29.7**

Q1. (a) The tree diagram is shown below.

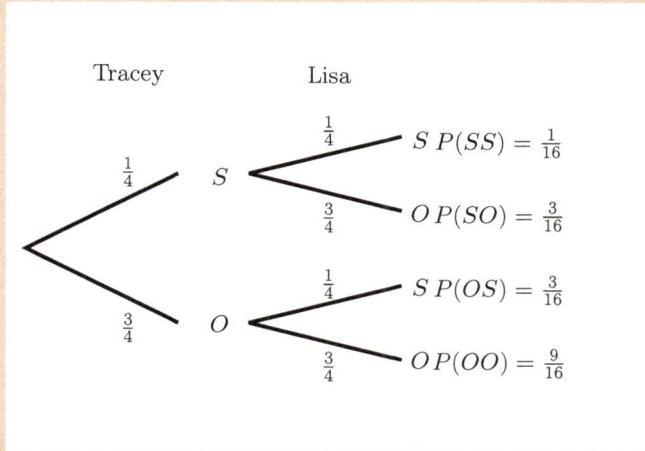

Tracey      Lisa

$\frac{1}{4}$   $S$

$\frac{1}{4}$   $S$   $P(SS) = \frac{1}{16}$

$\frac{3}{4}$   $O$   $P(SO) = \frac{3}{16}$

$\frac{3}{4}$   $O$

$\frac{1}{4}$   $S$   $P(OS) = \frac{3}{16}$

$\frac{3}{4}$   $O$   $P(OO) = \frac{9}{16}$

(b) $\frac{3}{8}$ (or equivalent);

(c) $\frac{9}{16}$

Q2. (a) The tree diagram is shown below.

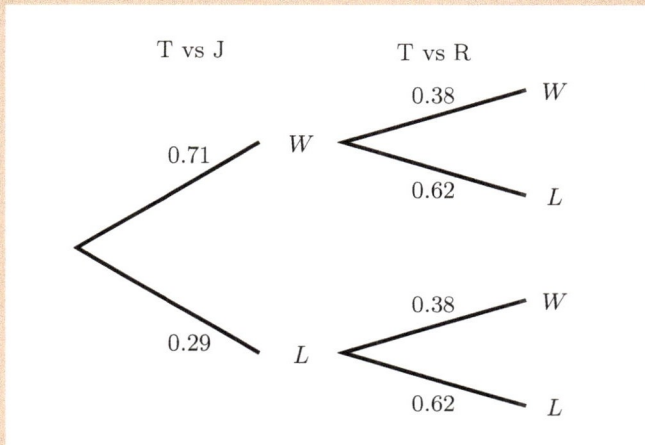

T vs J      T vs R

0.71   $W$

0.38   $W$

0.62   $L$

0.29   $L$

0.38   $W$

0.62   $L$

(b) 0.2698;

(c) 0.8202.

Q3. (a) The tree diagram is shown below.

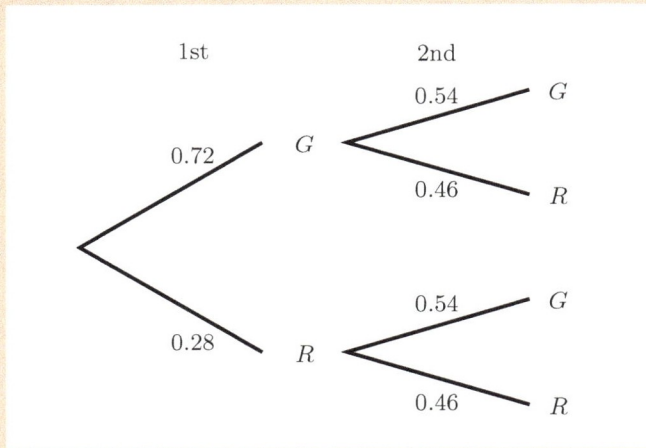

1st      2nd

0.72 — G

0.54 — G
0.46 — R

0.28 — R

0.54 — G
0.46 — R

(b) 0.3888;
(c) 0.4824.

**Answers — Exercise 29.8**
Q1. (a) 0.53;
(b) 0.88;
(c) 0.0567;
(d) The Venn diagram is below.

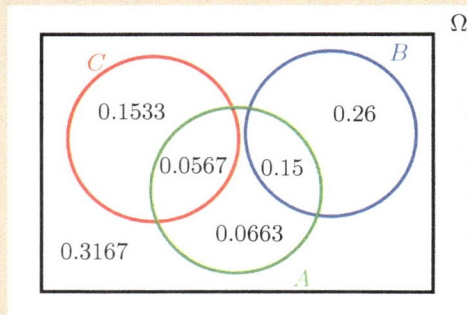

Ω

C    B

0.1533    0.26

0.0567   0.15

0.0663

0.3167

A

(e) (i) 0.62; (ii) 0.21.
Q2. (a) (i) 0.49; (ii) 0.215.

Q3. (a) The Venn diagram is shown below (a)

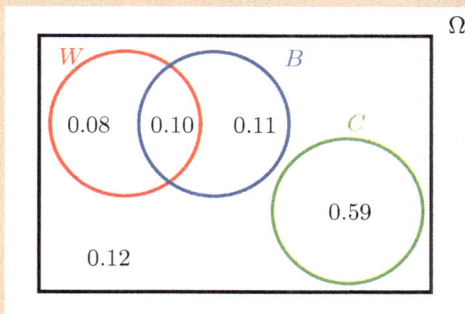

(b) 0.11;
(c) 0.8;
(d) Not statistically independent;
(e) $B$ and $C$ mutually exclusive, $W$ and $C$ mutually exclusive.

Q4. (a) $b = w = 0.11$;
(b) 1.

Q5. (a) $a = 0$;
(b) $b = 0.18$;
(c) $c = 0.08$;
(d) (i) 0.08; (ii) 0.82.

**Answers — Exercise 30.1**

Q1. (a) $\{HHHH\}, \{HHHT\}, \{HHTH\}, \{HHTT\},$
$\{HTHH\}, \{HTHT\}, \{HTTH\}, \{HTTT\},$
$\{THHH\}, \{THHT\}, \{THTH\}, \{THTT\},$
$\{TTHH\}, \{TTHT\}, \{TTTH\}, \{TTTT\}.$
(b) $\frac{1}{2}$
(c) $\mathbb{P}(X = 0) = \frac{1}{16}$, $\mathbb{P}(X = 1) = \frac{7}{16}$, $\mathbb{P}(X = 2) = \frac{5}{16}$, $\mathbb{P}(X = 3) = \frac{1}{8}$ and $\mathbb{P}(X = 4) = \frac{1}{16}$.

Q2. (a) $\{1, 1\}, \{1, 2\}, \{1, 3\}, \{1, 4\}, \{1, 5\}, \{1, 6\},$
$\{2, 1\}, \{2, 2\}, \{2, 3\}, \{2, 4\}, \{2, 5\}, \{2, 6\},$
$\{3, 1\}, \{3, 2\}, \{3, 3\}, \{3, 4\}, \{3, 5\}, \{3, 6\},$
$\{4, 1\}, \{4, 2\}, \{4, 3\}, \{4, 4\}, \{4, 5\}, \{4, 6\},$
$\{5, 1\}, \{5, 2\}, \{5, 3\}, \{5, 4\}, \{5, 5\}, \{5, 6\},$
$\{6, 1\}, \{6, 2\}, \{6, 3\}, \{6, 4\}, \{6, 5\}, \{6, 6\}.$
(b) $\frac{5}{9}$
(c) Let $X$ be your score. Then $\mathbb{P}(X = 1) = \frac{1}{6}$, $\mathbb{P}(X = 2) = \frac{1}{18}$, $\mathbb{P}(X = 3) = \frac{1}{9}$, $\mathbb{P}(X = 4) = \frac{1}{6}$, $\mathbb{P}(X = 5) = \frac{2}{9}$, and $\mathbb{P}(X = 6) = \frac{5}{18}$. The expectation is $4\frac{1}{18}$ compared to an expectation of $3\frac{1}{2}$ for rolling a single die.

Q3. (a) $\mathbb{P}(\text{top}) = \frac{1}{3}$, $\mathbb{P}(\text{middle}) = \frac{1}{3}$, $\mathbb{P}(\text{bottom}) = \frac{1}{3}$.
(b) $\mathbb{P}(\text{middle}) = \frac{1}{2}$, $\mathbb{P}(\text{bottom}) = \frac{1}{2}$.

Q4. (a) $\frac{1}{32}$
(b) $\frac{2}{11}$

(c) $\mathbb{P}(Y = 3) = \frac{1}{4}$, $\mathbb{P}(Y = 4) = \frac{1}{3}$ and $\mathbb{P}(Y = 5) = \frac{5}{12}$.

(d) $\frac{33}{192}$

Q5. Let $X$ be the colour of the first ball, and $Y$ the colour of the second.

(a) $\mathbb{P}(X = \text{red}) = \frac{2}{3}$, $\mathbb{P}(X = \text{green}) = \frac{1}{3}$

(b) $\mathbb{P}(Y = \text{red}|X = \text{red}) = \frac{7}{11}$, $\mathbb{P}(Y = \text{green}|X = \text{red}) = \frac{4}{11}$

(c) $\mathbb{P}(Y = \text{red}) = \frac{2}{3}$, $\mathbb{P}(Y = \text{green}) = \frac{1}{3}$

## Answers — Exercise 30.2

Q1. Given as fractions.

(a) $\frac{1}{8}$

(b) $\frac{1}{8}$

(c) $\frac{1}{8}$

(d) $\frac{3}{8}$

Q2. Given to 3 decimal places.

(a) 0.198

(b) 0.012

(c) 0.395

## Answers — Exercise 30.3

Q1. (a) 0.058

(b) 0.136

(c) 0.745

(d) 0.561

Q2. $Y \sim B(7, 0.5)$

(a) 0.055

(b) 0.164

(c) 0.164

(d) 0.5

Q3. (a) Rolling one six with a single throw is most likely to succeed, with a probability of 0.665. The chance of rolling three or more sixes with three throws is 0.597.

(b) With biased dice the probability of rolling one six in one throw goes up to 0.822. However, the chance of rolling at least three sixes with three throws has increased to 0.865, and this is now most likely to succeed.

Q4. (a) 0.315

(b) 0.401

(c) 0.215

(d) 0.370

## Answers — Exercise 30.4

Q1. (a) 0.050

(b) 0.801

(c) 0.388

Q2. (a) 0.323

(b) 48

(c) A Poisson distribution with parameter $\lambda = 48$.

(d) 0.567

Q3. 0.137

Q4. 0.042

Q5. The random variable $X$ has a Poisson distribution with parameter $\lambda$.

$$\mathbb{E}(X) = \sum_{k=0}^{\infty} k\mathbb{P}(X = k)$$

$$= \sum_{k=0}^{\infty} \frac{k\lambda^k e^{-\lambda}}{k!}$$

$$= \lambda e^{-\lambda} \sum_{k=0}^{\infty} \frac{k\lambda^{k-1}}{k!} = \lambda e^{-\lambda} \sum_{k=1}^{\infty} \frac{k}{k!}\lambda^{k-1}$$

$$= \lambda e^{-\lambda} \sum_{k=1}^{\infty} \frac{\lambda^{k-1}}{(k-1)!} = \lambda e^{-\lambda} \sum_{k=0}^{\infty} \frac{\lambda^k}{k!} = \lambda e^{-\lambda} e^{\lambda} = \lambda.$$

## Answers — Exercise 30.5

Q1. (a) 0.4

(b) 0.36

(c) 0.1296

(d) 0.36

Q2. Let $X$ be the number of strawberries scanned before the first small one is found.

(a) $\mathbb{P}(X = k) = \frac{1}{10}(\frac{9}{10})^k$ for $k \in \{0, 1, 2, \ldots\}$

(b) 0.271

(c) 0.09

(d) 0.09

## Answers — Exercise 31.1

Q1. (a) Critical region: $X \leq 10$; acceptance region: $X \geq 11$; reject $H_0$.

(b) Critical region is $X \leq 6$, $X \geq 14$; acceptance region: $7 \leq X \leq 13$; do not reject $H_0$.

(c) Critical region: $X \geq 35$; acceptance region: $X \leq 34$ reject $H_0$.

Q2. (a) $H_0 : p = 0.1$, $H_1 : p > 0.1$.

(b) $H_0 : p = 0.2$.

(c) $H_0 : p = 0.7$, $H_1 : p > 0.7$.

Q3. (a) Q2(a): 1% critical region: $X \geq 11$, 5% critical region $X \geq 10$, 10% critical region: $X \geq 9$.

(b) Q2(b): 1% critical region: $X \geq 10$, 5% critical region: $X = 0$, $X \geq 9$, 10%

critical region: $X = 0$, $X \geq 8$.
   (c) Q2(c):1% critical region: $X \geq 66$, 5% critical region $X \geq 64$, 10% critical region: $X \geq 62$.
Q4. (a) 0.0302.
   (b) The new critical value is less than 4.
   (c) Reject $H_0$ at both levels.
Q5. (a) $H_0 : p = 0.85$, $H_1 : p < 0.85$; critical value: critical region: $X \leq 78$.
   (b) (i) Do not reject $H_0$.
       (ii) Reject $H_0$.
Q6. Critical region: $510 \leq X \leq 602$. Reject $H_0$.

**Answers — Exercise 31.2**
Q1. (a) 0.2293.
   (b) 0.0579.
   (c) 0.4729.
   (d) 0.0010.
Q2. (a) p-value: 0.1036. Do not reject $H_0$.
   (b) p-value: 0.0032. Reject $H_0$.
   (c) p-value 0.0002. Reject $H_0$ .
Q3. (a) 0.0729.
   (b) Do not reject $H_0$.
   (c) p-value: 0.0126. Reject $H_0$.
   (d) p-value: 0.0011. Reject $H_0$.
Q4. (c) $H_0 : p = 0.25$, $H_1 : p > 0.25$.
   (d) p-value: 0.0002. Reject $H_0$.
Q5. (a) $H_0 : p = 0.1$, $H_1 : p < 0.1$.
   (b) $k = 1$.
Q6. p-value: 0.2089. Do not reject $H_0$.
Q7. (a) p-value: 0.0628. Do not reject $H_0$.
   (b) p-value: 0.0931. Reject $H_0$.
Q8. p-value: 0.1330. Do not reject $H_0$.
Q9. p-value: 0.1266. Do not reject $H_0$.
Q10. The null and alternative hypotheses are
   $H_0 : p = 0.5$, $H_1 : p > 0.5$, where $p$ is the proportion of coin tosses that land on heads.
   (a) (i) p-value: 0.0898; do not reject $H_0$.
       (ii) p-value: 0.0392; reject $H_0$
   (b) 18.

## 33. Notation

### 33.1 Set Notation

| | | | |
|---|---|---|---|
| $\in$ | is an element of | $\mathbb{Z}_0^+$ | the set of nonnegative integers, $\{0, 1, 2, 3, \ldots\}$ |
| $\notin$ | is not an element of | | |
| $\subseteq$ | is a subset of | $\mathbb{R}$ | the set of real numbers |
| $\subset$ | is a proper subset of | $\mathbb{Q}$ | the set of rational numbers |
| $\{x_1, x_2\}$ | the set with elements $x_1$, $x_2$ | | $\left\{\frac{p}{q} : p \in \mathbb{Z}, q \in \mathbb{Z}\right\}$ |
| $x :$ | the set of all $x$ such that | $\cup$ | union |
| $n(A)$ | the number of elements in set $A$ | $\cap$ | intersection |
| | | $(x, y)$ | the ordered pair $x$, $y$ |
| $\emptyset$ | the empty set | $[a, b]$ | the closed interval $\{x : a \le x \le b\}$ |
| $\mathcal{E}$ | the universal set | | |
| $A'$ | the complement of the set A | $[a, b)$ | the interval $\{x : a \le x < b\}$ |
| $\mathbb{N}$ | the set of natural numbers, $\{1, 2, 3, \ldots\}$ | $(a, b]$ | the interval $\{x : a < x \le b\}$ |
| | | $(a, b)$ | the open interval $\{x : a < x < b\}$ |
| $\mathbb{Z}$ | the set of integers, $\{0, \pm 1, \pm 2, \pm 3, \ldots\}$ | | |
| $\mathbb{Z}^+$ | the set of positive integers, $\{1, 2, 3, \ldots\}$ | | |

### 33.2 Miscellaneous Symbols

| | | | |
|---|---|---|---|
| $=$ | is equal to | $\approx$ | is approximately equal to |
| $\neq$ | is not equal to | $\infty$ | infinity |
| $\equiv$ | is identical to or is congruent to | $\propto$ | is proportional to |

| | | | |
|---|---|---|---|
| $\therefore$ | therefore | $\geq, \geqslant$ | is greater than or equal to, is |
| $\because$ | because | | not less than |
| $<$ | is less than | $p \Rightarrow q$ | $p$ implies $q$ (if $p$ then $q$) |
| $\leq, \leqslant$ | less than or equal to, is not | $p \Leftarrow q$ | $p$ is implied by $q$ (if $q$ then $p$) |
| | greater than | $p \Leftrightarrow q$ | $p$ implies and is implied by $q$ |
| $>$ | is greater than | | ($p$ is equivalent to $q$) |

## 33.3  Sequences

| | | | |
|---|---|---|---|
| $a$ | first term for an arithmetic or geometric sequence | $r$ | common ratio for a geometric sequence |
| $l$ | last term for an arithmetic sequence | $S_n$ | sum to $n$ terms of a sequence |
| | | $S_\infty$ | sum to infinity of a sequence |
| $d$ | common difference for an arithmetic sequence | | |

## 33.4  Operations

| | | | |
|---|---|---|---|
| $a + b$ | $a$ plus $b$ | $|a|$ | the modulus of $a$ |
| $a - b$ | $a$ minus $b$ | $n!$ | $n$ factorial: |
| $a \times b, ab, a \cdot b$ | $a$ multiplied by $b$ | | $n! = n \times (n-1) \times \ldots 2 \times 1$, |
| $a \div b, \frac{a}{b}, a/b$ | $a$ divided by $b$ | | $n \in \mathbb{N};\ 0! = 1$ |
| $\displaystyle\sum_{i=1}^{n} a_i$ | $a_1 + a_2 + \ldots + a_n$ | $\displaystyle\binom{n}{r},\ {}^nC_r,$ | the binomial coefficient |
| | | ${}_nC_r$ | $\frac{n!}{r!(n-r)!}$ for $n, r \in \mathbb{Z}_0^+$ , $r \leq n$ |
| $\displaystyle\prod_{i=1}^{n} a_i$ | $a_1 \times a_2 \times \ldots \times a_n$ | | or $\frac{n(n-1)\ldots(n-r+1)}{r!}$ for $n \in \mathbb{Q}$, |
| $\sqrt{a}$ | the nonnegative square root of $a$ | | $r \in \mathbb{Z}_0^+$ |

## 33.5  Functions

| | | | |
|---|---|---|---|
| $f(x)$ | the value of the function $f$ at $x$ | $\dfrac{\mathrm{d}^n y}{\mathrm{d}x^n}$ | the $n$th derivative of $y$ with respect to $x$ |
| $f : x \mapsto y$ | the function $f$ maps the element $x$ to the element $y$ | $f'(x), f''(x),$ $\ldots, f^{(n)}(x)$ | the first, second, ..., $n$th derivatives of $f(x)$ with respect to $x$ |
| $f^{-1}$ | the inverse function of the function $f$ | | |
| $gf$ | the composite function of $f$ and $g$ which is defined by $gf(x) = g(f(x))$ | $\dot{x}, \ddot{x}, \ldots$ | the first, second, ... derivatives of $x$ with respect to $t$ |
| $\displaystyle\lim_{x \to a} f(x)$ | the limit of $f(x)$ as $x$ tends to $a$ | $\int y\, \mathrm{d}x$ | the indefinite integral of $y$ with respect to $x$ |
| $\Delta x, \delta x$ | an increment of $x$ | $\int_a^b y\, \mathrm{d}x$ | the definite integral of $y$ with respect to $x$ between the limits $x = a$ and $x = b$ |
| $\dfrac{\mathrm{d}y}{\mathrm{d}x}$ | the derivative of $y$ with respect to $x$ | | |

## 33.6 Exponential and Logarithmic Functions

| | | | |
|---|---|---|---|
| e | base of natural logarithms | $\log_a(x)$ | logarithm to the base $a$ of $x$ |
| $e^x$, $\exp(x)$ | exponential function of $x$ | $\ln(x)$, $\log_e(x)$ | natural logarithm of $x$ |

## 33.7 Trigonometric Functions

| | | | |
|---|---|---|---|
| $\sin, \cos,$ $\tan, \csc,$ $\sec, \cot$ | the trigonometric functions | $\sin^{-1}, \cos^{-1},$ $\tan^{-1}, \arcsin,$ $\arccos, \arctan$ | the inverse trigonometric functions |
| $^\circ$ | degrees | rad | radians |

## 33.8 Vectors

| | | | |
|---|---|---|---|
| $\boldsymbol{a}$, $\underline{a}$, $\underset{\sim}{a}$ | the vector $\boldsymbol{a}$, $\underline{a}$, $\underset{\sim}{a}$; these alternatives apply throughout this section. | $|\boldsymbol{a}|$, $a$ | the magnitude of $\boldsymbol{a}$ |
| | | $|\overrightarrow{AB}|$, $AB$ | the magnitude of $\overrightarrow{AB}$ |
| $\overrightarrow{AB}$ | the vector represented in magnitude and direction by the directed line segment $AB$ | $\begin{pmatrix} a \\ b \end{pmatrix}$, $a\boldsymbol{i} + b\boldsymbol{j}$ | column vector and corresponding unit vector notation |
| | | $\boldsymbol{r}$ | position vector |
| $\hat{a}$ | a unit vector in the direction of $\boldsymbol{a}$ | $\boldsymbol{s}$ | displacement vector |
| | | $\boldsymbol{v}$ | velocity vector |
| $\boldsymbol{i}$, $\boldsymbol{j}$, $\boldsymbol{k}$ | unit vectors in the directions of the Cartesian coordinate axes | $\boldsymbol{a}$ | acceleration vector |

## 33.9 Probability and Statistics

| | | | |
|---|---|---|---|
| $A, B, C, etc$ | events | $p_1, p_2, \ldots$ | probabilities of the values $x_1$, $x_2, \ldots$ of the discrete random variable $X$ |
| $A \cup B$ | union of the events $A$ and $B$ | | |
| $A \cap B$ | intersection of the events $A$ and $B$ | | |
| $P(A)$ | probability of the event $A$ | $E(X)$ | expectation of the random variable $X$ |
| $A'$ | complement of the event $A$ | $\mathrm{Var}(X)$ | variance of the random variable $X$ |
| $P(A|B)$ | probability of the event $A$ conditional on the event $B$ | | |
| $X, Y, R, etc$ | random variables | $\sim$ | has the distribution |
| $x, y, r, etc$ | values of the random variables $X$, $Y$, $R, etc$ | $B(n, p)$ | binomial distribution with parameters $n$ and $p$, where $n$ is the number of trials and $p$ is the probability of success in a trial |
| $x_1, x_2, \ldots$ | values of observations | | |
| $f_1, f_2, \ldots$ | frequencies with which the observations $x_1, x_2, \ldots$ occur | $q$ | $q = 1 - p$ for binomial distribution |
| $p(x)$, $P(X = x)$ | probability function of the discrete random variable $X$ | | |

| $N(\mu, \sigma^2)$ | Normal distribution with mean $\mu$ and variance $\sigma^2$ | $\sigma$ | population standard deviation |
|---|---|---|---|
| $Z \sim N(0, 1)$ | standard Normal distribution | $\bar{x}$ | sample mean |
| $\phi$ | probability density function of the standardised Normal variable with distribution $N(0, 1)$ | $s^2$ | sample variance |
| | | $s$ | sample standard deviation |
| | | $H_0$ | Null hypothesis |
| | | $H_1$ | Alternative hypothesis |
| $\Phi$ | corresponding cumulative distribution function | $r$ | product moment correlation coefficient for a sample |
| $\mu$ | population mean | $\rho$ | product moment correlation coefficient for a population |
| $\sigma^2$ | population variance | | |

## 33.10   Mechanics

| kg | kilograms | $t$ | time |
|---|---|---|---|
| m | metres | $s$ | displacement |
| km | kilometres | $u$ | initial velocity |
| $m\,s^{-1}$, $ms^{-1}$ | metres per second (velocity) | $v$ | velocity or final velocity |
| $m/s^2$, $ms^{-2}$ | metres per second per second (acceleration) | $a$ | acceleration |
| | | $g$ | acceleration due to gravity |
| $F$ | Force or resultant force | $\mu$ | coefficient of friction |
| N | Newton | | |
| N m | Newton metre (moment of a force) | | |

# Index

Lightning Source UK Ltd.
Milton Keynes UK
UKOW07f0827200817

307571UK00004B/20/P